# Power Electronics and Power Systems

More information about this series at http://www.springer.com/series/6403

Haifeng Wang · Wenjuan Du

# Analysis and Damping Control of Power System Low-frequency Oscillations

Springer

Haifeng Wang
Beijing
China

Wenjuan Du
Beijing
China

ISSN 2196-3185 ISSN 2196-3193 (electronic)
Power Electronics and Power Systems
ISBN 978-1-4899-7694-9 ISBN 978-1-4899-7696-3 (eBook)
DOI 10.1007/978-1-4899-7696-3

Library of Congress Control Number: 2016931424



Printed on acid-free paper

This Springer imprint is published by SpringerNature
The registered company is Springer Science+Business Media LLC New York

# Preface

Power system electromechanical low-frequency oscillations threaten the stable operation of power systems. A great effort has been spent by many researchers for over half-century in order to understand and solve this complicated engineering problem. There are two main methods for the analysis and damping control of power system low-frequency oscillations. They are the damping torque analysis and modal analysis based on the linearized models of power systems for the study of small-signal angular stability. In this book, they are named as linearized methods or small-signal methods. The damping torque analysis is simple and of clear physical meaning. The phase compensation method developed on the basis of damping torque analysis for the design of power system stabilizers (PSSs) has been widely used in practice. The modal analysis is based on the modal control theory and has been the mostly used method in large-scale multi-machine power systems. This book is devoted to the introduction of those two linearized methods and their applications in the design of conventional PSSs and more recently appeared FACTS stabilizers. It is written as a textbook for postgraduate research students and a reference book for power system researchers. In addition to the introduction of modal analysis, great attention has been paid in the book to elaborate the principle and applications of the damping torque analysis, which we have applied in our research on the subject in recent 20 years.

We have used the main materials in this book in teaching and tutoring our research students for years. Some examples in the book are the results of their exercise work before starting their research projects. We would like to acknowledge their contributions to the preparation of examples in the book. They are Yifu Lin, Xiangfeng Wang, Yan Zhang, Shen Yang, Qianhui Wan, Ximin Li (Sect. 2.4), Chen Lv (Sect. 3.3), Yi Ge (Sect. 4.4), Tianyu Su (Sect. 5.3), Zhijin Zhen (Sect. 6.3), and Xiaobo Hu (Sect. 7.4.2).

Beijing
October 2015

Haifeng Wang
Wenjuan Du

# Contents

# Chapter 1
# Introduction

## 1.1 Power System Low-frequency Oscillations

Power system low-frequency oscillations are the oscillations of active power delivered along particular transmission corridors in a power system with the oscillation frequency from 0.1 Hz up to a couple of Hz. Once started, the oscillations can continue for a while and then disappear, or grow continuously to cause power system collapse. In this book, they are called power system oscillations or power oscillations for short.

Manifestation of a power oscillation is the oscillation of relative movement of angular positions of generators in the power system. The oscillation can be triggered by severe faults, such as a three-phase to-earth short circuit along or tripping of a transmission line. It can also occur under normal operating conditions when the power system is only subject to small disturbances. Hence if the power system collapse is caused by the power oscillation, it could belong to the problem of power system large-signal rotor angle (angular) stability or small-signal angular stability [1].

This book is devoted to the introduction of the linearized (small-signal) methods for the analysis and damping control of the power oscillation. Hence strictly speaking, the power oscillation discussed in the book should belong to the problem of power system small-signal angular stability. If the power oscillation exhibits increasing magnitude, the power oscillation is said to have negative damping and obviously the power system is unstable in terms of power system small-signal angular stability. If the power oscillation shows a sustained constant magnitude, or continues for a certain period of time (over several to tens of seconds) and sets down eventually, it is reckoned that the oscillation is of poor damping though the power system is still stable. If the oscillation subsides quickly within several seconds with a damping ratio greater than 0.1 (sometimes it could be 0.05 for large-scale power systems), it is considered that the power oscillation is well damped.



H. Wang and W. Du, *Analysis and Damping Control of Power System Low-frequency Oscillations*, Power Electronics and Power Systems,
DOI 10.1007/978-1-4899-7696-3_1

Power system oscillation was first observed in the Northern American power network in Oct. 1964 during a trial interconnection of the Northwest Power Pool and the Southwest Power Pool [2]. The power oscillation occurred on the tie line and was of a frequency of 0.1 Hz. Since then, many incidents of power system oscillations have been reported in power transmission networks around the world. Examples are as follows:

1. Between the late 1970s and early 1980s, power oscillations were observed in the power transmission corridor from Scotland to England in the Great Britain power network. Operational experience indicated that those oscillations were related to the relatively high loading level of power transmission lines from Scotland to England. A series of tests were carried out between 1980 and 1985. Those tests demonstrated that the oscillations occurred when the power transfer from Scotland to England reached a certain level and the typical oscillation frequency was around 0.5 Hz [3].
2. Sustained power oscillations were reported in the Taiwan power network in 1984. Those oscillations happened typically when a large amount of power was transferred along some particular high-voltage transmission lines. It was found that reduction of the amount of power delivered along the high-voltage transmission lines can effectively improve the damping of power oscillations. Installation of the PSSs (power system stabilizers) at selected locations worked successfully to suppress the power oscillations. Further investigation indicated that other factors, such as the gain value of automatic voltage regulator (AVR) and characteristics of load, also affected the damping of the power oscillations [4].
3. An outage of the Northern American Western Systems Coordinating Council (WSCC) network on August 10 1996 was directly due to the power oscillations. The power oscillations (with frequency between 0.2 and 0.3 Hz) were triggered by the loss of a 500-kV line when the system operated under depressed conditions (with overloaded lines and at low voltage level). The incident spread quickly, tripping more lines, and generating units, which eventually led to the separation of the WSCC network into four islands. The outage affected 7.5 million customers for up to 9 h, causing considerable economic loss [5].

Over the last half-century, many power system researchers and engineers have worked on and contributed to the understanding and solution of the problem of power system oscillations. It is now well recognized that the main cause of power system oscillations is the poor damping of the so-called electromechanical oscillation modes of the power system. Poor damping could be brought about by the (1) large amount of long-distance power transmission, (2) weak interconnection of large power sub-networks, and/or (3) negative damping due to the fast-acting high-gain AVRs. Power system low-frequency oscillations can be classified, according to the electromechanical oscillation modes of the power system, into two types: (1) local power oscillations (associated with local oscillation modes) and (2) inter-area power oscillations (associated with inter-area oscillation modes).

Local power oscillations associated with the local oscillation modes normally are the oscillations of one or a group of local generators against a large power network. Local generators send power over a long distance to the large power network. Frequency of the oscillations often is about one or several Hz. Inter-area power oscillations related to the inter-area oscillation modes are the oscillations between two or more sub-networks in a large-scale power system. A typical inter-area oscillation is the tie-line power oscillation between two weakly connected areas in the power system. The inter-area power oscillation could involve many sub-networks to oscillate against each other (which sometimes is referred to as an intra-area oscillation). Normally, the oscillation frequency is lower, from 0.1 to 1 Hz.

A power oscillation could be engaged by one oscillation mode only (local or inter-area oscillation mode). This is the case referred to as the "single-mode power oscillation" when there is only one electromechanical oscillation mode in the power system being poorly or negatively damped. In the case that there are multiple poorly or negatively damped electromechanical oscillation modes in the power system, the power oscillation is participated by multiple oscillation modes. The power oscillation in this case is called the "multi-mode power oscillation".

## 1.2 Linearized Methods for the Analysis and Damping Control of Power System Oscillations

Study of power system oscillations is in order to investigate the phenomena and to develop prevention measures from the occurrence of the problem. This is to understand why the oscillations happen and to recommend how they can be effectively suppressed. There are two main categories of linearized methods (small-signal methods) for the analysis and damping control of power system oscillations: the damping torque analysis and modal analysis.

The technique of damping torque analysis (DTA) was well explained in [6] for a single-machine infinite-bus power system to investigate the effect of excitation control on power system small-signal angular stability. Reference [6] is the milestone contribution by deMello and Concordia to the field. Proposal of the DTA is based on the linearized Heffron–Phillips model of the single-machine infinite-bus power system [7, 8]. It is used to examine the electric torque contribution from a particular source in the power system to the so-called electromechanical oscillation loop of a generator, which is in fact the linearized swing equation of rotor motion of the generator. The electric torque is decomposed into two components, synchronizing torque and damping torque. Synchronizing torque is in phase with rotor angle deviation and damping torque in phase with rotor speed deviation of the generator. Synchronizing torque is the force to keep the generator in synchronism with the rest of the power system. Lack of damping torque or contribution of negative damping torque may lead to poor damping of power system oscillations or even power system instability.

The DTA is established on the physical concept of electric torque and classic control theory with the description of the system in frequency and Laplace s domain. It is easy to be understood and simple to be applied. Often analytical conclusions can be obtained under certain reasonable assumptions. In particular later, an effective method, the phase compensation method, for the design of a power system stabilizer (PSS) was developed on the basis of the DTA [9–11], which has made the DTA a well-accepted method for the analysis and damping control of power system oscillations. Since the 1970s, considerable effort has been spent to extend the DTA for the application in general multi-machine power systems.

Modal analysis (MA) is based on the modal control theory, a special branch in modern control theory. It is established on the state-space representation of a power system, i.e. so-called ABCD state-space model [12]. Applications of modal analysis in the study of power system oscillations include the following: (1) computation of eigenvalues and associated eigenvectors (so-called eigensolution) of the state matrix of the system; (2) computation of controllability index, observability index, and their product, residue. There is no much difference in the procedure to apply the MA in a single-machine infinite-bus and a multi-machine power system.

Eigensolution is a very specific topic in mathematics. Examples of applying eigensolution in studying power system oscillations are computation of power system oscillation modes (i.e. computation of eigenvalues and identification of oscillation modes among all eigenvalues), eigenvalue sensitivity, and participation factor [13, 14]. For a large-scale power system, dimension of the state matrix could become very high. This could lead to the extremely high computational complexity and even numerical difficulty to calculate the eigenvalues of the state matrix. Hence, it has been a special research topic to compute eigenvalues of interests of a high-dimensional state matrix considering the features of the large-scale power system.

Computation of controllability index, observability index, and residue is mainly applied for the selection of installing locations and feedback signals of stabilizers in the multi-machine power system. Because computation of controllability index, observability index, and residue usually needs to perform eigensolution, various methods to reduce computational cost have been proposed in the literature. Examples are the selective modal analysis [13–15], partial modal analysis [16], and eigensolution-free modal analysis [17].

The MA is a method based on the results of numerical computation. It can tell whether a power system is stable at given system operating conditions. However, it is difficult to be used to draw general conclusions by performing numerical eigensolution. Hence in many occasions, effort has to be spent to establish the connections between the computational results from the MA and the physical explanations in order to understand and explain the results of numerical computation.

## 1.3 FACTS and Grid-Connected ESS

Flexible AC transmission systems (FACTS) are the technologies emerged around the end of last century to reinforce power systems. Foundation of the FACTS is the application of power-electronic based control in high-voltage power transmission systems. It is the result of rapid advance in high-power electronics technology over the last several decades. Though the FACTS controllers are still considered expensive compared to the conventional power system controllers, they have been installed in many real power systems in the world due to their superior control performance.

The basic operational principle of the FACTS controllers is shown in Fig. 1.1 [18, 19]. The active power delivered along the transmission line is as follows:

$$P_{12} = \frac{V_1 V_2}{x} \sin \theta \tag{1.1}$$

where $\theta$ is the phase difference between the voltage at the sending and receiving end of the transmission line, $\bar{V}_1$ and $\bar{V}_2$. A FACTS controller is to change in real time the line impedance x, and phase and magnitude of line voltage $\bar{V}_1$ and $\bar{V}_2$ individually or simultaneously to regulate the power flow along the transmission line, such that a great flexibility of power flow regulation is achieved.

In fact, the operational principle of the FACTS was known as early as in the 1920s. Since then, mechanically controlled series and shunt capacitors have been installed in power systems to control line power flow and regulate system voltage profile. Power electronics applications in high-voltage power systems were dated from early 1970s when converters for high-voltage direct-current systems began using thyristors to replace mercury arc valves. This was the time when the FACTS technology really emerged and was applied in power systems. The concept of the FACTS was formerly introduced by Prof. Hingarani [18] when work to use power electronic switching devices to replace thyristors began [19]. Hence normally, the FACTS controllers are classified into two groups: (1) the conventional thyristor-based FACTS controllers and (2) new generation of converter (power electronic switching devices)-based FACTS controllers. The following are three main types of thyristor-based FACTS controllers which have been applied in power systems:

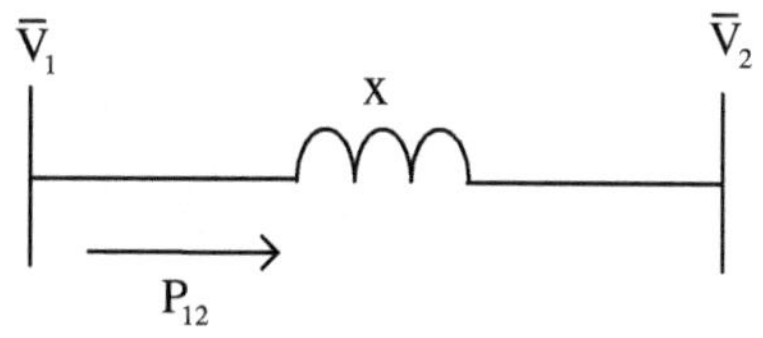


**Fig. 1.1** Illustration on the operational principle of FACTS controllers

1. Thyristor-controlled static VAR compensator (SVC), which controls the magnitude of line voltage;
2. Thyristor-controlled series compensator (TCSC), which changes the equivalent impedance of transmission line and hence can be used for power flow control;
3. Thyristor-controlled phase shifter (TCPS), which shifts the phase of line voltage and normally can be used for power flow control.

New generation of FACTS controllers employs self-commutated, voltage-soured switching power converters to realize rapidly controllable, static synchronous AC voltage or current sources. A new-generation FACTS controller is constructed mainly on the synchronous voltage source (SVS) which is an ideal machine with no rotation and inertia. The magnitude and phase of the SVS can be controlled instantaneously to generate reactive power (both capacitive and inductive) and/or to regulate active power flow independent of power system variables (voltage and current). For example, Fig. 1.2 shows an SVS constructed by a GTO-based voltage source converter (VSC), which employs the algorithm of the pulse width modulation (PWM). Through controlling the modulation ratio m and phase $\phi$, the magnitude and phase of the AC SVS can be controlled to realize various functions of power system regulation.

There are three main types of VSC-based FACTS controllers:

1. The static synchronous compensator (STATCOM). This is the controller realized by installing the VSC-based SVS in shunt along a transmission line.
2. The static synchronous series compensator (SSSC). This is the VSC-based SVS installed in series in the transmission line.
3. The unified power flow controller (UPFC). This is the most powerful FACTS controller, a combined device of a STATCOM and a SSSC with a common DC capacitor.

The development of power systems in the recent years is towards meeting the requirement of clean power generation. This will lead to the fundamental changes

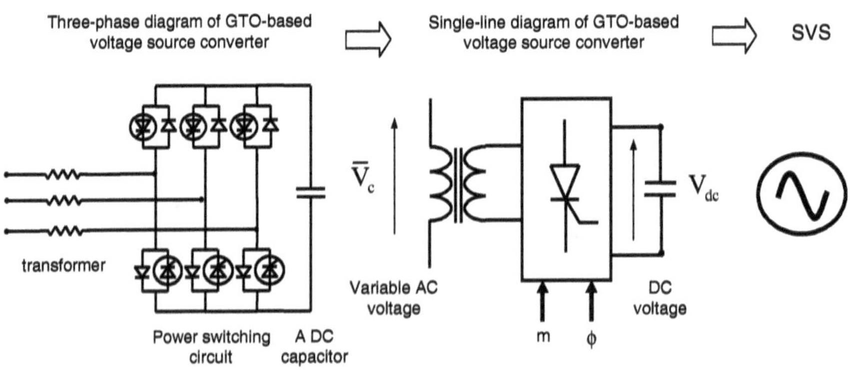


**Fig. 1.2** An SVS realized by a VSC circuit

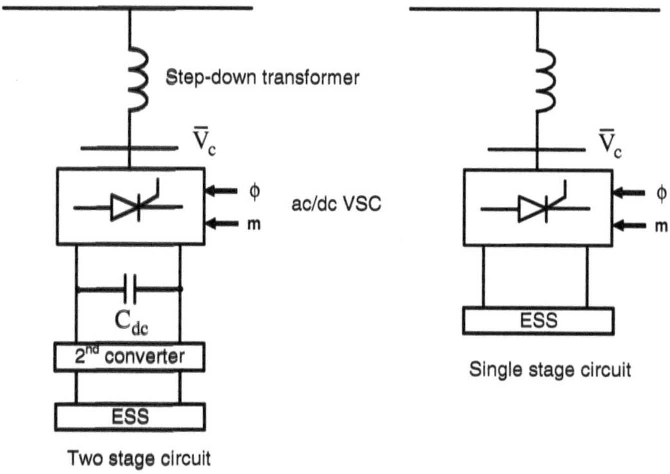


**Fig. 1.3** VSC-based power electronic circuit for grid-connected ESS or renewable generation plant

of the systems in the near future with a fairly large percentage of grid-connected renewable power generation. It can be foreseen that the high-power electronic VSC will play even more important role in shaping the future power systems. Not only the VSC will be used for the grid connection of many types of renewable power generation, such as wind, solar, and fuel cell, but also it can be applied for the energy storage systems (ESSs) to assist power system operation and control in accommodating variable grid-connected renewable power generation. Hence in future power systems, the VSC could become one of the mostly installed power components like transformers to meet various purposes of power regulation, storage, and generation.

Figure 1.3 shows the configuration of grid connection of an ESS device or a renewable power generation plant by use of the VSC-based power electronics circuits. The ESS device can be a battery ESS (BESS). The configuration is also applicable for the grid connection of a renewable generation source, such as a photovoltaic (PV) power plant. In the two-stage circuit, the second converter can be a DC/DC converter to connect to a fuel cell (FC) power plant or a DC/AC converter to a wind farm. In both circuits, the key component is the AC/DC converter (VSC) which is connected to the high-voltage busbar via a step-down transformer.

## 1.4 Controllers to Damp Power System Oscillations

To increase the damping of power system oscillations, installation of a supplementary damping controller, the PSS (power system stabilizer), on the side of excitation system of a generator is a simple, effective and economical method.

Through the effort of and experience accumulated by power engineers and researchers in the last half-century, the PSS has become a standard device installed in the generator to supply extra damping to power system oscillations.

However, it has been found that installation of the PSSs cannot always provide satisfactory solution to the problem of power system oscillations. Typical example is the inter-area power oscillations on the tie lines or along a particular long-distance power transmission corridor in a large-scale power system. If there is no power plant close to the tie lines and the transmission corridor where power oscillations are observed, it may need complex coordinated design of multiple PSSs installed at different remote power plants to jointly suppress the power oscillations. There is no guarantee that the coordinated design of those multiple PSSs is always able to supply sufficient damping to the power oscillations. Hence, any new option to the PSS for the effective suppression of power system oscillations is always welcome.

With the advent of FACTS technology, FACTS-based stabilizers (referred to as FACTS stabilizers in this book) have been investigated as an alternative type of controllers to damp power system oscillations. Main difference of the FACTS stabilizers to PSSs is that the FACTS devices, which the FACTS stabilizers are attached to, usually are installed at key locations of transmission network in power systems, such as that close to the tie lines and long-distance transmission corridors. Hence in certain cases, FACTS stabilizers could be simpler, easier to be implemented, and more effective to suppress power system oscillations.

In fact, the idea of applying variable series compensation to damp power system oscillations was proposed many decades ago [20, 21], though at that time, variable compensation was assumed to be realized by switching in or out the series capacitors using mechanical circuit breakers. Only with the thyristor-controlled high-power electronic circuits being available after the 1980s, high-speed variable reactive power compensation has become a feasible technology in power system application to suppress power system oscillations. Between the late 1980s and 1990s, case study of numerical calculation, computer simulation, and field tests confirmed the capability of thyristor-based FACTS stabilizers to damp power system oscillations. Theoretical analysis was carried out to obtain insight into and to understand the basic principles of thyristor-based FACTS stabilizers. Similarity of the thyristor-based FACTS stabilizers to the PSSs has helped considerably the understanding on why and how the thyristor-based FACTS stabilizers can suppress power system oscillations.

A FACTS stabilizer is superimposed on the normal control function of a FACTS controller. For example, Fig. 1.4 shows a TCR (thyristor-controlled reactor)-FC (fixed capacitor) type of SVC with a normal function of voltage control. A damping controller (the SVC stabilizer) is superimposed on the SVC voltage control function. In Fig. 1.4, the variable inductive current $\bar{I}_l$ is controlled by the thyristor firing circuit. Hence, the variable equivalent reactance of the TCR-FC SVC is as follows

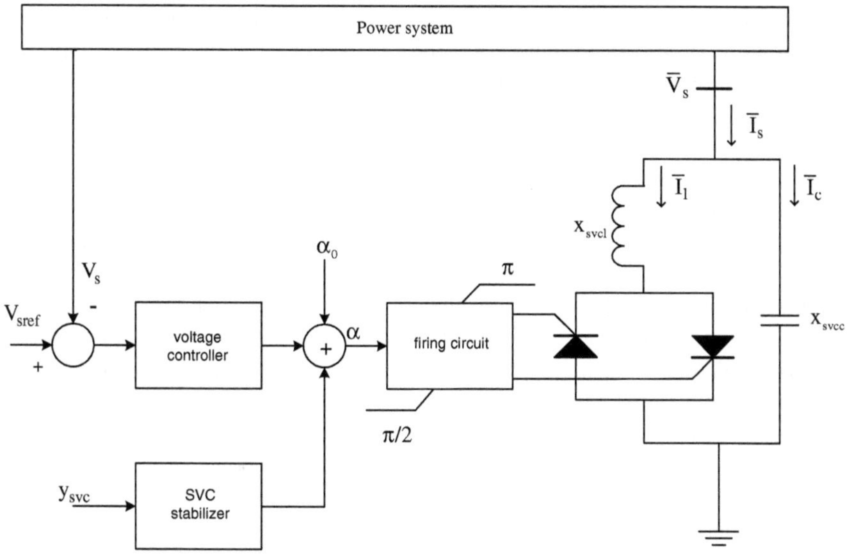


**Fig. 1.4** An SVC stabilizer superimposed on the SVC voltage control function

$$jx_{svc} = \frac{\bar{V}_s}{\bar{I}_s} = \frac{\bar{V}_s}{(\bar{I}_l + \bar{I}_c)} \tag{1.2}$$

where and in Fig. 1.4, $\bar{V}_s$ is the voltage measurement at the location where the SVC is installed, $\bar{V}_{sref}$ the reference signal of voltage control, $jx_{svc}$ the equivalent reactance of the SVC, and $y_{svc}$ the feedback signal of the SVC stabilizer. The SVC voltage controller is to maintain the magnitude of voltage at the SVC installing location close to $\bar{V}_{sref}$ by controlling the variable equivalent reactance of the SVC. Attachment of the SVC stabilizer contributes the control of variable equivalent reactance $jx_{svc}$, hence the exchange of reactive power between the SVC and rest of the power system. Thus, it affects the variations of active power flow along the transmission line where the SVC locates. If designed properly, the SVC stabilizer can help to suppress power oscillations. Attachment of the SVC stabilizer to SVC normal voltage control function is very similar to the arrangement of a PSS being superimposed on the voltage control loop of an AVR.

It was well reckoned that VSC-based FACTS stabilizers work as effectively as the thyristor-based FACTS stabilizers in damping power system oscillations, as confirmed by case study of numerical calculation, computer simulation, and field tests [22, 23]. There is no fundamental difference between the mechanism of the thyristor-based and VSC-based FACTS stabilizers, though examination of the VSC-based FACTS stabilizers usually is sometimes more complicated as it involves the VSC functions.

Occurrence of power system oscillations is essentially due to the response of generators to system disturbances in order to maintain the balance of supply and consumption of active power. Because generators need time to either accelerate or decelerate in responding to disturbances and they are at different geographic positions in the power system, power oscillations occur as the result of the relatively slow correcting actions of generators to keep the balance of active power. The slower the correcting actions are, the worse the power oscillations. That is why in the power system, the power oscillations are of low oscillation frequency and hence called low-frequency oscillations. Gird-connected ESSs do not have rotating components involved for the exchange of active power with the rest of power system. They can respond instantaneously to system disturbances with correcting actions to maintain the balance of active power in the power system. Hence in principle, applications of the ESSs in the power system can effectively provide extra damping to power system oscillations.

Research on the capability of ESS controllers to damp power system oscillations has been mainly about two types of energy storage devices, the superconducting magnetic energy storage (SMES) and the battery energy storage systems (BESS) that are also classified as FACTS devices [24]. The function and effectiveness of a SMES to damp power system oscillations have been well investigated since several decades ago. By regulating the active and reactive power exchange of the SMES with the power system, the SMES can effectively suppress power system oscillations, which have been demonstrated by simulation and laboratory experiment [25, 26].

In [27], a case of field application of the BESS in improving power oscillation damping was reported. A 10-MW BESS unit was installed in 1988 at the Chino substation owned by the Southern California Edison (SCE), USA. Six years later, an energy source power system stabilizer (ESPSS) was installed by taking the advantage of very fast response rate of the installed BESS to provide extra damping to the power system. The ESPSS was designed to modulate the active power output from the BESS to damp power system oscillations rather than through controlling voltage or reactive power as a conventional stabilizer does.

## 1.5 Design of Damping Controllers to Suppress Power System Oscillations

The prime specification of designing a damping controller (a PSS or a FACTS stabilizer) is the maximum effectiveness of the stabilizer at the minimum control cost. Two main issues in the design are as follows: (1) selection of installing locations and feedback signals and (2) parameters setting of the stabilizer.

In a large-scale multi-machine power system, there are many power stations that can be or have been equipped with the PSSs. To damp a particular oscillation mode of interests, it would be impractical if all the PSSs in the power system are set accordingly. Hence, the selection of installing locations of the PSSs can mean to choose the locations to install PSSs or to choose the PSSs among those installed to be set, in order to damp the particular oscillation mode. As far as a FACTS stabilizer is concerned, justification of installing an expensive FACTS device usually is based on other applications rather than the damping function. Hence, the selection of installing location of the FACTS stabilizer is carried out among the installed FACTS devices and/or FACTS normal control functions to attach the damping function. Even if a location in the power system is the best to have the FACTS stabilizer, it often needs further justification to consider the installation of the FACTS stabilizer there.

Leaving the practical consideration about the candidate locations aside, there is no difference in the strategy to select installing locations for the PSSs and FACTS stabilizers. On the other hand, selection of feedback signals for the PSSs and FACTS stabilizers normally is to determine the most effective and locally available feedback signals for the design of stabilizers. Those signals can be the rotor speed deviation of generators, active power, line current, etc. In the case that remote signals can be used, of course, it also includes the choice of signals at different locations in the power system.

Methods which have been proposed and developed so far for the selection of installing locations and feedback signals are mostly based on the DTA and modal analysis (MA). Examples are the induced damping and synchronizing torque coefficients (IDSTCs) index [28], damping torque coefficient (DTC) index [29], simple index [30], and damping index [31] derived by using the DTA. Sensitivity calculation [32], the participation factors [13–15], the LIED (location index for effective damping) [33], residue index [34, 35], the partial modal analysis [16], and eigensolution-free modal analysis [17] are established by use of the modal analysis.

Methods to set parameters of multiple stabilizers in a multi-machine power system can be classified into two categories, sequence setting and simultaneous tuning. Sequence setting is the strategy to set stabilizers one by one in a sequence. For example, suppose two stabilizers, stabilizers A and B, to be set. In sequence setting, stabilizer A is installed in the multi-machine power system firstly and its parameters are set. Afterwards, stabilizer B is installed with its parameters being set. The strategy of sequence setting fits the practice of installing stabilizers in the multi-machine power system. It is simple and easy to be used. The well-known phase compensation method [9–11] is considered to be an approach of sequential setting.

However, it was found that this strategy of sequential setting can cause a problem of so-called "eigenvalue drifting" [36] due to the dynamic interactions of multiple stabilizers."Eigenvalue drifting" is that, for example, setting of stabilizer B affects the setting results of stabilizer A which is designed previously. The consequence is that the sequential setting cannot complete the required setting of two stabilizers, A and B, to move oscillation modes to their required positions simultaneously. This phenomenon of "eigenvalue drifting" is also referred to as the problem of dynamic interactions between multiple stabilizers.

In order to solve the problem of "eigenvalue drifting" caused by interactions between multiple stabilizers, the simultaneous tuning strategy has been proposed and developed for the design of multiple stabilizers. In the simultaneous tuning, parameters of all the stabilizers are tuned simultaneously. Hence, once the tuning is completed, all the oscillation modes involved in the tuning are at the target positions. Simultaneous tuning is often called the coordinated design of multiple stabilizers. So far there have been many methods of simultaneous tuning for different types of stabilizers proposed and developed. Those methods to simultaneously tune parameters of multiple stabilizers can be classified into two groups. The first group are the analytical methods developed either from the DTA technique or by applying the modern control theory. The second group are the methods of parameter optimization. They usually convert the simultaneous tuning to the solution of an optimization problem by establishing an objective function in the space of stabilizers' parameters. Then, an optimization algorithm is employed to find the solution of the objective function. Normally, the second group of methods heavily relies on the eigensolution of the power system.

Other two important issues in the design of stabilizers are the robustness of and interactions between multiple stabilizers. The robustness is about the effectiveness of stabilizers when power system operating conditions vary. Because the linearized model of a power system is established at a chosen operating condition for the design of stabilizers, the effectiveness of stabilizers is ensured only at the operating point by the design. Hence, the robustness of stabilizers to the variations of power system operating conditions is an issue which needs special consideration in order to guarantee the effectiveness of the stabilizers over a certain range of system operating conditions. The interactions are those among multiple stabilizers themselves and with other control functions on which they are superimposed. The well-known examples are the "eigenvalue drifting" and interaction between a PSS and the AVR it is attached to.

## 1.6 Organization of the Book

Linearized methods for the analysis and damping control of power system oscillations covered by this book are based on the linearized models of power systems. Hence, a comprehensive introduction on establishing linearized models of power systems is presented in the book. This includes the introduction of the Heffron–Phillips model, a well-known and relatively simpler linearized model of power system for the study of power system oscillations. In addition, the establishment of a more general linearized model of power systems is introduced where the full-scale mathematical model of synchronous generators is used.

Two groups of linearized methods for the analysis and damping control of power system oscillations are introduced in the book. They are the damping torque

**Table 1.1** Chapter arrangement of the book

| Type of stabilizers / Type of power systems | Single-machine infinite-bus power systems | Multi-machine power systems |
|---|---|---|
| The PSSs | Chapter 2 | Chapter 5 |
| The thyristor-based FACTS stabilizers | Chapter 3 | Chapter 6 |
| The VSC-based FACTS stabilizers | Chapter 4 | Chapter 7 |

analysis (DTA) and the modal analysis (MA). The basic concepts and theory of the DTA and the MA are presented firstly for a simpler case of power systems, a single-machine infinite-bus power system. The introduction of the applications and extensions of the DTA and the MA in a more complicated case of a multi-machine power system is then followed in the book.

Power system damping controllers are classified into three groups in the book, the power system stabilizer (PSS), the thyristor-based FACTS stabilizers, and the VSC-based stabilizers. Introduction of linearized methods for the analysis and damping control of power system oscillations in the book is organized according to the classification of damping controllers (stabilizers) and their applications in the simple single-machine infinite-bus and complex multi-machine power system. The organization is illustrated in Table 1.1.

## References

1. IEEE/CIGRE Joint Task Force on Stability Terms and Definitions (2004) Definition and classification of power system stability. IEEE Trans Power Syst 19(2):1387–1401
2. Schleif FR, White JH (1966) Damping for the northwest-southwest tieline oscillations—an analogue study. IEEE Trans Power Appar Syst 85(12):1239–1247
3. Gibson CM (1988) Application of power system stabilizers on the Anglo-Scottish interconnection—programme of system proving tests and operational experience. IEE Proc Part C 135(3):255–260
4. Hsu Y-Y, Shyue S-W, Su C-C (1987) Low frequency oscillations in longitudinal power systems: experience with dynamic stability of Taiwan power system. IEEE Trans Power Syst 2 (1):92–98
5. Western Electricity Coordinating Council (1996) Western systems coordinating council disturbance report. www.bpa.biz
6. deMello FP, Concordia C (1969) Concepts of synchronous machine stability as affected by excitation control. IEEE Trans Power Appar Syst 88(4):316–329
7. Heffron WG, Phillips RA (1952) Effect of modern amplidyne voltage regulators on underexcited operation of large turbine generators. AIEE Trans (Power Appar Syst) 71: 692–697
8. deMello FP, Laskowski TF (1975) Concepts of power system dynamic stability. IEEE Trans Power Appar Syst 94(3):827–833
9. Larsen EV, Swann DA (1981) Applying power system stabilizers part I: general concepts. IEEE Trans Power Appar Syst 100(6):3017–3024

10. Larsen EV, Swann DA (1981) Applying power system stabilizers part II: performance objectives and tuning concepts. IEEE Trans Power Appar Syst 100(6):3025–3033
11. Larsen EV, Swann DA (1981) Applying power system stabilizers part III: practical considerations. IEEE Trans Power Appar Syst 100(6):3034–3046
12. Rogers G (2000) Power system oscillations. Kluwer Academic Publisher
13. Perez-Arriaga IJ, Verghese GC, Schweppe FC (1982) Selective modal analysis with applications to electric power systems, Part I. IEEE Trans Power Appar Syst 101(9):3117–3125
14. Perez-Arriaga IJ, Verghese GC, Schweppe FC (1982) Selective modal analysis with applications to electric power systems, part II. IEEE Trans Power Appar Syst 101(9):3126–3134
15. Sancha JL, Perez-Arriaga IJ (1988) Selective modal analysis of electric power system oscillatory instability. IEEE Trans Power Syst 3(2):429–438
16. Larsen EV, Sanchez-Gasca JJ, Chow JH (1995) Concept for design of FACTS controllers to damp power swings. IEEE Trans Power Syst 10(2):948–956
17. Wang HF (1999) Selection of robust installing locations and feedback signals of FACTS-based stabilizers in multi-machine power systems. IEEE Trans Power Syst 14 (2):569–574
18. Hingorani NG (1988, Aug) High power electronics and flexible ac transmission systems. IEEE Power Eng Rev 3–4
19. Gyugyi L (1992) Unified power-flow control; concept for flexible AC transmission systems. IEE Proc Part C 139(4):323–331
20. Smith OJM (1969) Power system transient control by capacitor switching. IEEE Trans Power Appar Syst 88(1):28–35
21. Webster RH, Mane AP, Smith OJM (1971) Series capacitor switching to quench electromechanical transient in power systems. IEEE Trans Power Appar Syst 99(2):427–433
22. Song YH, Johns AT (1999) Flexible AC transmission systems. IEE Press
23. Hingorani NG, Gyugyi L (1999) Understanding FACTS. IEEE Press
24. CIGRE TF 30-01-08 Report (1999) Modelling of power electronics equipment (FACTS) in load flow and stability programs
25. Hsu CS, Lee WJ (1993) Superconducting magnetic energy storage for power system applications. IEEE Trans Ind Appl 29(5):990–996
26. Xue XD, Cheng KWE, Sutanto D (2005) Power system applications of superconducting magnetic energy storage systems. In: Fourtieth IAS annual conference meeting on industry applications, vol 2, Oct 2005, pp 1524–1529
27. Bhargava B, Dishaw G (1998) Application of an energy source power system stabilizer on the 10 MW battery energy storage system at Chino substation. IEEE Trans Power Syst 13(1): 145–151
28. Pourbeik P, Gibbard MJ (1996) Damping and synchronizing torques induced on generators by FACTS stabilizers in multimachine power systems. IEEE Trans Power Syst 11(4):1920–1925
29. Swift FJ, Wang HF (1995) Static Var compensator to damp system oscillations in multi-machine power systems. In: Proceedings of IEE APSCOM conference, Hong Kong
30. Swift FJ, Wang HF (1996) Application of the controllable series compensator in damping power system oscillations. IEE Proc Part C 143(4):359–364
31. Wang HF, Swift FJ (1997) The indexes for selecting the best locations of PSS or FACTS-based stabilizers in multi-machine power systems: a comparison study. IEE Proc Part C 144(2):155–159
32. Abdalla OH, Hassan SA, Tweig NT (1984) Co-ordinated stabilisation of a multimachine power system. IEEE Trans Power Appar Syst 103(3):483–494
33. Okamoto H, Kurita A, Sekine Y (1995) A method for identification of effective locations of variable impedance apparatus on enhancement of steady-state stability in large-scale power systems. IEEE Trans Power Syst 10(3):1401–1407

34. Martins N, Lima LTG (1990) Determination of suitable locations for power system stabilizers and static VAR compensators for damping electromechanical oscillations in large power systems. IEEE Trans Power Syst 5(4):1455–1469
35. Yang X, Feliachi A (1994) Stabilization of inter-area oscillation modes through excitation systems. IEEE Trans Power Syst 9(1):494–502
36. Gooi HB, Hill EF, Mobarak MA, Thonne DH, Lee TH (1981) Coordinated multimachine stabilizer setting without eigenvalue drift. IEEE Trans Power Appar Syst 100(12):3879–3887

# Chapter 2
# A Single-Machine Infinite-Bus Power System Installed with a Power System Stabilizer

## 2.1 Linearized Model of a Single-Machine Infinite-Bus Power System Installed with a Power System Stabilizer

### *2.1.1 General Linearized Mathematical Model*

#### 2.1.1.1 Full Mathematical Model of a Synchronous Generator

Fundamental equations describing the dynamics of a synchronous generator are the well-known Park's voltage equations. They are given based on a coordinate system consisting of a d-axis (direct axis) fixed on the field winding axis of synchronous generator and a q-axis (quadrature axis). After Park's transformation, three armature phase windings a, b, and c on the stator of synchronous generator are replaced by two equivalent armature phase windings, d and q. Two damper windings on the rotor, D and Q, are permanently short-circuited. Field winding f is DC excited. Park's voltage equations of those five windings have the simplest form as follows:

$$\begin{aligned}
\dot{\psi}_d &= \omega_0(v_{td} + r_a i_d + \omega\psi_q) \\
\dot{\psi}_q &= \omega_0(v_{tq} + r_a i_q - \omega\psi_d) \\
\dot{\psi}_f &= \omega_0(v_f - r_f i_f) \\
\dot{\psi}_D &= -\omega_0 r_D i_D \\
\dot{\psi}_Q &= -\omega_0 r_Q i_Q
\end{aligned} \tag{2.1}$$

where $\psi_d, \psi_q, \psi_f, \psi_D, \psi_Q$, $i_d, i_q, i_f, i_D, i_Q$, $v_{td}, v_{tq}, v_f$, and $r_a, r_f, r_D, r_Q$ are the magnetic flux linkage, current, voltage, and resistance of each corresponding winding, respectively, $\omega_0$ is the synchronous speed, and $\omega$ is the rotor speed in per unit (p.u.)



H. Wang and W. Du, *Analysis and Damping Control of Power System Low-frequency Oscillations*, Power Electronics and Power Systems,
DOI 10.1007/978-1-4899-7696-3_2

of synchronous generator. Since there is no static coupling between any d-axis winding and q-axis winding, the relationship between the magnetic flux linkage and the current of those windings is as follows:

$$\begin{bmatrix} \psi_d \\ \psi_f \\ \psi_D \end{bmatrix} = \begin{bmatrix} x_d & x_{ad} & x_{ad} \\ x_{ad} & x_f & x_{ad} \\ x_{ad} & x_{ad} & x_D \end{bmatrix} \begin{bmatrix} -i_d \\ i_f \\ i_D \end{bmatrix}$$
$$\begin{bmatrix} \psi_q \\ \psi_Q \end{bmatrix} = \begin{bmatrix} x_q & x_{aq} \\ x_{aq} & x_Q \end{bmatrix} \begin{bmatrix} -i_q \\ i_Q \end{bmatrix} \tag{2.2}$$

where $x_d, x_q, x_f, x_D, x_Q$ are the self-reactance of winding d, q, f, D, and Q, respectively. In Eq. (2.2), it is assumed that the mutual reactance of all windings on a common axis is same, being $x_{ad}$ or $x_{aq}$ respectively.

The rotor motion equation of synchronous generator is as follows:

$$\begin{aligned} \dot{\delta} &= \omega_0(\omega - 1) \\ \dot{\omega} &= \frac{1}{M}[T_m - T_t - D(\omega - 1)] \end{aligned} \tag{2.3}$$

where M is the inertia of the rotor, D the damping coefficient of the rotor motion, $\delta$ the rotor angular position of synchronous generator to a reference axis, and $T_m$ and $T_t$ the mechanical torque and electric torque applied on the rotor of generator, respectively.

#### 2.1.1.2 Excitation System and the Automatic Voltage Regulator (AVR)

Voltage control in a power system is closely related to the regulation of reactive power flow. The objectives of voltage and reactive power control in the power system are as follows:

1. To maintain the voltage at terminals of all equipment in the power system within acceptable limits. As far as the system voltage profile is concerned, a constraint of system voltage deviations is imposed normally to be within ±5 % of the nominal value of voltage.
2. To improve system stability in order to maximize the utilization of the transmission system.
3. To minimize reactive power flow so as to reduce transmission losses.

Power system voltage control has a hierarchy structure with multiple levels. At the primary level, control devices attempt to compensate the rapid and random voltage variations by maintaining their output variables close to the setting reference values. The highest level of voltage control uses global information of power

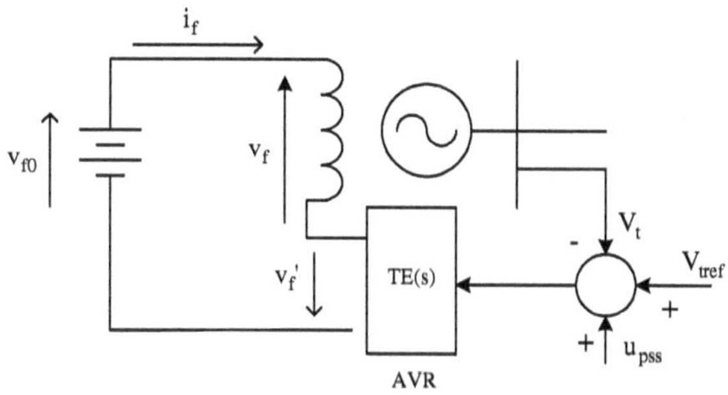


**Fig. 2.1** Arrangement of an AVR

system and is implemented by solving some optimization problems. A synchronous generator can generate or absorb reactive power depending on its excitation, which is controlled by a voltage control device, the automatic voltage regulator (AVR). Figure 2.1 shows simple illustration on the arrangement of the AVR on the synchronous generator.

In Fig. 2.1, TE(s) denotes the transfer function of the AVR. The excitation provided to the field winding of synchronous generator is $v_f$ which is from two sources. One is a constant excitation, $v_{f0}$, and another is the output from the AVR, $v_f'$. The AVR measures the terminal voltage of synchronous generator, $V_t$, and compares it to a reference setting value $V_{tref}$. The AVR responds to the deviation of terminal voltage to change the excitation of generator and hence regulates the reactive power supply or absorption by generator. By doing so, the terminal voltage of generator is effectively maintained close to the reference setting value. Therefore, the AVR works at the bottom of the hierarchy of voltage control of power system to eliminate voltage variations at generator's terminal. On the other hand, the reference setting value of the AVR is provided by the result of voltage and/or reactive control optimization of the whole power system such that the voltage profile of the system is kept within required constraints and the transmission losses are minimized. Hence, $V_{tref}$ is from the higher level of hierarchy of voltage control of the power system.

Historically, the role of excitation system to improve power system performance has been growing. Early excitation systems were for the maintenance of constant voltage level at generator's terminals and were implemented manually. Gradually, fast-acting AVRs were installed in many generators not only providing a satisfactory voltage control performance, but also considerably improving power system steady-state and transient stability. However, during 1960s to 1970s, it was found that fast-acting AVRs have an adverse effect in providing negative damping to power system oscillations in some occasions. This results in a conflict requirement in the design of the AVRs. Subsequently, the power system stabilizer (PSS) was introduced into excitation control to overcome the problem. Nowadays, it is quite common to have a combination of a fast-acting AVR and a PSS for the excitation control of synchronous generators. Stabilizing signal, $u_{pss}$, is superimposed on that of the AVR as shown in Fig. 2.1.

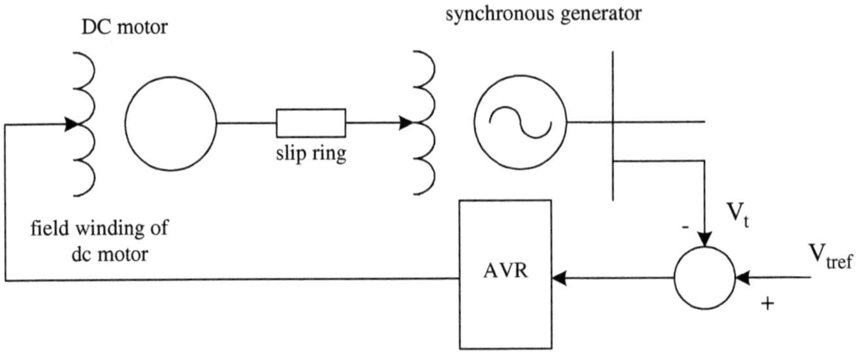


**Fig. 2.2** Arrangement of a DC excitation system

Based on the difference of the excitation power sources used, excitation systems can be classified into three major types.

1. DC excitation systems

A DC excitation system uses a DC generator as the source of excitation power to provide field current and is connected to field winding through slip rings. The exciter may be driven by a motor or by the generator itself. The DC excitation system represents the early application of excitation control. Figure 2.2 shows the arrangement of the DC excitation system.

2. AC excitation systems

An AC excitation system uses an AC machine (alternator) as the source of excitation power to provide field current. The AC excitation current is rectified to provide the DC excitation to the synchronous generator. Usually, the AC exciter is on the same shaft of synchronous generator. There are two major types of AC excitation systems, depending on the difference of the arrangement of AC excitation from either the stationary or rotating armature winding of the exciter, as shown in Figs. 2.3 and 2.4, respectively. In a rotating rectifier excitation system, the armature windings of the AC exciter and the diode rectifiers rotate with the synchronous generator field. Thus, the need for slip rings and brushes is eliminated. Hence, such a system is also called a brushless excitation system.

3. Static excitation systems

In a static excitation system, power supply is from the synchronous generator. The DC excitation is provided to the field of generator through slip rings after being rectified. There are three major types of static excitation systems. They are potential-source controlled-rectifier systems, compound-source rectifier systems, and compound-controlled rectified excitation systems. Figure 2.5 shows the arrangement of a potential-source controlled-rectifier excitation system.

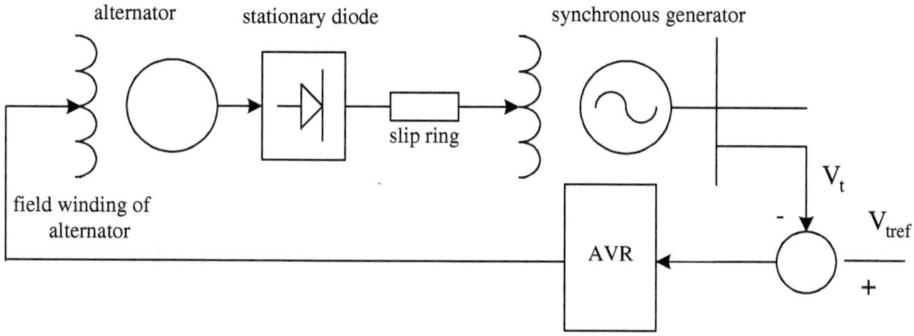


**Fig. 2.3** Arrangement of a stationary rectifier excitation system

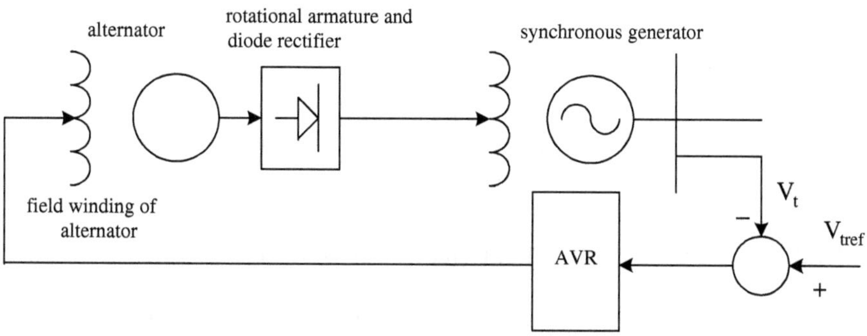


**Fig. 2.4** Arrangement of a rotating rectifier excitation system

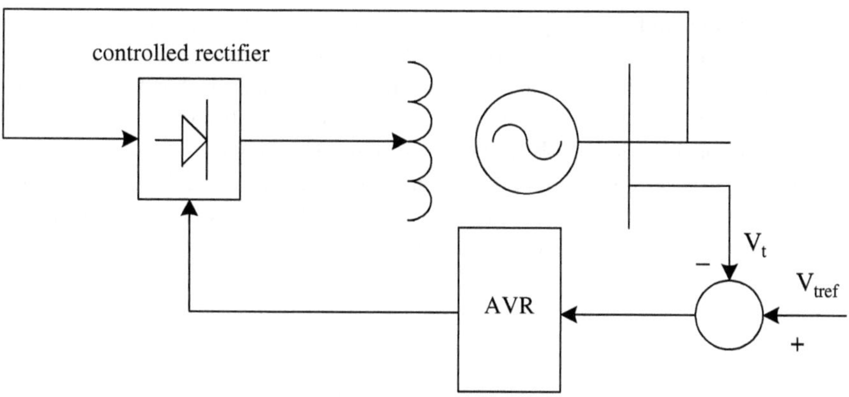


**Fig. 2.5** Arrangement of a potential-source controlled-rectifier excitation system

From Fig. 2.1, it can have

$$v_f = v_{f0} + TE(s)(V_{tref} - V_t + u_{pss}) \tag{2.4}$$

where $v_{f0}$ is the constant excitation, $u_{pss}$ the stabilizing signal of the PSS, and $V_t$ and $V_{tref}$ the terminal voltage of generator and its reference setting value, respectively. Various forms of transfer function have been recommended for different types of excitation systems and the AVR. In this book, the following simplest form is used for the purpose of simple presentation,

$$TE(s) = \frac{K_A}{1 + sT_A} \tag{2.5}$$

where $K_A$ is the gain and $T_A$ the time constant of the AVR. Hence, from Eqs. (2.4) and (2.5), mathematical model of the AVR can be written as

$$\begin{aligned} v_f &= v_{f0} + v_f' \\ \dot{v}_f' &= -\frac{1}{T_A} v_f' + \frac{K_A}{T_A}(V_{tref} - V_t + u_{pss}) \end{aligned} \tag{2.6}$$

#### 2.1.1.3 A Single-Machine Infinite-Bus Power System

Figure 2.6 shows the configuration of a power system where a generator sends power to a large network. Capacity of the large network is much greater than that of the generator such that operation of the large network is not affected at all by any changes in the part of the power system on the left-hand side of busbar b in Fig. 2.6. This effectively means that the voltage and frequency at busbar b are constant when the focus of the study is the part of the left-hand side of the power system. Thus, from the point of view of operation of the part of left-hand side of the power system, capacity of the large network is "infinite". Hence, busbar b is called the "infinite busbar", and the part of the power system on the left-hand side of busbar b is a "single-machine infinite-bus" power system. The single-machine infinite-bus power system is an approximate representation of a kind of real power systems, where a power plant with a generator or a group of generators are connected by transmission lines to a very large power network.

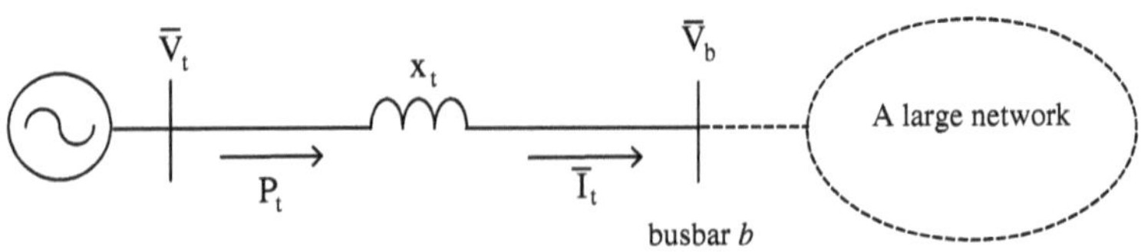


**Fig. 2.6** A single-machine infinite-bus power system

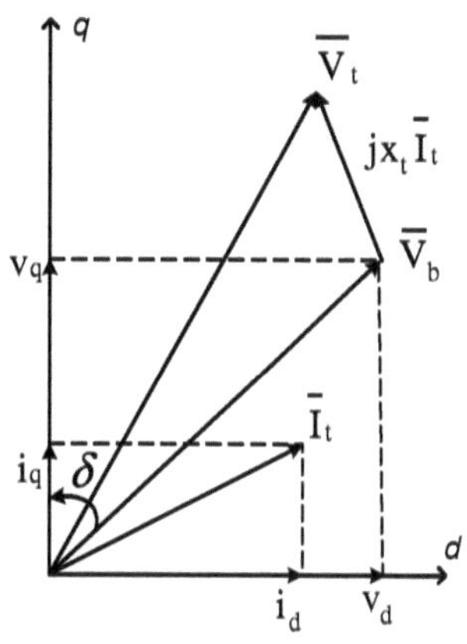


**Fig. 2.7** d–q coordinate of generator

For the single-machine infinite-bus power system shown in Fig. 2.6, it can have

$$\bar{V}_t = jx_t\bar{I}_t + \bar{V}_b \tag{2.7}$$

In the d–q coordinate of the generator shown in Fig. 2.7, Eq. (2.7) can be written as follows:

$$v_{td} + jv_{tq} = jx_t(i_d + ji_q) + v_d + jv_q \tag{2.8}$$

where $v_{td}, v_{tq}$ and $i_d, i_q$ $v_d, v_q$ are the d and q components of terminal voltage of generator, $\bar{V}_t$, line current, $\bar{I}_t$, and voltage at the infinite busbar, $\bar{V}_b$, respectively. Comparing the real and imaginary part on the both sides of Eq. (2.8), it can have

$$\begin{aligned} v_{td} &= -x_t i_q + v_d \\ v_{tq} &= x_t i_d + v_q \end{aligned} \tag{2.9}$$

where $v_d = V_b \sin\delta$, $v_q = V_b \cos\delta$ and the terminal voltage of generator is as follows:

$$V_t = \sqrt{v_{td}^2 + v_{tq}^2} \tag{2.10}$$

In per unit, the mechanical and electric torque, $T_m$ and $T_t$, in Eq. (2.3), is equal to the mechanical power input from the prime mover to the electric power supplied by the generator, respectively, i.e. $T_m = P_m$ and $T_t = P_t$. While $P_t$ is equal to the electric power received at the infinite busbar, that is

$$P_t = v_{dt}i_d + v_{qt}i_q = v_d i_d + v_q i_q \tag{2.11}$$

Equations (2.1)–(2.3), (2.6) and (2.9)–(2.11) are the complete dynamic model of the single-machine infinite-bus power system shown in Fig. 2.6 where $V_b$ and $P_m$ are constant.

#### 2.1.1.4 Linearized Model of Single-Machine Infinite-Bus Power System

Linearization of Eqs. (2.1) and (2.2) is as follows:

$$
\begin{aligned}
\Delta\dot{\psi}_d &= \omega_0(\Delta v_{td} + r_a\Delta i_d + \omega_0\Delta\psi_q + \psi_{q0}\Delta\omega) \\
\Delta\dot{\psi}_q &= \omega_0(\Delta v_{tq} + r_a\Delta i_q - \omega_0\Delta\psi_d - \psi_{d0}\Delta\omega) \\
\Delta\dot{\psi}_f &= \omega_0(\Delta v_f - r_f\Delta i_f) \\
\Delta\dot{\psi}_D &= -\omega_0 r_D\Delta i_D \\
\Delta\dot{\psi}_Q &= -\omega_0 r_Q\Delta i_Q
\end{aligned}
\tag{2.12}
$$

$$
\begin{aligned}
\begin{bmatrix}\Delta i_d\\ \Delta i_f\\ \Delta i_D\end{bmatrix} &= \begin{bmatrix} x_d & x_{ad} & x_{ad}\\ x_{ad} & x_f & x_{ad}\\ x_{ad} & x_{ad} & x_D\end{bmatrix}^{-1}\begin{bmatrix}\Delta\psi_d\\ \Delta\psi_f\\ \Delta\psi_D\end{bmatrix}\\
\begin{bmatrix}\Delta i_q\\ \Delta i_Q\end{bmatrix} &= \begin{bmatrix} x_q & x_{aq}\\ x_{aq} & x_Q\end{bmatrix}^{-1}\begin{bmatrix}\Delta\psi_q\\ \Delta\psi_Q\end{bmatrix}
\end{aligned}
\tag{2.13}
$$

where prefix, $\Delta$, and subscript 0 are used to denote small increment of a variable (linearized variable) and value of the variable at the power system steady-state operating condition where the linearization is carried out, respectively. This notation will be used throughout this book.

Linearization of Eqs. (2.10) and (2.11) is as follows:

$$
\Delta V_t = \frac{v_{td0}}{V_{t0}}\Delta v_{td} + \frac{v_{tq0}}{V_{t0}}\Delta v_{tq} \tag{2.14}
$$

$$
\Delta P_t = v_{dt0}\Delta i_d + v_{qt0}\Delta i_q + i_{d0}\Delta v_{dt} + i_{q0}\Delta v_{qt} \tag{2.15}
$$

By using Eqs. (2.14) and (2.15), linearization of Eqs. (2.3) and (2.6) can be obtained to be

$$
\begin{aligned}
\Delta\dot{\delta} &= \omega_0\Delta\omega \\
\Delta\dot{\omega} &= -\frac{1}{M}(\Delta P_t + D\Delta\omega) \\
&= -\frac{1}{M}(v_{dt0}\Delta i_d + v_{qt0}\Delta i_q + i_{d0}\Delta v_{dt} + i_{q0}\Delta v_{qt} + D\Delta\omega)
\end{aligned}
\tag{2.16}
$$

$$
\begin{aligned}
\Delta v_f &= \Delta v'_f \\
\Delta \dot{v}'_f &= -\frac{1}{T_A}\Delta v'_f + \frac{K_A}{T_A}\left(-\frac{v_{td0}}{V_{t0}}\Delta v_{td} - \frac{v_{tq0}}{V_{t0}}\Delta v_{tq} + \Delta u_{pss}\right)
\end{aligned}
\tag{2.17}
$$

Arranging Eqs. (2.12), (2.13), (2.16), and (2.17) in matrix form with all linearized current variables be cancelled, it can have

$$
\begin{aligned}
s\mathbf{\Delta X_{g-dq}} &= \mathbf{A_{g-dq}\Delta X_{g-dq}} + \mathbf{B_{g-dq}\Delta V_{dq}} + \mathbf{b_{pss}}\Delta u_{pss} \\
\mathbf{\Delta I_{dq}} &= \mathbf{C_{g-dq}\Delta X_{g-dq}}
\end{aligned}
\tag{2.18}
$$

where

$$
\begin{aligned}
\mathbf{\Delta X_{g-dq}} &= \begin{bmatrix} \Delta\delta & \Delta\omega & \Delta v_f & \Delta\psi_d & \Delta\psi_q & \Delta\psi_f & \Delta\psi_D & \Delta\psi_Q \end{bmatrix}^T, \\
\mathbf{\Delta V_{dq}} &= \begin{bmatrix} \Delta v_{td} & \Delta v_{tq} \end{bmatrix}^T, \mathbf{\Delta I_{dq}} = \begin{bmatrix} \Delta i_d & \Delta i_q \end{bmatrix}^T
\end{aligned}
$$

For the single-machine infinite-bus power system, the network voltage equation is Eq. (2.9). Its linearization is as follows:

$$
\begin{aligned}
\Delta v_{td} &= -x_t\Delta i_q + \Delta v_d = -x_t\Delta i_q + V_b\cos\delta_0\Delta\delta \\
\Delta v_{tq} &= x_t\Delta i_d + \Delta v_q = x_t\Delta i_d - V_b\sin\delta_0\Delta\delta
\end{aligned}
\tag{2.19}
$$

In matrix form, the above equation can be written as follows:

$$
\mathbf{\Delta V_{dq}} = \mathbf{F_{dq1}\Delta I_{dq}} + \mathbf{F_{dq2}\Delta X_{g-dq}}
\tag{2.20}
$$

where

$$
\mathbf{F_{dq1}} = \begin{bmatrix} 0 & -x_t \\ x_t & 0 \end{bmatrix}, \mathbf{F_{dq2}} = \begin{bmatrix} V_b\cos\delta_0 & \mathbf{0} \\ -V_b\sin\delta_0 & \mathbf{0} \end{bmatrix}
$$

Substituting Eq. (2.20) into (2.18), state-equation model of the single-machine infinite-bus power system is obtained to be

$$
s\mathbf{\Delta X_{g-dq}} = \mathbf{A_{gc-dq}\Delta X_{g-dq}} + \mathbf{b_{pss}}\Delta u_{pss}
\tag{2.21}
$$

where $\mathbf{A_{gc-dq}} = \mathbf{A_{g-dq}} + \mathbf{B_{g-dq}F_{dq1}C_{g-dq}} + \mathbf{B_{g-dq}F_{dq2}}$.

## 2.1.2 *Heffron–Phillips Model*

### 2.1.2.1 Simplification

For the study of power system oscillations, full mathematical model of synchronous generator of Eqs. (2.1)–(2.2) can be simplified based on the following considerations:

1. Effect of damper windings is not considered or directly included in the damping coefficient D in the rotor motion equation in Eq. (2.3). Thus, Eq. (2.1) is simplified to be

$$\begin{aligned}\dot{\psi}_d &= \omega_0(v_{td} + r_a i_d + \omega\psi_q)\\ \dot{\psi}_q &= \omega_0(v_{tq} + r_a i_q - \omega\psi_d)\\ \dot{\psi}_f &= \omega_0(v_f - r_f i_f)\end{aligned} \tag{2.22}$$

2. Effect of fast transient and the resistance of d and q armature windings are neglected. Equation (2.22) is further simplified to be

$$\begin{aligned}0 &= v_{td} + \omega\psi_q\\ 0 &= v_{tq} - \omega\psi_d\\ \dot{\psi}_f &= \omega_0(v_f - r_f i_f)\end{aligned} \tag{2.23}$$

3. In small-signal power oscillations, variation of rotor speed is very small, $\omega \approx 1$. Hence, the first two equations in Eq. (2.23) become

$$\begin{aligned}v_{td} &= -\psi_q\\ v_{tq} &= \psi_d\end{aligned} \tag{2.24}$$

To transform the third equation in Eq. (2.23) into a different form, it is defined that

$$E'_q = \frac{x_{ad}}{x_f}\psi_f,\ E_q = x_{ad} i_f,\ E_{fd} = \frac{x_{ad} v_f}{r_f} \tag{2.25}$$

where $E'_q$ is called the q-axis transient excitation voltage, $E_q$ the q-axis excitation voltage, and $E_{fd}$ the excitation voltage. Multiplying both sides of the third equation in Eq. (2.23) by $\frac{x_{ad}}{r_f}$, it can have

$$T'_{d0}\dot{E}'_q = E_{fd} - E_q \tag{2.26}$$

where $T'_{d0} = \frac{x_f}{\omega_0 r_f}$, which is the time constant of the field winding.

Equation (2.26) together with Eq. (2.3) forms the simplified third-order model of synchronous generator. Equation (2.2) becomes

$$\begin{bmatrix}\psi_d \\ \psi_f\end{bmatrix} = \begin{bmatrix} x_d & x_{ad} \\ x_{ad} & x_f \end{bmatrix}\begin{bmatrix} -i_d \\ i_f \end{bmatrix} \qquad \psi_q = -x_q i_q \tag{2.27}$$

From Eqs. (2.24), (2.25), and (2.27), it can have

$$\begin{aligned} v_{td} &= -\psi_q = x_q i_q \\ v_{tq} &= \psi_d = x_{ad} i_f - x_d i_d = E_q - x_d i_d \end{aligned} \tag{2.28}$$

From Eqs. (2.25) and (2.27), it can be obtained that

$$E'_q = \frac{x_{ad}}{x_f}\psi_f = \frac{x_{ad}}{x_f}(x_f i_f - x_{ad} i_d) = E_q - \frac{x_{ad}^2}{x_f} i_d = E_q - (x_d - x'_d) i_d \tag{2.29}$$

where $x'_d = x_d - \frac{x_{ad}^2}{x_f}$, which is called the transient d-axis reactance. Thus, Eq. (2.26) becomes

$$T'_{d0}\dot{E}'_q = E_{fd} - E'_q - (x_d - x'_d) i_d \tag{2.30}$$

#### 2.1.2.2 A Simplified Model of Single-Machine Infinite-Bus Power System

For the single-machine infinite-bus power system shown in Fig. 2.6, from Eqs. (2.9) and (2.28), it can have

$$\begin{aligned} v_{td} &= v_d - x_t i_q = x_q i_q \\ v_{tq} &= v_q + x_t i_d = E_q - x_d i_d = E'_q - x'_d i_d \end{aligned} \tag{2.31}$$

Thus,

$$\begin{aligned} v_d &= (x_t + x_q) i_q = x_{q\Sigma} i_q \\ v_q &= E_q - (x_d + x_t) i_d = E'_q - (x'_d + x_t) i_d = E_q - x_{d\Sigma} i_d = E'_q - x'_{d\Sigma} i_d \end{aligned} \tag{2.32}$$

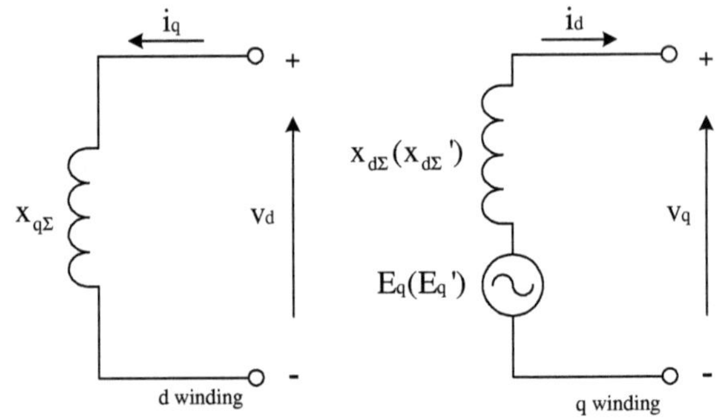


Fig. 2.8 Equivalent circuit model of single-machine infinite-bus power system

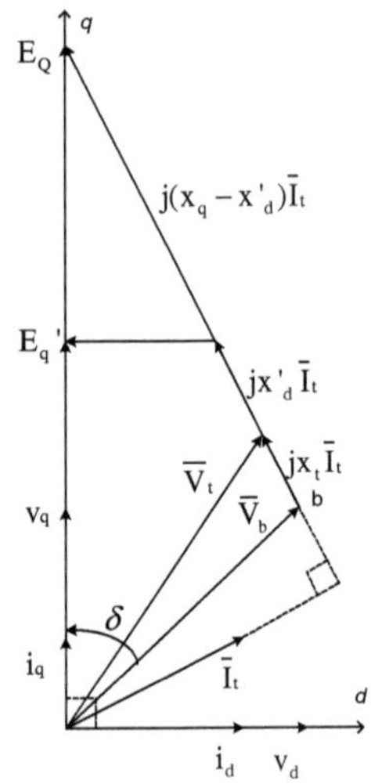


Fig. 2.9 Phasor diagram of single-machine infinite-bus power system

The single-machine infinite-bus power system can be represented by a circuit model of Fig. 2.8. Figure 2.9 shows the phasor diagram of the system on the d–q coordinate.

From Eq. (2.32) or Fig. 2.8, it can have

$$
\begin{aligned}
i_d &= \frac{E_q' - V_b \cos\delta}{x_{d\Sigma}'} \\
i_q &= \frac{V_b \sin\delta}{x_{q\Sigma}}
\end{aligned}
\tag{2.33}
$$

By substituting Eqs. (2.9) and (2.33) into Eq. (2.11), the electric power supplied by the generator can be expressed as follows:

$$
\begin{aligned}
P_t &= V_b \cos\delta \frac{V_b \sin\delta}{x_{q\Sigma}} + V_b \sin\delta \frac{E_q' - V_b \cos\delta}{x_{d\Sigma}'} \\
&= \frac{E_q' V_b}{x_{d\Sigma}'} \sin\delta - \frac{V_b^2}{2} \frac{(x_q - x_d')}{x_{d\Sigma}' x_{q\Sigma}} \sin 2\delta
\end{aligned}
\tag{2.34}
$$

From Eqs. (2.29) and (2.33), it can be obtained that

$$
\begin{aligned}
E_q &= E_q' + (x_d - x_d') i_d = E_q' + (x_d - x_d') \frac{E_q' - V_b \cos\delta}{x_{d\Sigma}'} \\
&= \frac{E_q' x_{d\Sigma}}{x_{d\Sigma}'} - \frac{(x_d - x_d') V_b \cos\delta}{x_{d\Sigma}'}
\end{aligned}
\tag{2.35}
$$

where $x_{d\Sigma} = x_d + x_t$. From Eqs. (2.31) and (2.33), it can have

$$
\begin{aligned}
v_{td} &= V_b \sin\delta - x_t i_q = V_b \sin\delta - x_t \frac{V_b \sin\delta}{x_{q\Sigma}} = \frac{x_q V_b \sin\delta}{x_{q\Sigma}}, \\
v_{tq} &= V_b \cos\delta + x_t i_d = V_b \cos\delta + x_t \frac{E_q' - V_b \cos\delta}{x_{d\Sigma}'} = \frac{x_t\, E_q'}{x_{d\Sigma}'} + \frac{V_b x_d' \cos\delta}{x_{d\Sigma}'}
\end{aligned}
\tag{2.36}
$$

Hence, the simplified model of single-machine infinite-bus power system is as follows:

$$
\begin{aligned}
\dot{\delta} &= \omega_0(\omega - 1) \\
\dot{\omega} &= \frac{1}{M}\left[P_m - P_t - D(\omega - 1)\right] \\
\dot{E}_q' &= \frac{1}{T_{do}'}(-E_q + E_{fd}) \\
\dot{E}_{fd}' &= -\frac{1}{T_A} E_{fd}' + \frac{K_A}{T_A}(V_{tref} - V_t + u_{pss})
\end{aligned}
\tag{2.37}
$$

where

$$
\begin{aligned}
P_t &= \frac{E_q' V_b}{x_{d\Sigma}'} \sin\delta - \frac{V_b^2}{2} \frac{(x_q - x_d')}{x_{d\Sigma}' x_{q\Sigma}} \sin 2\delta \\
E_q &= \frac{E_q' x_{d\Sigma}}{x_{d\Sigma}'} - \frac{(x_d - x_d') V_b \cos\delta}{x_{d\Sigma}'} \\
E_{fd} &= E_{fd0} + E_{fd}' \\
v_{td} &= \frac{x_q V_b \sin\delta}{x_{q\Sigma}}, \quad v_{tq} = \frac{x_t\, E_q'}{x_{d\Sigma}'} + \frac{V_b x_d' \cos\delta}{x_{d\Sigma}'}, \quad V_t = \sqrt{v_{td}^2 + v_{tq}^2}
\end{aligned}
\tag{2.38}
$$

The model is a group of 4 first-order differential equations plus 6 algebraic equations.

#### 2.1.2.3 Heffron–Phillips Model [1–3]

By linearizing Eqs. (2.37) and (2.38) at an operating point of power system, where $V_t = V_{t0}, V_{td} = V_{td0}, V_{tq} = V_{tq0}, \delta = \delta_0, \omega_0 = 1, E'_q = E'_{q0}, E_{fd} = E_{fd0}$, it can have

$$\begin{aligned}
\Delta\dot{\delta} &= \omega_o \Delta\omega \\
\Delta\dot{\omega} &= \frac{1}{M}(-\Delta P_t - D\Delta\omega) \\
\Delta\dot{E}'_q &= \frac{1}{T'_{do}}(-\Delta E_q + \Delta E'_{fd}) \\
\Delta\dot{E}'_{fd} &= -\frac{1}{T_A}\Delta E'_{fd} - \frac{K_A}{T_A}(\Delta V_t - \Delta u_{pss})
\end{aligned} \tag{2.39}$$

$$\begin{aligned}
\Delta P_t &= K_1 \Delta\delta + K_2 \Delta E'_q \\
\Delta E_q &= K_3 \Delta E'_q + K_4 \Delta\delta \\
\Delta V_t &= K_5 \Delta\delta + K_6 \Delta E'_q
\end{aligned} \tag{2.40}$$

where

$$\begin{aligned}
K_1 &= \frac{E'_{q0} V_b X'_{d\Sigma}}{\cos}\delta_0 - \frac{V_b^2 (X_q - X'_d)}{X'_{d\Sigma} X_{q\Sigma}}\cos 2\delta_0 \\
K_2 &= \frac{V_b}{X'_{d\Sigma}}\sin\delta_0 \\
K_3 &= \frac{X_{d\Sigma}}{X'_{d\Sigma}} \\
K_4 &= \frac{(X_d - X'_d) V_b \sin\delta_0}{X'_{d\Sigma}} \\
K_5 &= \frac{V_{td0}}{V_{t0}} \frac{X_q V_b \cos\delta_0}{X_{q\Sigma}} - \frac{V_{tq0}}{V_{t0}} \frac{V_{b0} X'_d \sin\delta_0}{X'_{d\Sigma}} \\
K_6 &= \frac{V_{tq0}}{V_{t0}} \frac{X_t}{X'_{d\Sigma}}
\end{aligned}$$

Substituting Eq. (2.40) into Eq. (2.39), it can be obtained that

$$\begin{aligned}
\Delta\dot{\delta} &= \omega_o \Delta\omega \\
\Delta\dot{\omega} &= \frac{1}{M}(-K_1\Delta\delta - K_2\Delta E'_q - D\Delta\omega) \\
\Delta\dot{E}'_q &= \frac{1}{T'_{do}}(-K_3\Delta E'_q - K_4\Delta\delta + \Delta E'_{fd}) \\
\Delta\dot{E}'_{fd} &= -\frac{1}{T_A}\Delta E'_{fd} - \frac{K_A}{T_A}(K_5\Delta\delta + K_6\Delta E'_q - \Delta u_{pss})
\end{aligned} \tag{2.41}$$

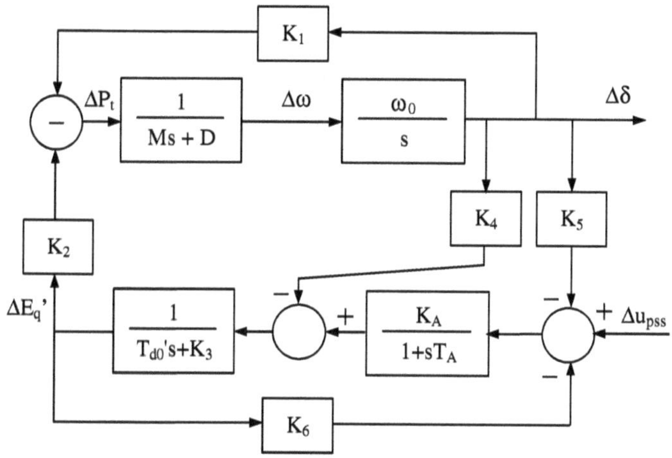


**Fig. 2.10** Heffron–Phillips model of a single-machine infinite-bus power system

Equation (2.41) is the so-called Heffron–Phillips model of single-machine infinite-bus power system, which is shown in Fig. 2.10.

The Heffron–Phillips model can be written in the form of state-space representation of Eq. (2.21) where

$$\mathbf{\Delta X_{g-dq}} = \begin{bmatrix} \Delta\delta \\ \Delta\omega \\ \Delta E'_q \\ \Delta E'_{fd} \end{bmatrix}, \mathbf{A_{gc-dq}} = \begin{bmatrix} 0 & \omega_o & 0 & 0 \\ -\frac{K_1}{M} & -\frac{D}{M} & -\frac{K_2}{M} & 0 \\ -\frac{K_4}{T'_{do}} & 0 & -\frac{K_3}{T'_{do}} & \frac{1}{T'_{do}} \\ -\frac{K_A K_5}{T_A} & 0 & -\frac{K_A K_6}{T_A} & -\frac{1}{T_A} \end{bmatrix}, \quad (2.42)$$

$$\mathbf{b_{pss}} = \begin{bmatrix} 0 \\ 0 \\ 0 \\ \frac{K_A}{T_A} \end{bmatrix}$$

## 2.2 Modal Analysis

### 2.2.1 Basis of Modal Analysis Theory

#### 2.2.1.1 Modal Decomposition

State-space representation of a linear system is as follows:

$$\begin{aligned} s\mathbf{X} &= \mathbf{A_o X} + \mathbf{b_o} u \\ y &= \mathbf{c_o^T X} \\ u &= H(s)y \end{aligned} \tag{2.43}$$

where $\mathbf{A_o}$, $\mathbf{b_o}$ and $\mathbf{c_o^T}$ is the state matrix, control vector, and output vector of open-loop system, respectively, and H(s) is the transfer function of feedback controller. Transfer function of open-loop system is as follow:

$$G(s) = \frac{y}{u} = \mathbf{c_o^T}(s\mathbf{I} - \mathbf{A_o})^{-1}\mathbf{b_o} \tag{2.44}$$

The system is shown in Fig. 2.11. Transfer function of closed-loop system is as follows:

$$T(s) = \frac{y}{w} = \frac{G(s)}{1 - G(s)H(s)} \tag{2.45}$$

Eigen solution is one of the basic techniques in the modal analysis, involving the computation of eigenvalues and eigenvectors of state matrix, $\mathbf{A_o}$. An eigenvalue of matrix $\mathbf{A_o}$, $\lambda$, is a scalar parameter, which satisfies the following equation

$$\mathbf{A_o v} = \lambda \mathbf{v} \tag{2.46}$$

with a non-trivial solution $(\mathbf{v} \neq \mathbf{0})$.

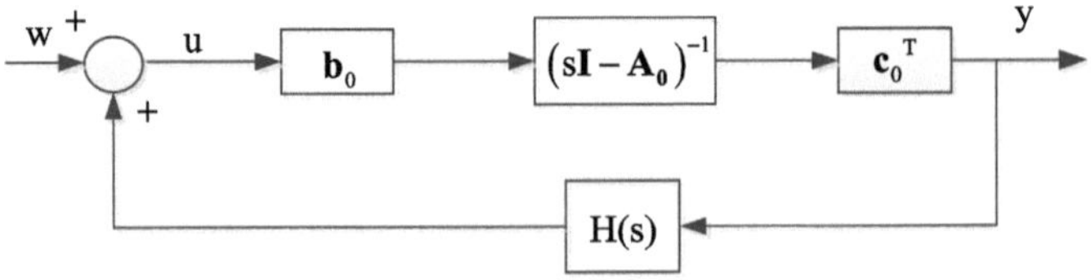


**Fig. 2.11** Block diagram of a closed-loop control system

Obviously, Eq. (2.46) can be written in the following form

$$(\mathbf{A_o} - \lambda \mathbf{I})\mathbf{v} = 0 \tag{2.47}$$

where **I** is an unity matrix. In order for Eq. (2.47) to have the non-trivial solution, it should have

$$|\mathbf{A_o} - \lambda \mathbf{I}| = 0 \tag{2.48}$$

Equation (2.48) is the following polynomial equation if $\mathbf{A_o}$ is an M × M matrix

$$(-1)^M \lambda^M + a_{M-1}\lambda^{M-1} + \cdots + a_1\lambda + a_0 = 0 \tag{2.49}$$

which is called the characteristic equation of state matrix $\mathbf{A_o}$. The characteristic equation should have M solutions; that is, matrix $\mathbf{A_o}$ has M eigenvalues, if the dimension of matrix is M.

For the ith eigenvalue of matrix $\mathbf{A_o}$, $\lambda_i$, if a nonzero vector $\mathbf{v_i}$ satisfies the equation

$$\mathbf{A}\mathbf{v_i} = \lambda_i \mathbf{v_i}, \quad i = 1, 2, \ldots, M \tag{2.50}$$

$\mathbf{v_i}$ is called the right eigenvector of matrix **A** associated with $\lambda_i$. Equation (2.50) can be arranged as follows:

$$\mathbf{A}[\mathbf{v_1} \quad \mathbf{v_2} \quad \ldots \quad \mathbf{v_M}] = [\mathbf{v_1} \quad \mathbf{v_2} \quad \ldots \quad \mathbf{v_M}] \begin{bmatrix} \lambda_1 & 0 & \mathbf{0} & 0 \\ 0 & \lambda_2 & \mathbf{0} & 0 \\ \mathbf{0} & \mathbf{0} & \ddots & \mathbf{0} \\ 0 & 0 & \mathbf{0} & \lambda_M \end{bmatrix} \tag{2.51}$$

That is

$$\mathbf{V}^{-1}\mathbf{A}\mathbf{V} = \mathbf{\Lambda} \tag{2.52}$$

where

$$\mathbf{V} = [\mathbf{v_1} \quad \mathbf{v_2} \quad \ldots \quad \mathbf{v_M}], \quad \mathbf{\Lambda} = \begin{bmatrix} \lambda_1 & 0 & \mathbf{0} & 0 \\ 0 & \lambda_2 & \mathbf{0} & 0 \\ \mathbf{0} & \mathbf{0} & \ddots & \mathbf{0} \\ 0 & 0 & \mathbf{0} & \lambda_M \end{bmatrix}$$

Denote

$$\mathbf{V}^{-1} = \begin{bmatrix} \mathbf{w}_1^T \\ \mathbf{w}_2^T \\ \vdots \\ \mathbf{w}_M^T \end{bmatrix} = [\mathbf{w}_1 \quad \mathbf{w}_2 \quad \ldots \quad \mathbf{w}_M]^T = \mathbf{W}^T \tag{2.53}$$

From Eqs. (2.52) and (2.53), obviously it can have

$$\mathbf{w}_i^T \mathbf{A} = \mathbf{w}_i^T \lambda_i, \quad i = 1, 2, \ldots, M \tag{2.54}$$

Hence, $\mathbf{w}_i^T$ is called the left eigenvector corresponding to eigenvalue $\lambda_i$.
If a new state variable vector $\mathbf{z}$ is introduced and defined to be

$$\mathbf{X} = \mathbf{VZ} \tag{2.55}$$

From Eqs. (2.43), (2.52), and (2.53), it can be obtained that

$$\begin{aligned} s\mathbf{Z} &= \mathbf{\Lambda Z} + \mathbf{W}^T \mathbf{b_o} u \\ y &= \mathbf{c_o^T V Z} \end{aligned} \tag{2.56}$$

That is

$$\begin{aligned} sz_i &= \lambda_i z_i, + \mathbf{w}_i^T \mathbf{b_o} u \quad i = 1, 2, \ldots, M \\ y &= \mathbf{c_o^T} \sum_{i=1}^{M} \mathbf{v}_i z_i \end{aligned} \tag{2.57}$$

According to Eq. (2.57), the system can also be shown in Fig. 2.12. This is the modal decomposition representation of state-space model of open-loop system.

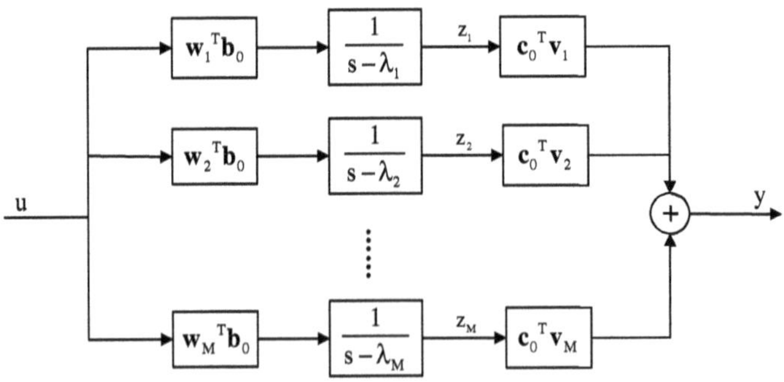


**Fig. 2.12** Block diagram of modal decomposition representation of state-space model of open-loop system

#### 2.2.1.2 Stability of Open-Loop System and Closed-Loop System

Considering the open-loop system when $u = 0$, the state-space representation of Eq. (2.57) is as follows:

$$sz_i = \lambda_i z_i \tag{2.58}$$

Solution of Eq. (2.58) is as follows:

$$z_i(t) = z_i(0)e^{\lambda_i t}, \quad i = 1, 2, \ldots, M \tag{2.59}$$

where $z_i(0)$ is the initial value of state variable $z_i(t)$, $i = 1, 2, \ldots, M$. From Eqs. (2.55) and (2.59), it can have

$$\mathbf{X} = \mathbf{V} \begin{bmatrix} z_1(0)e^{\lambda_1 t} \\ z_2(0)e^{\lambda_2 t} \\ \vdots \\ z_n(0)e^{\lambda_M t} \end{bmatrix} \tag{2.60}$$

Hence, time response of the kth state variable of the system, $x_k(t)$, $i = 1, 2, \ldots, M$, is as follows:

$$x_k(t) = v_{k1}z_1(0)e^{\lambda_1 t} + v_{k2}z_2(0)e^{\lambda_2 t} + \cdots + v_{kM}z_M(0)e^{\lambda_M t} = \sum_{i=1}^{M} v_{ki}z_i(0)e^{\lambda_i t} \tag{2.61}$$

Obviously, the time response of system state variables is decided by the eigenvalues of state matrix $\mathbf{A_o}$. If there is one or more eigenvalues on the right-hand half of the complex plan (the real part of eigenvalue is equal to or greater than zero), the system is unstable. If all the eigenvalues of $\mathbf{A_o}$ are on the left-hand side of the complex plane, the system is stable. Hence, eigenvalues of $\mathbf{A_o}$ determine the system stability. They often are called the modes of the system. If a pair of eigenvalues are conjugate complex number, i.e. $\bar{\lambda}_{i,i+1} = \xi_i \pm j\omega_i$, the corresponding component in the time response of the kth state variable of the system should be

$$\bar{v}_{ki}z_i(0)e^{\bar{\lambda}_i t} = \bar{v}_{ki}z_i(0)e^{(\xi_i + j\omega_i)t} = \bar{v}_{ki}z_i(0)e^{\xi_i t}[\cos \omega_i t + j \sin \omega_i t] \tag{2.62}$$

The component is oscillatory in respect of time. The oscillation angular frequency is $\omega_i$. The decaying and increasing of the oscillation are determined by the real part of the mode $\xi_i$. The pair of conjugate eigenvalues of state matrix $\mathbf{A_o}$ are often called the oscillation mode of the system.

The oscillation frequency $f_i$ (Hz) and damping $\zeta_i$ associated with $\bar{\lambda}_{i,i+1} = \xi_i \pm j\omega_i$ are normally defined as follows:

$$f_i = \frac{\omega_i}{2\pi}, \ \zeta_i = -\frac{\xi_i}{\sqrt{\xi_i^2 + \omega_i^2}} \tag{2.63}$$

From Eq. (2.59), it can be seen that $z_i(t),\ i = 1, 2, \ldots, M$ is related only with the ith mode of the system $\lambda_i$. Hence, $z_i(t),\ i = 1, 2, \ldots, M$ often is seen as the ith mode of the system. Equation (2.57) is often called the modal decomposition of state-space representation.

From Eq. (2.61), it can also be seen that the magnitude of $v_{ki}$ measures how much the ith mode $\lambda_i$ contributes to the kth state variable $x_k(t)$. Thus, $|v_{ki}|$ is a kind of measurement of the "observability" of the ith mode in the kth state variable.

On the basis of above discussion, from Fig. 2.12, it can be seen that $\mathbf{w_i^T b_o}$ is the weight on how much the control signal u affects the ith mode of the open-loop system, the so-called controllability index, whereas $\mathbf{c_o^T v_i}$ is the weight on how much the ith mode is observed in the system output, which is called the observability index. The product of controllability and observability index is called the residue. That is

$$R_i = \mathbf{w_i^T b_o c_o^T v_i} \tag{2.64}$$

From Eqs. (2.53) and (2.55), it can have

$$\mathbf{Z} = \mathbf{V}^{-1}\mathbf{X} = \mathbf{W}^{T}\mathbf{X} \tag{2.65}$$

or

$$z_i(t) = w_{1i}x_1(t) + w_{2i}x_2(t) + \cdots + w_{Mi}x_M(t) \tag{2.66}$$

$w_{ki}$ is the ith row kth column element of matrix $\mathbf{W}$. Equation (2.66) indicates that the magnitude of $w_{ki}$ measures the influence of the kth state variable $x_k(t)$ on the ith state variable $z_i(t)$, or the ith mode $\lambda_i$ of the system. It is a kind of measurement of "controllability" of the kth state variable on the ith mode.

Let the realization of the transfer function of feedback controller H(s) be

$$\begin{aligned} s\mathbf{X_f} &= \mathbf{A_f X_f} + \mathbf{b_f} y \\ u &= \mathbf{c_f^T X_f} \end{aligned} \tag{2.67}$$

That is $H(s) = \mathbf{c}_f^T(s\mathbf{I} - \mathbf{A}_f)^{-1}\mathbf{b}_f$. Thus, from Eqs. (2.43) and (2.67), the state-space representation of closed-loop system can be obtained to be

$$\begin{bmatrix} s\mathbf{X} \\ s\mathbf{X}_f \end{bmatrix} = \begin{bmatrix} \mathbf{A}_o & \mathbf{b}_o\mathbf{c}_f^T \\ \mathbf{b}_f\mathbf{c}_o^T & \mathbf{A}_f \end{bmatrix} \begin{bmatrix} \mathbf{X} \\ \mathbf{X}_f \end{bmatrix} = \mathbf{A}_c \begin{bmatrix} \mathbf{X} \\ \mathbf{X}_f \end{bmatrix} \tag{2.68}$$

where $\mathbf{A}_c$ is the state matrix of closed-loop system. Obviously, based on the discussion above, eigenvalues of $\mathbf{A}_c$ or modes of closed-loop system determine the stability of closed-loop system.

From Eq. (2.68), it can be obtained that

$$\mathbf{A}_c = \mathbf{A}_o + H(s)\mathbf{b}_o\mathbf{c}_o^T \tag{2.69}$$

Denote a variable parameter of feedback controller as $\alpha$. Thus, state matrix is a function of the parameter. Influence of the parameter on the ith mode of closed-loop system can be calculated by use of the following equation

$$\frac{\partial \lambda_i}{\partial \alpha} = \mathbf{w}_i^T \frac{\partial \mathbf{A}_c(\alpha)}{\partial \alpha} \mathbf{v}_i = \frac{\partial H(\lambda_i, \alpha)}{\partial \alpha} \mathbf{w}_i^T \mathbf{b}_o \mathbf{c}_o^T \mathbf{v}_i = R_i \frac{\partial H(\lambda_i, \alpha)}{\partial \alpha} \tag{2.70}$$

Hence, the residue measures how much the mode of closed-loop system is affected by the parameter of the controller.

## *2.2.2 Applications of Modal Analysis*

### 2.2.2.1 Modal Analysis for the AVR

Consider the simple case of a single-machine infinite-bus power system expressed by the Heffron–Phillips model shown in Fig. 2.10 without the PSS installed $(\Delta u_{pss} = 0)$. The upper part can be considered as the open-loop system and lower part the feedback controller. Thus, Eq. (2.41) can be rearranged as follows:

$$\begin{aligned} \begin{bmatrix} s\Delta\delta \\ s\Delta\omega \end{bmatrix} &= \begin{bmatrix} 0 & \omega_o \\ -\frac{K_1}{M} & -\frac{D}{M} \end{bmatrix} \begin{bmatrix} \Delta\delta \\ \Delta\omega \end{bmatrix} + \begin{bmatrix} 0 \\ -\frac{K_2}{M} \end{bmatrix} \Delta E'_q \\ \Delta\delta &= \begin{bmatrix} 1 & 0 \end{bmatrix} \begin{bmatrix} \Delta\delta \\ \Delta\omega \end{bmatrix} \\ \Delta E'_q &= F_{delta}(s)\Delta\delta \end{aligned} \tag{2.71}$$

In the above state-space model of the power system, the input to the open-loop system is $\Delta E'_q$, the output is $\Delta\delta$, and the transfer function of feedback controller is $F_{delta}(s)$. Obviously, the state-space realization of $F_{delta}(s)$ is as follows:

$$\begin{bmatrix} s\Delta E'_q \\ s\Delta E'_{fd} \end{bmatrix} = \begin{bmatrix} -\frac{K_3}{T'_{do}} & \frac{1}{T'_{do}} \\ -\frac{K_A K_6}{T_A} & -\frac{1}{T_A} \end{bmatrix} \begin{bmatrix} \Delta E'_q \\ \Delta E'_{fd} \end{bmatrix} + \begin{bmatrix} -\frac{K_4}{T'_{do}} \\ -\frac{K_A K_5}{T_A} \end{bmatrix} \Delta\delta$$
$$\Delta E'_q = \begin{bmatrix} 1 & 0 \end{bmatrix} \begin{bmatrix} \Delta E'_q \\ \Delta E'_{fd} \end{bmatrix} \tag{2.72}$$

According to Eq. (2.48), the modes of open-loop system can be found by solving the following characteristic equation

$$\left| \begin{bmatrix} 0 & \omega_o \\ -\frac{K_1}{M} & -\frac{D}{M} \end{bmatrix} - \lambda \begin{bmatrix} 1 & 0 \\ 0 & 1 \end{bmatrix} \right| = \left| \begin{bmatrix} -\lambda & \omega_o \\ -\frac{K_1}{M} & -\frac{D}{M} - \lambda \end{bmatrix} \right| = \lambda^2 + \frac{D}{M}\lambda + \frac{K_1}{M}\omega_o = 0 \tag{2.73}$$

Solution of Eq. (2.73) gives the oscillation mode of the single-machine infinite-bus power system when $\Delta E'_q = 0$, that is the case when the generator is modelled as a constant voltage source. The oscillation mode is as follows:

$$\bar{\lambda}_{1,2} = \frac{1}{2}\left[ -\frac{D}{M} \pm \sqrt{\left(\frac{D}{M}\right)^2 - \frac{4\omega_0 K_1}{M}} \right] = \xi_o \pm j\omega_{NOF} \tag{2.74}$$

The oscillation mode is related to the rotor motion of generator, i.e. state variables $\Delta\delta$ and $\Delta\omega$. It is often called the electromechanical oscillation mode of the power system.

From Eq. (2.71), it can have

$$\begin{bmatrix} s\Delta\delta \\ s\Delta\omega \end{bmatrix} = \begin{bmatrix} 0 & \omega_o \\ -\frac{K_1}{M} - \frac{K_2}{M} F_{delta}(s) & -\frac{D}{M} \end{bmatrix} \begin{bmatrix} \Delta\delta \\ \Delta\omega \end{bmatrix} = \mathbf{A_c} \begin{bmatrix} \Delta\delta \\ \Delta\omega \end{bmatrix} \tag{2.75}$$

where $\mathbf{A_c}$ is the state matrix of closed-loop system.

From Eq. (2.72), it can be obtained that

$$
\begin{aligned}
F_{delta}(s) &= [1 \quad 0]\begin{bmatrix} s+\frac{K_3}{T'_{do}} & -\frac{1}{T'_{do}} \\ \frac{K_A K_6}{T_A} & s+\frac{1}{T_A} \end{bmatrix}^{-1}\begin{bmatrix} -\frac{K_4}{T'_{do}} \\ -\frac{K_A K_5}{T_A} \end{bmatrix} \\
&= \frac{1}{\left(s+\frac{K_3}{T'_{do}}\right)\left(s+\frac{1}{T_A}\right)+\frac{K_A K_6}{T_A}\frac{1}{T'_{do}}}[1 \quad 0]\begin{bmatrix} s+\frac{1}{T_A} & \frac{1}{T'_{do}} \\ -\frac{K_A K_6}{T_A} & s+\frac{K_3}{T'_{do}} \end{bmatrix}\begin{bmatrix} -\frac{K_4}{T'_{do}} \\ -\frac{K_A K_5}{T_A} \end{bmatrix} \\
&= \frac{T'_{do} T_A}{(sT'_{do}+K_3)(T_A s+1)+K_A K_6}\begin{bmatrix} s+\frac{1}{T_A} & \frac{1}{T'_{do}} \end{bmatrix}\begin{bmatrix} -\frac{K_4}{T'_{do}} \\ -\frac{K_A K_5}{T_A} \end{bmatrix} \\
&= -\frac{(T_A s+1)K_4+K_A K_5}{(sT'_{do}+K_3)(T_A s+1)+K_A K_6}
\end{aligned}
\quad (2.76)
$$

Hence, according to Eq. (2.70), it can have

$$
\frac{\partial \bar{\lambda}_i}{\partial \alpha} = -R_i \frac{\partial \frac{(T_A \bar{\lambda}_i+1)K_4+K_A K_5}{(\bar{\lambda}_i T'_{do}+K_3)(T_A \bar{\lambda}_i+1)+K_A K_6}}{\partial \alpha} \quad (2.77)
$$

where $\alpha = K_A$ or $T_A$. By using Eq. (2.77), effect of the AVR on the damping of electromechanical oscillation, i.e. power oscillation, can be examined.

#### 2.2.2.2 Modal Analysis for the PSS

The general linearized model of the single-machine infinite-bus power system with the PSS installed is Eq. (2.21) which can be rearranged as follows:

$$
\begin{aligned}
\begin{bmatrix} s\Delta\delta \\ s\Delta\omega \\ s\Delta\mathbf{x}_{3-8} \end{bmatrix} &= \begin{bmatrix} 0 & \omega_0 & \mathbf{0} \\ \frac{-a_{21}}{M} & \frac{-a_{22}}{M} & \frac{-\mathbf{a}_{23-8}^T}{M} \\ \mathbf{a}_{13-8} & \mathbf{a}_{23-8} & \mathbf{A}_{33} \end{bmatrix}\begin{bmatrix} \Delta\delta \\ \Delta\omega \\ \Delta\mathbf{x}_{3-8} \end{bmatrix} + \begin{bmatrix} 0 \\ 0 \\ \mathbf{b}_{pss-3} \end{bmatrix}\Delta u_{pss} \\
y = \Delta\omega &= [0 \quad 1 \quad \mathbf{0}]\begin{bmatrix} \Delta\delta \\ \Delta\omega \\ \Delta\mathbf{x}_{3-8} \end{bmatrix} \\
\Delta u_{pss} &= T_{pss}(s)\Delta\omega
\end{aligned}
\quad (2.78)
$$

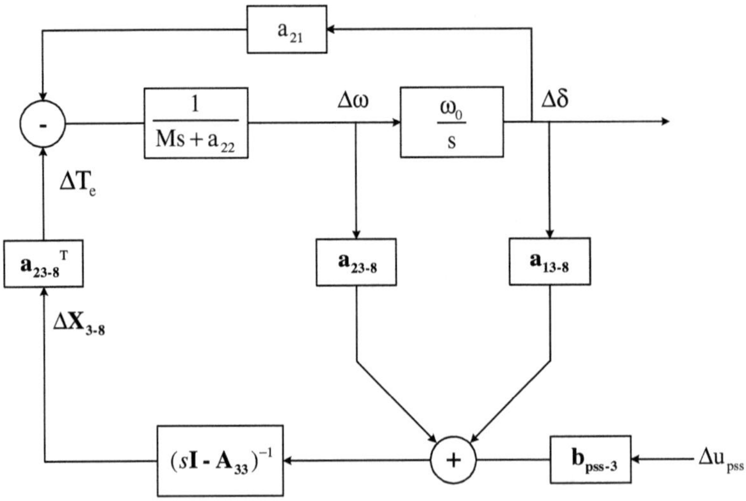


**Fig. 2.13** General linearized model of single-machine infinite-bus power system with PSS installed

The model is shown in Fig. 2.13.

Let the oscillation mode of the system without the PSS installed be $\bar{\lambda}_i$ and the corresponding left and right eigenvector be

$$\mathbf{w}_i^T = \begin{bmatrix} w_{i1} & w_{i2} & \mathbf{w}_{i3}^T \end{bmatrix}, \mathbf{v}_i = \begin{bmatrix} v_{i1} \\ v_{i2} \\ \mathbf{v}_{i3} \end{bmatrix} \tag{2.79}$$

That is

$$\begin{bmatrix} w_{i1} & w_{i2} & \mathbf{w}_{i3}^T \end{bmatrix} \begin{bmatrix} 0 & \omega_0 & \mathbf{0} \\ \frac{-a_{21}}{M} & \frac{-a_{22}}{M} & \frac{-\mathbf{a}_{23-8}^T}{M} \\ \mathbf{a}_{13-8} & \mathbf{a}_{23-8} & \mathbf{A}_{33} \end{bmatrix} = \bar{\lambda}_i \begin{bmatrix} w_{i1} & w_{i2} & \mathbf{w}_{i3}^T \end{bmatrix} \tag{2.80}$$

It can have

$$\begin{aligned} -\bar{w}_{i2}\frac{a_{21}}{M} + \bar{\mathbf{w}}_{i3}^T \mathbf{a}_{13-8} &= \bar{\lambda}_i \bar{w}_{i1} \\ \omega_0 \bar{w}_{i1} - \bar{w}_{i2}\frac{a_{22}}{M} + \bar{\mathbf{w}}_{i3}^T \mathbf{a}_{23-8} &= \bar{\lambda}_i \bar{w}_{i2} \\ -\bar{w}_{i2}\frac{\mathbf{a}_{23-8}^T}{M} + \bar{\mathbf{w}}_{i3}^T \mathbf{A}_{33} &= \bar{\lambda}_i \bar{\mathbf{w}}_{i3}^T \end{aligned} \tag{2.81}$$

Hence,

$$\bar{\mathbf{w}}_{\mathbf{i3}}^{\mathrm{T}} = -\bar{\mathrm{w}}_{\mathrm{i2}} \frac{\mathbf{a}_{\mathbf{23-8}}^{\mathrm{T}}}{\mathrm{M}} (\bar{\lambda}_{\mathrm{i}}\mathbf{I} - \mathbf{A}_{33})^{-1} \tag{2.82}$$

From Eqs. (2.64), (2.78), and (2.82), the residue can be obtained to be

$$\begin{aligned} \mathrm{R}_{\mathrm{i}} &= \begin{bmatrix} \mathrm{w}_{\mathrm{i1}} & \mathrm{w}_{\mathrm{i2}} & \bar{\mathbf{w}}_{\mathbf{i3}}^{\mathrm{T}} \end{bmatrix} \begin{bmatrix} 0 \\ 0 \\ \mathbf{b}_{\mathbf{pss-3}} \end{bmatrix} \begin{bmatrix} 0 & 1 & \mathbf{0} \end{bmatrix} \begin{bmatrix} \bar{\mathrm{v}}_{\mathrm{i1}} \\ \bar{\mathrm{v}}_{\mathrm{i2}} \\ \bar{\mathbf{v}}_{\mathbf{i3}} \end{bmatrix} \\ &= \mathbf{w}_{\mathbf{i3}}^{\mathrm{T}} \mathbf{b}_{\mathbf{pss-3}} \bar{\mathrm{v}}_{\mathrm{i2}} \\ &= -\bar{\mathrm{w}}_{\mathrm{i2}} \bar{\mathrm{v}}_{\mathrm{i2}} \frac{\mathbf{a}_{\mathbf{23-8}}^{\mathrm{T}}}{\mathrm{M}} (\bar{\lambda}_{\mathrm{i}}\mathbf{I} - \mathbf{A}_{33})^{-1} \mathbf{b}_{\mathbf{pss-3}} \end{aligned} \tag{2.83}$$

#### 2.2.2.3 Design of PSS by Pole Assignment

If the feedback signal and transfer function of the PSS to be designed is y and $\mathrm{T}_{\mathrm{PSS}}(\mathrm{s})$ respectively, Eq. (2.78) can be written more generally as follows:

$$\begin{aligned} \mathrm{s}\mathbf{X} &= \mathbf{A}\mathbf{X} + \mathbf{b}\Delta \mathrm{u}_{\mathrm{pss}} \\ \mathrm{y} &= \mathbf{c}^{\mathrm{T}}\mathbf{X} \\ \Delta \mathrm{u}_{\mathrm{pss}} &= \mathrm{T}_{\mathrm{pss}}(\mathrm{s})\mathrm{y} \end{aligned} \tag{2.84}$$

Transfer function of open-loop system is as follows:

$$\mathrm{G}(\mathrm{s}) = \mathbf{c}^{\mathrm{T}}(\mathrm{s}\mathbf{I} - \mathbf{A})^{-1}\mathbf{b} \tag{2.85}$$

Characteristic equation of closed-loop control system is as follows:

$$1 + \mathrm{G}(\mathrm{s})\mathrm{T}_{\mathrm{pss}}(\mathrm{s}) = 0 \tag{2.86}$$

If design of the PSS is to assign the electromechanical oscillation mode of the single-machine infinite-bus power system to a target position, $\bar{\lambda}_{\mathrm{c}} = -\xi_{\mathrm{c}} \pm \mathrm{j}\omega_{\mathrm{c}}$, $\bar{\lambda}_{\mathrm{c}}$ must be the solution of the characteristic equation of closed-loop control system of Eq. (2.86). Hence, it should have

$$1 + \mathrm{G}(\bar{\lambda}_{\mathrm{c}})\mathrm{T}_{\mathrm{pss}}(\bar{\lambda}_{\mathrm{c}}) = 0 \tag{2.87}$$

By separating the real and imaginary part of the above equation, two equations will be obtained which can be used to determine two parameters of the transfer function of PSS. If the transfer function of PSS adopts the following format of a lead–lag block,

$$T_{pss}(s) = K_{pss}\frac{(1+s\alpha T)^2}{(1+sT)^2} \tag{2.88}$$

With a predetermined T, parameters of the PSS, $K_{pss}$ and $\alpha$, can be set according to Eq. (2.87), thus completing the design of PSS via the pole assignment.

## 2.3 Damping Torque Analysis

### *2.3.1 Damping Torque and Synchronizing Torque*

#### 2.3.1.1 Damping Torque and Synchronizing Torque Derived from Heffron–Phillips Model

The damping torque analysis (DTA) was firstly introduced on the basis of the Heffron–Phillips model for a single-machine infinite-bus power system to examine the effect of excitation control, such as the AVR, on power system small-signal stability [1–3]. It was developed based on the understanding that the dynamic of the electromechanical oscillation loop of a generator decides the damping of power oscillations in the single-machine infinite-bus power system.

In the Heffron–Phillips model shown in Fig. 2.10, the upper part obviously is the linearized rotor motion equation and lower part is formed from the mathematical description of dynamic of the field winding of generator and the AVR. Figure 2.14 shows the upper part of the model which is called the electromechanical oscillation loop. Signal $\Delta T_e$ from the lower part in the Heffron–Phillips model is obviously an electric torque. Hence, from Fig. 2.14, it can have

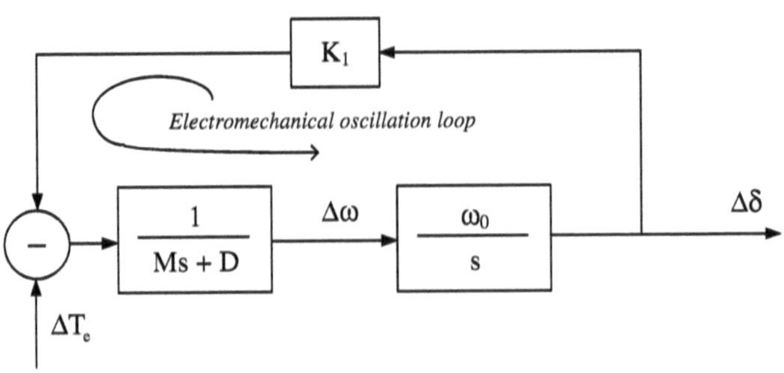


**Fig. 2.14** Electromechanical oscillation loop of generator

$$s^2\Delta\delta + \frac{D}{M}s\Delta\delta + \frac{\omega_0 K_1}{M}\Delta\delta + \frac{\omega_0}{M}\Delta T_e = 0 \tag{2.89}$$

If firstly the contribution from the lower part of Heffron–Phillips model, $\Delta T_e$, is not considered, the electromechanical oscillation loop of generator shown in Fig. 2.14 is described by the following second-order differential equation

$$s^2\Delta\delta + \frac{D}{M}s\Delta\delta + \frac{\omega_0 K_1}{M}\Delta\delta = 0 \tag{2.90}$$

Equation (2.90) in fact is the linearized model of the single-machine infinite-bus power system when the dynamic of exciter and the AVR are not considered. This is the case when the generator is modelled only by the rotor motion equation in Eq. (2.3).

Solution of Eq. (2.90) is as follows:

$$\Delta\delta(t) = a_0 e^{-\frac{D}{2M}t}\cos\omega_{NOF}t + b_0 \tag{2.91}$$

where $a_0$ and $b_0$ are two constants and $\omega_{NOF} = \frac{1}{2}\sqrt{\left(\frac{D}{M}\right)^2 - \frac{4\omega_0 K_1}{M}}$.

Equation (2.91) describes the behaviour of rotor motion, i.e. the acceleration and deceleration to store or release electric power. Hence, it determines the variations of active power supplied by the generator during dynamic transient (electromechanical transient), when the power system is subject to small disturbances. If $\frac{D}{2M}$ is small or negative, a poorly damped or magnitude-increasing power oscillations occur. This is the electromechanical oscillation associated with the rotor motion of synchronous generator, i.e. the power system low-frequency oscillation.

In Eq. (2.91), $\omega_{NOF}$ is called the angular frequency of natural oscillation. The angular oscillation frequency, $\omega_s$, of power oscillation in the single-machine infinite-bus power system is normally very close to the angular frequency of natural oscillation. Equation (2.91) indicates that the damping of power oscillation of the single-machine infinite-bus power system is determined by the coefficient of the first-order derivative in the second-order differential equation in Eq. (2.91) $\frac{D}{M}$.

At the angular oscillation frequency $\omega_s$, the electric torque contributed from the lower part of Heffron–Phillips model can be decomposed into two components

$$\Delta T_e = T_d\Delta\omega + T_s\Delta\delta \tag{2.92}$$

Equation (2.89) becomes

$$s^2\Delta\delta + \left(\frac{D}{M} + \frac{T_d}{M\omega_0}\right)s\Delta\delta + \left(\frac{\omega_0 K_1}{M} + \frac{\omega_0 T_s}{M}\right)\Delta\delta = 0 \tag{2.93}$$

Obviously, from the discussion on Eq. (2.91), it is easy to understand that the component in the decomposition of $\Delta T_e$, $T_d\Delta\omega$, contributes to the damping of power oscillation. This component is called the damping torque. In Eq. (2.92), $T_s\Delta\delta$ is called the synchronizing torque.

#### 2.3.1.2 Electric Torque Contributed from the PSS

From Fig. 2.10, it can be seen that the electric torque contributed from the lower part of Heffron–Phillips model is as follows:

$$\Delta T_{et} = F_{delta}(s)\Delta\delta + F_{pss}(s)\Delta u_{pss} \tag{2.94}$$

where $F_{delta}(s)$ and $F_{pss}(s)$ are the transfer function from $\Delta\delta$ and $\Delta u_{pss}$ respectively, to form the electric torque contribution to the electromechanical oscillation loop of generator.

The electric torque contribution from the PSS is as follows:

$$\Delta T_{pss} = F_{pss}(s)\Delta u_{pss} \tag{2.95}$$

Figure 2.15 shows that the PSS contributes the electric torque, $\Delta T_{pss}$, to the electromechanical oscillation loop of generator. Obviously, $F_{pss}(s)$ is the transfer function of forward path from the stabilizing signal of the PSS to the

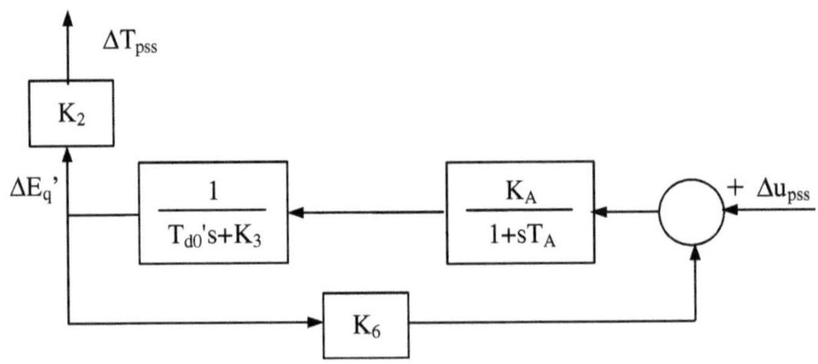


**Fig. 2.15** Forward path of stabilizing signal of the PSS

electromechanical oscillation loop of generator in the Heffron–Phillips model. From Fig. 2.15, it can be obtained that

$$F_{pss}(s) = K_2 \frac{\frac{1}{K_3 + sT'_{d0}} \frac{K_A}{1+sT_A}}{1 + K_6 \frac{1}{K_3 + sT'_{d0}} \frac{K_A}{1+sT_A}} = K_2 \frac{K_A}{(K_3 + sT'_{d0})(1+sT_A) + K_6 K_A} \quad (2.96)$$

Hence, at the angular oscillation frequency, $\omega_s$, the electric torque provided by the PSS to the electromechanical oscillation loop is as follows:

$$\Delta T_{pss} = \bar{F}_{pss}(j\omega_s)\Delta u_{pss}$$

For example, if the PSS is a pure-gain controller and takes the deviation of rotor speed of generator as the feedback signal, i.e. $\Delta u_{pss} = K_{pss}\Delta\omega$, the electric torque contributed from the PSS to the electromechanical oscillation loop of generator is as follows:

$$\Delta T_{pss} = K_{pss}\bar{F}_{pss}(j\omega_s)\Delta\omega = K_{pss}\mathrm{Re}[\bar{F}_{pss}(j\omega_s)]\Delta\omega + jK_{pss}\mathrm{Im}[\bar{F}_{pss}(j\omega_s)]\Delta\omega \quad (2.97)$$

where $\mathrm{Re}\left[\bar{F}_{pss}(j\omega_s)\right]$ and $\mathrm{Im}\left[\bar{F}_{pss}(j\omega_s)\right]$ denote the real and imaginary part of $\bar{F}_{pss}(j\omega_s)$, respectively (this notation will be used throughout the book). From the first equation in Eq. (2.41), it can have $s\Delta\delta = \omega_0\Delta\omega$, i.e.

$$\Delta\omega = \frac{j\omega_s}{\omega_0}\Delta\delta \quad (2.98)$$

By substituting Eqs. (2.98) into Eq. (2.97), it can be obtained that

$$\Delta T_{pss} = K_{pss}\mathrm{Re}[\bar{F}_{pss}(j\omega_s)]\Delta\omega - \frac{\omega_s}{\omega_0}K_{pss}\mathrm{Im}\left[\bar{F}_{pss}(j\omega_s)\right]\Delta\delta \quad (2.99)$$

Hence, the damping torque supplied by the PSS is $K_{pss}\mathrm{Re}[\bar{F}_{pss}(j\omega_s)]\Delta\omega$, and the synchronizing torque is $-\frac{\omega_s}{\omega_0}K_{pss}\mathrm{Im}[\bar{F}_{pss}(j\omega_s)]\Delta\delta$.

#### 2.3.1.3 Damping Torque and Synchronizing Torque Derived from the General Linearized Model

In Eq. (2.78) or Fig. 2.13, denote $\Delta T_e = -\mathbf{a}_{23-8}^T\Delta\mathbf{x}_{3-8}$. From Eq. (2.78) or Fig. 2.13, it can have

$$s^2\Delta\delta + \frac{a_{22}}{M}s\Delta\delta + \omega_0\frac{a_{21}}{M}\Delta\delta + \Delta T_e = 0 \quad (2.100)$$

Taking the similar procedure of DTA presented above in Sects. 2.3.1.1 and 2.3.1.2, transfer function of the forward path of stabilizing signal of the PSS can be obtained to be

$$F_{pss}(s) = -\mathbf{a}_{23-8}^{T}(s\mathbf{I} - \mathbf{A}_{33})^{-1}\mathbf{b}_{pss-3} \tag{2.101}$$

At the angular oscillation frequency, $\omega_s$, the electric torque contribution from the PSS is as follows:

$$\Delta T_{pss} = -\mathbf{a}_{23-8}^{T}(j\omega_s\mathbf{I} - \mathbf{A}_{33})^{-1}\mathbf{b}_{pss-3}\Delta u_{pss} \tag{2.102}$$

The electric torque can be decomposed into the damping and synchronizing torque. The damping torque contribution from the PSS determines its effect on the damping of power oscillation.

Assume that the installation of PSS brings about a change of damping coefficient $D_{pss}\Delta\omega$ in the electromechanical oscillation loop of generator. The state-space representation of power system with the PSS installed can be equivalently written as follows:

$$\begin{bmatrix} s\Delta\delta \\ s\Delta\omega \\ \Delta\mathbf{x}_{3-8} \end{bmatrix} = \begin{bmatrix} 0 & \omega_0 & 0 \\ -\frac{a_{21}}{M} & -\frac{a_{22}+D_{pss}}{M} & -\frac{\mathbf{a}_{23-8}^{T}}{M} \\ \mathbf{a}_{13-8} & \mathbf{a}_{23-8} & \mathbf{A}_{33} \end{bmatrix} \begin{bmatrix} \Delta\delta \\ \Delta\omega \\ \Delta\mathbf{x}_{3-8} \end{bmatrix} \tag{2.103}$$

From Eqs. (2.70) and (2.103), it can have

$$\begin{aligned} \frac{\partial\bar{\lambda}_i}{\partial D_{pss}} &= \bar{\mathbf{w}}_i^{T}\frac{\partial\mathbf{A}_c}{\partial D_{pss}}\bar{\mathbf{v}}_i \\ &= \begin{bmatrix} \bar{w}_{i1} & \bar{w}_{i2} & \bar{\mathbf{w}}_{i3}^{T} \end{bmatrix} \frac{\partial}{\partial D_{pss}} \begin{bmatrix} 0 & \omega_0 & 0 \\ -\frac{a_{21}}{M} & -\frac{a_{22}+D_{pss}}{M} & -\frac{\mathbf{a}_{23-8}^{T}}{M} \\ \mathbf{a}_{13-8} & \mathbf{a}_{23-8} & \mathbf{A}_{33} \end{bmatrix} \begin{bmatrix} \bar{v}_{i1} \\ \bar{v}_{i2} \\ \bar{\mathbf{v}}_{i3} \end{bmatrix} \\ &= -\frac{\bar{w}_{i2}\bar{v}_{i2}}{M} \end{aligned} \tag{2.104}$$

From Eqs. (2.83), (2.102), and (2.104), it can be seen that the residue in fact measures the effect of the PSS on the electromechanical oscillation mode of the system. At the complex frequency $\bar{\lambda}_i$, it is equal to the forward path of the PSS multiplied by the sensitivity of the mode to the damping torque contribution.

## 2.3.2 Damping Torque Analysis and Design of PSS by Phase Compensation

### 2.3.2.1 Theoretical Basis of the Damping Torque Analysis

This section explains the theoretical basis of the damping torque analysis by use of the Phillips–Heffron model as follows.

Firstly, the effect of PSS is not considered, i.e. $\Delta u_{pss} = 0$. From Fig. 2.10, it can have

$$\begin{aligned} &\left(Ms^2 + Ds + \omega_0 K_1\right)\Delta\delta(s) = -\omega_0 \Delta T(s) \\ &\Delta T(s) = F_{delta}(s)\Delta\delta(s) \end{aligned} \tag{2.105}$$

where $F_{delta}(s)$ is the transfer function from $\Delta\delta(s)$ to $\Delta T(s)$. Combining two equations above gives

$$\left[Ms^2 + Ds + \omega_0 K_1 + \omega_0 F_{delta}(s)\right]\Delta\delta(s) = 0 \tag{2.106}$$

Thus, characteristic equation of the system is as follows:

$$Ms^2 + Ds + \omega_0 K_1 + \omega_0 F_{delta}(s) = 0 \tag{2.107}$$

Solutions of the characteristic equation are the eigenvalues of state matrix of system model given by Eq. (2.42). One of the pair of complex solutions is called the electromechanical oscillation mode. Its real part defines the damping of power oscillation. Denote the mode as $\bar{\lambda}_s = \xi_s + j\omega_s$. In the complex frequency domain, it should have

$$M\bar{\lambda}_s^2 + D\bar{\lambda}_s + \omega_0 K_1 + \omega_0 \bar{F}_{delta}(\bar{\lambda}_s) = 0 \tag{2.108}$$

The second equation in Eq. (2.105) expressed in the complex frequency domain is as follows:

$$\Delta T(\bar{\lambda}_s) = \bar{F}_{delta}(\bar{\lambda}_s)\Delta\delta(\bar{\lambda}_s) \tag{2.109}$$

Also in the complex frequency domain, the first equation in Eq. (2.41) becomes

$$\Delta\omega(\bar{\lambda}_s) = \frac{\xi_s + j\omega_s}{\omega_0}\Delta\delta(\bar{\lambda}_s) = \frac{\xi_s}{\omega_0}\Delta\delta(\bar{\lambda}_s) + j\frac{\omega_s}{\omega_0}\Delta\delta(\bar{\lambda}_s) \tag{2.110}$$

Let the electric torque defined by Eq. (2.109) be decomposed as follows:

$$\Delta T(\bar{\lambda}_s) = T_{s1}\Delta\delta(\bar{\lambda}_s) + T_{d1}\Delta\omega(\bar{\lambda}_s) \tag{2.111}$$

From Eqs. (2.109), (2.110), and (2.111), it can be obtained that

$$\bar{F}_{delta}(\bar{\lambda}_s)\Delta\delta(\bar{\lambda}_s) = T_{s1}\Delta\delta(\bar{\lambda}_s) + T_{d1}\frac{\xi_s}{\omega_0}\Delta\delta(\bar{\lambda}_s) + jT_{d1}\frac{\omega_s}{\omega_0}\Delta\delta(\bar{\lambda}_s) \tag{2.112}$$

That is

$$\bar{F}_{delta}(\bar{\lambda}_s) = T_{s1} + T_{d1}\frac{\xi_s}{\omega_0} + jT_{d1}\frac{\omega_s}{\omega_0} \tag{2.113}$$

From Eq. (2.113), it can have

$$\begin{cases} T_{d1} = \frac{\omega_0}{\omega_s}\mathrm{Im}[\bar{F}_{delta}(\bar{\lambda}_s)] \\ T_{s1} = \mathrm{Re}[\bar{F}_{delta}(\bar{\lambda}_s)] - \frac{T_{d1}\xi_s}{\omega_0} \end{cases} \tag{2.114}$$

The above derivation indicates that in the complex frequency domain, the electric torque can be decomposed into damping and synchronizing torque according to Eq. (2.111). Substituting Eqs. (2.111) into (2.108), it can have

$$\begin{aligned} (M\bar{\lambda}_s^2 + D\bar{\lambda}_s + \omega_0 K_1)\Delta\delta(\bar{\lambda}_s) &= -\omega_0 T_{s1}\Delta\delta(\bar{\lambda}_s) - \omega_0 T_{d1}\Delta\omega(\bar{\lambda}_s) \\ &= -\omega_0 T_{s1}\Delta\delta(\bar{\lambda}_s) - T_{d1}\bar{\lambda}_s\Delta\delta(\bar{\lambda}_s) \end{aligned} \tag{2.115}$$

Thus,

$$M\bar{\lambda}_s^2 + (D + T_{d1})\bar{\lambda}_s + \omega_0 K_1 + \omega_0 T_{s1} = 0 \tag{2.116}$$

Solution of the above equation is as follows:

$$\xi_s = -\frac{D + T_{d1}}{2M} \tag{2.117}$$

Equation (2.117) indicates that the damping torque affects the real part of electromechanical oscillation mode, i.e. the damping of power oscillation.

#### 2.3.2.2 Graphical Explanation of the Damping Torque Analysis

Equations (2.16) and (2.39) include the following linearized rotor motion equation of generator

$$\begin{aligned} \Delta\dot{\delta} &= \omega_0\Delta\omega \\ \Delta\dot{\omega} &= -\frac{1}{M}(\Delta P_t + D\Delta\omega) \end{aligned} \tag{2.118}$$

Without referring to any linearized model of single-machine infinite-bus power system, assume $\Delta P_t$ being comprised of contributions from $\Delta\delta$ and stabilizing signal of the PSS, $\Delta u_{pss}$. That is to let

$$\begin{aligned}\Delta P_t &= \Delta P_t(\Delta\delta) + \Delta P_t(\Delta u_{pss}) \\ \Delta P_t(\Delta\delta) &= F_{delta}(s)\Delta\delta, \Delta P_t(\Delta u_{pss}) = F_{pss}(s)\Delta u_{pss}\end{aligned} \tag{2.119}$$

At a known angular oscillation frequency $\omega_s$, if $\Delta P_t$ is decomposed in the complex frequency domain, then

$$\begin{aligned}\Delta P_t(\Delta\delta) &= \bar{F}_{delta}(j\omega_s)\Delta\delta = C_{delta}\Delta\delta + D_{delta}\Delta\omega, \\ \Delta P_t(\Delta u_{pss}) &= \bar{F}_{pss}(j\omega_s)\Delta u_{pss} = C_{pss}\Delta\delta + D_{pss}\Delta\omega\end{aligned} \tag{2.120}$$

Obviously, $D_{delta}\Delta\omega$ and $D_{pss}\Delta\omega$ are the damping torque contributed to the electromechanical oscillation loop of generator from $\Delta\delta$ and $\Delta u_{pss}$' respectively. Function of the damping torque component in suppressing the power oscillation can be explained graphically by the use of the linearized equal-area criterion as follows.

Without affecting the conclusions of following discussion, it is assumed that in Eq. (2.120), $D_{delta} = 0$. When there is no PSS installed $\left(\Delta P_t(\Delta u_{pss}) = 0\right)$,

$$\Delta P_t = \Delta P_t(\Delta\delta) = C_{delta}\Delta\delta \tag{2.121}$$

The linearized $\Delta P_t - \Delta\delta$ curve is line a–f as shown in Fig. 2.16. In Fig. 2.16, the operating point of system at steady state is d $(P_{t0}, \delta_0)$ and it moves to point a $(P_{t1}, \delta_1)$ after the system is subject to a small disturbance. Hence, when the operating point moves down from the initial point a $(P_{t1}, \delta_1)$ along line a–f, it will stop at point f $(P_{t2'}, \delta_{2'})$ with area a–d–c being equal to area d–g–f. Obviously, in this case, $|P_{t1} - P_{t0}| = |P_{t2'} - P_{t0}|, |\delta_1 - \delta_0| = |\delta_{2'} - \delta_0|$, power oscillation is of unchanged magnitude and not damped at all.

When the PSS is installed to provide a pure positive damping torque,

$$\Delta P_t(\Delta u_{pss}) = D_{pss}\Delta\omega \text{ (assuming } D_{pss} > 0) \tag{2.122}$$

From Eqs. (2.120), (2.121), and (2.122), it should have

$$\Delta P_t = \Delta P_t(\Delta\delta) + \Delta P_t(\Delta u_{pss}) = C_{delta}\Delta\delta + D_{pss}\Delta\omega, \tag{2.123}$$

When the operating point moves down from point a $(P_{t1}, \delta_1)$, power angle of generator decreases and thus $\Delta\omega < 0$. $D_{pss}\Delta\omega < 0$ is added on $C_{delta}\Delta\delta$ as shown in Eq. (2.123). Hence, the operating point should move below line a–f along curve $\Delta P_t = C_{delta}\Delta\delta + D_{pss}\Delta\omega$,. When the operating point stops moving, $\Delta\omega = 0$. Thus, it should stop on line a–f at point c $(P_{t2}, \delta_2)$ with area $A_1$ being equal to area $A_2$.

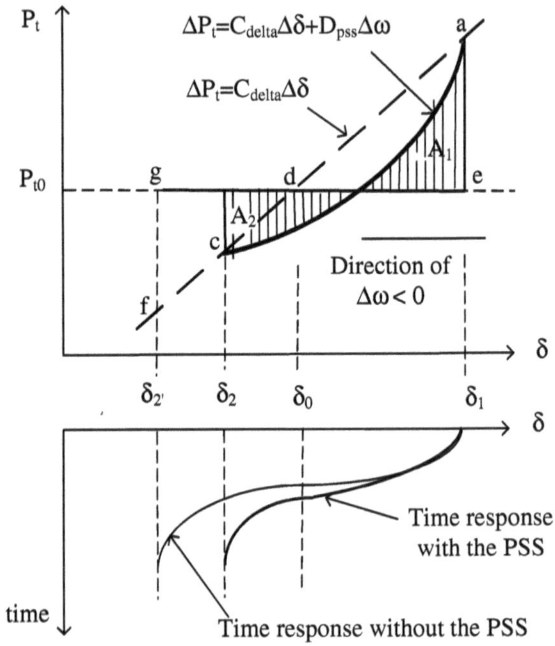


**Fig. 2.16** Graphical explanation of the DTA

Obviously, $|P_{t2} - P_{t0}| < |P_{t1} - P_{t0}|$, $|\delta_2 - \delta_0| < |\delta_1 - \delta_0|$, indicating extra positive damping is provided by the PSS to the power oscillation. A similar analysis can be carried out to examine the case when the operating point moves up from point c $(P_{t2}, \delta_2)$.

The above discussion explains the function of damping provided by the PSS in suppressing the power oscillation. It is important to note that the explanation relies only on the linearized rotor motion equation in Eqs. (2.118), (2.119), and (2.120) without referring to any particular type of model of power system. This means that for any type of linearized model of power system, including that of a multi-machine power system, if Eqs. (2.119) and (2.120) can be established on the basis of the model, the above procedure can be applied.

#### 2.3.2.3 Design of PSS by the Phase Compensation Method [4]

If the rotor speed of generator is taken as the feedback signal of the PSS, transfer function of the PSS is $T_{pss}(s)$, that is

$$\Delta u_{PSS} = T_{pss}(s)\Delta\omega \tag{2.124}$$

From Eqs. (2.95) or (2.102) and above equation, it can have

$$\Delta T_{pss} = F_{pss}(s)T_{pss}(s)\Delta\omega \tag{2.125}$$

At the angular oscillation frequency, $\omega_s$, the decomposition of the electric torque contributed by the PSS is as follows:

$$\begin{aligned}\Delta T_{pss} &= \bar{F}_{pss}(j\omega_s)\bar{T}_{pss}(j\omega_s)\Delta\omega \\ &= \mathrm{Re}\left[\bar{F}_{pss}(j\omega_s)\bar{T}_{pss}(j\omega_s)\right]\Delta\omega + j\mathrm{Im}[\bar{F}_{pss}(j\omega_s)\bar{T}_{pss}(j\omega_s)]\Delta\omega \\ &= \mathrm{Re}[\bar{F}_{pss}(j\omega_s)\bar{T}_{pss}(j\omega_s)]\Delta\omega - \frac{\omega_s}{\omega_0}\mathrm{Im}[\bar{F}_{pss}(j\omega_s)\bar{T}_{pss}(j\omega_s)]\Delta\delta \\ &= T_{pssd}\Delta\omega + T_{psss}\Delta\delta\end{aligned} \tag{2.126}$$

The damping and synchronizing torque provided by the PSS is $T_{pssd}\Delta\omega$ and $T_{psss}\Delta\delta$, respectively. In order to achieve the most efficient design, ideally the PSS should provide only the damping torque, that is,

$$\Delta T_{pss} = D_{pss}\Delta\omega, \quad D_{pss} > 0 \tag{2.127}$$

where $D_{pss}$ is the coefficient of the damping torque which needs to be provided by the PSS. Hence, from Eqs. (2.126) and (2.127), it can be seen that design of the PSS should satisfy that

$$D_{pss} = \bar{F}_{pss}(j\omega_s)\bar{T}_{pss}(j\omega_s) \tag{2.128}$$

According to Eq. (2.128), design of the PSS should set the phase of the PSS, $\angle\bar{T}_{pss}(j\omega_s)$, to be equal to the minus phase of the forward path, $\angle\bar{F}_{pss}(j\omega_s)$, that is to design the PSS such that it can compensate the phase lag of the forward path and ensure it to provide a pure positive damping torque. Hence, the method to design the PSS based on Eq. (2.128) is called the phase compensation method.

If it is denoted that

$$\bar{F}_{pss}(j\omega_s) = |F_{pss}|\angle\phi, \ \bar{T}_{pss}(j\omega_s) = |T_{pss}|\angle\gamma \tag{2.129}$$

The phase compensation method requires

$$\begin{aligned}T_{pssd} &= |F_{pss}T_{pss}|\cos(\phi+\gamma) = D_{pss} \\ T_{psss} &= |F_{pss}T_{pss}|\sin(\phi+\gamma) = 0\end{aligned} \tag{2.130}$$

This can be achieved by setting

$$\gamma = -\phi, \ |T_{pss}| = \frac{D_{pss}}{|F_{pss}|} \tag{2.131}$$

Often the PSS is constructed as a lead–lag block with its main part of transfer function to be

$$T_{pss}(s) = K_{pss}\frac{(1+sT_2)}{(1+sT_1)}\frac{(1+sT_4)}{(1+sT_3)} = K_{pss1}\frac{(1+sT_2)}{(1+sT_1)}K_{pss2}\frac{(1+sT_4)}{(1+sT_3)} \tag{2.132}$$

where $K_{pss} = K_{pss1}K_{pss2}$. Parameters of the PSS then can be set to satisfy

$$\begin{aligned} K_{pss1}\frac{(1+j\omega_s T_2)}{(1+j\omega_s T_1)} &= \frac{D_{pss}}{|F_{pss}|}\angle -\frac{\phi}{2} \\ K_{pss2}\frac{(1+j\omega_s T_4)}{(1+j\omega_s T_3)} &= 1.0\angle -\frac{\phi}{2} \end{aligned} \tag{2.133}$$

for the PSS to provide a positive damping torque $D_{pss}\Delta\omega$.

With the PSS installed, from Fig. 2.10, it can have

$$\begin{aligned} (Ms^2 + Ds + \omega_0 K_1)\Delta\delta(s) &= -\omega_0\Delta T_{delta}(s) - \omega_0\Delta T_{pss}(s) \\ \Delta T_{delta}(s) &= F_{delta}(s)\Delta\delta(s) \\ \Delta T_{pss}(s) &= F_{pss}(s)T_{pss}(s)\Delta\omega(s) \end{aligned} \tag{2.134}$$

Let $\bar{\lambda}_c = \xi_c + j\omega_c$ be the electromechanical oscillation mode of the closed-loop system with the PSS installed. According to the DTA discussed above in Sect. 2.3.2.1, in the complex frequency domain, it can have

$$\begin{aligned} \Delta T_{delta}(\bar{\lambda}_c) &= \bar{F}_{delta}\bar{\lambda}_c(s)\Delta\delta(\bar{\lambda}_c) = C_1\Delta\delta(\bar{\lambda}_c) + D_1\Delta\omega(\bar{\lambda}_c) \\ \Delta T_{pss}(\bar{\lambda}_c) &= \bar{F}_{pss}(\bar{\lambda}_c)\bar{T}_{pss}(\bar{\lambda}_c)\Delta\omega(\bar{\lambda}_c) = C_{pss}\Delta\delta(\bar{\lambda}_c) + D_{pss}\Delta\omega(\bar{\lambda}_c) \end{aligned} \tag{2.135}$$

Taking the similar procedure of discussion in Sect. 2.3.2.1, it can easily prove that $D_{pss}\Delta\omega(\bar{\lambda}_c)$ affects the damping of the electromechanical oscillation mode. Hence, the PSS can be designed to satisfy

$$F_{pss}(\bar{\lambda}_c)T_{pss}(\bar{\lambda}_c) = D_{pss} \tag{2.136}$$

When the PSS is being designed, if $\bar{\lambda}_c = \xi_c + j\omega_c$ is given, the PSS can be designed by using the phase compensation method defined by Eq. (2.136) to move the electromechanical oscillation mode strictly to the position at $\bar{\lambda}_c = \xi_c + j\omega_c$ in the complex plane. If only the amount of damping torque provision $D_{pss}\Delta\omega, D_{pss} > 0$ is given, the PSS can also be designed by the phase compensation method introduced above from Eqs. (2.126) to (2.133).

## 2.4 Examples

### *2.4.1 Linearized Mathematical Models of an Example Power System*

#### 2.4.1.1 Linearized Mathematical Model with Full Model of Generator Used

Parameters (in p.u.) of an example single-machine infinite-bus power system are as follows:

1. Generator:

$$x_d = 1.18, x_q = 0.78, x_{ad} = 1.0, x_{aq} = 0.6, x_D = 1.11, x_Q = 0.73, x_f = 1.13,$$
$$r_a = 0.005, r_f = 0.00075, r_D = 0.002, r_Q = 0.04, M = 7\text{ s}, D = 0, T'_{d0} = 5\text{ s}$$

2. AVR: $K_A = 100, T_A = 0.01$ s
3. Transmission line: $x_t = 0.15$
4. Steady-state operating point: $P_{t0} = 0.5, V_{t0} = 1.05, V_{b0} = 1.0$

At the steady-state operating point, the complex power received at the infinite busbar is as follows:

$$\bar{V}_{b0}\left(\frac{\bar{V}_{t0} - \bar{V}_{b0}}{jx_t}\right)^* = P_{t0} + jQ_{b0}$$

where $Q_{b0}$ is the reactive power received at the infinite busbar and * denotes the conjugate of a complex phasor. By choosing $\bar{V}_{b0}$ as the reference phasor, that is, $\bar{V}_{b0} = \bar{V}_{b0}^* = V_{b0}$, from the above equation, it can be obtained that

$$Q_{b0} = \frac{V_{b0}}{x_t}\left[\sqrt{V_{t0}^2 - \left(\frac{x_t P_{t0}}{V_{b0}}\right)^2} - V_{b0}\right] = 0.3155\text{ p.u.}$$

In the above, calculation is in p.u. Throughout the following calculation in p.u., p.u. is omitted.

The line current is as follows:

$$\bar{I}_{t0} = \frac{P_{t0} - jQ_{b0}}{V_{b0}} = 0.5 - j0.3155$$

Terminal voltage of generator is :

$$\bar{V}_{t0} = jx_t\bar{I}_{t0} + V_{b0} = 1.0473 + j0.075$$

At steady state, $i_{D0} = i_{Q0} = 0, \omega_0 = 1$, Eqs. (2.1) and (2.2) become

$$\begin{aligned}
0 &= v_{td0} + r_a i_{d0} + \psi_{q0} \\
0 &= v_{tq0} + r_a i_{q0} - \psi_{d0} \\
0 &= v_{f0} - r_f i_{f0} \\
\begin{bmatrix} \psi_{d0} \\ \psi_{f0} \end{bmatrix} &= \begin{bmatrix} x_d & x_{ad} \\ x_{ad} & x_f \end{bmatrix} \begin{bmatrix} -i_{d0} \\ i_{f0} \end{bmatrix} \\
\psi_{q0} &= -x_q i_{q0}
\end{aligned}$$

Thus, according to Eq. (2.28)

$$\begin{aligned}
v_{td0} &= -r_a i_{d0} - \psi_{q0} = -r_a i_{d0} + x_q i_{q0} \\
v_{tq0} &= -r_a i_{q0} + \psi_{d0} = -r_a i_{q0} + x_{ad} i_{f0} - x_d i_{d0} = -r_a i_{q0} + E_{q0} - x_d i_{d0}
\end{aligned}$$

Hence, in d–q coordinate of generator,

$$\begin{aligned}
\bar{V}_{t0} &= V_{td0} + jV_{tq0} = -r_a i_{d0} + x_q i_{q0} + j(-r_a i_{q0} + E_{q0} - x_d i_{d0}) \\
&= -r_a(i_{d0} + ji_{q0}) + x_q i_{q0} - jx_q i_{d0} + j(E_{q0} - x_d i_{d0} + x_q i_{d0})
\end{aligned}$$

Let an imaginary electromotive force (EMF) be

$$E_{Q0} = E_{q0} - x_d i_{d0} + x_q i_{d0} = E_{q0} - (x_d - x_q) i_{d0}$$

Thus,

$$\bar{V}_{t0} = -(r_a + jx_q)(i_{d0} + ji_{q0}) + \bar{E}_{Q0}$$

Thus, the q-axis of generator can be found by calculating the imaginary EMF $\bar{E}_Q$

$$\bar{E}_{Q0} = \bar{V}_{t0} + (r_a + jx_q)\bar{I}_{t0} = 1.2959 + j0.4634 = 1.38\angle 19.68^\circ$$

Hence, $\delta_0 = 19.68^\circ$. From the above equation about $E_{q0}$ and $E_{Q0}$, it can have

$$\begin{aligned}
E_{q0} &= E_{Q0} - (x_q - x_d) i_{d0} = 1.1506 \\
i_{f0} &= \frac{E_{q0}}{x_{ad}} = 1.5624
\end{aligned}$$

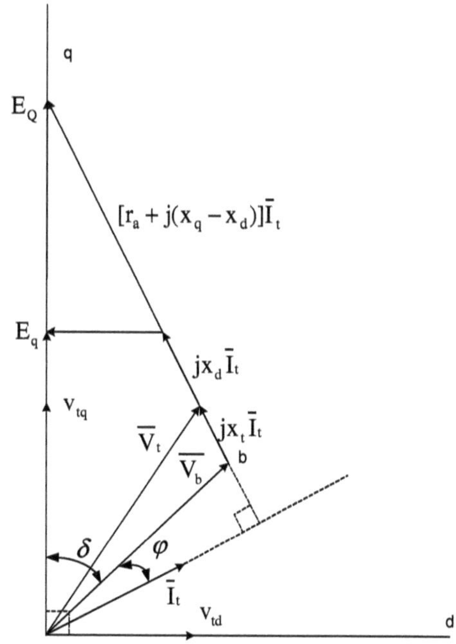


**Fig. 2.17** Phasor diagram for computing initial values of state variables

Above computation can be shown by the phasor diagram of Fig. 2.17 where subscript 0 is omitted. From Fig. 2.17, it can be seen that

$$i_{d0} = I_{t0}\sin(\delta_0 + \varphi_0) = 0.4654, i_{q0} = I_{t0}\cos(\delta_0 + \varphi_0) = 0.3646$$
$$v_{d0} = V_{b0}\sin\delta_0 = 0.3368, v_{q0} = V_{b0}\cos\delta_0 = 0.9416$$

Because

$$\bar{V}_{t0} = v_{td0} + jv_{tq0} = jx_t\bar{I}_{t0} + V_{b0} = jx_t(i_{d0} + ji_{q0}) + v_{d0} + jv_{q0}$$

thus

$$v_{td0} = v_{d0} - x_t i_{q0} = 0.2820$$
$$v_{tq0} = x_t i_{d0} + v_{q0} = 1.0114$$

From Eq. (2.23), it can have

$$i_{f0} = \frac{E_{q0}}{x_{ad}} = 1.5624$$

Because $i_{D0} = 0, i_{Q0} = 0$, thus according to Eq. (2.2), it can be obtained that

$$\begin{bmatrix} \psi_{d0} \\ \psi_{f0} \\ \psi_{D0} \end{bmatrix} = \begin{bmatrix} x_d & x_{ad} & x_{ad} \\ x_{ad} & x_f & x_{ad} \\ x_{ad} & x_{ad} & x_D \end{bmatrix} \begin{bmatrix} -i_{d0} \\ i_{f0} \\ i_{D0} \end{bmatrix} = \begin{bmatrix} 1.0132 \\ 1.3001 \\ 1.0970 \end{bmatrix}$$

$$\begin{bmatrix} \psi_{q0} \\ \psi_{Q0} \end{bmatrix} = \begin{bmatrix} x_q & x_{aq} \\ x_{aq} & x_Q \end{bmatrix} \begin{bmatrix} -i_{q0} \\ i_{Q0} \end{bmatrix} = \begin{bmatrix} -0.2844 \\ -0.2187 \end{bmatrix}$$

Denote

$$\begin{bmatrix} x_d & x_{ad} & x_{ad} \\ x_{ad} & x_f & x_{ad} \\ x_{ad} & x_{ad} & x_D \end{bmatrix}^{-1} = \begin{bmatrix} a_{11} & a_{12} & a_{13} \\ a_{21} & a_{22} & a_{23} \\ a_{31} & a_{32} & a_{33} \end{bmatrix} = \begin{bmatrix} -4.2331 & 1.8311 & 2.1640 \\ -1.8311 & 5.1570 & -2.9963 \\ -2.1640 & -2.9963 & 5.5498 \end{bmatrix}$$

$$\begin{bmatrix} x_q & x_{aq} \\ x_{aq} & x_Q \end{bmatrix}^{-1} = \begin{bmatrix} b_{11} & b_{12} \\ b_{13} & b_{14} \end{bmatrix} = \begin{bmatrix} -3.6842 & 2.8653 \\ -2.8653 & 3.7249 \end{bmatrix}$$

From Eqs. (2.12) and (2.13), it can have

$$\mathbf{A}_{g-dq} = \begin{bmatrix} 0 & \omega_0 & 0 & 0 & 0 & 0 & 0 & 0 \\ 0 & -\frac{D}{M} & 0 & -\frac{a_{11}V_{td0}}{M} & -\frac{b_{11}V_{tq0}}{M} & -\frac{a_{12}V_{td0}}{M} & -\frac{a_{13}V_{td0}}{M} & -\frac{b_{12}V_{tq0}}{M} \\ 0 & 0 & -\frac{1}{T_A} & 0 & 0 & 0 & 0 & 0 \\ 0 & \omega_0\psi_{q0} & 0 & a_{11}\omega_0 r_a & \omega_0\omega & a_{12}\omega_0 r_a & a_{13}\omega_0 r_a & 0 \\ 0 & \omega_0\psi_{d0} & 0 & -\omega_0\omega & b_{11}\omega_0 r_a & 0 & 0 & b_{12}\omega_0 r_a \\ 0 & 0 & \omega_0 & -a_{21}\omega_0 r_f & 0 & -a_{22}\omega_0 r_f & -a_{23}\omega_0 r_f & 0 \\ 0 & 0 & 0 & -a_{31}\omega_0 r_D & 0 & -a_{32}\omega_0 r_D & -a_{33}\omega_0 r_D & 0 \\ 0 & 0 & 0 & 0 & -b_{21}\omega_0 r_Q & 0 & 0 & -b_{22}\omega_0 r_Q \end{bmatrix}$$

$$= \begin{bmatrix} 0 & 314.16 & 0 & 0 & 0 & 0 & 0 & 0 \\ 0 & 0 & 0 & 0.1706 & 0.5037 & -0.0738 & -0.0872 & -0.4140 \\ 0 & 0 & -100 & 0 & 0 & 0 & 0 & 0 \\ 0 & -89.3377 & 0 & -6.6494 & 314.16 & 2.8762 & 3.3992 & 0 \\ 0 & 318.3167 & 0 & -314.16 & -5.4760 & 0 & 0 & 4.5008 \\ 0 & 0 & 314.16 & 0.4314 & 0 & -1.2151 & 0.7060 & 0 \\ 0 & 0 & 0 & 1.35797 & 0 & 1.8826 & -3.4871 & 0 \\ 0 & 0 & 0 & 0 & 36.0068 & 0 & 0 & -46.8088 \end{bmatrix}$$

$$\mathbf{B_{g-dq}} = \begin{bmatrix} 0 & 0 \\ -\frac{i_{d0}}{M} & -\frac{i_{q0}}{M} \\ -\frac{K_A V_{td0}}{T_A V_{t0}} & -\frac{K_A V_{tq0}}{T_A V_{t0}} \\ \omega_0 & 0 \\ 0 & \omega_0 \\ 0 & 0 \\ 0 & 0 \\ 0 & 0 \end{bmatrix} = \begin{bmatrix} 0 & 0 \\ -0.0665 & -0.0521 \\ -2686.1 & -9632.5 \\ 314.1593 & 0 \\ 0 & 314.1593 \\ 0 & 0 \\ 0 & 0 \\ 0 & 0 \end{bmatrix}$$

$$\mathbf{C_{g-dq}} = \begin{bmatrix} 0 & 0 & 0 & a_{11} & 0 & a_{12} & a_{13} & 0 \\ 0 & 0 & 0 & 0 & b_{11} & 0 & 0 & b_{12} \end{bmatrix}$$

$$= \begin{bmatrix} 0 & 0 & 0 & -4.2331 & 0 & 1.8331 & 2.1640 & 0 \\ 0 & 0 & 0 & 0 & -3.4862 & 0 & 0 & 2.8653 \end{bmatrix}$$

$$F_{dq\mathbf{1}} = \begin{bmatrix} 0 & -x_t \\ x_t & 0 \end{bmatrix} = \begin{bmatrix} 0 & -0.15 \\ 0.15 & 0 \end{bmatrix}$$

$$F_{dq\mathbf{2}} = \begin{bmatrix} V_b \cos\delta_0 & \mathbf{0} \\ -V_b \sin\delta_0 & \mathbf{0} \end{bmatrix} = \begin{bmatrix} 0.9416 & \mathbf{0} \\ -0.3367 & \mathbf{0} \end{bmatrix}$$

Thus, state matrix and control vector of linearized state-space model are obtained to be

$$\mathbf{A_{gc-dq}} = \mathbf{A_{g-dq}} + \mathbf{B_{g-dq}F_{dq1}C_{g-dq}} + \mathbf{B_{g-dq}F_{dq2}}$$

$$= \begin{bmatrix} 0 & 314.16 & 0 & 0 & 0 & 0 & 0 & 0 \\ -0.045 & 0 & 0 & 0.20 & 0.47 & -0.09 & -0.104 & -0.38 \\ 714.3 & 0 & -100 & 6116.3 & -1404.6 & -2646 & -3127 & 1154.5 \\ 295.8 & -89.34 & 0 & -6.64 & 478.4 & 2.88 & 3.39 & -135.03 \\ -105.78 & -318.3 & 0 & -513.64 & -5.48 & 86.29 & 101.9 & 4.50 \\ 0 & 0 & 314.2 & 0.43 & 0 & -1.22 & 0.70 & 0 \\ 0 & 0 & 0 & 1.36 & 0 & 1.88 & -3.48 & 0 \\ 0 & 0 & 0 & 0 & 36.0 & 0 & 0 & -46.8 \end{bmatrix}$$

$$\mathbf{b_{pss}} = \begin{bmatrix} 0 \\ 0 \\ \frac{K_A}{T_A} \\ 0 \\ 0 \\ 0 \\ 0 \\ 0 \end{bmatrix} = \begin{bmatrix} 0 \\ 0 \\ 10000 \\ 0 \\ 0 \\ 0 \\ 0 \\ 0 \end{bmatrix}$$

Eigenvalues of above state matrix are computed to be

$$\bar{\lambda}_1, \bar{\lambda}_2 = -24.14 \pm j971.56$$
$$\bar{\lambda}_3, \bar{\lambda}_4 = -36.33 \pm j359.63$$
$$\lambda_5 = -35.57$$
$$\bar{\lambda}_6, \bar{\lambda}_7 = -0.71 \pm j8.44$$
$$\lambda_8 = -5.71$$

According to Eq. (2.63), the oscillation frequency $f_i$ (Hz) and damping $\zeta_i$ for the electromechanical oscillation mode, $\bar{\lambda}_6$ and $\bar{\lambda}_7$, are as follows:

$$f_i = \frac{\omega_i}{2\pi} = 1.34 \ \text{Hz}$$
$$\zeta_i = -\frac{\xi_i}{\sqrt{\xi_i^2 + \omega_i^2}} = 0.0838, \quad i = 6, 7$$

### 2.4.1.2 Heffron–Phillips Model of Example Power System

With D and Q damping winding of generator being ignored, from the given parameters of above example power system and results of calculation, it can have

$$x_d' = x_d - \frac{x_{ad}^2}{x_f} = 0.2951$$
$$E_{q0}' = E_{Q0} - \left(x_q - x_d'\right) i_{td0} = 1.1506$$

From Eq. (2.40), it can be obtained that

$$K_1 = 1.5248, K_2 = 0.7602, K_3 = 2.9885,$$
$$K_4 = 0.6727, K_5 = -0.0027, K_6 = 0.3245$$

State equation of (2.41) can be obtained to be

$$\begin{bmatrix} \Delta\dot{\delta} \\ \Delta\dot{\omega} \\ \Delta\dot{E}_q' \\ \Delta\dot{E}_{fd}' \end{bmatrix} = \begin{bmatrix} 0 & 314.16 & 0 & 0 \\ -0.2178 & 0 & -0.1086 & 0 \\ -0.1346 & 0 & -0.5978 & 0.2 \\ 26.4729 & 0 & -3245 & -100 \end{bmatrix} \begin{bmatrix} \Delta\delta \\ \Delta\omega \\ \Delta E_q' \\ \Delta E_{fd}' \end{bmatrix} + \begin{bmatrix} 0 \\ 0 \\ 0 \\ 10000 \end{bmatrix} \Delta u_{pss}$$

Eigenvalues of state matrix are as follows:

$$\lambda_1 = -92.9741, \ \bar{\lambda}_{2,3} = -0.0114 \pm 8.2610, \ \lambda_4 = -7.6008$$

According to Eq. (2.63), oscillation frequency $f_i$ (Hz) and damping $\zeta_i$ for the electromechanical oscillation mode, $\bar{\lambda}_{2,3}$, are as follows:

$$f_i = \frac{\omega_i}{2\pi} = 1.315 \ \ \text{Hz}$$

$$\zeta_i = -\frac{\xi_i}{\sqrt{\xi_i^2 + \omega_i^2}} = 0.0014, \quad i = 2, 3$$

## 2.4.2 *Modal Analysis of Example Power System*

Heffron–Phillips model established above is used to demonstrate the modal analysis of example power system in this section.

### 2.4.2.1 Modal Decomposition and Stability of Example Power System

Right eigenvectors corresponding to each of eigenvalues of state matrix of Heffron–Phillips model are calculated to be

$$\mathbf{v}_1 = \begin{bmatrix} -8.4781 \times 10^{-6} \\ 2.5091 \times 10^{-6} \\ 0.0022 \\ -1 \end{bmatrix}, \bar{\mathbf{v}}_2 = \begin{bmatrix} 0.9014 \\ j0.0237 \\ -0.0050 + j0.0050 \\ 0.3868 - j0.1930 \end{bmatrix}$$

$$\bar{\mathbf{v}}_3 = \begin{bmatrix} 0.9014 \\ -j0.0237 \\ -0.0050 - j0.0050 \\ 0.3868 + j0.1930 \end{bmatrix}, \mathbf{v}_4 = \begin{bmatrix} -0.0077 \\ 0.0002 \\ 0.0284 \\ -0.9996 \end{bmatrix}$$

Hence,

$$\bar{\mathbf{V}} = \begin{bmatrix} \mathbf{v}_1 & \bar{\mathbf{v}}_2 & \bar{\mathbf{v}}_3 & \mathbf{v}_4 \end{bmatrix}$$
$$= \begin{bmatrix} -8.4781 \times 10^{-6} & 0.9014 & 0.9014 & -0.0077 \\ 2.5091 \times 10^{-6} & j0.0237 & -j0.0237 & 0.0002 \\ 0.0022 & -0.0050 + j0.0050 & -0.0050 - j0.0050 & 0.0284 \\ -1 & 0.3868 - j0.1930 & 0.3868 + j0.1930 & -0.9996 \end{bmatrix}$$

From Eq. (2.53),

$$\bar{\mathbf{W}}^{\mathrm{T}} = \begin{bmatrix} \mathbf{w}_1 & \bar{\mathbf{w}}_2 & \bar{\mathbf{w}}_3 & \mathbf{w}_4 \end{bmatrix}^{\mathrm{T}} = \bar{\mathbf{V}}^{-1}$$

$$= \begin{bmatrix} 0.2524 & 0.5554 - j0.0001 & 0.5554 + j0.0001 & 0.1773 \\ -0.8529 & -0.0312 - j21.1226 & -0.0312 + j21.1226 & -7.3269 \\ -38.0214 & 0.1626 + j0.1475 & 0.1626 - j0.1475 & 38.2206 \\ -1.0823 & 0.0003 + j0.0003 & 0.0003 - j0.0003 & 0.0827 \end{bmatrix}$$

Modal decomposition thus is obtained to be

$$\mathrm{s}z_i = \lambda_i z_i, + \mathbf{w}_i^{\mathrm{T}} \mathbf{b}_o \mathrm{u}, \quad i = 1, 2, 3, 4$$

Without considering the PSS $(\Delta u_{pss} = 0)$, solution of modal decomposition is obtained to be

$$z_i(t) = z_i(0)e^{\lambda_i t}, \quad i = 1, 2, 3, 4$$

From Eq. (2.55), it can have

$$\mathbf{Z}(0) = \mathbf{V}^{-1}\mathbf{X}(0) = \mathbf{W}^{\mathrm{T}}\mathbf{X}(0)$$

Time response of state variables can be written as follows:

$$x_k(t) = \sum_{i=1}^{4} v_{ki} z_i(0) e^{\lambda_i t} = \sum_{i=1}^{4} v_{ki} z_i(0) e^{\xi_i t} [\cos \omega_i t + j \sin \omega_i t]$$

It can be seen that when time approaches infinity $(t \to \infty)$,

$$\begin{aligned} \lim_{t\to\infty} x_k(t) &= \lim_{t\to\infty} \sum_{i=1}^{4} v_{ki} z_i(0) e^{\lambda_i t} = \lim_{t\to\infty} \sum_{i=1}^{4} v_{ki} z_i(0) e^{\xi_i t} [\cos \omega_i t + j \sin \omega_i t] \\ &= \lim_{t\to\infty} \{ e^{-92.9741t} v_{k1} z_1(0) + e^{-0.0114t} v_{k2} z_2(0) [\cos(8.2610t) + j \sin(8.2610t)] \\ &\quad + e^{-0.0114t} v_{k3} z_3(0) [\cos(8.2610t) - j \sin(8.2610t)] + e^{-7.6008t} v_{k4} z_4(0) \} = 0 \end{aligned}$$

Since $\mathbf{X}(t) = \begin{bmatrix} \Delta\delta(t) & \Delta\omega(t) & \Delta E'_q(t) & \Delta E'_{fd}(t) \end{bmatrix}^{\mathrm{T}}$ and

$$\begin{aligned} &\Delta\delta(t) = \delta(t) - \delta_0, \Delta\omega(t) = \omega(t) - \omega_0, \\ &\Delta E'_q(t) = E'_q(t) - E'_{q0}(t), \Delta E'_{fd}(t) = E'_{fd}(t) - E'_{fd0}(t) \end{aligned}$$

$\lim_{t\to\infty} x_k(t) = 0, \quad k = 1,2,3,4$ should give

$$\lim_{t\to\infty} \delta(t) = \delta_0, \lim_{t\to\infty} \omega(t) = \omega_0, \lim_{t\to\infty} E'_q(t) = E'_{q0}(t), \lim_{t\to\infty} E'_{fd}(t) = E'_{fd0}(t)$$

All the state variables return to their initial points $\mathbf{X}(0)$, the equilibrium point of the system. Hence, the system is stable in terms of small-signal stability.

#### 2.4.2.2 Modal Analysis of the AVR

State matrix, control vector, and output vector of Eq. (2.71) are as follows:

$$A_0 = \begin{bmatrix} 0 & w_0 \\ -\frac{K_1}{M} & -\frac{D}{M} \end{bmatrix} = \begin{bmatrix} 0 & 314.16 \\ -0.2178 & 0 \end{bmatrix}$$

$$b_0 = \begin{bmatrix} 0 \\ -\frac{K_2}{M} \end{bmatrix} = \begin{bmatrix} 0 \\ -0.1086 \end{bmatrix} \quad c_0^T = [1 \quad 0]$$

Its eigenvalues are calculated to be $\bar{\lambda}_{1,2} = \pm j8.2725$ with the corresponding eigenvectors to be

$$\bar{v}_1 = \begin{bmatrix} 0.9997 \\ j0.0263 \end{bmatrix}, \bar{v}_2 = \begin{bmatrix} 0.9997 \\ -j0.0263 \end{bmatrix}$$

Thus

$$\bar{\mathbf{W}}^T = [\bar{w}_1 \quad \bar{w}_2]^T = \bar{\mathbf{V}}^{-1} = [\bar{v}_1 \quad \bar{v}_2]^{-1}$$

$$== \begin{bmatrix} 0.9997 & 0.9997 \\ j0.0263 & -j0.0263 \end{bmatrix}^{-1} = \begin{bmatrix} 0.5002 & 0.5002 \\ -j18.9947 & j18.9947 \end{bmatrix}^T$$

Hence, the controllability and observability index, respectively, can be obtained to be

$$\bar{w}_1^T b_0 = [0.5002 \quad -j18.9914] \begin{bmatrix} 0 \\ -0.1086 \end{bmatrix} = j2.0628$$

$$\bar{w}_2^T b_0 = [0.5002 \quad j18.9914] \begin{bmatrix} 0 \\ -0.1086 \end{bmatrix} = -j2.0628$$

$$c_0^T \bar{v}_1 = [1 \quad 0] \begin{bmatrix} 0.9997 \\ j0.0263 \end{bmatrix} = 0.9997$$

$$c_0^T \bar{v}_2 = [1 \quad 0] \begin{bmatrix} 0.9997 \\ -j0.0263 \end{bmatrix} = 0.9997$$

The residue is as follows:

$$\bar{R}_1 = \bar{w}_1^T b_0 * c_0^T v_1 = j2.0622$$
$$\bar{R}_2 = \bar{w}_2^T b_0 * c_0^T v_2 = -j2.0622$$

From Eq. (2.77), when $\alpha = K_A$, it can have

$$\begin{aligned}\frac{\partial\bar{\lambda}_i}{\partial K_A} &= -R_i\frac{\partial\frac{(T_A\bar{\lambda}_i+1)K_4+K_AK_5}{(\bar{\lambda}_i T'_{do}+K_3)(T_A\bar{\lambda}_i+1)+K_AK_6}}{\partial K_A}\\ &= -R_i\frac{K_5[(\bar{\lambda}_i T'_{do}+K_3)(T_A\bar{\lambda}_i+1)+K_AK_6]-[(T_A\bar{\lambda}_i+1)K_4+K_AK_5]K_6}{[(\bar{\lambda}_i T'_{do}+K_3)(T_A\bar{\lambda}_i+1)+K_AK_6]^2}\\ &= 0.00018158 \pm j0.00005839 \quad i = 1,2\end{aligned}$$

When $\alpha = T_A$ it can have

$$\begin{aligned}\frac{\partial\bar{\lambda}_i}{\partial T_A} &= -R_i\frac{\partial\frac{(T_A\bar{\lambda}_i+1)K_4+K_AK_5}{(\bar{\lambda}_i T'_{do}+K_3)(T_A\bar{\lambda}_i+1)+K_AK_6}}{\partial T_A}\\ &= -R_i\frac{\bar{\lambda}_i K_4[(\bar{\lambda}_i T'_{do}+K_3)(T_A\bar{\lambda}_i+1)+K_AK_6]-[(T_A\bar{\lambda}_i+1)K_4+K_AK_5](\bar{\lambda}_i T'_{do}+K_3)\bar{\lambda}_i}{[(\bar{\lambda}_i T'_{do}+K_3)(T_A\bar{\lambda}_i+1)+K_AK_6]^2}\\ &= 0.0357 \pm j0.1533 \quad i = 1,2\end{aligned}$$

Above results indicate that increase of the gain value and time constant will move the electromechanical oscillation mode towards the right on the complex plane and hence is detrimental to the small-signal angular stability of the power system. It has been well known that the fast-acting high-gain AVR may be detrimental to the damping of power system electromechanical oscillation modes. This means that increase of gain value of the AVR could move the oscillation mode to the right. However, increase of the time constant (slower action of the AVR) should not.

In order to further clarify the results of derivative of the oscillation mode in respect to the parameters of the AVR obtained above, Fig. 2.18 presents the loci of movement of the electromechanical oscillation mode on the complex plane with the change of gain value and time constant of the AVR. In Fig. 2.18, $K_A$ increases from $K_A = 50$ to $K_A = 100$ with $T_A = 0.01$ and $T_A$ increases from $T_A = 0.01$ to $T_A = 0.1$ with $K_A = 100$. From Fig. 2.18, it can be seen that although at the point $K_A = 100, T_A = 0.01$ where the derivatives are calculated, the trend of loci movement is towards the right with the increase of the gain value and time constant, and the oscillation mode in fact moves towards left when the time constant of the AVR increases. The trend of the loci with the increase of the time constant of the AVR actually changes the direction at the point $K_A = 100, T_A = 0.01$.

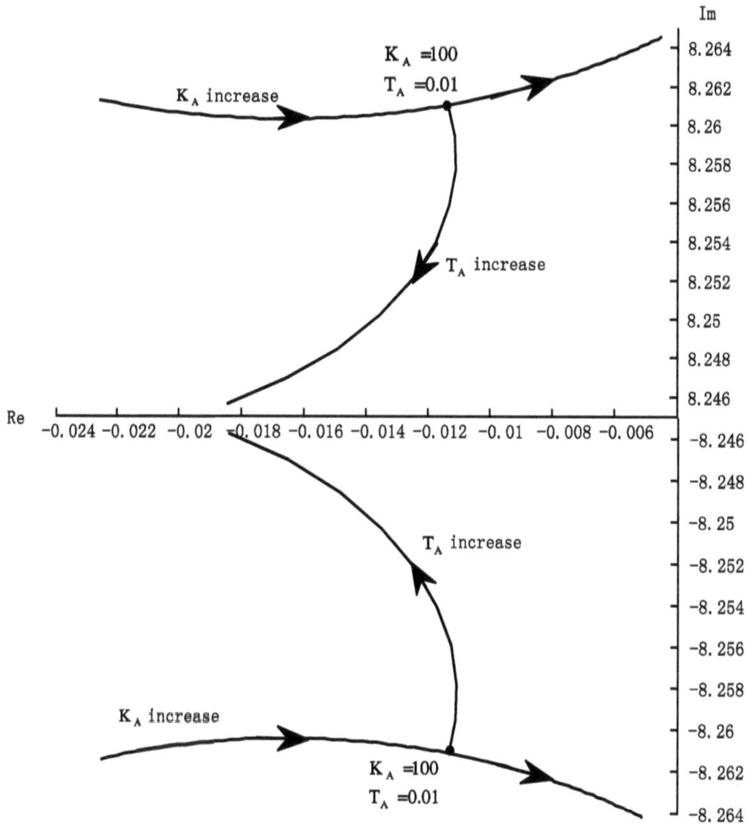


**Fig. 2.18** Loci of the movement of the oscillation mode on the complex plane with the changes of parameters of the AVR

#### 2.4.2.3 Design of the PSS by Pole Assignment for the Example Power System

A PSS can be installed to increase the damping of example power system by assigning the electromechanical oscillation mode, $\overline{\lambda}_{2,3} = -0.0114 \pm j8.2611$, to a target position on the complex plane, $\overline{\lambda}_c = -0.9 \pm j8.2611$, which is damping over 10 %.

Let the transfer function of the PSS be (see Eq. (2.88)) with $T = 0.1$ s

$$T_{pss}(s) = K_{pss}\frac{(1 + s\alpha T)^2}{(1 + sT)^2}$$

State-space model of example power system with the PSS to be installed is

$$\begin{bmatrix} \Delta\dot{\delta} \\ \Delta\dot{\omega} \\ \Delta\dot{E}'_q \\ \Delta\dot{E}'_{fd} \end{bmatrix} = \begin{bmatrix} 0 & 314.16 & 0 & 0 \\ -0.2178 & 0 & -0.1086 & 0 \\ -0.1346 & 0 & -0.5978 & 0.2 \\ 26.4729 & 0 & -3245 & -100 \end{bmatrix} \begin{bmatrix} \Delta\delta \\ \Delta\omega \\ \Delta E'_q \\ \Delta E'_{fd} \end{bmatrix} + \begin{bmatrix} 0 \\ 0 \\ 0 \\ 10000 \end{bmatrix} \Delta u_{pss}$$

$$\Delta\omega = \begin{bmatrix} 0 & 1 & 0 & 0 \end{bmatrix} \begin{bmatrix} \Delta\delta \\ \Delta\omega \\ \Delta E'_q \\ \Delta E'_{fd} \end{bmatrix}$$

$$\Delta u_{pss} = T_{pss}(s)\Delta\omega$$

Thus, the transfer function of open-loop system can be obtained to be

$$G(s) = \mathbf{c}^T(s\mathbf{I} - \mathbf{A})^{-1}\mathbf{b} = \frac{-238.8s}{1.1s^4 + 110.6s^3 + 854.5s^2 + 7563s + 53020}$$

From the characteristic equation of closed-loop control system of Eq. (2.85), for the electromechanical oscillation mode $\bar{\lambda}_c$, it should have

$$T_{pss}(\bar{\lambda}_c) = \frac{1}{G(\bar{\lambda}_c)} = \frac{1}{0.0753 - j0.0996}$$

That is

$$K_{pss}\frac{(1+\bar{\lambda}_c\alpha T)^2}{(1+\bar{\lambda}_c T)^2} = \frac{1}{0.0753 - j0.0996}$$

By solving the above equations, parameters of the PSS can be obtained to be

$$\begin{cases} K_{pss} = 2.6151 \\ \alpha = 2.4256 \end{cases}$$

In order to establish the state-space model of closed-loop system with the PSS installed, let

$$\Delta x_1 = \frac{(1+s\alpha T)}{(1+sT)}\Delta\omega$$

$$\Delta u_{pss} = K_{pss}\frac{(1+s\alpha T)}{(1+sT)}\Delta x_1$$

Thus, state equation of the PSS is as follows:

$$
\begin{aligned}
s\Delta x_1 &= \frac{1}{T}\Delta\omega + \alpha \cdot s\Delta\omega - \frac{1}{T}\Delta x_1 \\
&= -\frac{K_1}{M}\alpha\Delta\delta + \left(\frac{1}{T} - \frac{D}{M}\alpha\right)\Delta\omega - \frac{K_2}{M}\alpha\Delta E_q' - \frac{1}{T}\Delta x_1 \\
s\Delta u_{pss} &= \frac{1}{T}K_{pss}\Delta x_1 - \frac{1}{T}\Delta u_{pss} + K_{pss}\alpha \cdot s\Delta x_1 \\
&= -K_{pss}\frac{K_1}{M}\alpha^2\Delta\delta + K_{pss}\alpha\left(\frac{1}{T} - \frac{D}{M}\alpha\right)\Delta\omega - K_{pss}\alpha^2\frac{K_2}{M}\Delta E_q' \\
&\quad + (1-\alpha)\frac{1}{T}K_{pss}\Delta x_1 - \frac{1}{T}\Delta u_{pss}
\end{aligned}
$$

By writing the state equation of open-loop system and the PSS together, the state-space model of closed-loop system is obtained to be

$$
\begin{bmatrix} \Delta\dot{\delta} \\ \Delta\dot{\omega} \\ \Delta\dot{E}_q' \\ \Delta\dot{E}_{fd}' \\ \Delta\dot{x}_1 \\ \Delta\dot{u}_{pss} \end{bmatrix}
=
\begin{bmatrix}
0 & 314.159 & 0 & 0 & 0 & 0 \\
-0.218 & 0 & -0.109 & 0 & 0 & 0 \\
-0.134 & 0 & -0.597 & 0.200 & 0 & 0 \\
26.582 & 0 & -3245.044 & -100 & 0 & 10000 \\
-0.528 & 10 & -0.263 & 0 & -10 & 0 \\
3.352 & -63.432 & 1.671 & 0 & 37.281 & -10
\end{bmatrix}
\begin{bmatrix} \Delta\delta \\ \Delta\omega \\ \Delta E_q' \\ \Delta E_{fd}' \\ \Delta x_1 \\ \Delta u_{pss} \end{bmatrix}
$$

Eigenvalues of state matrix are calculated to be

$$
\begin{aligned}
\lambda_1 &= -93.4535 \\
\lambda_{2,3} &= -0.8995 \pm j5.2621 \\
\lambda_{4,5} &= -9.4136 \pm j5.0918 \\
\lambda_6 &= -6.5176
\end{aligned}
$$

Hence, the electromechanical oscillation mode is successfully assigned to the target position.

Figure 2.19 shows the simulation result of example power system without and with the PSS installed. At 1.0 s of simulation, a three-phase to-earth short circuit occurred on the transmission line which was cleared in 100 ms. From Fig. 2.19, it can be seen that the low-frequency oscillation is damped effectively by the PSS designed by use of the method of pole assignment.

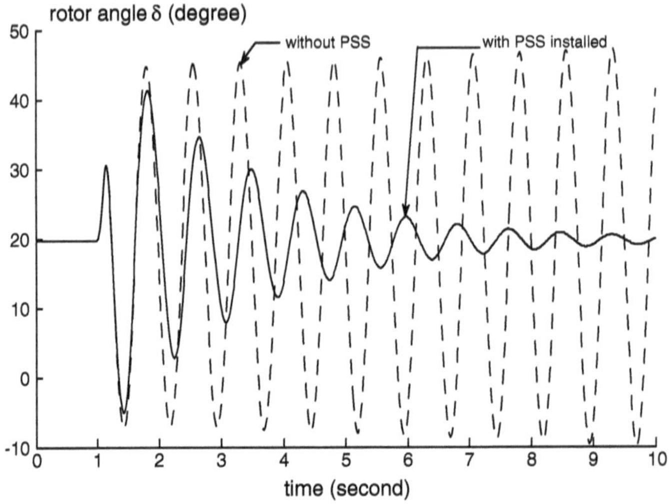


**Fig. 2.19** Simulation result of example power system without and with PSS installed

## *2.4.3 Damping Torque Analysis of Example Power System*

Heffron–Phillips model established is again used in this section to demonstrate the DTA of example power system.

### 2.4.3.1 Damping Torque Provided by the AVR in the Example Power System

From Fig. 2.10, it can be seen that the electric torque provided by the AVR to the electromechanical oscillation loop of generator is as follows:

$$\Delta T_{avr} = K_2 \frac{-K_5 \dfrac{1}{sT'_{do}+K_3}\dfrac{K_A}{sT_A+1}}{1+\dfrac{K_6}{sT'_{do}+K_3}\dfrac{K_A}{sT_A+1}}\Delta\delta$$

$$= -\frac{K_2K_5K_A}{K_6K_A+(sT'_{do}+K_3)(sT_A+1)}\Delta\delta = F_{avr}(s)\Delta\delta$$

At the complex frequency of the electromechanical oscillation with $\bar{\lambda}_s = \xi_s + j\omega_s = \bar{\lambda}_1 = -0.0114 + j8.2610$, let the decomposition of the electric torque provided by the AVR be

$$\Delta T_{avr}(\bar{\lambda}_s) = T_{savr}\Delta\delta(\bar{\lambda}_s) + T_{davr}\Delta\omega(\bar{\lambda}_s)$$

Because

$$\Delta\omega(\bar{\lambda}_s) = \frac{\xi_s + j\omega_s}{\omega_0}\Delta\delta(\bar{\lambda}_s) = \frac{\xi_s}{\omega_0}\Delta\delta(\bar{\lambda}_s) + j\frac{\omega_s}{\omega_0}\Delta\delta(\bar{\lambda}_s)$$

thus

$$\Delta T_{avr}(\bar{\lambda}_s) = T_{savr}\Delta\delta(\bar{\lambda}_s) + T_{davr}\frac{\xi_s}{\omega_0}\Delta\delta(\bar{\lambda}_s) + jT_{davr}\frac{\omega_s}{\omega_0}\Delta\delta(\bar{\lambda}_s)$$

That is

$$T_{davr}\frac{\omega_s}{\omega_0} = \mathrm{Im}F_{avr}(\bar{\lambda}_s)$$

Since

$$\begin{aligned}
F_{avr}(\bar{\lambda}_s) &= -\frac{K_2K_5K_A}{K_6K_A + (\bar{\lambda}_s T'_{do} + K_3)(\bar{\lambda}_s T_A + 1)} \\
&= -\frac{K_2K_5K_A}{K_6K_A + K_3 + (K_3T_A + T'_{d0})\bar{\lambda}_s + T'_{d0}T_A\bar{\lambda}_s^2} \\
&= -\frac{K_2K_5K_A}{K_6K_A + K_3 + (K_3T_A + T'_{d0})(\xi_s + j\omega_s) + T'_{d0}T_A(\xi_s + j\omega_s)^2} \\
&= -\frac{K_2K_5K_A}{a + jb} = -\frac{K_2K_5K_A}{a^2 + b^2}(a - jb) = -0.0024 + j0.0031
\end{aligned}$$

where

$$\begin{aligned}
a &= K_6K_A + K_3 + \left(K_3T_A + T'_{d0}\right)\xi_s + T'_{d0}T_A\left(\xi_s^2 - \omega_s^2\right) = 31.9695 \\
b &= \left(K_3T_A + T'_{d0}\right)\omega_s + 2T'_{d0}T_A\xi_s\omega_s = 41.5426
\end{aligned}$$

thus it can be obtained

$$\begin{aligned}
T_{davr} &= \frac{\omega_0}{\omega_s}\mathrm{Im}F_{avr}(\bar{\lambda}_s) = \frac{\omega_0}{\omega_s}\frac{K_2K_5K_A}{a^2 + b^2}b \\
&= 0.0988
\end{aligned}$$

Since

$$\frac{\partial a}{\partial K_A} = K_6 = 0.3245, \quad \frac{\partial a}{\partial T_A} = K_3\xi_s + T'_{d0}(\xi_s^2 - \omega_s^2) = -341.2561$$
$$\frac{\partial b}{\partial K_A} = 0, \quad \frac{\partial b}{\partial T_A} = K_3\omega_s + 2T'_{d0}\xi_s\omega_s = 23.7488$$

sensitivity of the damping torque provided by the AVR to its parameters can be obtained to be

$$\frac{\partial T_{davr}}{\partial K_A} = \frac{\omega_0}{\omega_s}\frac{K_2K_5\left[\left(b + K_A\frac{\partial b}{\partial K_A}\right)(a^2+b^2) - K_A b\left(2a\frac{\partial a}{\partial K_A} + 2b\frac{\partial b}{\partial K_A}\right)\right]}{(a^2+b^2)^2} = -0.0003$$
$$\frac{\partial T_{davr}}{\partial T_A} = \frac{\omega_0}{\omega_s}\frac{K_2K_5K_A\left[\frac{\partial b}{\partial T_A}(a^2+b^2) - b\left(2a\frac{\partial a}{\partial T_A} + 2b\frac{\partial b}{\partial T_A}\right)\right]}{(a^2+b^2)^2} = -0.9055$$

The above results indicate that (1) with the increase of the AVR gain, less damping torque will be provided by the AVR, detrimental to the system small-signal angular stability and (2) with the increase of the AVR time constant, less damping torque will be provided by the AVR, also detrimental to the damping of low-frequency power oscillations.

#### 2.4.3.2 Design of PSS Installed in the Example Power System by the Phase Compensation Method

The PSS to be designed is to provide a damping torque $\Delta T_{pss} = D_{pss}\Delta\omega, D_{pss} = 15$. From Eq. (2.96), the forward path of stabilizing signal of the PSS can be obtained to be $(j\omega_s = j8.44)$

$$F_{pss}(j\omega_s) = K_2\frac{K_A}{(K_3 + j\omega_s T'_{d0})(1 + j\omega_s T_A) + K_6K_A}$$
$$= 0.8598 - j1.1451 = 1.4320\angle{-53.0989^\circ}$$

The PSS adopts the deviation of rotor speed of generator as the feedback signal, and its transfer function is as follows:

$$T_{pss}(s) = K_1\frac{(1+sT_2)}{(1+sT_1)}K_2\frac{(1+sT_4)}{(1+sT_3)} \quad \text{with } T_1 = 0.09\text{ s}, T_3 = 0.09\text{ s}$$

According to Eq. (2.133), parameters of the PSS are set to compensate the phase of the forward path and thus obtained to be

$$K_{pss} = K_{pss1}K_{pss2} = 3.2271, T_2 = 0.2405\ s, T_4 = 0.2405\ s$$

Let

$$\Delta x_1 = K_{pss2}\frac{(1+sT_4)}{(1+sT_3)}\Delta\omega = (9.3706 + j4.6819)\Delta\omega$$
$$\Delta u_{pss} = K_{pss1}\frac{(1+sT_2)}{(1+sT_1)}\Delta x_1 = (0.8946 + j0.4470)\Delta x_1$$

Thus, state-space realization of the PSS is as follows:

$$\begin{aligned}
s\Delta x_1 &= -\frac{1}{T_3}\Delta x_1 + \frac{K_{pss2}}{T_3}(T_4 s\Delta\omega + \Delta\omega) \\
&= -\frac{1}{T_3}\Delta x_1 + \frac{K_{pss2}}{T_3}\left[\frac{T_4}{M}(-K_1\Delta\delta - K_2\Delta E'_q - D\Delta\omega) + \Delta\omega)\right] \\
&= -\frac{K_{pss2}T_4K_1}{T_3M}\Delta\delta + \frac{K_{pss2}}{T_3}\left(1 - \frac{T_4}{M}D\right)\Delta\omega - \frac{K_{pss2}T_4K_2}{T_3M}\Delta E'_q - \frac{1}{T_3}\Delta x_1 \\
&= -0.3230\Delta\delta + 6.1672\Delta\omega - 0.1224\Delta E'_q - 11.1111\Delta x_1
\end{aligned}$$

$$\begin{aligned}
s\Delta u_{pss} &= -\frac{1}{T_1}\Delta u_{pss} + \left(\frac{K_{pss1}}{T_1} - \frac{T_2}{T_3}\frac{K_{pss1}}{T_1}\right)\Delta x_1 + \frac{K_{pss1}T_2}{T_1}\left[-\frac{K_{pss2}T_4K_1}{T_3M}\Delta\delta\right. \\
&\quad \left. - \frac{K_{pss2}T_4K_2}{T_3M}\Delta E'_q + \frac{K_{pss2}}{T_3}\left(1 - \frac{T_4}{M}D\right)\Delta\omega\right] \\
&= -\frac{K_{pss}T_2T_4K_1}{T_1T_3M}\Delta\delta + \frac{K_{pss}T_2}{T_1T_3}\left(1 - \frac{T_4}{M}D\right)\Delta\omega - \frac{K_{pss}T_2T_4K_2}{T_1T_3M}\Delta E'_q \\
&\quad + \left(\frac{K_{pss1}}{T_1} - \frac{T_2}{T_3}\frac{K_{pss1}}{T_1}\right)\Delta x_1 - \frac{1}{T_1}\Delta u_{pss} \\
&= -5.0182\Delta\delta + 95.8026\Delta\omega - 2.5018\Delta E'_q - 108.0008\Delta x_1 - 11.1111\Delta u_{pss}
\end{aligned}$$

By writing the state equation of open-loop system and the PSS together, state matrix of closed-loop system is obtained to be

$$
\mathbf{A} = \begin{bmatrix}
0 & \omega_0 & 0 & 0 & 0 & 0 \\
-\frac{K_1}{M} & -\frac{D}{M} & -\frac{K_2}{M} & 0 & 0 & 0 \\
-\frac{K_4}{T'_{d0}} & 0 & -\frac{K_3}{T'_{d0}} & -\frac{1}{T'_{d0}} & 0 & 0 \\
-\frac{K_A K_5}{T_A} & 0 & -\frac{K_A K_6}{T_A} & -\frac{K_A}{T_A} & 0 & \frac{K_A}{T_A} \\
-\frac{K_{pss2} T_4 K_1}{T_3 M} & \frac{K_{pss2}}{T_3}\left(1 - \frac{T_4}{T_3 M}\right) & -\frac{K_{20} T_4 K_2}{T_3 M} K_2 & 0 & -\frac{1}{T_3} & 0 \\
-\frac{T_2 T_4 K_1}{T_1 T_3 M} K_{PSS} & \frac{T_2}{T_1 T_3}\left(1 - \frac{T_4 D}{M}\right) K_{PSS} & -\frac{T_2 T_4 K_2}{T_1 T_3 M} K_{PSS} & 0 & \frac{K_{pss1}}{T_1}\left(1 - \frac{T_2}{T_3}\right) & -\frac{1}{T_1}
\end{bmatrix}
$$

$$
= \begin{bmatrix}
0 & 314.16 & 0 & 0 & 0 & 0 \\
-0.2178 & 0 & -0.1086 & 0 & 0 & 0 \\
-0.1345 & 0 & -0.5977 & 0.2 & 0 & 0 \\
26.5821 & 0 & 3245 & -100 & 0 & 10000 \\
-0.3230 & 6.1672 & -0.1224 & 0 & -11.1111 & 0 \\
-5.0182 & 95.8026 & -2.5018 & 0 & -108.0008 & -11.1111
\end{bmatrix}
$$

Eigenvalues of state matrix are calculated to be

$$
\begin{aligned}
\lambda_1 &= -93.6838 \\
\bar{\lambda}_{2,3} &= -1.2125 \pm j8.0051 \\
\bar{\lambda}_{4,5} &= -8.7171 \pm j6.2080 \\
\lambda_6 &= -9.2771
\end{aligned}
$$

Hence, damping of the electromechanical oscillation mode is successfully increased by the PSS designed above via the phase compensation method.

Figure 2.20 shows the simulation result of the example power system without and with the PSS installed. At 1.0 s of the simulation, a three-phase to-earth short

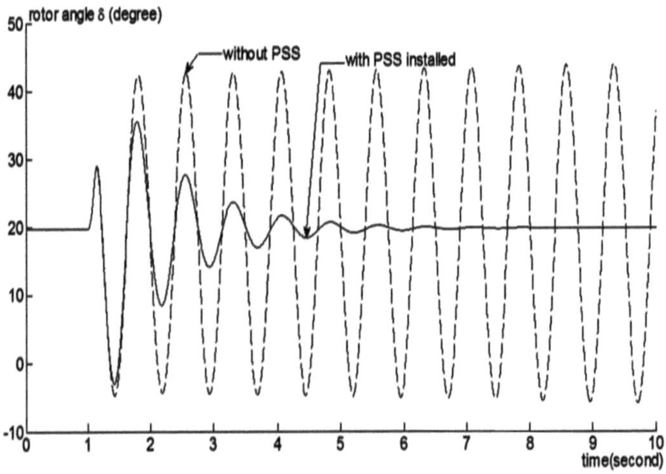


**Fig. 2.20** Simulation result of power system without and with the PSS designed by use of the phase compensation method

circuit occurred on the transmission line which was cleared in 100 ms. From Fig. 2.20, it can be seen that the low-frequency oscillation is damped effectively by the PSS designed by use of the phase compensation method.

#### 2.4.3.3 Theoretical Basis and Graphical Explanation of the Damping Torque Analysis

From Fig. 2.10 and Eq. (2.105), it can be obtained that

$$F_{delta}(s) = K_2\left(\frac{-K_4\frac{1}{sT'_{do}+K_3}}{1+\frac{K_6}{sT'_{do}+K_3}\frac{K_A}{sT_A+1}} + \frac{-K_5\frac{1}{sT'_{do}+K_3}\frac{K_A}{sT_A+1}}{1+\frac{K_6}{sT'_{do}+K_3}\frac{K_A}{sT_A+1}}\right)$$

$$= -\frac{K_2[K_4(sT_A+1)+K_5K_A]}{K_6K_A+(sT'_{do}+K_3)(sT_A+1)}$$

At the complex frequency of electromechanical oscillation with $\bar{\lambda}_s = \xi_s + j\omega_s = \overline{\lambda_1} = -0.0114 + j8.2610$,

$$\bar{F}_{delta}(\bar{\lambda}_s) = -0.0042 + j0.0042$$

According to Eq. (2.114)

$$\begin{cases} T_{d1} = \frac{\omega_0}{\omega_s}\mathrm{Im}\left[\bar{F}_{delta}(\bar{\lambda}_s)\right] = 0.1591 \\ T_{s1} = \mathrm{Re}\left[\bar{F}_{delta}(\bar{\lambda}_s)\right] - \frac{T_{d1}\xi_s}{\omega_0} = -0.0042 \end{cases}$$

Substituting the above result into Eq. (2.116), it can have

$$7\bar{\lambda}_s^2 + 0.1553\bar{\lambda}_s + 477.7128 = 0$$

Solution of above equation is $\bar{\lambda}_s = -0.0114 + j8.2610$. It is the electromechanical oscillation mode of example power system without the PSS being installed.

From Fig. 2.10, it can have

$$\Delta P_t = K_1\Delta\delta + F_{delta}(\bar{\lambda}_s)\Delta\delta = K_1\Delta\delta + T_{s1}\Delta\delta + T_{d1}\Delta\omega = 1.5207\Delta\delta + 0.1553\Delta\omega$$

This is the first equation in Eq. (2.120). Figure 2.21 presents the $P_t - \delta$ curve from simulation. At 1 s of simulation, the mechanical power input to the generator

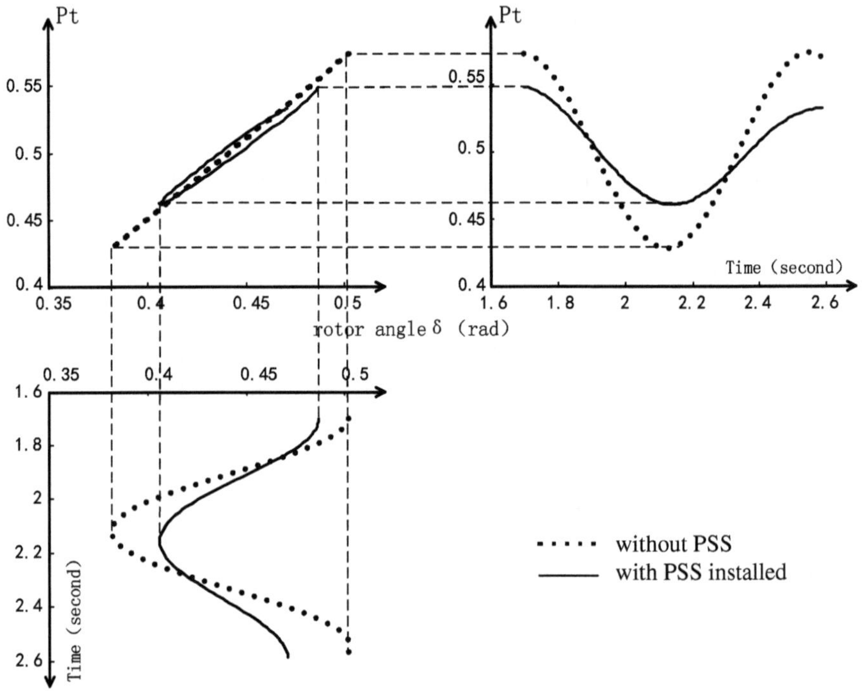


**Fig. 2.21** $P_t - \delta$ curve, corresponding variation of rotor angle and power output of the generator of example power system

increased by 1 % and then returned to its original value in 10 ms. $P_t - \delta$ curve depicted in Fig. 2.20 is the first cycle of power oscillation starting from its first peak. Dashed curve is the case without the PSS installed.

With the PSS designed in Sect. 2.4.3.2 being installed,

$$\begin{aligned}\Delta P_t &= K_1\Delta\delta + F_{delta}(\bar{\lambda}_c)\Delta\delta + \bar{F}_{pss}(\bar{\lambda}_c)\bar{T}_{pss}(\bar{\lambda}_c)\Delta\omega(\bar{\lambda}_c)\\ &= K_1\Delta\delta + T_{s1}\Delta\delta + T_{d1}\Delta\omega + D_{pss}\Delta\omega(\bar{\lambda}_c)\\ &= 1.5207\Delta\delta + 15.1553\Delta\omega\end{aligned}$$

where $\bar{\lambda}_c = -1.2279 + j8.0264$ is the electromechanical oscillation mode of example power system with the PSS being installed. Solid $P_t - \delta$ curve in Fig. 2.21 is the case that the PSS is installed in the example power system. Figure 2.21 confirms the graphical explanation of the DTA illustrated in Fig. 2.16.

### *2.4.4 Equivalence Between the Damping Torque and Modal Analysis*

It is concluded in Sect. 2.3.1.3 that at a complex frequency $\bar{\lambda}_i$, the residue is equal to the forward path of the PSS multiplied by the sensitivity of the mode to the damping torque contribution. In this section, this conclusion is to be demonstrated by example power system.

#### 2.4.4.1 Demonstration by Use of Heffron–Phillips Model of Example Power System

In Sect. 2.4.2.3, state-space model of the example power system is obtained to be

$$\begin{bmatrix} \Delta\dot{\delta} \\ \Delta\dot{\omega} \\ \Delta\dot{E}'_q \\ \Delta\dot{E}'_{fd} \end{bmatrix} = \begin{bmatrix} 0 & 314.16 & 0 & 0 \\ -0.2178 & 0 & -0.1086 & 0 \\ -0.1345 & 0 & -0.5977 & 0.2 \\ 26.5821 & 0 & -3245 & -100 \end{bmatrix} \begin{bmatrix} \Delta\delta \\ \Delta\omega \\ \Delta E'_q \\ \Delta E'_{fd} \end{bmatrix} + \begin{bmatrix} 0 \\ 0 \\ 0 \\ 10000 \end{bmatrix} \Delta u_{pss}$$

$$\Delta\omega = \begin{bmatrix} 0 & 1 & 0 & 0 \end{bmatrix} \begin{bmatrix} \Delta\delta \\ \Delta\omega \\ \Delta E'_q \\ \Delta E'_{fd} \end{bmatrix}$$

The electromechanical oscillation modes are as follows:

$$\bar{\lambda}_{2,3} = -0.0114 \pm j8.2610$$

In Sect. 2.4.2.1, matrix formed by right eigenvectors is as follows:

$$\bar{\mathbf{V}} = \begin{bmatrix} \mathbf{v}_1 & \bar{\mathbf{v}}_2 & \bar{\mathbf{v}}_3 & \mathbf{v}_4 \end{bmatrix}$$

Thus, the right eigenvectors related to the electromechanical oscillation modes are as follows:

$$\bar{\mathbf{v}}_2 = \begin{bmatrix} 0.9014 \\ j0.0237 \\ -0.0050 + j0.0050 \\ 0.3868 - j0.1930 \end{bmatrix} \quad \bar{\mathbf{v}}_3 = \begin{bmatrix} 0.9014 \\ -j0.0237 \\ -0.0050 - j0.0050 \\ 0.3868 + j0.1930 \end{bmatrix}$$

Matrix formed by left eigenvectors is as follows:

$$\bar{\mathbf{W}}^{\mathrm{T}} = [\mathbf{w}_1 \quad \bar{\mathbf{w}}_2 \quad \bar{\mathbf{w}}_3 \quad \mathbf{w}_4]^{\mathrm{T}} = \bar{\mathbf{V}}^{-1}$$

The left eigenvectors related to the electromechanical oscillation modes are as follows:

$$\bar{\mathbf{w}}_2^{\mathrm{T}} = [0.5554 - \mathrm{j}0.0001 \quad -0.0312 - \mathrm{j}21.1226 \quad 0.1626 + \mathrm{j}0.1475 \quad 0.0003 + \mathrm{j}0.0003]$$
$$\bar{\mathbf{w}}_3^{\mathrm{T}} = [0.5554 + \mathrm{j}0.0001 \quad -0.0312 + \mathrm{j}21.1226 \quad 0.1626 - \mathrm{j}0.1475 \quad 0.0003 - \mathrm{j}0.0003]$$

Thus, according to Eq. (2.64), for the electromechanical oscillation modes, the residue is calculated to be

$$\begin{aligned}\bar{\mathrm{R}}_2 &= \bar{\mathbf{w}}_2^{\mathrm{T}}\mathbf{b}_o\mathbf{c}_o^{\mathrm{T}}\bar{\mathbf{v}}_2 \\ &= [0.5554 - \mathrm{j}0.0001 \quad -0.0312 - \mathrm{j}21.1226 \quad 0.1626 + \mathrm{j}0.1476 \quad 0.0003 + \mathrm{j}0.0003] \\ &\quad \times \begin{bmatrix} 0 \\ 0 \\ 0 \\ 10000 \end{bmatrix} [0 \quad 1 \quad 0 \quad 0] \begin{bmatrix} 0.9014 \\ \mathrm{j}0.0237 \\ -0.0050 + \mathrm{j}0.0050 \\ 0.3868 - \mathrm{j}0.1930 \end{bmatrix} \\ &= -0.0633 + \mathrm{j}0.0822\end{aligned}$$

$$\begin{aligned}\bar{\mathrm{R}}_3 &= \bar{\mathbf{w}}_3^{\mathrm{T}}\mathbf{b}_o\mathbf{c}_o^{\mathrm{T}}\bar{\mathbf{v}}_3 \\ &= [0.5554 + \mathrm{j}0.0001 \quad -0.0312 + \mathrm{j}21.1226 \quad 0.1626 - \mathrm{j}0.1475 \quad 0.0003 - \mathrm{j}0.0003] \\ &\quad \times \begin{bmatrix} 0 \\ 0 \\ 0 \\ 10000 \end{bmatrix} [0 \quad 1 \quad 0 \quad 0] \begin{bmatrix} 0.9014 \\ -\mathrm{j}0.0237 \\ -0.0050 - \mathrm{j}0.0050 \\ 0.3868 + \mathrm{j}0.1930 \end{bmatrix} \\ &= -0.0633 - \mathrm{j}0.0822\end{aligned}$$

From Eq. (2.104), sensitivity of the electromechanical oscillation modes to the coefficient of damping torque provided by the PSS can be computed to be

$$\frac{\partial\bar{\lambda}_2}{\partial \mathrm{D}_{\mathrm{pss}}} = -\frac{\bar{\mathrm{w}}_{22}\bar{\mathrm{v}}_{22}}{\mathrm{M}} = -\frac{(-0.0312 - \mathrm{j}21.1226)(\mathrm{j}0.0237)}{7} = -0.0715$$
$$\frac{\partial\bar{\lambda}_3}{\partial \mathrm{D}_{\mathrm{pss}}} = -\frac{\bar{\mathrm{w}}_{32}\bar{\mathrm{v}}_{32}}{\mathrm{M}} = -\frac{(-0.0312 + \mathrm{j}21.1226)(-\mathrm{j}0.0237)}{7} = -0.0715$$

At the complex frequency $\bar{\lambda}_{2,3} = -0.0114 \pm j8.2610$, the forward path can be calculated from Eq. (2.96) as

$$\begin{aligned} F_{pss}(\bar{\lambda}_{2,3}) &= K_2 \frac{K_A}{(K_3 + \bar{\lambda}_{2,3} T'_{d0})(1 + \bar{\lambda}_{2,3} T_A) + K_6 K_A} \\ &= \frac{0.7602 \times 100}{[2.9885 + (-0.0114 \pm j8.2611) \times 5][1 + (-0.0114 \pm j8.2611) \times 0.01] + 0.3245 \times 100} \\ &= -0.8845 \pm j1.1493 \end{aligned}$$

Hence,

$$\begin{aligned} R_{2,3} &= -\frac{\partial \bar{\lambda}_{2,3}}{\partial D_{pss}} F_{pss}(\bar{\lambda}_{2,3}) = -(-0.0715) \times (-0.8845 \pm j1.1493) \\ &= -0.0633 \pm j0.0822 \end{aligned}$$

Thus, it is conformed that the residue is equal to the forward path of the PSS multiplied by the sensitivity of electromechanical oscillation modes to the coefficient of damping torque contribution from the PSS.

#### 2.4.4.2 Demonstration by Use of General Linearized Model of Example Power System

In Sect. 2.4.1.1, state matrix and control vector of state-space model of example power system are obtained to be

$$\mathbf{A}_{gc-dq} = \begin{bmatrix} 0 & 314.16 & 0 & 0 & 0 & 0 & 0 & 0 \\ -0.045 & 0 & 0 & 0.20 & 0.47 & -0.09 & -0.104 & -0.39 \\ 714.29 & 0 & -100 & 6116.3 & -1404.6 & -2646 & -3127 & 1154.5 \\ 295.81 & -89.35 & 0 & -6.65 & 478.4 & 2.88 & 3.40 & -135.03 \\ -105.79 & -318.31 & 0 & -513.64 & -5.48 & 86.29 & 101.98 & 4.50 \\ 0 & 0 & 314.16 & 0.43 & 0 & -1.22 & 0.71 & 0 \\ 0 & 0 & 0 & 1.36 & 0 & 1.88 & -3.49 & 0 \\ 0 & 0 & 0 & 0 & 36.0 & 0 & 0 & -46.8 \end{bmatrix}$$

$$\mathbf{b}_{pss} = \begin{bmatrix} 0 \\ 0 \\ 10000 \\ 0 \\ 0 \\ 0 \\ 0 \\ 0 \end{bmatrix}$$

Eigenvalues of state matrix are as follows:

$$\begin{aligned}\bar{\lambda}_{1,2} &= -24.12 \pm j971.56\\ \bar{\lambda}_{3,4} &= -36.32 \pm j359.63\\ \lambda_5 &= -35.57\\ \bar{\lambda}_{6,7} &= -0.71 \pm j8.44\\ \lambda_8 &= -5.71\end{aligned}$$

The pair of electromechanical oscillation modes are as follows:

$$\bar{\lambda}_{6,7} = -0.71 \pm j8.44$$

For each of eigenvalues, right eigenvector is calculated to be

$$\bar{\mathbf{v}}_1 = \begin{bmatrix} 0.000005 - j0.000011 \\ 0.000032 + j0.000016 \\ 0.9507 \\ -0.0001 + j0.0182 \\ -0.0369 + j0.0014 \\ -0.0072 - j0.3072 \\ -0.000569 + j0.000026 \\ j0.0014 \end{bmatrix} \quad \bar{\mathbf{v}}_2 = \begin{bmatrix} 0.000005 + j0.000011 \\ 0.000032 - j0.000016 \\ 0.9507 \\ -0.0001 - j0.0182 \\ -0.0369 - j0.0014 \\ -0.0072 + j0.3072 \\ -0.000569 - j0.000026 \\ -j0.0014 \end{bmatrix}$$

$$\bar{\mathbf{v}}_3 = \begin{bmatrix} -0.000215 - j0.000026 \\ 0.000055 - j0.000244 \\ 0.7251 \\ 0.0163 - j0.2196 \\ 0.1654 + j0.025 \\ -0.0615 - j0.6274 \\ -0.0041 + j0.0006 \\ 0.003 - j0.0165 \end{bmatrix} \quad \bar{\mathbf{v}}_4 = \begin{bmatrix} -0.000215 + j0.000026 \\ 0.000055 + j0.000244 \\ 0.7251 \\ 0.0163 + j0.2196 \\ 0.1654 - j0.025 \\ -0.0615 + j0.6274 \\ -0.0041 - j0.0006 \\ 0.003 + j0.0165 \end{bmatrix} \quad \mathbf{v}_5 = \begin{bmatrix} -0.0527 \\ 0.006 \\ -0.0539 \\ 0.1055 \\ 0.2563 \\ 0.4925 \\ -0.0334 \\ 0.8209 \end{bmatrix}$$

$$\bar{\mathbf{v}}_6 = \begin{bmatrix} 0.7004 \\ -0.0016 + j0.0188 \\ 0.000061 - j0.000914 \\ -0.1482 + j0.0043 \\ -0.5516 + j0.0283 \\ -0.0317 + j0.0041 \\ -0.0077 + j0.0284 \\ -0.413 + j0.0977 \end{bmatrix} \quad \bar{\mathbf{v}}_7 = \begin{bmatrix} 0.7004 \\ -0.0016 - j0.0188 \\ 0.000061 + j0.000914 \\ -0.1482 - j0.0043 \\ -0.5516 - j0.0283 \\ -0.0317 - j0.0041 \\ -0.0077 - j0.0284 \\ -0.413 - j0.0977 \end{bmatrix} \quad \mathbf{v}_8 = \begin{bmatrix} 0.000104 \\ -0.000002 \\ 0.0095 \\ -0.000116 \\ -0.000082 \\ -0.7636 \\ 0.6456 \\ -0.000072 \end{bmatrix}$$

They form the following matrix

$$\bar{\mathbf{V}} = [\,\bar{\mathbf{v}}_1 \quad \bar{\mathbf{v}}_2 \quad \ldots \quad \bar{\mathbf{v}}_8\,]$$

As $\bar{\mathbf{W}}^{\mathrm{T}} = [\,\bar{\mathbf{w}}_1 \quad \bar{\mathbf{w}}_2 \quad \ldots \quad \mathbf{w}_8\,]^{\mathrm{T}} = \bar{\mathbf{V}}^{-1}$, left eigenvectors corresponding to eigenvalues are calculated to be

$$\bar{\mathbf{w}}_1^{\mathrm{T}} = \begin{bmatrix} -1.2419 - j0.4483 \\ -0.0486 - j0.262 \\ 0.4494 + j0.0412 \\ -0.2153 - j3.8659 \\ -1.9686 + j0.7712 \\ -0.0188 + j1.3997 \\ -0.0288 + j1.6542 \\ 0.6025 - j0.5408 \end{bmatrix}^{\mathrm{T}} \quad \bar{\mathbf{w}}_2^{\mathrm{T}} = \begin{bmatrix} -1.2419 + j0.4483 \\ -0.0486 + j0.262 \\ 0.4494 - j0.0412 \\ -0.2153 + j3.8659 \\ -1.9686 - j0.7712 \\ -0.0188 - j1.3997 \\ -0.0288 - j1.6542 \\ 0.6025 + j0.5408 \end{bmatrix}^{\mathrm{T}}$$

$$\bar{\mathbf{w}}_3^{\mathrm{T}} = \begin{bmatrix} 1.6552 + j0.1633 \\ -0.1388 + j0.9214 \\ 0.1003 - j0.0038 \\ 0.2788 + j1.9563 \\ 2.5805 - j0.126 \\ 0.0246 + j0.1141 \\ 0.0277 + j0.1353 \\ -0.7412 - j0.2714 \end{bmatrix}^{\mathrm{T}} \quad \bar{\mathbf{w}}_4^{\mathrm{T}} = \begin{bmatrix} 1.6552 - j0.1633 \\ -0.1388 - j0.9214 \\ 0.1003 + j0.0038 \\ 0.2788 - j1.9563 \\ 2.5805 + j0.126 \\ 0.0246 - j0.1141 \\ 0.0277 - j0.1353 \\ -0.7412 + j0.2714 \end{bmatrix}^{\mathrm{T}} \quad \mathbf{w}_5^{\mathrm{T}} = \begin{bmatrix} 0.7411 \\ -6.8756 \\ -0.000219 \\ -0.0934 \\ -0.0106 \\ -0.000045 \\ -0.00006 \\ 1.3311 \end{bmatrix}^{\mathrm{T}}$$

$$\bar{\mathbf{w}}_6^{\mathrm{T}} = \begin{bmatrix} 0.7422 + j0.06 \\ -0.2586 - j27.632 \\ 0.000019 + j0.00094 \\ -0.0034 + j0.0146 \\ 0.0004 + j0.000131 \\ -0.000019 + j0.000298 \\ -0.000139 + j0.0002727 \\ 0.0499 + j0.2027 \end{bmatrix}^{\mathrm{T}} \quad \bar{\mathbf{w}}_7^{\mathrm{T}} = \begin{bmatrix} 0.7422 - j0.06 \\ -0.2586 + j27.632 \\ 0.000019 - j0.00094 \\ -0.0034 - j0.0146 \\ 0.0004 - j0.000131 \\ -0.000019 - j0.000298 \\ -0.000139 - j0.0002727 \\ 0.0499 - j0.2027 \end{bmatrix}^{\mathrm{T}}$$

$$\mathbf{w}_8^{\mathrm{T}} = \begin{bmatrix} 0.0803 \\ -2.7921 \\ 0.0021 \\ 0.003 \\ 0.0283 \\ 0.000637 \\ 1.5496 \\ 0.079 \end{bmatrix}^{\mathrm{T}}$$

According to Eq. (2.64), for the pair of electromechanical oscillation modes, the residue is calculated to be

$$\bar{R}_{6,7} = \bar{\mathbf{w}}_{6,7}^{T}\mathbf{b_o}\mathbf{c_o}^{T}\bar{\mathbf{v}}_{6,7}$$

$$= \begin{bmatrix} 0.7422 \pm j0.06 \\ -0.2586 \mp j27.632 \\ 0.000019 \pm j0.00094 \\ -0.0034 \pm j0.0146 \\ 0.0004 \pm j0.000131 \\ -0.000019 \pm j0.000298 \\ -0.000139 \pm j0.0002727 \\ 0.0499 + j0.2027 \end{bmatrix}^{T} \begin{bmatrix} 0 \\ 0 \\ 10000 \\ 0 \\ 0 \\ 0 \\ 0 \\ 0 \end{bmatrix}$$

$$\times \begin{bmatrix} 0 & 1 & 0 & 0 & 0 & 0 & 0 & 0 \end{bmatrix} \begin{bmatrix} 0.7004 \\ -0.0016 \pm j0.0188 \\ 0.000061 \mp j0.000914 \\ -0.1482 \pm j0.0043 \\ -0.5516 \pm j0.0283 \\ -0.0317 \pm j0.0041 \\ -0.0077 \pm j0.0284 \\ -0.413 \pm j0.0977 \end{bmatrix}$$

$$= -0.1772 \mp j0.0114$$

From Eq. (2.104), the sensitivity of electromechanical oscillation modes to the coefficient of damping torque provided by the PSS can be computed to be

$$\frac{\partial \bar{\lambda}_{6,7}}{\partial D_{pss}} = -\frac{\bar{w}_{62,72}\bar{v}_{62,72}}{M} = -\frac{(-0.2586 \mp j27.632)(-0.0016 \pm j0.0188)}{7}$$

$$= -0.0743 \mp j0.0056$$

At the complex frequency $\bar{\lambda}_{6,7} = -0.71 \pm j8.44$, the forward path can be calculated to be

$$
\begin{aligned}
F_{pss}(\bar{\lambda}_{6,7}) &= -\mathbf{a}_{23-8}^{T}(s\mathbf{I}-\mathbf{A}_{33})^{-1}\mathbf{b}_{pss-3} \\
&= \begin{bmatrix} 0 & 1.425 & 3.283 & -0.617 & -0.729 & -2.698 \end{bmatrix} \\
&\quad \left( (-0.71 \pm j8.44) \begin{bmatrix} 1 & & & & & \\ & 1 & & & & \\ & & 1 & & & \\ & & & 1 & & \\ & & & & 1 & \\ & & & & & 1 \end{bmatrix} - \begin{bmatrix} -100 & 6116.3 & -1404.6 & -2646 & -3127 & 1154.5 \\ 0 & -6.65 & 478.4 & 2.88 & 3.40 & -135.03 \\ 0 & 513.64 & -5.48 & 86.29 & 101.98 & 4.50 \\ 314.16 & 0.43 & 0 & -1.22 & 0.71 & 0 \\ 0 & 1.36 & 0 & 1.88 & -3.49 & 0 \\ 0 & 0 & 36.0 & 0 & 0 & -46.8 \end{bmatrix} \right)^{-1} \begin{bmatrix} 10000 \\ 0 \\ 0 \\ 0 \\ 0 \\ 0 \end{bmatrix} \\
&= -2.3823 \pm j0.0253
\end{aligned}
$$

Hence,

$$
\begin{aligned}
R_{6,7} &= -\frac{\partial\bar{\lambda}_{6,7}}{\partial D_{pss}} F_{pss}(\bar{\lambda}_{6,7}) = -(-0.0743 \mp j0.0056)(-2.3823 \pm j0.0253) \\
&= -0.1772 \mp j0.0114
\end{aligned}
$$

It is thus confirmed that the residue is equal to the forward path of the PSS multiplied by the sensitivity of oscillation modes to the coefficient of damping torque contribution from the PSS.

## References

1. Demello FP, Concordia C (1969) Concepts of synchronous machine stability as affected by excitation control. IEEE Trans Power Appar Syst 88(4):316–329
2. Heffron WG, Phillips RA (1952) Effect of modern amplidyne voltage regulators on underexcited operation of large turbine generators. AIEE Trans Power Apparatus Syst 71:692–697
3. Demello FP, Laskowski TF (1975) Concepts of power system dynamic stability. IEEE Trans Power Apparatus Syst 94(3):827–833
4. Larsen EV, Swann DA (1981) Applying power system stabilizers part I: general concepts. IEEE Trans Power Appar Syst 100(6):3017–3024

# Chapter 3
# Damping Torque Analysis of Thyristor-Based FACTS Stabilizers Installed in Single-Machine Infinite-Bus Power Systems

## 3.1 A Single-Machine Infinite-Bus Power System Installed with an SVC Stabilizer

### *3.1.1 Extended Heffron—Phillips Model of a Single-Machine Infinite-Bus Power System Installed with an SVC Stabilizer*

#### 3.1.1.1 Nonlinear Mathematical Model of a Single-Machine Infinite-Bus Power System Installed with an SVC Stabilizer

Installation of a shunt thyristor-controlled reactive power compensator, the static VAR compensator (SVC), is for the voltage support at a key location in a power system. An additional stabilizing signal can be superimposed on the voltage control loop of the SVC to provide extra damping to the power system. This supplementary damping controller is named the SVC stabilizer in this book. Figure 3.1 shows the configuration of a thyristor-controlled reactor and fixed capacitor (TCR-FC) type of SVC with an additional damping control loop, i.e., the SVC stabilizer.

Figure 3.2 shows the configuration of a single-machine infinite-bus power system installed with an SVC, where $b_{svc}$ is the equivalent admittance of the SVC. For the TCR-FC, SVC [1]

$$jb_{svc} = [1 - C(\alpha)]/jx_{svcl} - 1/jx_{svcc} \tag{3.1}$$

where $C(\alpha) = \frac{2\alpha - \sin 2\alpha}{2\pi} - 1$ and $\alpha$ is the firing angle.

From Fig. 3.2, it can have

$$\begin{aligned} \bar{I}_{sb} &= \bar{I}_{ts} - \bar{I}_s = \bar{I}_{ts} - jb_{svc}\bar{V}_s \\ \bar{V}_s &= jx_{sb}\bar{I}_{sb} + \bar{V}_b = jx_{sb}\bar{I}_{ts} + x_{sb}b_{svc}\bar{V}_s + \bar{V}_b \end{aligned} \tag{3.2}$$



H. Wang and W. Du, *Analysis and Damping Control of Power System Low-frequency Oscillations*, Power Electronics and Power Systems,
DOI 10.1007/978-1-4899-7696-3_3

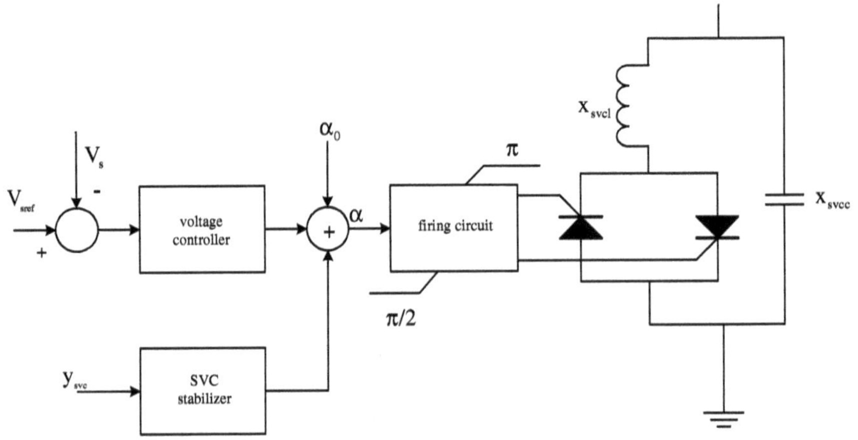


**Fig. 3.1** Configuration of a TCP-FC SVC with damping control function

**Fig. 3.2** A single-machine infinite-bus power system installed with an SVC

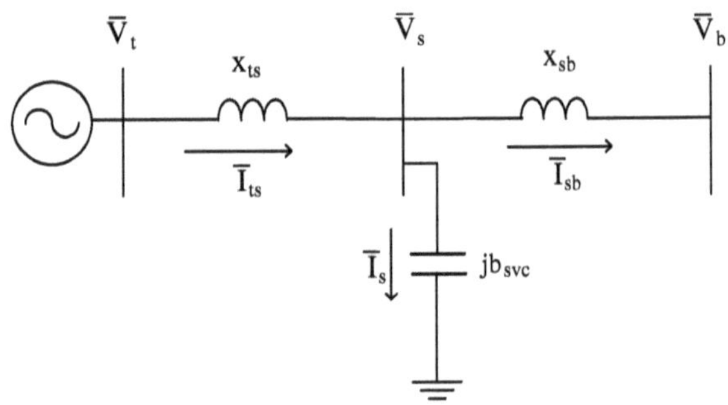


$$\bar{V}_t = jx_{ts}\bar{I}_{ts} + \bar{V}_s \tag{3.3}$$

From Eq. (3.2), it can be obtained that

$$\bar{V}_s = \frac{jx_{sb}\bar{I}_{ts} + \bar{V}_b}{c_{svc}} \tag{3.4}$$

where $c_{svc} = 1 - x_{sb}b_{svc}$.

By substituting Eq. (3.4) into Eq. (3.3) it can have

$$\bar{V}_t = jx_{1\Sigma}\bar{I}_{ts} + \bar{V}_b/c_{svc} \tag{3.5}$$

where $x_{1\Sigma} = x_{ts} + x_{sb}/c_{svc}$.

Equation (3.5) indicates that the power system installed with the SVC is equivalent to a power system without the SVC installed. The equivalent system is of a line impedance $x_{1\Sigma}$ connected to an infinite bus with a voltage being $\frac{\bar{V}_b}{c_{svc}}$.

Therefore, the mathematical model of the single-machine infinite-bus power system installed with the SVC as shown in Fig. 3.2 can be obtained simply by modifying the model of the system without the SVC of Eqs. (2.37) and (2.38). The modification is to replace $x_t$ and $V_b$ in Eqs. (2.37) and (2.38) by $x_{l\Sigma}$ and $\frac{V_b}{C_{svc}}$, respectively, to obtain that

$$\begin{aligned}\dot{\delta} &= \omega_o(\omega - 1)\\ \dot{\omega} &= \frac{1}{M}[P_m - P_t - D(\omega - 1)]\\ \dot{E}'_q &= \frac{1}{T'_{do}}(-E_q + E_{fd})\\ \dot{E}'_{fd} &= -\frac{1}{T_A}E'_{fd} + \frac{K_A}{T_A}(V_{tref} - V_t)\end{aligned} \tag{3.6}$$

$$\begin{aligned}P_t &= \frac{E'_q V_b}{c_{svc} x'_{d\Sigma}}\sin\delta - \frac{V_b^2}{c_{svc}^2 2}\frac{(x_q - x'_d)}{x'_{d\Sigma} x_{q\Sigma}}\sin 2\delta\\ E_q &= \frac{E'_q x_{d\Sigma}}{x'_{d\Sigma}} - \frac{(x_d - x'_d)\,V_b\cos\delta}{c_{svc} x'_{d\Sigma}}\\ E_{fd} &= E_{fd0} + E'_{fd}\\ v_{td} &= \frac{x_q V_b \sin\delta}{c_{svc} x_{q\Sigma}},\quad v_{tq} = \frac{x_t\,E'_q}{x'_{d\Sigma}} + \frac{V_b x'_d\cos\delta}{c_{svc} x'_{d\Sigma}},\quad V_t = \sqrt{v_{td}^2 + v_{tq}^2}\end{aligned} \tag{3.7}$$

where $x'_{d\Sigma} = x'_d + x_{l\Sigma}$, $x_{d\Sigma} = x_d + x_{l\Sigma}$, $x_{q\Sigma} = x_q + x_{l\Sigma}$.

Similar modification of Eq. (2.33) gives

$$\begin{aligned}i_{tsd} &= \frac{E'_q}{x'_{d\Sigma}} - \frac{V_b\cos\delta}{c_{svc} x'_{d\Sigma}}\\ i_{tsq} &= \frac{V_b\sin\delta}{c_{svc} x_{q\Sigma}}\end{aligned} \tag{3.8}$$

where $i_{tsd}$ and $i_{tsq}$ are the d and q component of line current $\bar{I}_{ts}$, respectively.

#### 3.1.1.2 Extended Heffron–Phillips Model

By linearizing the first two equations in Eq. (3.7), it can have

$$\begin{aligned}\Delta P_t &= K_1\Delta\delta + K_2\Delta E'_q + K_p\Delta b_{svc}\\ \Delta E_q &= K_3\Delta E'_q + K_4\Delta\delta + K_q\Delta b_{svc}\end{aligned} \tag{3.9}$$

where

$$K_1 = \frac{E'_{q0}V_b}{c_{svc0}x'_{d\Sigma}}\cos\delta_0 - \frac{V_b^2(x_q - x'_d)}{c_{svc0}^2 x'_{d\Sigma}x_{q\Sigma}}\cos 2\delta_0$$
$$K_2 = \frac{V_b}{c_{svc0}x'_{d\Sigma}}\sin\delta_0$$

$$K_p = \frac{\partial P_t}{\partial b_{svc}}\bigg|_0 = \frac{V_b E'_{q0}\sin\delta_0 x_{sb}(x_{ts} + x'_d)}{(c_{svc0}x'_{d\Sigma})^2} - \frac{V_b^2(x_q - x'_d)\sin(2\delta_0)x_{sb}(x_{ts} + x'_d)}{2c_{svc0}^3(x'_{d\Sigma})^2 x_{q\Sigma}}$$
$$- \frac{V_b^2(x_q - x'_d)\sin(2\delta_0)x_{sb}(x_{ts} + x_q)}{2c_{svc0}^3 x'_{d\Sigma}(x_{q\Sigma})^2}$$
$$= \frac{x_{sb}(x_{ts} + x'_d)}{c_{svc0}x'_{d\Sigma}}\left[\frac{V_b E'_{q0}\sin\delta_0}{c_{svc0}x'_{d\Sigma}} - \frac{V_b^2(x_q - x'_d)\sin(2\delta_0)}{2c_{svc0}^2 x'_{d\Sigma}x_{q\Sigma}}\right]$$
$$- \frac{V_b^2(x_q - x'_d)\sin(2\delta_0)}{2c_{svc0}^2 x'_{d\Sigma}x_{q\Sigma}}\frac{x_{sb}(x_{ts} + x_q)}{c_{svc0}x_{q\Sigma}}$$

$$K_3 = \frac{x_{d\Sigma}}{x'_{d\Sigma}}$$
$$K_4 = \frac{(x_d - x'_d)V_b\sin\delta_0}{c_{svc0}x'_{d\Sigma}}$$
$$K_q = -\frac{E'_{q0}x_{d\Sigma}}{(x'_{d\Sigma})^2}\frac{x_{sb}^2}{c_{svc0}^2} + \frac{E'_{q0}}{x'_{d\Sigma}}\frac{x_{sb}^2}{c_{svc0}^2} - \frac{(x_d - x'_d)V_b\cos\delta_0}{(c_{svc0}x'_{d\Sigma})^2}x_{sb}(x_{ts} + x'_d)$$

In deriving $K_p$ and $K_q$ above, following equations are used

$$\begin{aligned} c_{svc}x'_{d\Sigma} &= c_{svc}(x'_d + x_{ts} + x_{sb}/c_{svc}) = c_{svc}(x'_d + x_{ts}) + x_{sb} \\ &= (1 - x_{sb}b_{svc})(x'_d + x_{ts}) + x_{sb} \\ c_{svc}x_q &= (1 - x_{sb}b_{svc})(x_q + x_{ts}) + x_{sb} \end{aligned} \tag{3.10}$$

By linearizing the last three equations in Eq. (3.7), it can have

$$\begin{aligned} \Delta v_{td} &= \frac{x_q V_b\cos\delta_0}{c_{svc0}x_{q\Sigma}}\Delta\delta + \frac{x_q V_b\sin\delta_0}{(c_{svc0}x_{q\Sigma})^2}x_{sb}(x_{sb} + x_q)\Delta b_{svc} \\ \Delta v_{tq} &= -\frac{V_b x'_d\sin\delta_0}{c_{svc0}x'_{d\Sigma}}\Delta\delta + \frac{x_{L\Sigma}}{x'_{d\Sigma}}\Delta E'_q \\ &+ \left[-\frac{x_{l\Sigma}E'_{q0}}{(x'_{d\Sigma})^2}\frac{x_{sb}^2}{c_{svc0}^2} + \frac{E'_{q0}}{x'_{d\Sigma}}\frac{x_{sb}^2}{c_{svc0}^2} + \frac{V_b x'_d\cos\delta_0}{(c_{svc0}x'_{d\Sigma})^2}x_{sb}(x_{sb} + x'_d)\right]\Delta b_{svc} \end{aligned} \tag{3.11}$$

Hence,

$$\Delta V_t = \frac{v_{td0}}{V_{t0}}\Delta v_{td} + \frac{v_{tq0}}{V_{t0}}\Delta v_{tq} = K_5\Delta\delta + K_6\Delta E'_q + K_v\Delta b_{svc} \tag{3.12}$$

where

$$\begin{aligned}
K_5 &= \frac{v_{td0}}{V_{t0}}\frac{x_q V_b \cos\delta_0}{c_{svc0}x_{q\Sigma}} - \frac{v_{tq0}}{V_{t0}}\frac{V_b x'_d \sin\delta_0}{c_{svc0}x'_{d\Sigma}} \\
K_6 &= \frac{v_{tq0}}{V_{t0}}\frac{x_{L\Sigma}}{x'_{d\Sigma}} \\
K_v &= \frac{v_{td0}}{V_{t0}}\frac{x_q V_b \sin\delta_0}{(c_{svc0}x_{q\Sigma})^2}x_{sb}(x_{sb}+x_q) \\
&\quad + \frac{v_{tq0}}{V_{t0}}\left[-\frac{x_{l\Sigma}}{(x'_{d\Sigma})^2}\frac{E'_{q0}}{}\frac{x_{sb}^2}{c_{svc0}^2} + \frac{E'_{q0}}{x'_{d\Sigma}}\frac{x_{sb}^2}{c_{svc0}^2} + \frac{V_b x'_d \cos\delta_0}{(c_{svc0}x'_{d\Sigma})^2}x_{sb}(x_{sb}+x'_d)\right]
\end{aligned}$$

By substituting Eqs. (3.9) and (3.12) into the linearized equation of Eq. (3.6), it can have

$$\begin{aligned}
\Delta\dot{\delta} &= \omega_o\Delta\omega \\
\Delta\dot{\omega} &= \frac{1}{M}(-K_1\Delta\delta - D\Delta\omega - K_2\Delta E'_q - K_p\Delta b_{svc}) \\
\Delta\dot{E}'_q &= \frac{1}{T'_{d0}}(-K_3\Delta E'_q - K_4\Delta\delta - K_q\Delta b_{svc} + \Delta E_{fd}) \\
\Delta\dot{E}'_{fd} &= -\frac{1}{T_A}\Delta\dot{E}'_{fd} - \frac{K_A}{T_A}(K_5\Delta\delta + K_6\Delta E'_q + K_v\Delta b_{svc})
\end{aligned} \tag{3.13}$$

Equation (3.13) is the extended Heffron–Phillips model of the single-machine infinite-bus power system installed with the SVC, which is shown in Fig. 3.3.

#### 3.1.1.3 Extended Heffron–Phillips Model with Both the SVC Voltage and Damping Control Function Included

By substituting $\bar{I}_{ts} = i_{tsd} + ji_{tsq}$ into Eq. (3.4), it can have

$$\begin{aligned}
v_{sd} &= \frac{V_b \sin\delta - i_{tsq}x_{sb}}{c_{svc}} \\
vsq &= \frac{V_b \cos\delta + i_{tsd}x_{sb}}{c_{svc}}
\end{aligned} \tag{3.14}$$

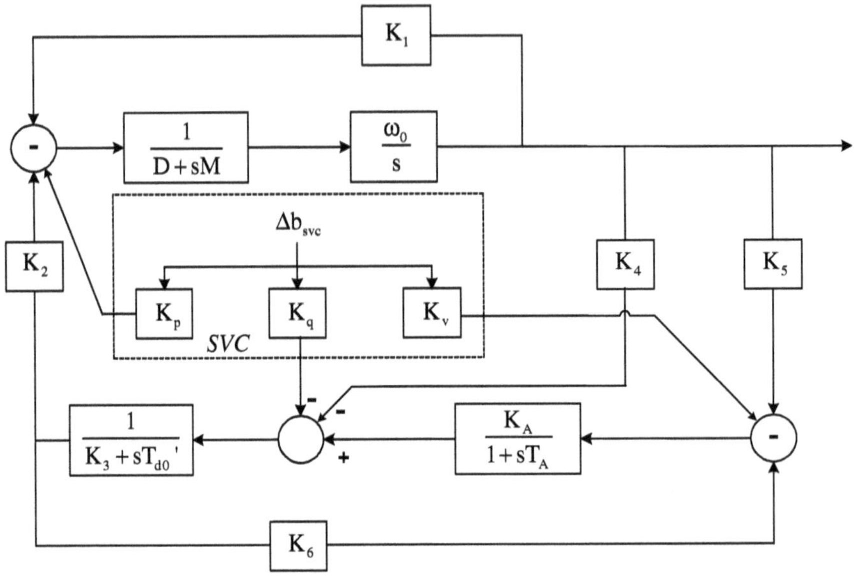


**Fig. 3.3** Extended Heffron–Phillips model of the single-machine infinite-bus power system installed with an SVC stabilizer

where $v_{sd}$ and $v_{sq}$ are the d and q component of voltage $\bar{V}_s$, respectively. By using Eq. (3.8), from Eq. (3.14), it can be obtained that

$$
\begin{aligned}
v_{sd} &= \frac{V_b \sin\delta - \frac{V_b \sin\delta}{C_{svc} X_{q\Sigma}} x_{sb}}{c_{svc}} = \frac{1}{c_{svc}}\left(V_b \sin\delta - \frac{x_{sb} V_b \sin\delta}{C_{svc} X_{q\Sigma}}\right) \\
vsq &= \frac{1}{c_{svc}}\left(V_b \cos\delta + \frac{x_{sb} E'_q}{x'_{d\Sigma}} - \frac{V_b \cos\delta}{c_{svc} x'_{d\Sigma}} x_{sb}\right) \\
V_s &= \sqrt{v_{sd}^2 + v_{sq}^2}
\end{aligned}
\tag{3.15}
$$

Linearization of Eq. (3.15) is as follows:

$$
\Delta V_s = C_1 \Delta\delta + C_2 \Delta E'_q + C_3 \Delta b_{svc} \tag{3.16}
$$

where

$$
\begin{aligned}
C_1 &= \left.\frac{\partial V_s}{\partial \delta}\right|_0 = \frac{v_{sd0}}{V_{s0}} \left.\frac{\partial v_{sd}}{\partial \delta}\right|_0 + \frac{v_{sq0}}{V_{s0}} \left.\frac{\partial v_{sq}}{\partial \delta}\right|_0 \\
&= \frac{v_{sd0}}{V_{s0}} \frac{1}{c_{svc0}} \left(V_{b0} \cos\delta_0 - \frac{x_{sb} V_b \cos\delta_0}{c_{svc0} x_{q\Sigma}}\right) + \frac{v_{sq0}}{V_{s0}} \frac{1}{c_{svc0}} \left(-V_{b0} \sin\delta_0 + \frac{V_b \sin\delta_0}{c_{svc0} x'_{d\Sigma}} x_{sb}\right)
\end{aligned}
$$

$$
\begin{aligned}
C_2 &= \left.\frac{\partial V_s}{\partial E'_q}\right|_0 = \left.\frac{v_{sd0}}{V_{s0}}\frac{\partial v_{sd}}{\partial E'_q}\right|_0 + \left.\frac{v_{sq0}}{V_{s0}}\frac{\partial v_{sq}}{\partial E'_q}\right|_0 = \frac{v_{sq0}}{V_{s0}}\frac{1}{c_{svc0}}\frac{x_{sb}}{x'_{d\Sigma}} \\
C_3 &= \left.\frac{\partial V_s}{\partial b_{svc}}\right|_0 = \left.\frac{v_{sd0}}{V_{s0}}\frac{\partial v_{sd}}{\partial b_{svc}}\right|_0 + \left.\frac{v_{sq0}}{V_{s0}}\frac{\partial v_{sq}}{\partial b_{svc}}\right|_0 \\
&= \frac{v_{sd0}}{V_{s0}}\left[\frac{x_{sb}v_{sd0}}{c_{svc0}} - \frac{x_{sb}^2 V_b \sin\delta_0}{c_{svc0}(c_{svc0}x_{q\Sigma})^2}(x_q + x_{ts})\right] \\
&\quad + \frac{v_{sq0}}{V_{s0}}\left\{\frac{x_{sb}}{c_{svc0}^2}\left(V_b\cos\delta_0 + \frac{x_{sb}E'_{q0}}{x'_{d\Sigma}} - \frac{V_b\cos\delta_0}{c_{svc0}x'_{d\Sigma}}x_{sb}\right)\right. \\
&\quad \left. + \frac{1}{c_{svc0}}\left[-\frac{E'_{q0}}{x'^2_{d\Sigma}}\frac{x_{sb}^3}{c_{svc0}^2} - \frac{V_b\cos\delta_0}{c_{svc0}^2 x'^2_{d\Sigma}}x_{sb}^2(x'_d + x_{ts})\right]\right\}
\end{aligned}
$$

For the simplicity of discussion, it is assumed that both the SVC voltage and damping controller adopt the proportional control law with a gain $C_V$ and $C_S$, respectively. If the feedback signal of the SVC stabilizer is the rotor speed of generator, from Fig. 3.1, it can have

$$\Delta\alpha = -C_V\Delta V_s + C_S\Delta\omega \tag{3.17}$$

From Eq. (3.1), it can be obtained that

$$b_{svc} = \frac{1}{x_{svcc}} - \frac{1 - C(\alpha)}{x_{svcl}} \tag{3.18}$$

Hence,

$$
\begin{aligned}
\Delta b_{svc} &= \frac{\dot{C}(\alpha_0)}{x_{svcl}}\Delta\alpha = \frac{\dot{C}(\alpha_0)}{x_{svcl}}(C_V\Delta V_s + C_S\Delta\omega) \\
&\text{where } \frac{4}{\pi} > \dot{C}(\alpha_0) = \frac{(2 - 2\cos 2\alpha_0)}{\pi} > 0
\end{aligned}
\tag{3.19}
$$

By substituting Eq. (3.16) into Eq. (3.19), the following equation can be obtained

$$\Delta b_{svc} = \frac{\dot{C}(\alpha_0)}{X_{svcl}}(-C_V C_1\Delta\delta - C_V C_2\Delta E'_q - C_V C_3\Delta b_{svc} + C_S\Delta\omega) \tag{3.20}$$

Thus,

$$\Delta b_{svc} = \frac{-C_V C_1\Delta\delta - C_V C_2\Delta E'_q + C_S\Delta\omega}{\frac{X_{svcl}}{\dot{C}(\alpha_0)} + C_V C_3} \tag{3.21}$$

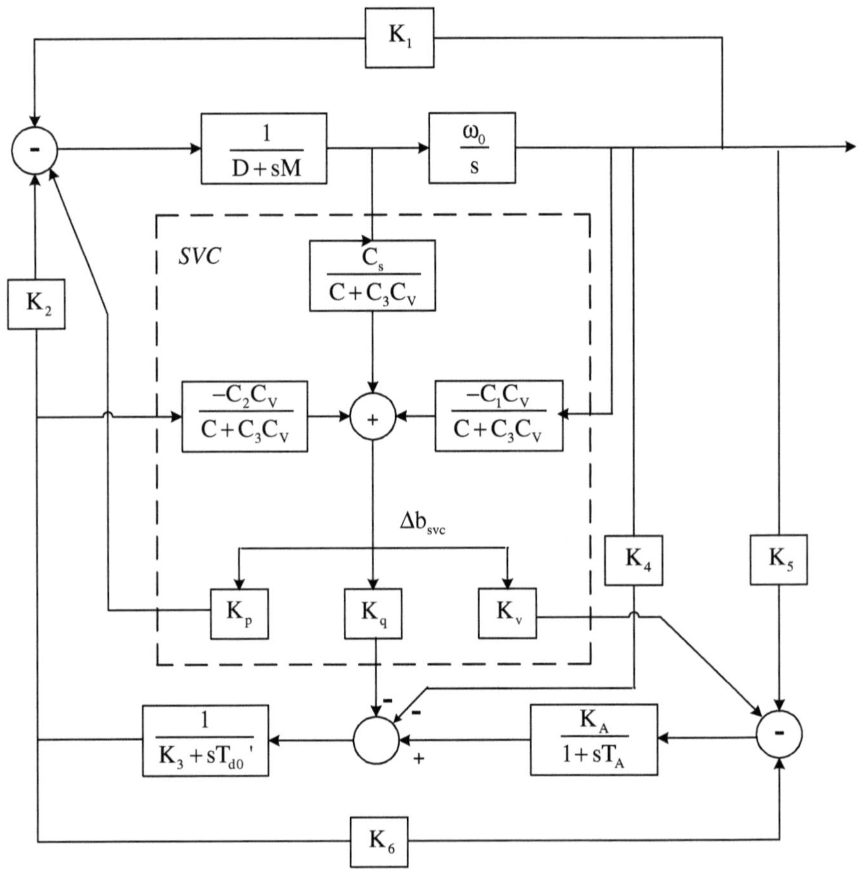


**Fig. 3.4** Extended Philip–Heffron model with SVC voltage and damping control function included

By taking account of Eq. (3.21), Fig. 3.3 can be further extended to include the SVC voltage and damping control function as shown in Fig. 3.4, where $C = \frac{X_{svcl}}{\dot{C}(\alpha_0)}$.

#### 3.1.1.4 Calculation of Initial Compensation

This section demonstrates how the initial compensation of the SVC installed in the single-machine infinite-bus power system shown in Fig. 3.2 can be calculated to satisfy the requirement to maintain the voltage profile of the system. The initial conditions to compute the initial compensation, $b_{svc0}$, of the SVC are the magnitude of busbar voltage, $V_{t0}$, $V_{b0}$, $V_{s0}$, and the active power supplied by the generator, $P_{t0}$.

Figure 3.5 shows the circuit model of the system as shown in Fig. 3.2. From Fig. 3.5, it can be obtained that

$$\frac{\bar{V}_{s0} - \bar{V}_b}{jx_{sb}}\bar{V}_{s0}^* = P_{sb0} - jQ_{sb0} \tag{3.22}$$

Since the resistance of the transmission lines is ignored, $P_{sb0} = P_{t0}$. By choosing $\bar{V}_{s0}$ as the reference phasor on x-axis, it can have $\bar{V}_{s0} = \bar{V}_{s0}^* = V_{s0}$. From Eq. (3.14), the following equation can be obtained

$$\bar{V}_b = V_{s0} - \frac{x_{sb}}{V_{s0}}(Q_{sb0} + jP_{sb0}) = V_{s0} - \frac{x_{sb}}{V_{s0}}Q_{sb0} - j\frac{x_{sb}}{V_b}P_{sb0}$$

The above equation gives

$$V_b^2 = \left(V_{s0} - \frac{x_{sb}}{V_{s0}}Q_{sb0}\right)^2 + \left(\frac{x_{sb}}{V_{s0}}P_{sb0}\right)^2 \tag{3.23}$$

Solution of Eq. (3.23) is as follows

$$Q_{sb0} = \frac{V_{s0}}{x_{sb}}\left[V_{s0} - \sqrt{V_b^2 - \left(\frac{x_{sb}}{V_{s0}}P_{sb0}\right)^2}\right] \tag{3.24}$$

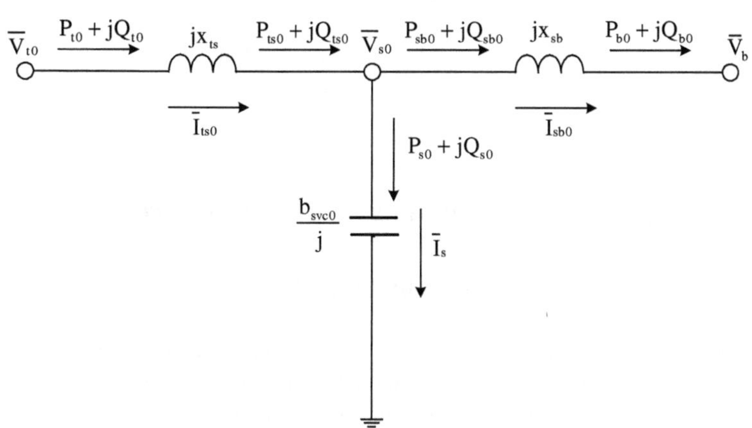


**Fig. 3.5** Circuit model of the single-machine infinite-bus power system installed with SVC

Similarly, from the following equation

$$\frac{\bar{V}_{t0}-\bar{V}_{s0}}{jx_{ts}}\bar{V}_{s0}^{*}=P_{ts0}-jQ_{ts0}$$

it can have

$$V_{t0}^{2}=\left(V_{s0}+\frac{x_{ts}}{V_{s0}}Q_{ts0}\right)^{2}+\left(\frac{x_{ts}}{V_{s0}}P_{ts0}\right)^{2} \tag{3.25}$$

Solution of Eq. (3.25) is as follows:

$$Q_{ts0}=\frac{V_{s0}}{x_{ts}}\left[\sqrt{V_{t0}^{2}-\left(\frac{x_{ts}}{V_{s0}}P_{ts0}\right)^{2}}-V_{s0}\right] \tag{3.26}$$

Hence,

$$P_{s0}+jQ_{s0}=0.0+j(Q_{ts0}-Q_{sb0})$$

Since

$$\bar{V}_{s0}jb_{svc0}=\bar{I}_{s0}$$

it can have

$$\bar{V}_{s0}\bar{V}_{s0}^{*}jb_{svc0}=\bar{I}_{s0}\bar{V}_{s0}^{*}=P_{s0}-jQ_{s0}$$

Finally from the above equation, the initial compensation is calculated to be

$$b_{svc0}=\frac{(P_{s0}-jQ_{s0})}{jV_{s0}^{2}}=-\frac{Q_{s0}}{V_{s0}^{2}}=\frac{Q_{sb0}-Q_{ts0}}{V_{s0}^{2}} \tag{3.27}$$

## *3.1.2 Damping Torque Analysis of SVC Stabilizer*

### 3.1.2.1 Electric Torque Provided by the SVC Stabilizer

From Fig. 3.4, the forward path of the SVC stabilizer to the electromechanical oscillation loop of generator can be established as shown in Fig. 3.6.

From Fig. 3.6, it can be seen that the SVC stabilizer contributes the electric torque to the electromechanical oscillation loop through two paths. The electric

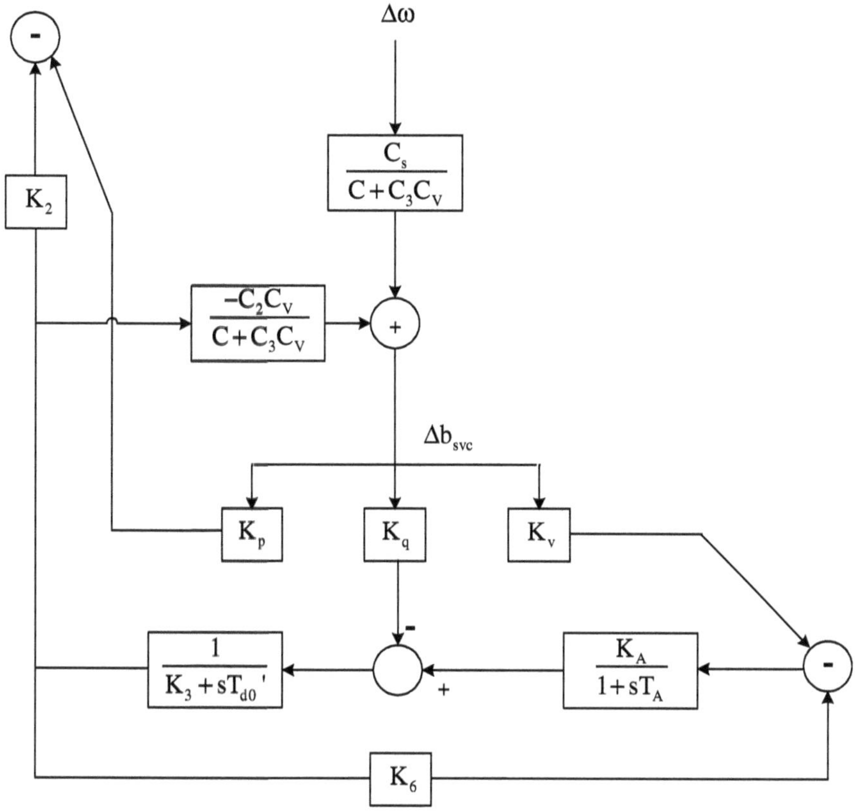


**Fig. 3.6** Forward path of the SVC stabilizer

torque through the block, $K_P$, is called the direct electric torque, denoted by $\Delta T_{ed}$. That through the blocks, $K_q$ and $K_V$, is named the indirect electric torque, denoted by $\Delta T_{ei}$. According to the damping torque analysis introduced in Chap. 2, electric torque can be decomposed into a damping torque and a synchronizing torque at the angular oscillation frequency, $\omega_s$, that is

$$\begin{aligned} \Delta T_{ed} &= \Delta T_{sd} + \Delta T_{dd} = T_{sd}\Delta\delta + T_{dd}\Delta\omega \\ \Delta T_{ei} &= \Delta T_{si} + \Delta T_{di} = T_{si}\Delta\delta + T_{di}\Delta\omega \end{aligned} \tag{3.28}$$

where $T_{sd}, T_{si}, T_{dd}, T_{di}$ are direct, indirect synchronizing torque and direct, indirect damping torque coefficients, respectively. From Fig. 3.6, the transfer function of forward path of the SVC stabilizer to the electromechanical oscillation loop of generator can be derived to be

$$
\begin{aligned}
F_{svc}(s) &= K_p - K_2K_q\frac{\frac{1}{K_3+sT'_{d0}}}{1+K_6\frac{1}{K_3+sT'_{d0}}\frac{K_A}{1+sT_A}} - K_2K_v\frac{\frac{1}{K_3+sT'_{d0}}\frac{K_A}{1+sT_A}}{1+K_6\frac{1}{K_3+sT'_{d0}}\frac{K_A}{1+sT_A}} \\
&= K_p - \frac{K_2[K_q(1+sT_A)+K_vK_A]}{(K_3+sT'_{d0})(1+sT_A)+K_6K_A}
\end{aligned}
\tag{3.29}
$$

Hence, the electric torque provided by the SVC stabilizer is as follows

$$
\Delta T_{et} = F_{svc}(j\omega_s)\Delta b_{svc} \tag{3.30}
$$

From Fig. 3.6, it can be seen that the signals are attenuated by the exciter and the AVR before they form the indirect damping toque. Hence, normally $T_{dd} \gg T_{di}$. From Fig. 3.6 and Eqs. (3.29) and (3.30), it can be seen that the coefficient $K_p$ is the dominant weight parameter in determining the amount of damping torque contribution from the SVC stabilizer. Hence, in the following section, how the effectiveness of the SVC stabilizer is affected by various factors is examined by checking the dominant weight parameter $K_P$.

#### 3.1.2.2 Damping Control as Affected by the Load Conditions

From Eq. (3.7), it can have

$$
P_{t0} = \frac{E'_{q0}V_b}{c_{svc0}x'_{d\Sigma}}\sin\delta_0 - \frac{V_b^2}{c_{svc0}^2 2}\frac{(x_q-x'_d)}{x'_{d\Sigma}x_{q\Sigma}}\sin 2\delta_0 = P_{t10} - P_{t20} \tag{3.31}
$$

where $P_{t10} = \frac{E'_{q0}V_b}{c_{svc0}x'_{d\Sigma}}\sin\delta_0$, $P_{t20}=\frac{V_b^2}{c_{svc0}^2 2}\frac{(x_q-x'_d)}{x'_{d\Sigma}x_{q\Sigma}}\sin 2\delta_0$.

From Eq. (3.9) and by using Eq. (3.31), the weight parameter $K_p$ can be obtained to be

$$
\begin{aligned}
K_p &= \frac{x_{sb}(x_{ts}+x'_d)}{c_{svc0}x'_{d\Sigma}}\left[\frac{V_bE'_{q0}\sin\delta_0}{c_{svc0}x'_{d\Sigma}} - \frac{V_b^2(x_q-x'_d)\sin(2\delta_0)}{2c_{svc0}^2x'_{d\Sigma}x_{q\Sigma}}\right] \\
&\quad - \frac{V_b^2(x_q-x'_d)\sin(2\delta_0)}{2c_{svc0}^2x'_{d\Sigma}x_{q\Sigma}}\frac{x_{sb}(x_{ts}+x_q)}{c_{svc0}x_{q\Sigma}} \\
&= \frac{x_{sb}}{c_{svc0}}\left[\frac{(x_{ts}+x'_d)}{x'_{d\Sigma}}P_{t0} - \frac{(x_{ts}+x_q)}{x_{q\Sigma}}P_{t20}\right]
\end{aligned}
\tag{3.32}
$$

From Fig. 3.7 and Eq. (3.32), it can be seen that at a higher load condition, difference between $P_{t0}$ and $P_{t20}$ is bigger. Thus, the higher the load condition is, the bigger $K_p$ is, the more damping torque is provided by the SVC stabilizer. Hence, effectiveness of the SVC stabilizer should increase with system load conditions.

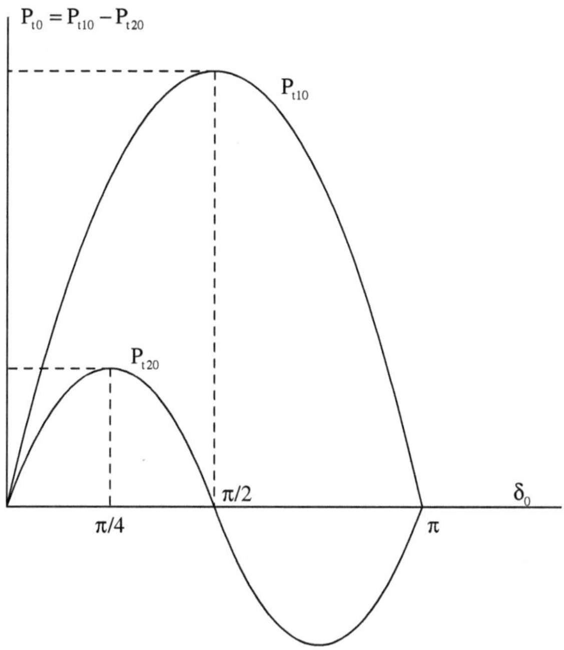


**Fig. 3.7** $P_t - \delta$ curve of the power system

#### 3.1.2.3 Influence of Parameters of the Generator

Since $x_q > x'_d$, it should have

$$\frac{(x_{ts} + x'_d)}{x'_{d\Sigma}} = \frac{1}{1 + \frac{X_{sb}}{C_{svc0}(X_{ts} + X'_d)}} < \frac{1}{1 + \frac{X_{sb}}{C_{svc0}(X_{ts} + X_q)}} = \frac{(x_{ts} + x_q)}{x_{q\Sigma}} \tag{3.33}$$

Equations (3.32) and (3.33) indicate that $K_p$ is affected not only by the load conditions of the power system, but also by the parameters of the generator, $x'_d$ and $x_q$, such that:

1. If the difference between the values of $x'_d$ and $x_q$ is small (for example, for the generator in a hydropower station, value of $x'_d$ is around 0.2–0.35 and that of $x_q$ is around 0.45–0.7) or if the system operates at a higher load condition ($P_{t0} \gg P_{t20}$), $\frac{(X_{ts} + X'_d)}{X'_{d\Sigma}} P_{t0} > \frac{(X_{ts} + X_q)}{X_{q\Sigma}} P_{t20}$ such that $K_p > 0$, the SVC stabilizer will provide the power system with positive damping.
2. If $x_q$ is much greater than $x'_d$ such that $\frac{(X_{ts} + X'_d)}{X'_{d\Sigma}} \ll \frac{(X_{ts} + X_q)}{X_{q\Sigma}}$, (for example, for a generator in a thermal power plant, $x'_d$ is around 0.15 to 0.24 and $x_q$ is around

1.2–2.2) and when the system operates at a lower load condition such that the difference between $P_{t0}$ and $P_{t20}$ is small, there is a possibility that $\frac{(X_{ts}+X'_d)}{X'_{d\Sigma}}P_{t0} < \frac{(X_{ts}+X_q)}{X_{q\Sigma}}P_{t20}$ such that $K_p < 0$. This means that the SVC stabilizer which supplies positive damping torque at a higher load condition may provide the power system with negative damping torque at a lower load condition.

3. Around the operating point where $\frac{(X_{ts}+X'_d)}{X'_{d\Sigma}}P_{t0} = \frac{(X_{ts}+X_q)}{X_{q\Sigma}}P_{t20}$ such that $K_p = 0$, the damping torque provided by the SVC stabilizer is zero. At this operating point, the SVC stabilizer will lose its capability to suppress power system oscillation. This load condition is called the "dead point" of the SVC stabilizer's function.

#### 3.1.2.4 Electric Length of the Transmission Line

Define z to be an index of the length of the transmission line and take $z_0 = 1$. The reactance of the transmission line is proportional to its electric length, i.e., the index z. $K_p$ can be written as a function z as (see Eq. (3.32))

$$K_p(z) = \frac{zx_{sb}}{c_{svc0}}\left[\frac{(zx_{ts}+x'_d)}{x'_d+z(x_{ts}+x_{sb}/c_{svc0})}P_{t0} - \frac{(zx_{ts}+x_q)}{x_q+z(x_{ts}+x_{sb}/c_{svc0})}P_{t20}\right] \quad (3.34)$$

For the simplicity of discussion, it is assumed that at the steady-state operation, $b_{svc0} = 0$ such that $c_{svc0} = 1$. Hence,

$$K_p(z) = zx_{sb}\left[\frac{(zx_{ts}+x'_d)}{x'_d+z(x_{ts}+x_{sb})}P_{t0} - \frac{(zx_{ts}+x_q)}{x_q+z(x_{ts}+x_{sb})}P_{t20}\right] \quad (3.34)$$

It can have

$$\begin{aligned}\frac{\partial}{\partial z}K_p(z) = {} & x_{sb}\left[\frac{(zx_{ts}+x'_d)}{x'_d+z(x_{ts}+x_{sb})}P_{t0} - \frac{(zx_{ts}+x_q)}{x_q+z(x_{ts}+x_{sb})}P_{t20}\right] \\ & + zx_{sb}\left\{\frac{x_{ts}}{x'_d+z(x_{ts}+x_{sb})}P_{t0} - \frac{(zx_{ts}+x'_d)(x_{ts}+x_{sb})}{[x'_d+z(x_{ts}+x_{sb})]^2}P_{t0}\right. \\ & \left. - \frac{x_{ts}}{x_q+z(x_{ts}+x_{sb})}P_{t20} + \frac{(zx_{ts}+x_q)(x_{ts}+x_{sb})}{[x_q+z(x_{ts}+x_{sb})]^2}P_{t20}\right\}\end{aligned}$$

Hence,

$$
\begin{aligned}
\left.\frac{\partial K_p(z)}{\partial z}\right|_{z=z_0=1} &= x_{sb}\left[\frac{(x_{ts}+x'_d)}{x'_d+(x_{ts}+x_{sb})}P_{to}-\frac{(x_{ts}+x_q)}{x_q+(x_{ts}+x_{sb})}P_{t02}\right] \\
&\quad + x_{sb}\left\{\frac{x_{ts}}{x'_d+(x_{ts}+x_{sb})}P_{t0}-\frac{(x_{ts}+x'_d)(x_{ts}+x_{sb})}{[x'_d+(x_{ts}+x_{sb})]^2}P_{t20}\right. \\
&\quad \left. -\frac{x_{ts}}{x_q+(x_{ts}+x_{sb})}P_{t20}+\frac{(x_{ts}+x_q)(x_{ts}+x_{sb})}{[x_q+(x_{ts}+x_{sb})]^2}P_{t20}\right\} \\
&= \frac{x_{sb}(x_{ts}+x'_d)}{x'_d+(x_{ts}+x_{sb})}P_{t0}+\frac{x_{sb}x_{ts}}{x'_d+(x_{ts}+x_{sb})}P_{t0}-\frac{x_{sb}(x_{ts}+x'_d)(x_{ts}+x_{sb})}{(x'_d+x_{ts}+x_{sb})^2}P_{t0} \\
&\quad -\frac{x_{sb}(x_{ts}+x_q)}{x_q+(x_{ts}+x_{sb})}P_{t20}-\frac{x_{sb}x_{ts}}{x_q+(x_{ts}+x_{sb})}P_{t20}+\frac{x_{sb}(x_{ts}+x_q)(x_{ts}+x_{sb})}{(x_q+x_{ts}+x_{sb})^2}P_{t20} \\
&= \frac{x_{sb}(x_{ts}+x'_d)}{x'_d+(x_{ts}+x_{sb})}P_{t0}+\frac{x_{sb}x_{ts}}{x'_d+(x_{ts}+x_{sb})}P_{t0}-\frac{x_{sb}(x_{ts}+x'_d)(x_{ts}+x_{sb})}{(x'_d+x_{ts}+x_{sb})^2}P_{t0} \\
&\quad -\frac{x_{sb}(x_{ts}+x_q)}{x_q+(x_{ts}+x_{sb})}P_{t20}-\frac{x_{sb}x_{ts}}{x_q+(x_{ts}+x_{sb})}P_{t20}+\frac{x_{sb}(x_{ts}+x_q)(x_{ts}+x_{sb})}{(x_q+x_{ts}+x_{sb})^2}P_{t20} \\
&= \frac{x_{sb}(x_{ts}+x'_d)x'_d+x_{sb}x_{ts}(x'_d+x_{ts}+x_{sb})}{(x'_d+x_{ts}+x_{sb})^2}P_{t0} \\
&\quad -\frac{x_{sb}(x_{ts}+x_q)x_q+x_{sb}x_{ts}(x_q+x_{ts}+x_{sb})}{(x_q+x_{ts}+x_{sb})^2}P_{t20} \\
&= \frac{x_{sb}x'_d(x_{ts}+x'_d)}{(x'_d+x_{ts}+x_{sb})^2}P_{t0}+\frac{x_{sb}x_{ts}}{x'_d+x_{ts}+x_{sb}}P_{t0}-\frac{x_{sb}x_q(x_{ts}+x_q)}{(x_q+x_{ts}+x_{sb})^2}P_{t20}-\frac{x_{sb}x_{ts}}{x_q+x_{ts}+x_{sb}}P_{t20} \\
&= \frac{x_{sb}x'_d(x_{ts}+x'_d)}{x'^2_{d\Sigma}}P_{t0}-\frac{x_{sb}x_q(x_{ts}+x_q)}{x^2_{q\Sigma}}P_{t20}+\frac{x_{sb}x_{ts}}{x'_{d\Sigma}}P_{t0}-\frac{x_{sb}x_{ts}}{x_{q\Sigma}}P_{t20} \\
&= \frac{x_{sb}x'_d(x_{ts}+x'_d)}{x'^2_{d\Sigma}}P_{t0}-\frac{x_{sb}x_q(x_{ts}+x_q)}{x^2_{q\Sigma}}P_{t20}+x_{sb}x_{ts}\left(\frac{P_{t0}}{x'_{d\Sigma}}-\frac{P_{t20}}{x_{q\Sigma}}\right)
\end{aligned}
\tag{3.35}
$$

Thus,

$$
\begin{aligned}
\left.\frac{\partial K_p(z)}{\partial z}\right|_{z=z_0=1} &= \frac{x_{sb}x'_d(x_{ts}+x'_d)+x_{sb}x_{ts}x'_{d\Sigma}}{X'^2_{d\Sigma}}P_{t0}-\frac{x_{sb}x_q(x_{ts}+x_q)+x_{sb}x_{ts}x_{q\Sigma}}{X^2_{q\Sigma}}P_{t20} \\
&= \frac{x_{sb}(x_{ts}+x'_d)^2+x_{ts}x^2_{sb}}{x'^2_{d\Sigma}}P_{t0}-\frac{x_{sb}(x_{ts}+x_q)^2+x_{ts}x^2_{sb}}{x^2_{q\Sigma}}P_{t20}
\end{aligned}
\tag{3.36}
$$

When $K_p > 0$

$$
\frac{(x_{ts}+x'_d)}{x'_{d\Sigma}}P_{t0} > \frac{(x_{ts}+x_q)}{x_{q\Sigma}}P_{t20} \tag{3.37}
$$

Hence, normally it should have

$$\left[\frac{(x_{ts}+x'_d)}{x'_{d\Sigma}}\right]^2 P_{t0} > \left[\frac{(x_{ts}+x_q)}{x_{q\Sigma}}\right]^2 P_{t20} \tag{3.38}$$

Using Eqs. (3.36) and (3.38), from Eq. (3.35), it can be obtained that

$$\frac{\partial K_p(z)}{\partial z} > 0 \tag{3.39}$$

Equation (3.39) means that for the SVC stabilizer designed to supply positive damping to power system oscillations, it will be more effective when the electric length of transmission lines increases.

#### 3.1.2.5 Installing Location of the SVC

If it is assumed that the total length of the transmission line is fixed, that is, $x_{ts} - x_{sb} = x_t$, from Eq. (3.32), it can be obtained that (assuming $c_{svc0} = 1$)

$$K_p(x_{sb}) = \left[\frac{x_{sb}(x_t - x_{sb} + x'_d)}{x_t + x'_d} P_{t0} - \frac{x_{sb}(x_t - x_{sb} + x_q)}{x_q + x_t} P_{t20}\right] \tag{3.40}$$

Define the following two functions of $x_{sb}$

$$\begin{aligned} g(x_{sb}) &= \frac{x_{sb}(x_t - x_{sb} + x'_d)}{x'_d + x_t} P_{t0} \\ h(x_{sb}) &= \frac{x_{sb}(x_t - x_{sb} + x_q)}{x_q + x_t} P_{t20} \end{aligned} \tag{3.41}$$

Thus,

$$K_p(x_{sb}) = g(x_{sb}) - h(x_{sb}) \tag{3.42}$$

By solving the following two equations

$$\begin{aligned} \frac{\partial}{\partial x_{sb}} g(x_{sb}) &= \frac{x_t - 2x_{sb} + x'_d}{x'_d + x_t} P_{t0} = 0 \\ \frac{\partial}{\partial x_{sb}} h(x_{sb}) &= \frac{x_t - 2x_{sb} + x_q}{x_q + x_t} P_{t20} = 0 \end{aligned} \tag{3.43}$$

It can have $x_{sb} = \frac{(x'_d + x_t)}{2} = x_{sb1}$ and $x_{sb} = \frac{(x_q + x_t)}{2} = x_{sb2}$. Hence, $g(x_{sb})$ and $h(x_{sb})$ achieve its maximum value at $x_{sb} = \frac{(x'_d + x_t)}{2} = x_{sb1}$ and $x_{sb} = \frac{(x_q + x_t)}{2} = x_{sb2}$, respectively.

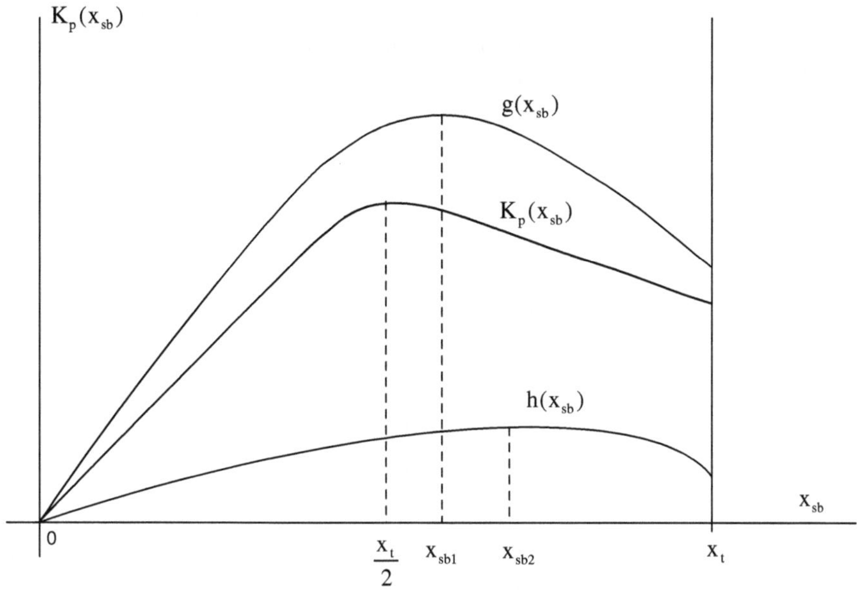


**Fig. 3.8** Illustration on the most effective installing location of the SVC stabilizer

From Eqs. (3.42) and (3.43), the curve of $K_p(x_{sb})$ can be sketched as shown in Fig. 3.8 ($P_{t0} \gg P_{t20}$). Figure 3.8 demonstrates that when $x_{sb} \approx \frac{x_t}{2}$, $K_p(x_{sb})$ achieves its maximum value. This means that when the SVC is installed near the middle point of the transmission line, the SVC stabilizer contributes the most amount of damping torque and hence is most effective in suppressing power system oscillations.

## 3.2 A Single-Machine Infinite-Bus Power System Installed with a TCSC or TCPS Stabilizer

### *3.2.1 Extended Heffron–Phillips Model of a Single-Machine Infinite-Bus Power System Installed with a TCSC or TCPS Stabilizer*

#### 3.2.1.1 Extended Heffron–Phillips Model of a Single-Machine Infinite-Bus Power System Installed with a TCSC

It was discovered decades ago that power oscillations along transmission lines can be effectively suppressed by switching series capacitors in and out from the lines after a fault [2, 3]. Nowadays, the high-speed thyristor-controlled series compensator

(TCSC) has made it practically possible. A TCSC can be simply represented as a capacitor whose capacitance is adjustable within a certain range as shown in Fig. 3.9, which is a single-machine infinite-bus power system installed with a TCSC. Obviously, installation of the TCSC only changes the equivalent reactance of the transmission line connecting the generator and the infinite busbar. Therefore, the mathematical model of the system can be established by simply modifying Eqs. (2.37) and (2.38). The modification is just to change the expressions of the equivalent reactance of the transmission line as follows

$$\begin{aligned}
\dot{\delta} &= \omega_o(\omega - 1) \\
\dot{\omega} &= \frac{1}{M}[P_m - P_t - D(\omega - 1)] \\
\dot{E}'_q &= \frac{1}{T'_{do}}(-E_q + E_{fd}) \\
\dot{E}'_{fd} &= -\frac{1}{T_A}E'_{fd} + \frac{K_A}{T_A}(V_{tref} - V_t)
\end{aligned} \tag{3.44}$$

$$\begin{aligned}
P_t &= \frac{E'_q V_b}{x'_{d\Sigma}}\sin\delta - \frac{V_b^2}{2}\frac{(x_q - x'_d)}{x'_{d\Sigma}x_{q\Sigma}}\sin 2\delta \\
E_q &= \frac{E'_q x_{d\Sigma}}{x'_{d\Sigma}} - \frac{(x_d - x'_d)V_b\cos\delta}{x'_{d\Sigma}} \\
E_{fd} &= E_{fd0} + E'_{fd} \\
v_{td} &= \frac{x_q V_b \sin\delta}{x_{q\Sigma}}, \quad v_{tq} = \frac{x_t E'_q}{x'_{d\Sigma}} + \frac{V_b x'_d \cos\delta}{x'_{d\Sigma}}, \quad V_t = \sqrt{v_{td}^2 + v_{tq}^2}
\end{aligned} \tag{3.45}$$

where

$$\begin{aligned}
x_{d\Sigma} &= x_d + x_L - x_{tcsc}, \; x'_{d\Sigma} = x'_d + x_t - x_{tcsc} \\
x_{q\Sigma} &= x_q + x_L - x_{tcsc}
\end{aligned} \tag{3.46}$$

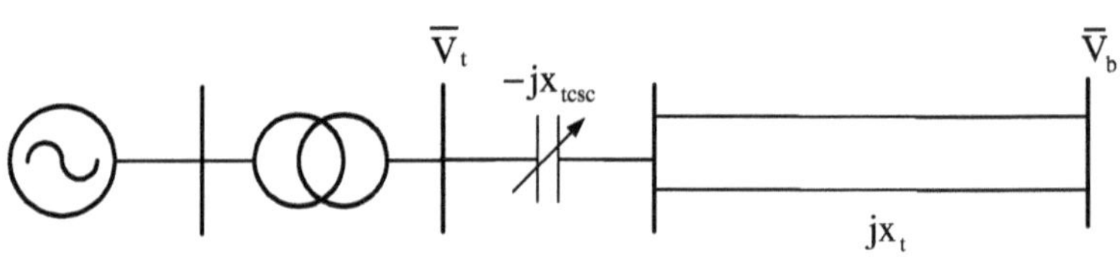


**Fig. 3.9** A single-machine infinite-bus power system installed with a TCSC

By linearizing Eq. (3.45), it can be obtained that

$$
\begin{aligned}
\Delta P_t &= K_1\Delta\delta + K_2\Delta E'_q + K_p\Delta x_{tcsc} \\
\Delta E_q &= K_3\Delta E'_q + K_4\Delta\delta + K_q\Delta x_{tcsc} \\
\Delta V_t &= K_5\Delta\delta + K_6\Delta E'_q + K_v\Delta x_{tcsc}
\end{aligned}
\tag{3.47}
$$

where

$$
\begin{aligned}
K_1 &= \frac{E'_{q0}V_b}{x'_{d\Sigma}}\cos\delta_0 - \frac{V_b^2\left(x_q - x'_d\right)}{x'_{d\Sigma}x_{q\Sigma}}\cos 2\delta_0 \\
K_2 &= \frac{V_b}{x'_{d\Sigma}}\sin\delta_0 \\
K_p &= \left.\frac{\partial P_t}{\partial x_{tcsc}}\right|_0 = \frac{V_bE'_{q0}\sin\delta_0}{x'^2_{d\Sigma}} - \frac{V_b^2(x_q - x'_d)\sin 2\delta_0}{2x'^2_{d\Sigma}x^2_{q\Sigma}}\left(x'_{d\Sigma} + x_{q\Sigma}\right) \\
K_3 &= \frac{x_{d\Sigma}}{x'_{d\Sigma}} \\
K_4 &= \frac{\left(x_d - x'_d\right)V_b\sin\delta_0}{x'_{d\Sigma}} \\
K_q &= \left.\frac{\partial E_q}{\partial x_{tcsc}}\right|_0 = \frac{E'_{q0}x_{d\Sigma}}{x'^2_{d\Sigma}} - \frac{\left(x_d - x'_d\right)V_b\cos\delta_0}{x'^2_{d\Sigma}} \\
K_5 &= \frac{v_{td0}}{V_{t0}}\frac{x_qV_b\cos\delta_0}{x_{q\Sigma}} - \frac{v_{tq0}}{V_{t0}}\frac{V_bx'_d\sin\delta_0}{x'_{d\Sigma}} \\
K_6 &= \frac{v_{tq0}}{V_{t0}}\frac{x_{L\Sigma}}{x'_{d\Sigma}} \\
K_v &= \left.\frac{\partial V_t}{\partial x_{tcsc}}\right|_0 = \frac{v_{td0}}{V_{t0}}\frac{x_qV_b\sin\delta_0}{x^2_{q\Sigma}} + \frac{v_{tq0}}{V_{t0}}\left[\left(\frac{x_L}{x'^2_{d\Sigma}} - \frac{1}{x'_{d\Sigma}}\right)E'_{q0} + \frac{V_bx'_d\cos\delta_0}{x'^2_{d\Sigma}}\right]
\end{aligned}
$$

By substituting Eq. (3.46) into the linearized equation of Eq. (3.44), it can have

$$
\begin{aligned}
\Delta\dot{\delta} &= \omega_o\Delta\omega \\
\Delta\dot{\omega} &= \frac{1}{M}\left(-K_1\Delta\delta - D\Delta\omega - K_2\Delta E'_q - K_p\Delta x_{tcsc}\right) \\
\Delta\dot{E}'_q &= \frac{1}{T'_{d0}}\left(-K_3\Delta E'_q - K_4\Delta\delta - K_q\Delta x_{tcsc} + \Delta E'_{fd}\right) \\
\Delta E'_{fd} &= -\frac{1}{T_A}\Delta E'_{fd} - \frac{K_A}{T_A}\left(K_5\Delta\delta + K_6\Delta E'_q + K_v\Delta x_{tcsc}\right)
\end{aligned}
\tag{3.48}
$$

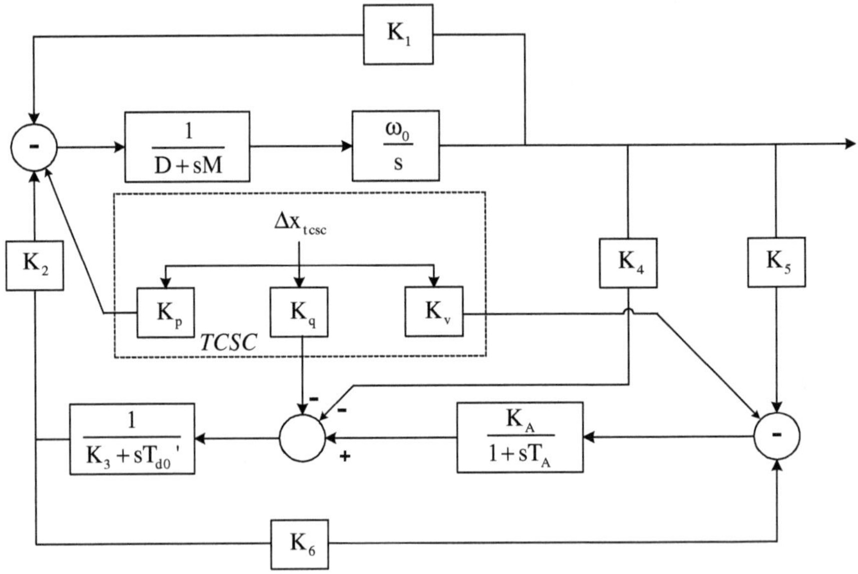


**Fig. 3.10** Extended Heffron–Phillips model of the single-machine infinite-bus power system installed with a TCSC stabilizer

Equation (3.48) is the extended Heffron–Phillips model of the single-machine infinite-bus power system installed with the TCSC stabilizer, which is shown in Fig. 3.10.

#### 3.2.1.2 Extended Heffron–Phillips Model of a Single-Machine Infinite-Bus Power System Installed with a TCPS

The phase shifters have been used to regulate the steady-state power flow in power systems. The idea of applying online control of phase shifters was proposed decades ago. However, the low-speed mechanical tap changers precluded the use of online control of phase shifters to improve power system dynamic performance. With the advances in power electronics, high-power high-speed electronic switches make it possible to realize real-time control of a phase shifter. The feasibility of a thyristor-controlled phase shifter (TCPS) as a means to adjust the value of quadrature component of line voltage has been well recognized.

Figure 3.11 illustrates the arrangement of a TCPS installed along a transmission line. Its function can be simply expressed as a phase shifting of line voltage as shown in Fig. 3.11. Hence, for a single-machine infinite-bus power system installed with a TCPS as shown in Fig. 3.12, system dynamic Eqs. (2.37) and (2.38) need to be modified only in the phase relationship as follows.

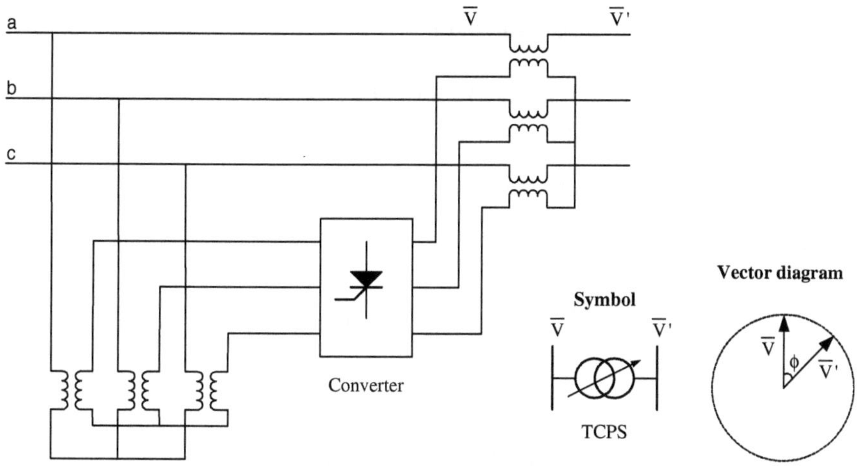


**Fig. 3.11** Arrangement of a TCPS and its simple representation

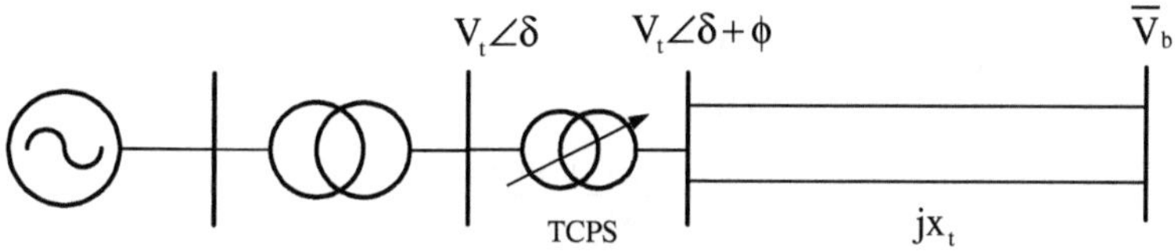


**Fig. 3.12** The single-machine infinite-bus power system installed with a TCPS

$$\begin{aligned}
\dot{\delta} &= \omega_0(\omega - 1) \\
\dot{\omega} &= \frac{1}{M}[P_m - P_t - D(\omega - 1)] \\
\dot{E}'_q &= \frac{1}{T'_{do}}(-E_q + E_{fd}) \\
\dot{E}'_{fd} &= -\frac{1}{T_A}E'_{fd} + \frac{K_A}{T_A}(V_{tref} - V_t)
\end{aligned} \tag{3.49}$$

$$\begin{aligned}
P_t &= \frac{E'_q V_b}{x'_{d\Sigma}}\sin(\delta+\phi) - \frac{V_b^2}{2}\frac{(x_q - x'_d)}{x'_{d\Sigma}x_{q\Sigma}}\sin 2(\delta+\phi) \\
E_q &= \frac{E'_q x_{d\Sigma}}{x'_{d\Sigma}} - \frac{(x_d - x'_d)V_b\cos(\delta+\phi)}{x'_{d\Sigma}} \\
v_{td} &= \frac{x_q V_b \sin(\delta+\phi)}{x_{q\Sigma}} \\
v_{tq} &= \frac{x_L E'_q}{x'_{d\Sigma}} + \frac{V_b x'_d \cos(\delta+\phi)}{x'_{d\Sigma}}
\end{aligned} \tag{3.50}$$

By linearizing Eq. (3.50), it can have

$$\begin{aligned} \Delta P_t &= K_1\Delta\delta + K_2\Delta E_q' + K_p\Delta\phi \\ \Delta E_q &= K_3\Delta E_q' + K_4\Delta\delta + K_q\Delta\phi \\ \Delta V_t &= K_5\Delta\delta + K_6\Delta E_q' + K_v\Delta\phi \end{aligned} \tag{3.51}$$

where

$$\begin{aligned}
K_1 &= \frac{E_{q0}' V_b}{x_{d\Sigma}'}\cos\delta_0 - \frac{V_b^2(x_q - x_d')}{x_{d\Sigma}' x_{q\Sigma}}\cos 2\delta_0 \\
K_2 &= \frac{V_b}{x_{d\Sigma}'}\sin\delta_0 \\
K_p &= \left.\frac{\partial P_t}{\partial\phi}\right|_0 = \frac{E_{q0}' V_b}{x_{d\Sigma}'}\cos\delta_0 - \frac{V_b^2(x_q - x_d')}{x_{d\Sigma}' x_{q\Sigma}}\cos 2\delta_0 \\
K_3 &= \frac{x_{d\Sigma}}{x_{d\Sigma}'} \\
K_4 &= \frac{(x_d - x_d')V_b\sin\delta_0}{x_{d\Sigma}'} \\
K_q &= \left.\frac{\partial E_q}{\partial\phi}\right|_0 = \frac{(x_d - x_d')V_b\sin\delta_0}{x_{d\Sigma}'} \\
K_5 &= \frac{v_{td0}}{V_{t0}}\frac{x_q V_b\cos\delta_0}{x_{q\Sigma}} - \frac{v_{tq0}}{V_{t0}}\frac{V_b x_d'\sin\delta_0}{x_{d\Sigma}'} \\
K_6 &= \frac{v_{tq0}}{V_{t0}}\frac{x_{L\Sigma}}{x_{d\Sigma}'} \\
K_v &= \left.\frac{\partial V_t}{\partial\phi}\right|_0 = \frac{v_{td0}}{V_{t0}}\frac{x_q V_b\cos\delta_0}{x_{q\Sigma}} - \frac{v_{tq0}}{V_{t0}}\frac{V_b x_d'\sin\delta_0}{x_{d\Sigma}'}
\end{aligned}$$

By substituting Eq. (3.51) into the linearized equation of Eq. (3.49), it can be obtained that

$$\begin{aligned}
\Delta\dot{\delta} &= \omega_o\Delta\omega \\
\Delta\dot{\omega} &= \frac{1}{M}(-K_1\Delta\delta - D\Delta\omega - K_2\Delta E_q' - K_p\Delta\phi) \\
\Delta\dot{E}_q' &= \frac{1}{T_{d0}'}(-K_3\Delta E_q' - K_4\Delta\delta - K_q\Delta\phi) \\
\Delta E_{fd}' &= -\frac{1}{T_A}\Delta E_{fd}' - \frac{K_A}{T_A}(K_5\Delta\delta + K_6\Delta E_q' + K_v\Delta\phi)
\end{aligned} \tag{3.52}$$

Equation (3.52) is the extended Heffron–Phillips model of the single-machine infinite-bus power system installed with the TCSC stabilizer, which is shown in Fig. 3.13.

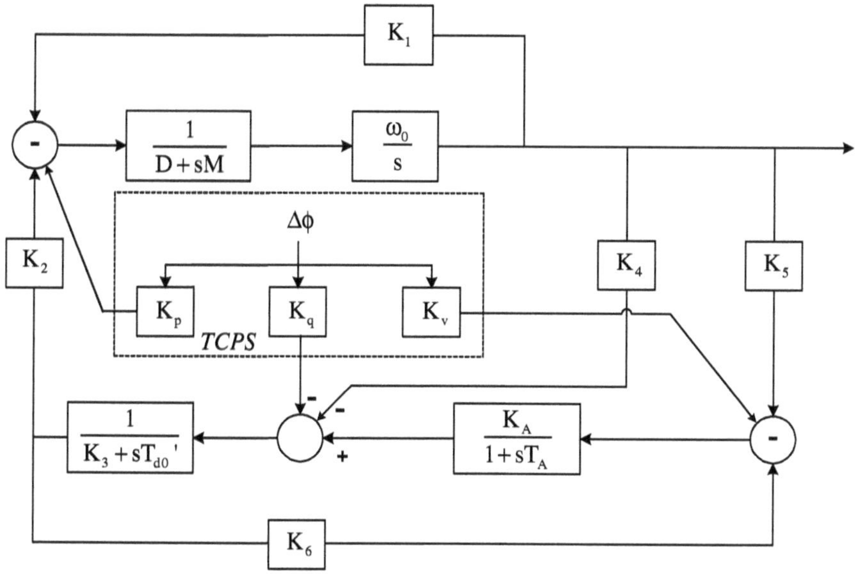


**Fig. 3.13** Extended Heffron–Phillips model of the single-machine infinite-bus power system installed with a TCPS stabilizer

## 3.2.2 *Damping Torque Analysis of TCSC and TCPS Stabilizers*

### 3.2.2.1 General Expression of Damping Torque Contributed by the TCSC and TCPS Stabilizers

For the simplicity of discussion, it is assumed that both the TCSC and TCPS stabilizers adopt the deviation of rotor speed of the generator as the feedback signal. Denote the transfer function of the stabilizers by $K_{tc}T_{tc}(s)$. The output control signal of the TCSC and TCPS stabilizers is as follows

$$\Delta u_{tc} = K_{tc}T_{tc}(s)\Delta\omega \tag{3.53}$$

where $\Delta u_{tc}$ is $\Delta x_{tcsc}$ or $\Delta\phi$ for the TCSC stabilizer or TCPS stabilizer, respectively. From Figs. 3.10 and 3.13, the forward path of the stabilizer can be obtained and shown in Fig. 3.14.

From Fig. 3.14, it can be seen that the direct damping torque contribution, $\Delta T_{dd}$, is through $K_p$. It should be much greater than the indirect damping torque, $\Delta T_{di}$,

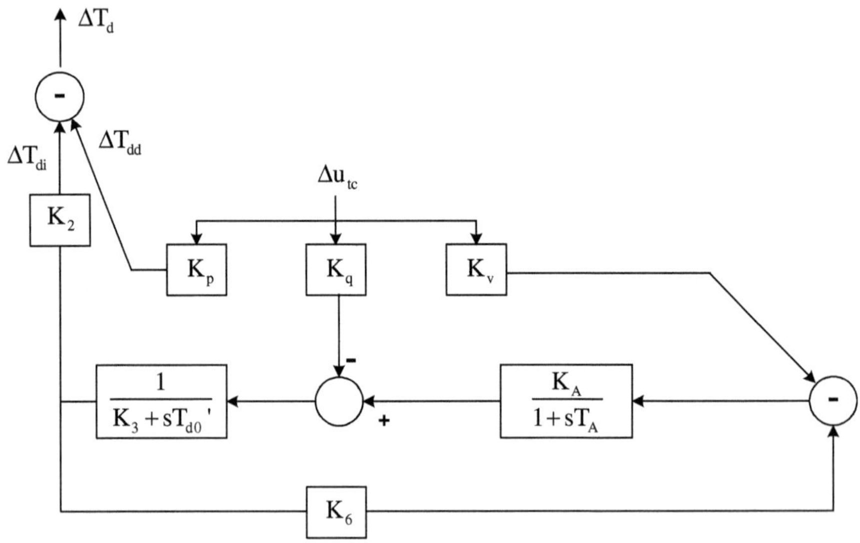


**Fig. 3.14** Forward path of the TCSC or TCPS stabilizer

because $\Delta u_{tc}$ is attenuated by two first-order lag blocks before it forms the indirect damping torque. That is

$$\Delta T_d \approx T_{dd}\Delta\omega = \mathrm{Re}[K_p K_{tc} T_{tc}(j\omega_s)]\Delta\omega \tag{3.54}$$

From Eq. (3.54), it can be seen that coefficient $K_p$ weighs the amount of damping torque contributed by the TCSC and TCPS stabilizers to the power system.

#### 3.2.2.2 Damping Torque Provided by the TCSC Stabilizer [4]

From Eq. (3.45), it can have

$$P_{t0} = \frac{E'_{q0}V_b}{x'_{d\Sigma}}\sin\delta_0 - \frac{V_b^2}{2}\frac{(x_q - x'_d)}{x'_{d\Sigma}x_{q\Sigma}}\sin 2\delta_0 = P_{t10} - P_{t20} \tag{3.55}$$

where

$$\begin{aligned} P_{t10} &= \frac{E'_{q0}V_b}{x'_{d\Sigma}}\sin\delta_0 \\ P_{t20} &= \frac{V_b^2}{2}\frac{(x_q - x'_d)}{x'_{d\Sigma}x_{q\Sigma}}\sin 2\delta_0 \end{aligned} \tag{3.56}$$

Hence, for the TCSC stabilizer, from Eq. (3.47), it can be obtained that

$$
\begin{aligned}
K_p &= \left.\frac{\partial P_t}{\partial x_{tcsc}}\right|_0 = \frac{V_b E'_{q0}\sin\delta_0}{x'^2_{d\Sigma}} - \frac{V_b^2(x_q - x'_d)\sin(2\delta_0)}{2x'^2_{d\Sigma}X^2_{q\Sigma}}(x'_{d\Sigma} + x_{q\Sigma}) \\
&= \frac{1}{x'_{d\Sigma}}\left[\frac{V_b E'_{q0}\sin\delta_0}{x'_{d\Sigma}} - \frac{V_b^2(x_q - x'_d)\sin(2\delta_0)}{2x'_{d\Sigma}x_{q\Sigma}}\right] - \frac{1}{x_{q\Sigma}}\frac{V_b^2(x_q - x'_d)\sin(2\delta_0)}{2x'_{d\Sigma}x_{q\Sigma}} \\
&= \frac{1}{x'_{d\Sigma}}P_{t10} - \frac{1}{x_{q\Sigma}}P_{t20}
\end{aligned}
\tag{3.57}
$$

Since $x_{q\Sigma} > x'_{d\Sigma}$ and $P_{t0} > P_{t20}$ as shown by Fig. 3.15,

$$K_p > 0 \tag{3.58}$$

Equation (3.58) indicates that when the load conditions of the power system change, a properly designed TCSC stabilizer can always provide positive damping torque to the power system.

At a higher load condition, difference between $P_{t0}$ and $P_{t20}$ is greater as shown in Fig. 3.15 and hence $K_p$ is bigger as shown in Eq. (3.57). This means that

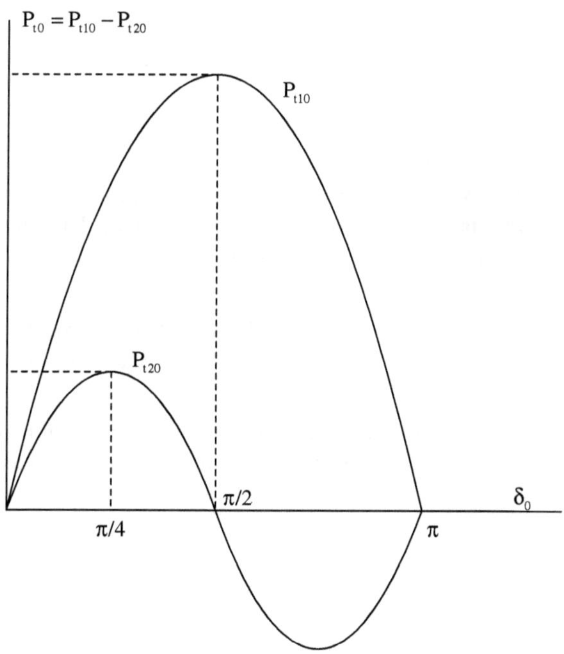


**Fig. 3.15** $P_t - \delta$ curve of the power system installed with a TCSC

effectiveness of the TCSC stabilizer in damping power system oscillations increases with the increase in system load conditions.

From Eq. (3.57), it can have

$$\frac{\partial K_p}{\partial x_t} = -\left(\frac{P_{t0}}{x'^2_{d\Sigma}} - \frac{P_{t20}}{x^2_{q\Sigma}}\right) < 0 \tag{3.59}$$

Equation (3.59) indicates that the TCSC stabilizer is less effective when the equivalent reactance of transmission line increases if the load condition keeps unchanged.

#### 3.2.2.3 Damping Torque Provided by the TCPS Stabilizer [5]

For the TCPS stabilizer, from Eq. (3.51), it can have

$$K_p = \left.\frac{\partial P_t}{\partial \phi}\right|_0 = K_{p1} - K_{p2} \tag{3.60}$$

where

$$\begin{aligned} K_{p1} &= \frac{V_b E'_{q0} \cos \delta_0}{x'_{d\Sigma}} \\ K_{p2} &= \frac{V_b^2 \left(x_q - x'_d\right) \cos 2\delta_0}{2 x'_{d\Sigma} x_{q\Sigma}} \end{aligned} \tag{3.61}$$

$K_{p1}$ and $K_{p2}$ are shown in Fig. 3.16. From Fig. 3.16, it can be seen that $K_p > 0$. Hence, when the load conditions of the power system change, a properly designed TCPS stabilizer can always supply positive damping torque to the power system.

From Eqs. (3.60) and (3.61), it can have

$$\frac{\partial K_p}{\partial x_t} = -\left(\frac{K_{p1}}{x'^2_{d\Sigma}} - \frac{K_{p2}}{x^2_{q\Sigma}}\right) < 0 \tag{3.62}$$

Equation (3.62) indicates that the TCPS stabilizer is less effective when the equivalent reactance of the transmission line increases if the loading condition is unchanged.

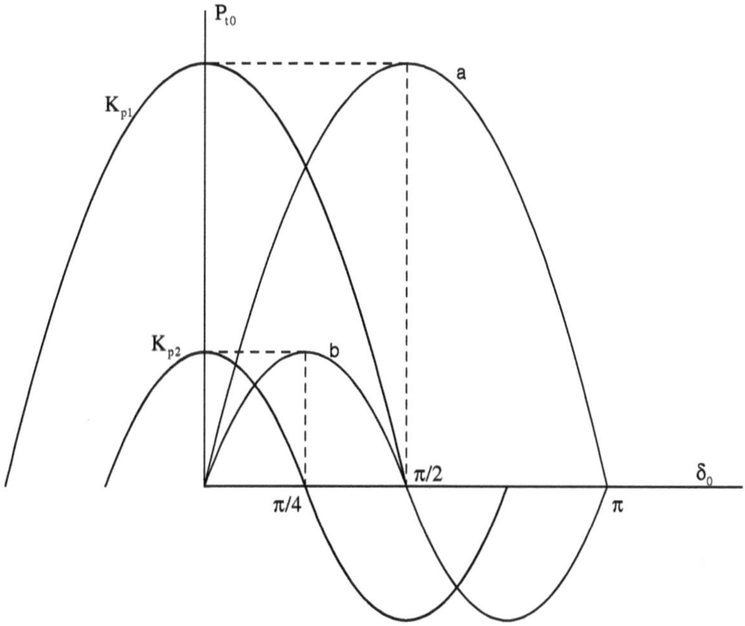


$$a = \frac{V_b E_{q0}' \sin \delta_0}{x_{d\Sigma}'}, \quad b = \frac{V_b^2 (x_q - x_d') \sin 2\delta_0}{2 x_{d\Sigma}' x_{q\Sigma}}$$

**Fig. 3.16** Examination of the damping effectiveness of the TCPS stabilizer

## 3.3 An Example Power System Installed with an SVC Stabilizer

Parameters of an example single-machine infinite-bus power system with an SVC installed in Fig. 3.2 are (in p.u. except indicated)

Generator: $x_d = 1.0, x_q = 0.8, x_d' = 0.15, M = 6.0\,\text{s.}, D = 0$, $T_{d0}' = 5.044\,\text{s.}$
The AVR: $K_A = 20.0$, $T_A = 0.01\,\text{s.}$
Transmission line: $X_{ts} = 0.3, X_{sb} = 0.3.$
The SVC: $x_{svcl} = 1, x_{svcc} = 1, K_{vp} = 1, K_{vi} = 8.$
Steady-state operating condition: $P_{t0} = 0.5, V_{t0} = 1.0, V_{s0} = 1.0, V_b = 1.0$

### 3.3.1 Linearized Model of Example Power System

#### 3.3.1.1 Linearized Model with the Transfer Function of the SVC Voltage Controller Included

The SVC is equipped with a proportional–integral (PI) voltage controller. The added damping controller adopts the locally available deviation of line active power as the feedback signal and the transfer function of a conventional PSS. That is (see Fig. 3.1)

$$\alpha = \alpha_0 + \left(K_{vp} + \frac{K_{vi}}{s}\right)(V_{sref} - V_s) + u_{svc-s} \tag{3.63}$$

where $u_{svc\text{-}s}$ is the output stabilizing signal of the SVC stabilizer, i.e.,

$$u_{svc-s} = K_s \frac{(1+sT_2)(1+sT_4)}{(1+sT_1)(1+sT_3)}(P_{tref} - P_t)$$

Linearization of Eq. (3.63) is as follows

$$\Delta\alpha = -\left(K_{vp} + \frac{K_{vi}}{s}\right)\Delta V_s + \Delta u_{svc-s} \tag{3.64}$$

From Eq. (3.19) with $\Delta b_{svc} = \frac{\dot{C}(\alpha_0)}{X_{svcl}}\Delta\alpha$, the above equation can be converted to

$$\Delta b_{svc} = -\frac{\dot{C}(\alpha_0)}{X_{svcl}}\left(K_{vp} + \frac{K_{vi}}{s}\right)\Delta V_s + \frac{\dot{C}(\alpha_0)}{X_{svcl}}\Delta u_{svc-s} \tag{3.65}$$

Substituting Eq. (3.16) into the above equation, it can have

$$\Delta b_{svc} = -\frac{\dot{C}(\alpha_0)}{X_{svcl}}\left[K_{vp} + \frac{K_{vi}}{s}\right]\left[C_1\Delta\delta + C_2\Delta E'_q + C_3\Delta b_{svc}\right] + \frac{\dot{C}(\alpha_0)}{X_{svcl}}\Delta u_{svc-s} \tag{3.66}$$

The above equation can give

$$\Delta b_{svc} = F_{svc1}(s)(C_1\Delta\delta + C_2\Delta E'_q) + F_{svc2}(s)\Delta u_{svc-s} \tag{3.67}$$

where

$$F_{svc1}(s) = \frac{-\frac{\dot{C}(\alpha_0)}{X_{svcl}}\left(K_{vp} + \frac{K_{vi}}{s}\right)}{1 + C_3\frac{\dot{C}(\alpha_0)}{X_{svcl}}\left(K_{vp} + \frac{K_{vi}}{s}\right)}, \quad F_{svc2}(s) = \frac{1}{1 + C_3\frac{\dot{C}(\alpha_0)}{X_{svcl}}\left(K_{vp} + \frac{K_{vi}}{s}\right)}\frac{\dot{C}(\alpha_0)}{X_{svcl}}$$

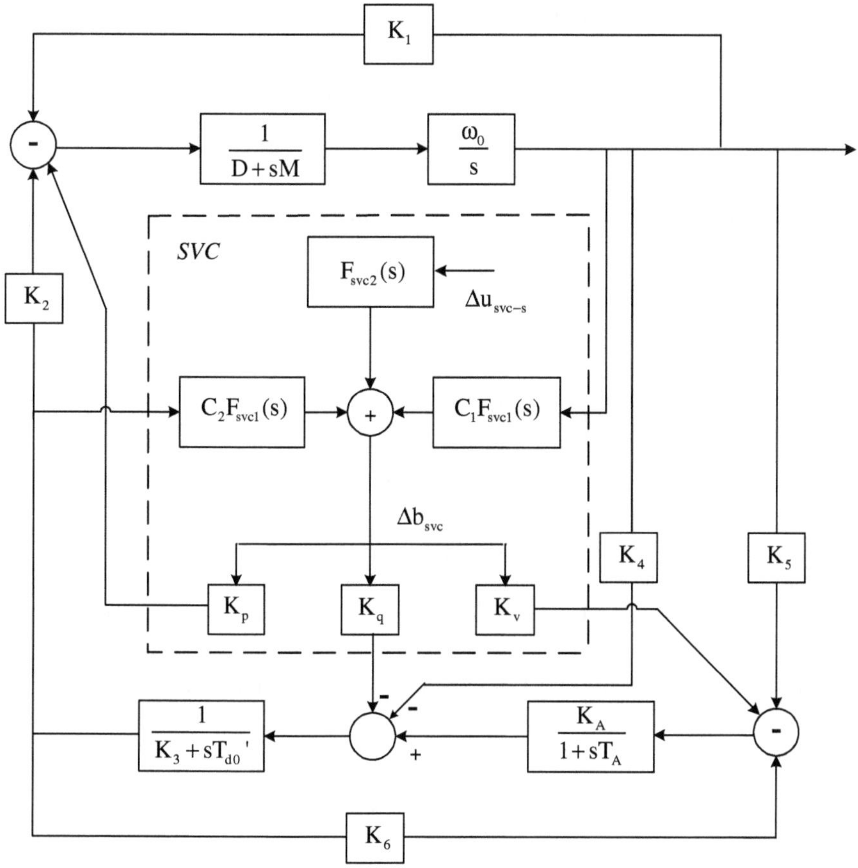


**Fig. 3.17** Linearized model with SVC voltage control function included

From Eq. (3.67) and Fig. 3.3, linearized Heffron–Phillips model of the power system with the SVC installed is shown in Fig. 3.17, where the SVC voltage controller adopts a PI controller and $u_{svc-s}$ is the stabilizing signal of the SVC-based stabilizer.

In Eq. (3.66), denote $\Delta z_{svc} = -\frac{\dot{C}(\alpha_0)}{X_{svcl}}\frac{K_{vi}}{s}(C_1\Delta\delta + C_2\Delta E'_q + C_3\Delta b_{svc})$, that is

$$\Delta \dot{z}_{svc} = -\frac{\dot{C}(\alpha_0)}{X_{svcl}} K_{vi}(C_1\Delta\delta + C_2\Delta E'_q + C_3\Delta b_{svc}) \tag{3.68}$$

Equation (3.66) becomes

$$\Delta b_{svc} = -\frac{\dot{C}(\alpha_0)}{X_{svcl}} K_{vp}\left(C_1\Delta\delta + C_2\Delta E'_q + C_3\Delta b_{svc}\right) + \Delta z + \frac{\dot{C}(\alpha_0)}{X_{svcl}}\Delta u_{svc-s} \tag{3.69}$$

The above equation can give

$$\Delta b_{svc} = \frac{-\frac{\dot{C}(\alpha_0)}{X_{svcl}} K_{vp}\left(C_1\Delta\delta + C_2\Delta E'_q\right) + \Delta z + \frac{\dot{C}(\alpha_0)}{X_{svcl}} \Delta u_{svc-s}}{1 + \frac{\dot{C}(\alpha_0)}{X_{svcl}} K_{vp} C_3} \tag{3.70}$$

Substituting Eq. (3.70) into Eq. (3.68), it can have

$$\Delta\dot{z}_{svc} = C_A\Delta\delta + C_E\Delta E'_q + C_Z\Delta z + C_U\Delta u_{svc-s} \tag{3.71}$$

where

$$C_A = -\frac{\dot{C}(\alpha_0)}{x_{svcl}} K_{vi} C_1 \left(1 - \frac{\frac{\dot{C}(\alpha_0)}{X_{svcl}} K_{vp} C_3}{1 + \frac{\dot{C}(\alpha_0)}{X_{svcl}} K_{vp} C_3}\right), \quad C_Z = -\frac{\frac{\dot{C}(\alpha_0)}{X_{svcl}} K_{vi} C_3}{1 + \frac{\dot{C}(\alpha_0)}{X_{svcl}} K_{vp} C_3},$$

$$C_E = -\frac{\dot{C}(\alpha_0)}{x_{svcl}} K_{vi} C_2 \left(1 - \frac{\frac{\dot{C}(\alpha_0)}{X_{svcl}} K_{vp} C_3}{1 + \frac{\dot{C}(\alpha_0)}{X_{svcl}} K_{vp} C_3}\right), \quad C_U = -\frac{\left[\frac{\dot{C}(\alpha_0)}{X_{svcl}}\right]^2 K_{vi} C_3}{1 + \frac{\dot{C}(\alpha_0)}{X_{svcl}} K_{vp} C_3}$$

Substituting Eq. (3.70) into Eq. (3.13), it can be obtained that

$$\begin{aligned}
\Delta\dot{\delta} &= \omega_0\Delta\omega \\
\Delta\dot{\omega} &= \frac{1}{M}(-K'_1\Delta\delta - D\Delta\omega - K'_2\Delta E'_q - K_{pz}\Delta z - K'_p\Delta u_{svc-s}) \\
\Delta\dot{E}'_q &= \frac{1}{T'_{d0}}(-K'_3\Delta E'_q - K'_4\Delta\delta - K_{qz}\Delta z - K'_q\Delta u_{svc-s} + \Delta E_{fd}) \\
\Delta\dot{E}'_{fd} &= -\frac{1}{T_A}\Delta\dot{E}'_{fd} - \frac{K_A}{T_A}(K'_5\Delta\delta + K'_6\Delta E'_q + K_{vz}\Delta z + K'_v\Delta u_{svc-s})
\end{aligned} \tag{3.72}$$

where

$$K'_1 = K_1 + K_p\frac{-\frac{\dot{C}(\alpha_0)}{X_{svcl}} K_{vp} C_1}{1 + \frac{\dot{C}(\alpha_0)}{X_{svcl}} K_{vp} C_3}, \quad K'_2 = K_2 + K_p\frac{-\frac{\dot{C}(\alpha_0)}{X_{svcl}} K_{vp} C_2}{1 + \frac{\dot{C}(\alpha_0)}{X_{svcl}} K_{vp} C_3},$$

$$K_{pz} = \frac{K_p}{\frac{1}{1 + \frac{\dot{C}(\alpha_0)}{X_{svcl}} K_{vp} C_3}}, \quad K'_p = \frac{K_p \frac{\dot{C}(\alpha_0)}{X_{svcl}}}{\frac{1}{1 + \frac{\dot{C}(\alpha_0)}{X_{svcl}} K_{vp} C_3}}$$

$$K_3' = K_3 + K_q \frac{-\frac{\dot{C}(\alpha_0)}{X_{svcl}} K_{vp} C_2}{1 + \frac{\dot{C}(\alpha_0)}{X_{svcl}} K_{vp} C_3}, \quad K_4' = K_4 + K_q \frac{-\frac{\dot{C}(\alpha_0)}{X_{svcl}} K_{vp} C_1}{1 + \frac{\dot{C}(\alpha_0)}{X_{svcl}} K_{vp} C_3},$$

$$K_{qz} = \frac{K_q}{\frac{1}{1 + \frac{\dot{C}(\alpha_0)}{X_{svcl}} K_{vp} C_3}}, \quad K_q' = \frac{K_q \frac{\dot{C}(\alpha_0)}{X_{svcl}}}{\frac{1}{1 + \frac{\dot{C}(\alpha_0)}{X_{svcl}} K_{vp} C_3}}$$

$$K_5' = K_5 + K_v \frac{-\frac{\dot{C}(\alpha_0)}{X_{svcl}} K_{vp} C_1}{1 + \frac{\dot{C}(\alpha_0)}{X_{svcl}} K_{vp} C_3}, \quad K_6' = K_6 + K_v \frac{-\frac{\dot{C}(\alpha_0)}{X_{svcl}} K_{vp} C_2}{1 + \frac{\dot{C}(\alpha_0)}{X_{svcl}} K_{vp} C_3},$$

$$K_{vz} = \frac{K_v}{\frac{1}{1 + \frac{\dot{C}(\alpha_0)}{X_{svcl}} K_{vp} C_3}}, \quad K_v' = \frac{K_v \frac{\dot{C}(\alpha_0)}{X_{svcl}}}{\frac{1}{1 + \frac{\dot{C}(\alpha_0)}{X_{svcl}} K_{vp} C_3}}$$

By writing Eqs. (3.71) and (3.72) together in matrix form, the state equation of the system can be obtained to be

$$\dot{\mathbf{X}} = \mathbf{A}\mathbf{X} + \mathbf{b}\Delta u_{svc-s} \tag{3.73}$$

where

$$\mathbf{X} = \begin{bmatrix} \Delta\delta \\ \Delta\omega \\ \Delta E_q' \\ \Delta E_{fd}' \\ \Delta z_{svc} \end{bmatrix}, \mathbf{b} = \begin{bmatrix} 0 \\ -\frac{K_p'}{M} \\ -\frac{K_q'}{T_{do}'} \\ -\frac{K_A K_v'}{T_A} \\ C_U \end{bmatrix},$$

$$\mathbf{A} = \begin{bmatrix} 0 & \omega_0 & 0 & 0 & 0 \\ -\frac{K_1'}{M} & -\frac{D}{M} & -\frac{K_2'}{M} & 0 & -\frac{K_{pz}}{M} \\ -\frac{K_4'}{T_{do}'} & 0 & -\frac{K_3'}{T_{do}'} & \frac{1}{T_{do}'} & -\frac{K_{qz}}{T_{do}'} \\ -\frac{K_A K_5'}{T_A} & 0 & -\frac{K_A K_6'}{T_A} & -\frac{1}{T_A} & -\frac{K_A K_{vz}}{T_A} \\ C_A & 0 & C_E & 0 & C_Z \end{bmatrix}$$

Linearization Eq. (3.63) is

$$\begin{aligned}\Delta u_{svc-s} &= K_s \frac{(1+sT_2)(1+sT_4)}{(1+sT_1)(1+sT_3)}(-\Delta P_t) = T_{svc-s}(s)(-\Delta P_t) \\ &= T_{svc-s}(s)(K_1'\Delta\delta + K_2'\Delta E_q' + K_{pz}\Delta z + K_p'\Delta u_{svc-s})\end{aligned} \quad (3.74)$$

Hence, the output equation of the system is as follows:

$$y = \mathbf{cX} + d\Delta u_{svc-s} \quad (3.75)$$

where

$$y = -\Delta P_t, \mathbf{c} = \begin{bmatrix} K_1' & 0 & K_2' & 0 & K_{pz} \end{bmatrix}, d = K_p'$$

#### 3.3.1.2 Linearized Model of the Example Power System

At the given steady-state operating condition, the voltage at the busbar where the SVC is installed is $V_{s0} = 1.0\,\text{p.u.}$ This is achieved by an initial compensation from the SVC at the operating condition, $b_{svc0}$. This initial compensation can be calculated as follows as introduced in Sect. 3.1.1.4.

From Eqs. (3.24) and (3.26), the reactive power injection at the busbar with the SVC installed is as follows

$$Q_{sb0} = \frac{V_{s0}}{x_{sb}}\left[V_{s0} - \sqrt{V_b^2 - \left(\frac{x_{sb}}{V_{s0}}P_{sb0}\right)^2}\right] = 0.0377$$

$$Q_{ts0} = \frac{V_{s0}}{x_{ts}}\left[\sqrt{V_{t0}^2 - \left(\frac{x_{ts}}{V_{s0}}P_{ts0}\right)^2} - V_{s0}\right] = -0.0377$$

Hence, the initial compensation is calculated to be

$$b_{svc0} = \frac{Q_{sb0} - Q_{ts0}}{V_{s0}^2} = 0.0754$$

This gives $c_{svc0} = 1 - x_{sb}b_{svc0} = 0.9774$.
From Eq. (3.1), it can have

$$-b_{svc0} = [1 - C(\alpha_0)]/X_{svcl} - 1/X_{svcc}$$

Thus, it can be obtained that at the steady-state operating condition,

$$C(\alpha_0) = \frac{2\alpha_0 - \sin 2\alpha_0}{2\pi} - 1 = 0.0754$$

Hence, $\alpha_0 = 42.0727°$.

From the circuit model of the power system installed with the SVC in Fig. 3.5, it can be calculated that

$$\bar{I}_{s0} = V_{s0} jb_{SVC0} = -j0.0754, \quad \bar{I}_{ts0} = \frac{(P_{t0} - jQ_{ts0})}{V_{s0}} = 0.5 + j0.0377$$
$$\bar{I}_{sb0} = \bar{I}_{ts0} - \bar{I}_{s0} = 0.5 + j0.0377, \quad \bar{V}_{b0} = \bar{V}_{s0} - j\bar{I}_{sb0} x_{sb} = 0.9887 - j0.15$$
$$P_{sb0} + jQ_{b0} = \bar{I}_{sb0} \bar{V}_{b0} = 0.4887 - j0.1123$$

With the reference voltage, recalculation of the value of various variables is as follows

$$\bar{I}_{sb0} = \frac{(P_{t0} - jQ_{b0})}{\bar{V}_{b0}} = 0.5 + j0.0377, \quad \bar{V}_{s0} = \bar{V}_{b0} + jX_{sb}\bar{I}_{sb0} = 0.9887 + j0.15$$
$$\bar{I}_{s0} = \bar{V}_{s0} jb_{svc0} = -0.0113 + j0.0746, \quad \bar{I}_{ts0} = \bar{I}_{sb0} + \bar{I}_{s0} = 0.4887 + j0.1123$$
$$\bar{V}_{t0} = \bar{V}_{s0} + j\bar{I}_{ts0} x_{ts} = 0.955 + j0.2966, \quad \bar{E}_{Q0} = \bar{V}_{t0} + jx_q \bar{I}_{ts0} = 0.8652 + j0.6876$$
$$\delta_0 = 38.4741°$$
$$I_{d0} = (\sin\delta_0 + j\cos\delta_0) I_{ts0} = 0.2161, \quad E_{q0} = E_{Q0} + (x_d - x_q) I_{d0} = 1.1483$$
$$E'_{q0} = E_{Q0} + (x'_d - x_q) I_{d0} = 0.9646, \quad I_{sbd0} = (\sin\delta_0 + j\cos\delta_0) I_{sb0} = 0.2816$$
$$I_{sbq0} = (\cos\delta_0 - j\sin\delta_0) I_{sb0} = 0.4149$$

From Eq. (3.7) it can have

$$x_{l\Sigma} = x_{ts} + \frac{x_{ts}}{c_{SVC0}} = 0.6069, \quad x'_{d\Sigma} = x'_d + x_{l\Sigma} = 0.7569,$$
$$x_{d\Sigma} = x_d + x_{l\Sigma} = 1.6069, \quad x_{q\Sigma} = x_q + x_{l\Sigma} = 1.4069$$

Thus, from Eqs. (3.9) and (3.12), parameters of the Heffron–Phillips model of the example power system can be calculated to be

$$K_1 = 0.8765, \quad K_2 = 0.8410, \quad K_3 = 2.1229, \quad K_4 = 0.7148, \quad K_5 = 0.0473,$$
$$K_6 = 0.7475, \quad K_p = 0.0166, \quad K_q = -0.2990, \quad K_v = 0.0806$$

From Eqs. (3.71) and (3.72), it can have

$$
\begin{aligned}
&C_A = 0.0358, \quad C_Z = -0.4259, \quad C_E = -0.7615, \quad C_U = -0.1218, \\
&K_1' = 0.8766, \quad K_2' = 0.8394, \quad K_{pz} = 0.0175, \quad K_p' = 0.0050, \quad K_3' = 2.1514, \\
&K_4' = 0.7134, \quad K_{qz} = -0.3158, \quad K_q' = -0.0903, \quad K_5' = 0.0476, \\
&K_6' = 0.7398, \quad K_{vz} = 0.0852, \quad K_v' = 0.0243.
\end{aligned}
$$

State matrix of Eq. (3.73) thus is obtained to be

$$
\mathbf{A} = \begin{bmatrix} 0 & 314.2 & 0 & 0 & 0 \\ -0.1461 & 0 & -0.1399 & 0 & -0.0029 \\ -0.1415 & 0 & -0.4299 & 0.1983 & 0.0626 \\ -95.2539 & 0 & -1479.6 & -100.0 & -170.3097 \\ 0.0358 & 0 & -0.7614 & 0 & -0.4259 \end{bmatrix}
$$

Eigenvalues of the state matrix are calculated to be

$$
\begin{aligned}
&\lambda_1 = -96.9595, \quad \bar{\lambda}_{2,3} = -0.1349 \pm j6.7166 \\
&\bar{\lambda}_4 = -3.2758, \quad \bar{\lambda}_5 = -0.3474
\end{aligned}
$$

Hence, the electromechanical oscillation mode of the power system is $\bar{\lambda}_{2,3} = -0.1349 \pm j6.7166$.

## 3.3.2 *Design of SVC-Based Stabilizer*

### 3.3.2.1 SVC Stabilizer Designed by Using the Phase Compensation Method

From Fig. 3.17 and Eq. (3.74), the forward path of the stabilizing signal of the SVC-based stabilizer, $\Delta\ u_{svc\text{-}s}$, can be drawn as shown in Fig. 3.18. Transfer function of the forward path of the SVC-based stabilizer can be obtained to be

$$
F_{svc-s}(s) = -F_{svc2}(s)\left\{K_p - \frac{\left(K_q + K_v\frac{K_A}{1+sT_A}\right)\left[C_2F_{svc1}(s)K_p + K_2\right]}{\left(K_3 + sT'_{d0} + K_6\frac{K_A}{1+sT_A}\right) + C_2F_{svc1}(s)\left(K_q + K_v\frac{K_A}{1+sT_A}\right)}\right\} \tag{3.76}
$$

Hence, from Fig. 3.17, the electric torque contribution from the SVC-based stabilizer can be obtained to be

$$
\Delta T_{et} = F_{svc-s}(s)T_{svc-s}(s)(-\Delta P_t) = F_{svc-s}(s)T_{svc-s}(s)(D + sM)\Delta\omega \tag{3.77}
$$

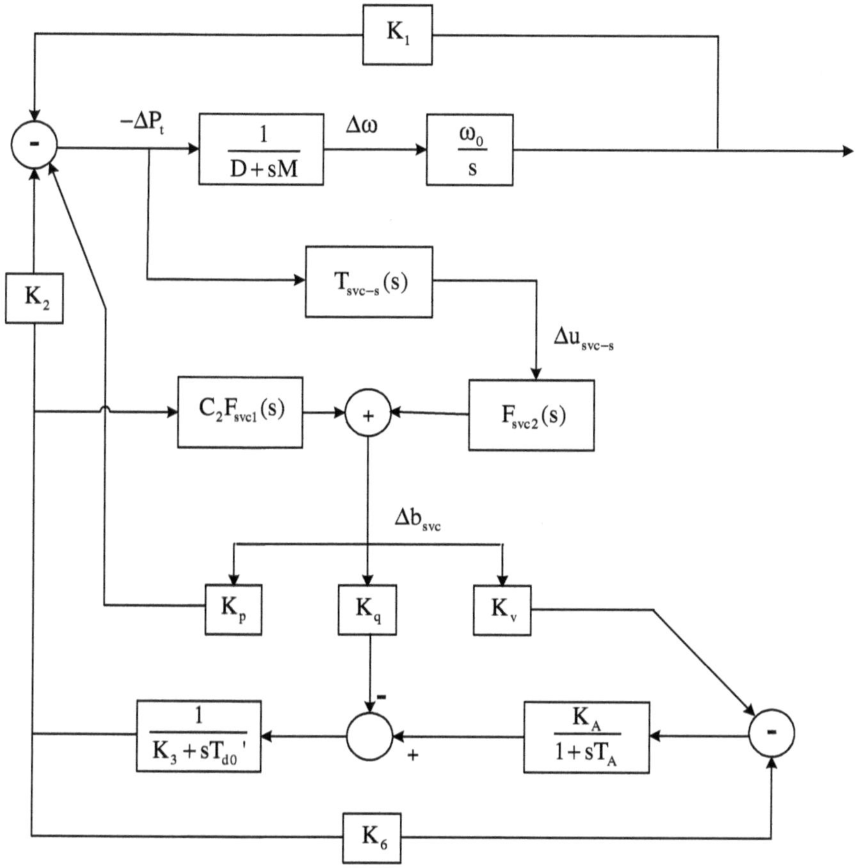


**Fig. 3.18** Forward path of the SVC-based stabilizer

Oscillation modes of the power system are $\bar{\lambda}_{2,3} = -0.1349 \pm j6.7166$. By taking $\omega_s = 6.7166$, from Eqs. (3.67) and (3.76), it can have

$$\bar{F}_{svc1}(j\omega_s) = -0.2899 + j0.3040, \quad \bar{F}_{svc2}(j\omega_s) = 0.2696 + j0.0171,$$
$$\bar{F}_{svc-s}(j\omega_s) = 0.0912 - j0.0114$$

Take the transfer function of the SVC stabilizer to be

$$T_{svc-s}(s) = K_s \frac{(1 + sT_2)(1 + sT_4)}{(1 + sT_1)(1 + sT_3)}, \quad T_1 = 0.09\,\text{s.}, \;\; T_3 = 0.09\,\text{s.}$$

The SVC stabilizer is to be designed to provide a damping torque $D_{svc}\Delta\omega = 1.6\Delta\omega$, i.e., $D_{svc} = 1.6$. Parameters of the SVC-based stabilizer can be set by using the phase compensation method to provide a pure damping torque

$$\Delta T_{et} = F_{svc-s}(j\omega_s)(D + j\omega_s M)T_{svc-s}(j\omega_s)\Delta\omega = D_{svc}\Delta\omega \tag{3.78}$$

The results of computation by the use of Eq. (3.78) are as follows

$$K_s = K_1K_2 = 5.8540, \quad T_2 = 1.7369\,\mathrm{s}, \quad T_4 = 1.7369\,\mathrm{s}$$

Let

$$\Delta x_1 = \frac{1 + sT_4}{1 + sT_3}\Delta\omega, \quad \Delta u_{svc-s} = K\frac{1 + sT_2}{1 + sT_1}\Delta x_1$$

Thus,

$$\begin{aligned} s\Delta x_1 &= -\tfrac{1}{T_3}\Delta x_1 + \tfrac{1}{T_3}\Big[-\tfrac{T_4K_1'}{M}\Delta\delta - \Big(\tfrac{DT_4}{M} - 1\Big)\Delta\omega \\ &\quad - \tfrac{T_4K_2'}{M}\Delta E_q' - \tfrac{T_4K_{pz}}{M}\Delta z_{svc} - \tfrac{T_4K_p'}{M}\Delta u_{svc-s}\Big] \\ s\Delta u_{svc} - s &= \tfrac{K_s(1 + sT_2)}{1 + sT_1}\Delta x_1 \end{aligned} \tag{3.79}$$

From Eqs. (3.73) and (3.79), state equation of the closed-loop system can be obtained to be

$$\begin{bmatrix} \Delta\dot{\delta} \\ \Delta\dot{\omega} \\ \Delta\dot{E}_q' \\ \Delta\dot{E}_{fd}' \\ \Delta\dot{z}_{svc} \\ \Delta\dot{x}_1 \\ \Delta\dot{u}_{svc-s} \end{bmatrix} = \mathbf{A}_c \begin{bmatrix} \Delta\delta \\ \Delta\omega \\ \Delta E_q' \\ \Delta E_{fd}' \\ \Delta z_{svc} \\ \Delta x_1 \\ \Delta u_{svc-s} \end{bmatrix} \tag{3.80}$$

where

$$\mathbf{A}_c = \begin{bmatrix} 0 & 314.1593 & 0 & 0 & 0 & 0 & 0 \\ -0.1461 & 0 & -0.1399 & 0 & -0.0029 & 0 & -0.0008 \\ -0.1415 & 0 & -0.4265 & 0.1983 & 0.0626 & 0 & 0.0179 \\ -95.2539 & 0 & -1479.6 & -100 & -170.3097 & 0 & -48.6868 \\ 0.0358 & 0 & -0.7615 & 0 & -0.4259 & 0 & -0.1218 \\ 0.7364 & 524.6004 & 0.2133 & 0.3170 & -0.0005 & -1.0411 & 0.0205 \\ 7.4877 & 5333.9 & 2.1683 & 3.2233 & -0.0056 & -4.7312 & -0.7913 \end{bmatrix}$$

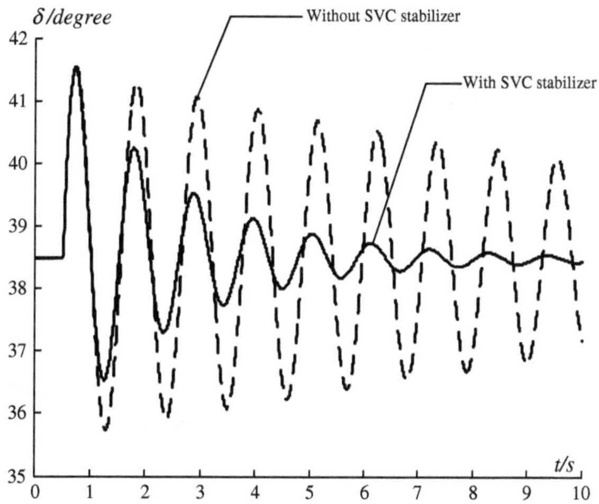


**Fig. 3.19** Simulation of the example power system without and with the SVC stabilizer installed

Eigenvalues of the state matrix of the closed-loop system are as follows

$$\lambda_1 = -95.2487, \quad \bar{\lambda}_{2,3} = -0.6870 \pm j6.6669, \quad \lambda_4 = -3.7484,$$
$$\lambda_5 = -1.2281, \quad \lambda_6 = -0.2937, \quad \lambda_7 = -0.7921$$

The electromechanical oscillation mode of the power system is now $\bar{\lambda}_{2,3} = -0.6870 \pm j6.6669$ with sufficient damping.

Figure 3.19 gives the simulation results of the power system without and with the SVC stabilizer installed. At 0.5 s of the simulation, a three-phase to-earth short circuit occurred on the transmission line of the single-machine infinite-bus power system in Fig. 3.2. The fault was cleared in 0.1 s. From Fig. 3.19, it can be seen that the SVC stabilizer effectively suppresses the power oscillation.

#### 3.3.2.2 Damping Control Effectiveness of the SVC Stabilizer as Affected by Various Factors

Effectiveness of the SVC stabilizer installed is affected by various factors as discussed in Sect. 3.1.2. Figure 3.20 shows the computational results of the damping torque provided by the SVC stabilizer when the load condition of the power system varies. From Fig. 3.20, it can be seen that

1. The direct damping torque provided by the SVC stabilizer is much greater than the indirect damping torque;
2. When the active power supplied by the generator increases, more damping torque is provided by the SVC stabilizer.

Table 3.1 shows the computational result of electromechanical oscillation mode of the power system with the SVC stabilizer installed at three different load

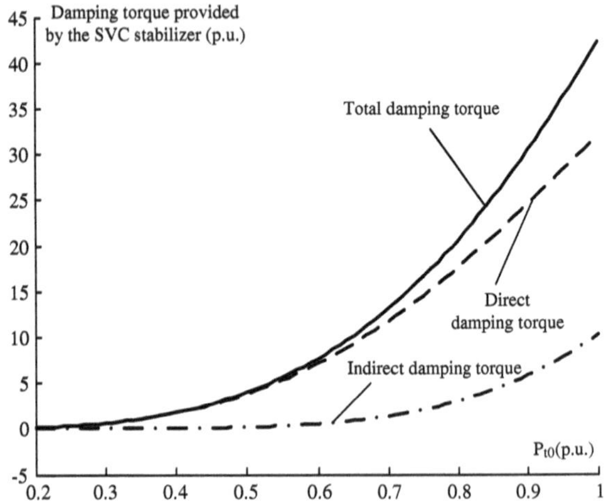


**Fig. 3.20** Computational results of the damping torque provided by the SVC stabilizer installed

**Table 3.1** Electromechanical oscillation mode

| Load condition $P_{t0}$ (p.u.) | 0.3 | 0.5 | 0.7 |
|---|---|---|---|
| Electromechanical oscillation mode | $-0.1732 \pm j5.4744$ | $-0.6870 \pm j6.6669$ | $-1.0831 \pm j6.0391$ |

conditions. Figures 3.21 and 3.22 show the simulation results. All those results in Figs. 3.20, 3.21, and 3.22 and in Table 3.1 confirm the analytical conclusion that the effectiveness of the SVC stabilizer increases with the load condition of power system.

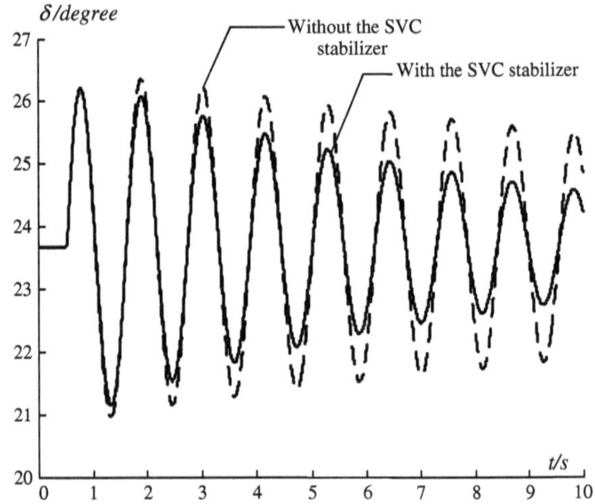


**Fig. 3.21** Simulation results at $P_{t0} = 0.3$

Fig. 3.22 Simulation results at $P_{t0} = 0.7$

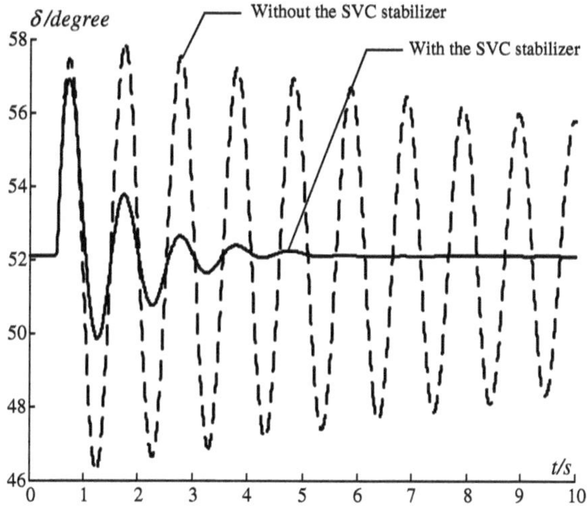


Discussions in Sect. 3.1.2 conclude that the effectiveness of the SVC stabilizer is also affected by the length of the transmission line and the installing locations of the SVC stabilizer. To confirm the conclusion, Fig. 3.23 presents the computational results of the damping torque provided by the SVC stabilizer at variable load condition when the system operates with single and double transmission lines connecting the generator and the infinite busbar. From Fig. 3.23, it can be seen that when the electric length of the line is longer (single transmission line connecting the

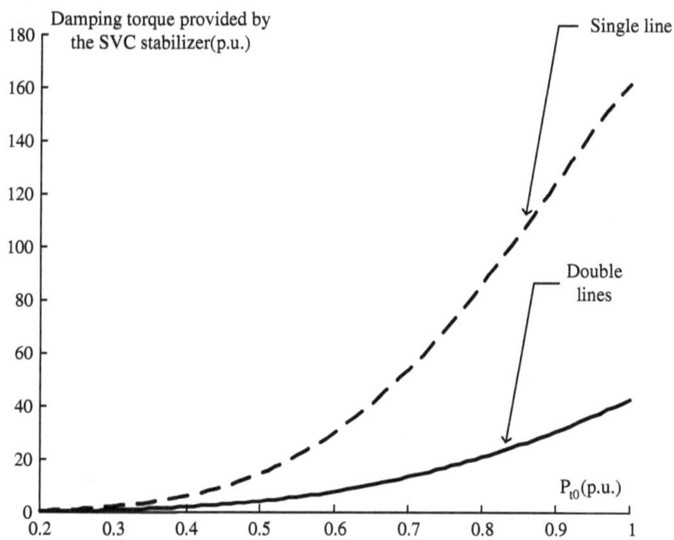


Fig. 3.23 Computational results of the damping torque provided by the SVC stabilizer

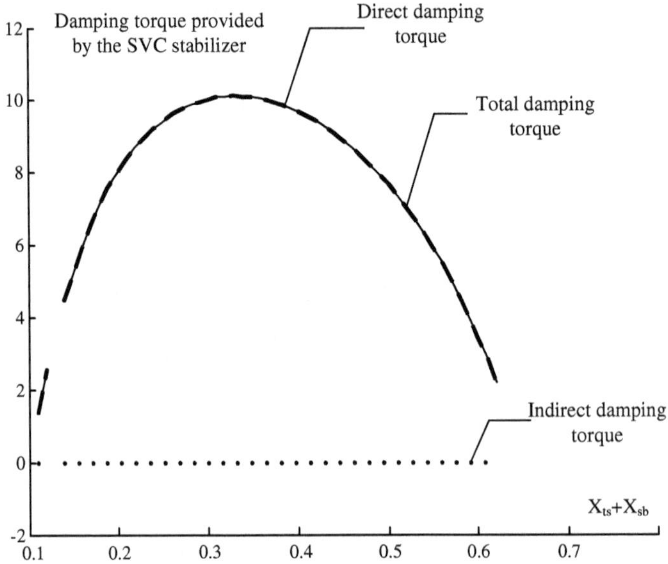


**Fig. 3.24** Computational results of the damping torque provided by the SVC stabilizer installed at various locations

generator and the infinite busbar), the SVC stabilizer provides more damping torque and hence is more effective in suppressing the power oscillation.

Keeping the operating condition of the example power system unchanged at $P_{t0} = 0.5, V_{b0} = 1.0, V_{t0} = 1.0, V_{s0} = 1.0$, the computational results of the damping torque provided by the SVC stabilizer installed at various locations along the transmission line are presented in Fig. 3.24. From Fig. 3.24, it can be seen that the SVC stabilizer is most effective when it is installed around the middle point of the line as it provides the most damping torque at the point.

## References


1. Wang HF, Swift FJ (1996) Capability of the static VAr compensator in damping power system oscillations. IEE Proc Part C 143(4):353–358
2. Smith OJM (1969) Power system transient control by capacitor switching. IEEE Trans Power Apparatus Syst 88(1):28–35
3. Smith OJM, Webster RH (1971) Series capacitor switching to quench electromechanical transients in power systems. IEEE Trans Power Apparatus Syst 90(2):427–433
4. Wang HF, Swift FJ (1996) Application of the Controllable Series Compensator in damping power system oscillations. IEE Proc Part C 143(4):359–364
5. Wang HF, Swift FJ (1997) Analysis of thyristor-controlled phase shifter applied in damping power system oscillations. Int J Electr Power Energ Syst 19(1):1–9

# Chapter 4
# Single-Machine Infinite-Bus Power Systems Installed with VSC-Based Stabilizers

## 4.1 Damping Torque Analysis of a Shunt VSC-Based Stabilizer Installed in a Single-Machine Infinite-Bus Power System

### *4.1.1 Extended Heffron–Phillips Model of a Single-Machine Infinite-Bus Power System Installed with a Shunt VSC-Based Stabilizer*

#### 4.1.1.1 A Shunt VSC Installed in a Single-Machine Infinite-Bus Power System

Figure 4.1 shows the configuration of a shunt voltage source converter (VSC)-based unit connected to high-voltage transmission line through a step-down transformer. $x_s$ is the equivalent reactance of the step-down transformer. The unit can be an energy storage system (ESS), such as a battery energy storage system (BESS) or a superconducting magnetic energy storage (SMES) device. It can also be a renewable power generation plant, such as a grid-connected photovoltaic (PV) or fuel cell (FC) power plant. In case that $I_{dc2} = 0$, it is a static synchronous compensator (STATCOM) [1].

There are two commonly used algorithms to control the voltage at the AC terminal of the converter $\overline{V}_c$, the pulse width modulation (PWM), and the pulse amplitude modulation (PAM). When the PWM is used, $V_{dc}$ is kept constant by controlling the modulation phase, $\phi$ (angle between $\overline{V}_c$ and $\overline{V}_s$), to charge or discharge the capacitor. The amplitude of voltage at the AC terminal is regulated by



H. Wang and W. Du, *Analysis and Damping Control of Power System Low-frequency Oscillations*, Power Electronics and Power Systems,
DOI 10.1007/978-1-4899-7696-3_4

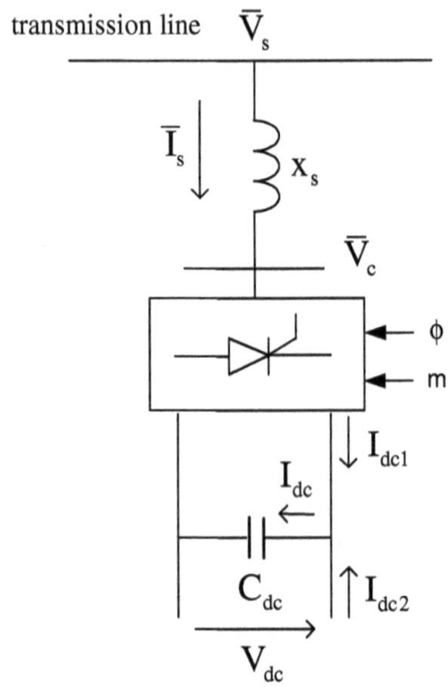


**Fig. 4.1** Configuration of a shunt VSC-based unit

the modulation ratio m, i.e. $V_c = mkV_{dc}$, where k is the ratio between the AC and DC voltages of the converter (including the ratio of the step-down transformer), dependent of the structure of the converter circuit.

Figure 4.2 shows the configuration of a single-machine infinite-bus power system, where a shunt VSC-based unit is installed at a location along the transmission line. From Fig. 4.2, it can be obtained that

$$\begin{aligned} \overline{V}_t &= jx_{ts}\overline{I}_{ts} + \overline{V}_s \\ \overline{V}_s &= jx_s\overline{I}_s + \overline{V}_c \\ \overline{V}_s - \overline{V}_b &= jx_{sb}(\overline{I}_{ts} - \overline{I}_s) \end{aligned} \quad (4.1)$$

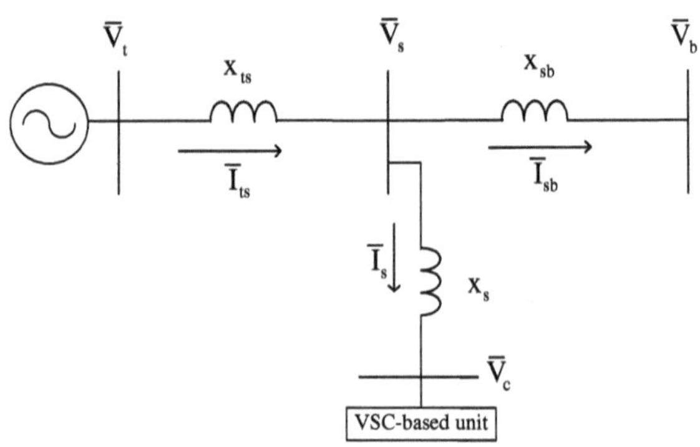


**Fig. 4.2** A single-machine infinite-bus power system installed with a VSC-based unit

Hence

$$\begin{aligned} & jx_s\bar{I}_s + \bar{V}_c - \bar{V}_b = jx_{sb}(\bar{I}_{ts} - \bar{I}_s) \\ & \bar{V}_t = jx_{ts}\bar{I}_{ts} + jx_{sb}(\bar{I}_{ts} - \bar{I}_s) + \bar{V}_b \end{aligned} \tag{4.2}$$

In the d–q coordinate of generator, as shown in Fig. 4.3, Eq. (4.2) gives

$$\begin{aligned} & jx_s(i_{sd} + ji_{sq}) + V_c(\cos\psi + j\sin\psi) - V_b(\sin\delta + j\cos\delta) \\ & \quad = jx_{sb}[(i_{tsd} - i_{sd}) + j(i_{tsq} - i_{sq})] \\ & v_{td} + jv_{tq} \\ & \quad = j(x_{ts} + x_{sb})(i_{tsd} + ji_{tsq}) - jx_{sb}(i_{sd} + ji_{sq}) + V_b(\sin\delta + j\cos\delta) \end{aligned} \tag{4.3}$$

By equating the real and imaginary parts on both sides of Eq. (4.3), it can have

$$\begin{aligned} & -x_s i_{sq} + V_c\cos\psi - V_b\sin\delta = -x_{sb}(i_{tsq} - i_{sq}) \\ & v_{td} = -(x_{ts} + x_{sb})i_{tsq} + x_{sb}i_{sq} + V_b\sin\delta \\ & x_s i_{sd} + V_c\sin\psi - V_b\cos\delta = x_{sb}(i_{tsd} - i_{sd}) \\ & v_{tq} = (x_{ts} + x_{sb})i_{tsd} - x_{sb}i_{sd} + V_b\cos\delta \end{aligned} \tag{4.4}$$

Because

$$\begin{aligned} & v_{td} = x_q i_{tsq} \\ & v_{tq} = E'_q - x'_d i_{tsd} \end{aligned} \tag{4.5}$$

**Fig. 4.3** Phasor diagram of system shown in Fig. 4.2

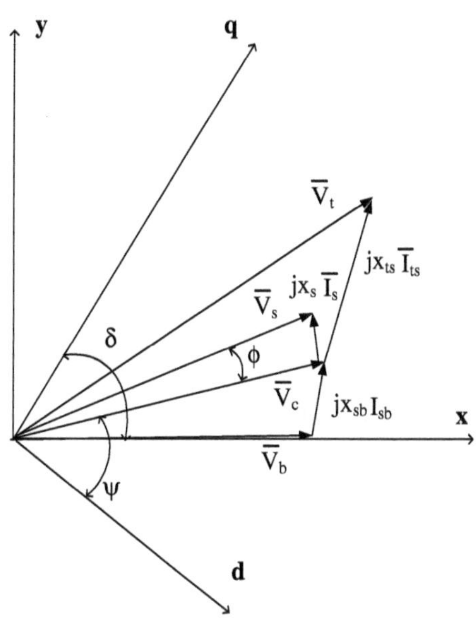

From Eqs. (4.4) and (4.5), it can be obtained that

$$
\begin{aligned}
&\begin{bmatrix} x_{sb} & -x_s - x_{sb} \\ x_q + x_{ts} + x_{sb} & -x_{sb} \end{bmatrix} \begin{bmatrix} i_{tsq} \\ i_{sq} \end{bmatrix} = \begin{bmatrix} -V_c \cos\psi + V_b \sin\delta \\ V_b \sin\delta \end{bmatrix} \\
&\begin{bmatrix} x_{sb} & -x_s - x_{sb} \\ x'_d + x_{ts} + x_{sb} & -x_{sb} \end{bmatrix} \begin{bmatrix} I_{tsd} \\ I_{sd} \end{bmatrix} = \begin{bmatrix} V_c \sin\psi - V_b \cos\delta \\ E'_q - V_b \cos\delta \end{bmatrix}
\end{aligned} \tag{4.6}
$$

The above equation gives

$$
\begin{aligned}
&\begin{bmatrix} i_{tsq} \\ i_{sq} \end{bmatrix} = \begin{bmatrix} x_{sb} & -x_s - x_{sb} \\ x_q + x_{ts} + x_{sb} & -x_{sb} \end{bmatrix}^{-1} \begin{bmatrix} -V_c \cos\psi + V_b \sin\delta \\ V_b \sin\delta \end{bmatrix} \\
&\begin{bmatrix} i_{tsd} \\ i_{sd} \end{bmatrix} = \begin{bmatrix} x_{sb} & -x_s - x_{sb} \\ x'_d + x_{ts} + x_{sb} & -x_{sb} \end{bmatrix}^{-1} \begin{bmatrix} V_c \sin\psi - V_b \cos\delta \\ E'_q - V_b \cos\delta \end{bmatrix}
\end{aligned} \tag{4.7}
$$

The active power supplied by the generator can be expressed as

$$
\begin{aligned}
P_t &= v_{td} i_{tsd} + v_{tq} i_{tsq} = x_q i_{tsq} i_{tsd} + (E'_q - x'_q i_{tsd}) i_{tsq} \\
&= E'_q i_{tsq} + (x_q - x'_q) i_{tsd} i_{tsq}
\end{aligned} \tag{4.8}
$$

From Eqs. (2.32) and (4.5), it can have

$$
\begin{aligned}
E_q &= E'_q + (x_d - x'_q) i_{tsd} \\
V_t &= \sqrt{v_{td}^2 + v_{tq}^2} = \sqrt{(x_q i_{tsq})^2 + (E'_q - x'_q i_{tsd})^2}
\end{aligned} \tag{4.9}
$$

Magnitude of $\overline{V}_c$ is $mkV_{dc}$. In the d–q coordinate of generator as shown in Fig. 4.3, its phase is $\psi$. Hence,

$$
\overline{V}_c = mkV_{dc}(\cos\psi + j\sin\psi) = mkV_{dc}\angle\psi \tag{4.10}
$$

Active power received by the DC capacitor of VSC-based unit is

$$
V_{dc} I_{dc1} = i_{sd} v_{cd} + i_{sq} v_{cq} = i_{sd} mkV_{dc} \cos\psi + i_{sq} mkV_{dc} \sin\psi \tag{4.11}
$$

The above equation gives

$$
I_{dc} = i_{sd} mk \cos\psi + i_{sq} mk \sin\psi \tag{4.12}
$$

Dynamic equation on the DC side of the converter is

$$
\dot{V}_{dc} = \frac{1}{C_{dc}}(I_{dc1} + I_{dc2}) = \frac{1}{C_{dc}}(i_{sd} mk \cos\psi + i_{sq} mk \sin\psi) + \frac{I_{dc2}}{C_{dc}} \tag{4.13}
$$

From the phasor diagram of Fig. 4.3, it can have

$$\psi = \tan^{-1}\frac{v_{sq}}{v_{sq}} - \phi \tag{4.14}$$

Equations (4.7)–(4.9), (4.13), and (4.14), together with the differential equations of synchronous generator of Eq. (2.37), form the mathematical model of the single-machine infinite-bus power system with the shunt VSC-based unit installed. In Eqs. (4.10) and (4.14), the modulation ratio and phase of PWM algorithm, m and $\phi$, are regulated by two controllers. Hence, m and $\phi$ are two control variables in the mathematical model of the power system.

#### 4.1.1.2 Extended Heffron–Phillips Model

Linearization of Eq. (4.7) is

$$\begin{aligned}
\begin{bmatrix}\Delta i_{tsq}\\ \Delta i_{sq}\end{bmatrix} &= \begin{bmatrix}c_{11} & c_{12}\\ c_{21} & c_{22}\end{bmatrix}\\
&\times\begin{bmatrix}-m_0k\cos\psi_0\Delta V_{dc} - kV_{dc0}\cos\psi_0\Delta m + m_0kV_{dc0}\sin\psi_0\Delta\psi + V_b\cos\delta_0\Delta\delta\\ V_b\cos\delta_0\Delta\delta\end{bmatrix}\\
\begin{bmatrix}\Delta i_{tsd}\\ \Delta i_{sd}\end{bmatrix} &= \begin{bmatrix}d_{11} & d_{12}\\ d_{21} & d_{22}\end{bmatrix}\\
&\times\begin{bmatrix}m_0k\sin\psi_0\Delta V_{dc} + kV_{dc0}\sin\psi_0\Delta m + m_0kV_{dc0}\cos\psi_0\Delta\psi + V_b\sin\delta_0\Delta\delta\\ \Delta E'_q + V_b\sin\delta_0\Delta\delta\end{bmatrix}
\end{aligned} \tag{4.15}$$

The second equation in Eq. (4.1) is

$$v_{sd} + jv_{sq} = jx_s(i_{sd} + ji_{sq}) + mkV_{dc}\cos\psi + jmkV_{dc}\sin\psi$$

Hence, it can have

$$\begin{aligned}
v_{sd} &= -x_s i_{sq} + mkV_{dc}\cos\psi\\
v_{sq} &= x_s i_{sd} + mkV_{dc}\sin\psi
\end{aligned}$$

By using Eq. (4.15), linearization of above equations can be obtained as

$$\begin{aligned}
\Delta v_{sd} &= (x_s+1)m_0k\cos\psi_0\Delta V_{dc} + (x_s+1)kV_{dc0}\cos\psi_0\Delta m\\
&\quad - (x_s+1)m_0kV_{dc0}\sin\psi_0\Delta\psi - x_sV_b\cos\delta_0\Delta\delta\\
&= a_{d1}\Delta V_{dc} + a_{d2}\Delta m + a_{d3}\Delta\psi + a_{d4}\Delta\delta\\
\Delta v_{sq} &= (x_s+1)m_0k\sin\psi_0\Delta V_{dc} + (x_s+1)kV_{dc0}\sin\psi_0\Delta m\\
&\quad + (x_s+1)m_0kV_{dc0}\cos\psi_0\Delta\psi + V_b\sin\delta_0\Delta\\
&= a_{q1}\Delta V_{dc} + a_{q2}\Delta m + a_{q3}\Delta\psi + a_{q4}\Delta\delta
\end{aligned}$$

By using above equations, linearization of Eq. (4.14) can be obtained as

$$\Delta\psi = a'_1 \Delta V_{dc} + a'_2 \Delta m + a'_3 \Delta\psi + a'_4 \Delta\delta - \Delta\phi$$

where

$$\begin{aligned}
a'_1 &= \frac{1}{v_{sd0}^2 + v_{sq0}^2}(v_{sd0} a_{1q} - v_{sq0} a_{1d}) \\
a'_2 &= \frac{1}{v_{sd0}^2 + v_{sq0}^2}(v_{sd0} a_{2q} - v_{sq0} a_{2d}) \\
a'_3 &= \frac{1}{v_{sd0}^2 + v_{sq0}^2}(v_{sd0} a_{3q} - v_{sq0} a_{3d}) \\
a'_4 &= \frac{1}{v_{sd0}^2 + v_{sq0}^2}(v_{sd0} a_{4q} - v_{sq0} a_{4d})
\end{aligned}$$

That is

$$\begin{aligned}
\Delta\psi &= \frac{a'_1}{1 - a'_3} \Delta V_{dc} + \frac{a'_2}{1 - a'_3} \Delta m + \frac{a'_4}{1 - a'_3} \Delta\delta - \frac{1}{1 - a'_3} \Delta\phi \\
&= a_1 \Delta V_{dc} + a_2 \Delta m + a_3 \Delta\delta + a_4 \Delta\phi
\end{aligned} \tag{4.16}$$

By using Eqs. (4.15) and (4.16), linearization of Eqs. (4.8) and (4.9) can be obtained as

$$\begin{aligned}
\Delta P_t &= K'_1 \Delta\delta + K_2 \Delta E'_q + K'_{pdc} \Delta V_{dc} + K'_{pm} \Delta m + K'_{p\psi} \Delta\psi \\
&= (K'_1 - a_3 K'_{p\psi}) \Delta\delta + K_2 \Delta E'_q + (K'_{pdc} - a_1 K'_{p\psi}) \Delta V_{dc} \\
&\quad + (K'_{pm} - a_2 K'_{p\psi}) \Delta m + a_4 K'_{p\psi} \Delta\phi \\
&= K_1 \Delta\delta + K_2 \Delta E'_q + K_{pdc} \Delta V_{dc} + K_{pm} \Delta m + K_{p\psi} \Delta\phi \\
\Delta E_q &= K'_4 \Delta\delta + K_3 \Delta E'_q + K'_{qdc} \Delta V_{dc} + K'_{qm} \Delta m + K'_{q\psi} \Delta\psi \\
&= (K'_4 - a_3 K'_{q\psi}) \Delta\delta + K_3 \Delta E'_q + (K'_{qdc} - a_1 K'_{q\psi}) \Delta V_{dc} \\
&\quad + (K'_{qm} - a_2 K'_{q\psi}) \Delta m + a_4 K'_{q\psi} \Delta\phi \\
&= K_4 \Delta\delta + K_3 \Delta E'_q + K_{qdc} \Delta V_{dc} + K_{qm} \Delta m + K_{q\psi} \Delta\phi \\
\Delta V_t &= K'_5 \Delta\delta + K_6 \Delta E'_q + K'_{vdc} \Delta V_{dc} + K'_{vm} \Delta m + K'_{v\psi} \Delta\psi \\
&= (K_5 - a_3 K'_{v\psi}) \Delta\delta + K_6 \Delta E'_q + (K'_{vdc} - a_1 K'_{v\psi}) \Delta V_{dc} \\
&\quad + (K'_{vm} - a_2 K'_{v\psi}) \Delta m + a_4 K'_{v\psi} \Delta\phi \\
&= K_5 \Delta\delta + K_6 \Delta E'_q + K_{vdc} \Delta V_{dc} + K_{vm} \Delta m + K_{v\psi} \Delta\phi
\end{aligned} \tag{4.17}$$

where

$$
\begin{aligned}
C_0 &= m_0 k \\
K_1' &= \left[E_{q0}' - i_{tsd0}\cdot\left(x_d' - x_q\right)\right](c_{11}+c_{12})V_b\cos\delta_0 \\
&\quad - i_{tsq0}\cdot\left(x_d' - x_q\right)(d_{11}+d_{12})V_b\sin\delta_0 \\
K_2 &= i_{tsq0} - i_{tsq0}\cdot\left(x_d' - x_q\right)d_{12} \\
K_{pdc}' &= -c_{11}C_0\cos\psi_0\left[E_{q0}' - i_{tsd0}\cdot\left(x_d' - x_q\right)\right] - d_{11}C_0\sin\psi_0 I_{tsq0}\cdot\left(x_d' - x_q\right) \\
K_{pm}' &= -c_{11}kV_{dc0}\cos\psi_0\left[E_{q0}' - i_{tsd0}\cdot\left(x_d' - x_q\right)\right] - d_{11}kV_{dc0}\sin\psi_0 I_{tsq0}\cdot\left(x_d' - x_q\right) \\
K_{p\psi}' &= c_{11}C_0V_{dc0}\sin\psi_0\left[E_{q0}' - i_{tsd0}\cdot\left(x_d' - x_q\right)\right] - d_{11}C_0V_{dc0}\cos\psi_0 I_{tsq0}\cdot\left(x_d' - x_q\right) \\
K_4 &= (d_{11}+d_{12})V_b\sin\delta_0(x_d - x_d') \\
K_3' &= d_{12}(x_d - x_d') + 1 \\
K_{qdc}' &= d_{11}C_0\sin\psi_0(x_d - x_d') \\
K_{qm}' &= d_{11}kV_{dc0}\sin\psi_0(x_d - x_d') \\
K_{q\psi}' &= d_{11}C_0V_{dc0}\cos\psi_0(x_d - x_d') \\
K_5' &= \frac{\left[v_{td0}X_q(c_{11}+c_{12})V_b\cos\delta_0 - v_{tq0}x_d'(d_{11}+d_{12})V_b\sin\delta_0\right]}{V_{t0}} \\
K_6 &= \frac{v_{tq0}\left(1 - x_d' d_{12}\right)}{V_{t0}} \\
K_{vdc}' &= \frac{\left(-v_{td0}x_q c_{11}C_0\cos\psi_0 - v_{tq0}x_d' d_{11}C_0\sin\psi_0\right)}{V_{t0}} \\
K_{vm}' &= \frac{\left(-v_{td0}x_q kc_{11}V_{dc0}\cos\psi_0 - v_{tq0}x_d' d_{11}kV_{dc0}\sin\psi_0\right)}{V_{t0}} \\
K_{v\psi}' &= \frac{\left(v_{td0}x_q c_{11}C_0V_{dc0}\sin\psi_0 - v_{tq0}x_d' d_{11}C_0V_{dc0}\cos\psi_0\right)}{V_{t0}}
\end{aligned}
$$

Linearization of Eq. (4.13) is

$$
\begin{aligned}
C_{dc}\Delta\dot{V}_{dc} &= (i_{sd0}k\cos\psi_0 + i_{sq0}k\sin\psi_0)\Delta m \\
&\quad + m_0k(-i_{sd0}\sin\psi_0 + i_{sq0}\cos\psi_0)\Delta\psi \\
&\quad + m_0k(\cos\psi_0\Delta i_{sd} + \sin\psi_0\Delta i_{sq}) + \Delta I_{dc2}
\end{aligned}
$$

By using Eqs. (4.15) and (4.16), it can have

$$\begin{aligned}\Delta\dot{V}_{dc} &= K_7'\Delta\delta + K_8\Delta E_q' + K_9'\Delta V_{dc} + K_{dm}'\Delta m + K_{d\psi}'\Delta\psi + \frac{1}{C_{dc}}\Delta I_{dc2}\\ &= (K_7' - a_3K_{d\psi}')\Delta\delta + K_8\Delta E_q' + (K_9' - a_1K_{d\psi}')\Delta V_{dc}\\ &\quad + (K_{dm} - a_2K_{d\psi}')\Delta m + a_4K_{d\psi}'\Delta f + \frac{1}{C_{dc}}\Delta I_{dc2}\\ &= K_7\Delta\delta + K_8\Delta E_q' + K_9\Delta V_{dc} + K_{dm}\Delta m + K_{d\psi}\Delta\phi + \frac{1}{C_{dc}}\Delta I_{dc2}\end{aligned} \tag{4.18}$$

where

$$\begin{aligned}K_7' &= \frac{[m_0k\cos\psi_0(d_{21}+d_{22})V_b\sin\delta_0 + m_0k\sin\psi_0(c_{21}+c_{22})V_b\cos\delta_0]}{C_{dc}}\\ K_8 &= \frac{m_0k\cos\psi_0 d_{22}}{C_{dc}}\\ K_9' &= \frac{(m_0^2k^2\cos\psi_0 d_{21}\sin\psi_0 - m_0^2k^2\sin\psi_0 c_{21}\cos\psi_0)}{C_{dc}}\\ K_{dm}' &= \frac{1}{C_{dc}}(ki_{sd0}\cos\psi_0 + ki_{sq0}\sin\psi_0 + m_0k^2\cos\psi_0 d_{21}V_{dc0}\sin\psi_0\\ &\quad - m_0k^2\sin\psi_0 c_{21}V_{dc0}\cos\psi_0)\\ K_{d\psi}' &= \frac{1}{C_{dc}}[-m_0k(i_{sd0}\sin\psi_0 - i_{sq0}\cos\psi_0) + m_0^2k^2\cos\psi_0 d_{21}V_{dc0}\cos\psi_0\\ &\quad + m_0^2k^2\sin\psi_0 c_{21}V_{dc0}\sin\psi_0]\end{aligned}$$

By substituting Eq. (4.17) into the linearized differential equation of synchronous generator of Eq. (2.37) (for the simplicity of expression, the PSS is not considered), the extended Heffron–Phillips model of the single-machine infinite-bus power system installed with the shunt VSC-based unit can be obtained as

$$\begin{aligned}\Delta\dot{\delta} &= \omega_o\Delta\omega\\ \Delta\dot{\omega} &= \frac{1}{M}(-K_1\Delta\delta - D\Delta\omega - K_2\Delta E'_q - K_{pdc}\Delta V_{dc} - K_{pm}\Delta m - K_{p\psi}\Delta\phi)\\ \Delta\dot{E}_q' &= \frac{1}{T_{d0}'}(-K_4\Delta\delta - K_3\Delta E'_q + \Delta E_{fd}' + K_{qdc}\Delta V_{dc} + K_{qm}\Delta m + K_{q\psi}\Delta\phi)\\ \Delta\dot{E}_{fd}' &= -\frac{1}{T_A}\Delta E_{fd}' - \frac{K_A}{T_A}(K_5\Delta\delta + K_6\Delta E'_q + K_{vdc}\Delta V_{dc} + K_{vm}\Delta m + K_{v\psi}\Delta\phi)\\ \Delta\dot{V}_{dc} &= K_7\Delta\delta + K_8\Delta E'_q + K_9\Delta V_{dc} + K_{dm}\Delta m + K_{d\psi}\Delta\phi + \frac{1}{C_{dc}}\Delta I_{dc2}\end{aligned} \tag{4.19}$$

The model is shown in Figs. 4.4 and 4.5.

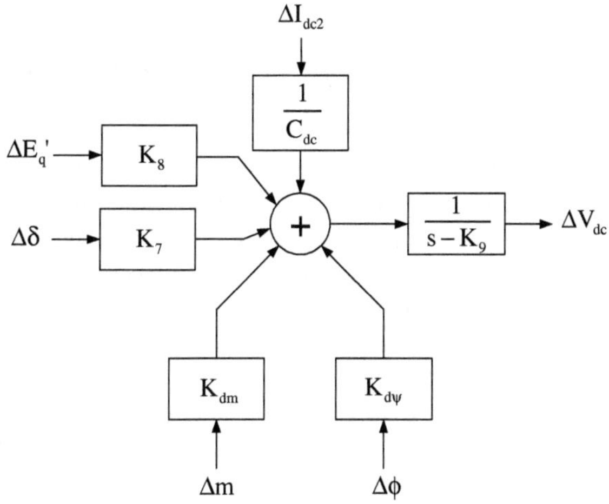


**Fig. 4.4** Extended Heffron–Phillips model—part of the VSC-based unit

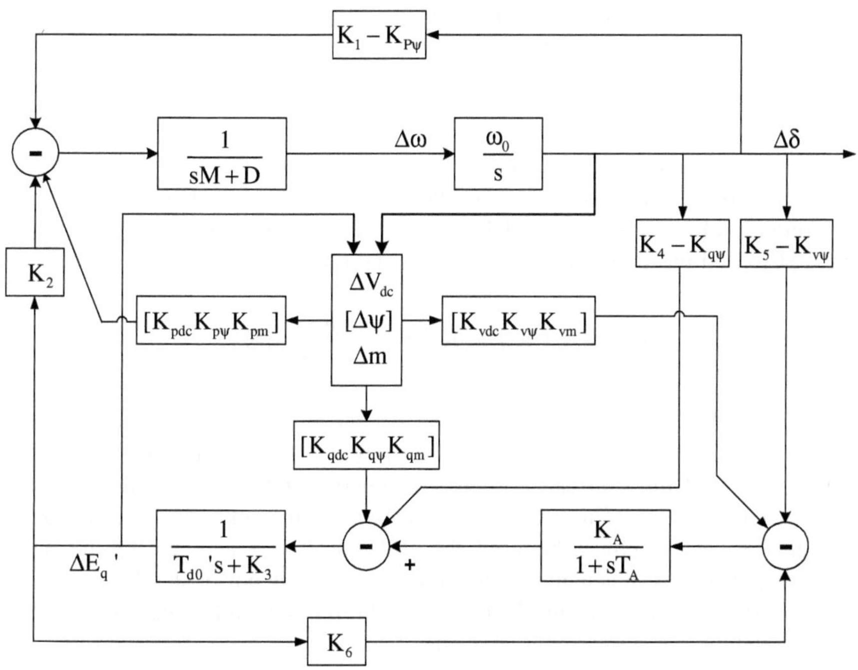


**Fig. 4.5** Extended Heffron–Phillips model—part of the power system

## 4.1.2 Damping Torque Analysis of Shunt VSC-Based Stabilizer Installed in Single-Machine Infinite-Bus Power System

### 4.1.2.1 Damping Torque Contributed by the Shunt VSC-Based Stabilizer [2, 3]

A damping control function can be superimposed on the modulation ratio m of the PWM algorithm. This forms a VSC-based stabilizer which functions through regulating the exchange of reactive power between the VSC-based unit and rest of the power system. In this book, it is named as the VSC-based reactive power stabilizer for the convenience of discussion. If it is assumed that the feedback signal and transfer function of the stabilizer are y and $T_{vsc}(s)$, respectively, and only the stabilizer's function is considered, it can have $\Delta m = T_{vsc}(s)\Delta y$. At the angular oscillation frequency of the power system, $\omega_s$, stabilizing signal of the VSC-based reactive power stabilizer $\Delta m = T_{vsc}(j\omega_s)\Delta y$ can always be decomposed into a synchronizing and damping torque, that is

$$\Delta m = T_{vsc}(j\omega_s)\Delta y = K_{vsc-d}\Delta\omega + K_{vsc-s}\Delta\delta \tag{4.20}$$

For the simplicity of discussion and presentation, it is assumed that the feedback signal of the stabilizer is the rotor speed of generator and the stabilizer adopts a proportional control law, it has

$$\Delta m = K_{rs}\Delta\omega \tag{4.21}$$

From Eq. (4.20), it can be seen that the above assumption will not change the following discussion and conclusions about the effectiveness of stabilizer.

From Eq. (4.21), Figs. 4.4, and 4.5, the direct damping torque contribution from the VSC-based reactive power stabilizer can be obtained as

$$\Delta T_{rs-d} = \left(K_{pm} - K_{pdc}K_{dm}\frac{K_9}{\omega_s^2 + K_9^2}\right)K_{rs}\Delta\omega \tag{4.22}$$

If the shunt VSC-based unit (such as an energy storage system or a renewable power generation plant) can exchange active power with the rest of power system, a damping control function can also be superimposed on the modulation phase of PWM algorithm, $\phi$. This VSC-based stabilizer functions by controlling the exchange of active power between the VSC-based unit and rest of the power system. In this book, it is named as the VSC-based active power stabilizer. Again, it can be assumed that the feedback signal of stabilizer is the rotor speed of generator and the stabilizer is a pure-gain controller. That is

$$\Delta\phi = K_{as}\Delta\omega \tag{4.23}$$

From Eq. (4.23), Figs. 4.4, and 4.5, the direct damping torque contribution from the VSC-based active power stabilizer can be obtained as

$$\Delta T_{as-d} = \left(K_{p\psi} - K_{pdc}K_{d\psi}\frac{K_9}{\omega_s^2 + K_9^2}\right)K_{as}\Delta\omega \tag{4.24}$$

In the following section, difference between the damping effectiveness of the VSC-based reactive and active power stabilizers is examined using Eqs. (4.22) and (4.24).

#### 4.1.2.2 Difference of Damping Control Effectiveness Between the VSC-Based Reactive and Active Power Stabilizers

From Fig. 4.2, it can have

$$\overline{V}_S = \frac{jx_{sb}}{1+\frac{x_{sb}}{x_s}}\overline{I}_{ts} + \frac{x_{sb}}{x_s(1+\frac{x_{sb}}{x_s})}\overline{V}_c + \frac{\overline{V}_b}{1+\frac{x_{sb}}{x_s}} \tag{4.25}$$

That gives

$$\begin{aligned}\overline{V}_t &= jx_{ts}\overline{I}_{ts} + \overline{V}_s \\ &= j\left(x_{ts} + \frac{x_s x_{sb}}{x_s + x_{sb}}\right)\overline{I}_{ts} + \frac{x_{sb}}{x_s + x_{sb}}\overline{V}_c + \frac{x_s}{x_s + x_{sb}}\overline{V}_b = jx\overline{I}_{ts} + \overline{V}_a\end{aligned} \tag{4.26}$$

where

$$x = \left(x_{ts} + \frac{x_s x_{sb}}{x_s + x_{sb}}\right)$$

$$\overline{V}_a = \frac{x_{sb}}{x_s + x_{sb}}\overline{V}_c + \frac{x_s}{x_s + x_{sb}}\overline{V}_b = a\overline{V}_c + b\overline{V}_b$$

Hence, from Eq. (4.26), the active power supplied by the generator can be expressed as

$$P_t = \frac{E_q' V_a}{x_{d\Sigma}'}\sin\delta' - \frac{V_a^2}{2}\frac{(x_q - x_d')}{x_{d\Sigma}' x_{q\Sigma}}\sin 2\delta' \tag{4.27}$$

where $\delta'$ is the angle between $E_q'$ and $\overline{V}_a$ and $x_{d\Sigma}' = x + x_d'$, $x_{q\Sigma} = x + x_q$. From the phasor diagram of Fig. 4.6, it can have

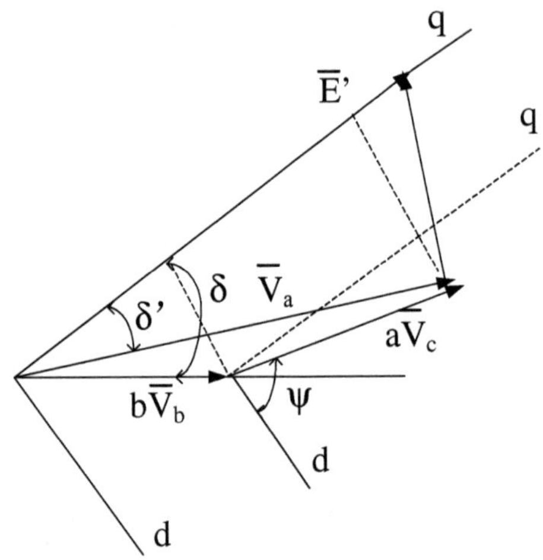


**Fig. 4.6** Phasor diagram for Eq. (4.27)

$$V_a \sin\delta' = bV_b \sin\delta + aV_c \cos\psi$$
$$V_a \cos\delta' = bV_b \cos\delta + aV_c \sin\psi \tag{4.28}$$

Thus, from Eqs. (4.27) and (4.28), the active power supplied by the generator can be expressed alternatively to be

$$\begin{aligned} P_t &= \frac{E'_q}{x'_{d\Sigma}}(bV_b \sin\delta + aV_c \cos\psi) \\ &\quad - \frac{(x_q - x'_d)}{x'_{d\Sigma}x_{q\Sigma}}(bV_b \sin\delta + aV_c \cos\psi)(bV_b \cos\delta + aV_c \sin\psi) \\ &\approx \frac{E'_q}{x'_{d\Sigma}}(bV_b \sin\delta + aV_c \cos\psi) \end{aligned} \tag{4.29}$$

where $V_c = mkV_{dc}$. Equation (4.29) is the explicit expression of the active power supplied by the generator in the single-machine infinite-bus power system installed with the shunt VSC-based unit. By using Eqs. (4.16) and (4.17), from Eq. (4.29), it can be obtained that

$$\begin{aligned} K_{pm} &= a_4 \frac{\partial P_t}{\partial m}\bigg|_0 = a_4 akV_{dc0} \cos\psi_0 \\ K_{p\psi} &= a_4 \frac{\partial P_t}{\partial \psi}\bigg|_0 = -a_4 akm_0 V_{dc0} \sin\psi_0 \\ K_{pdc} &= a_4 \frac{\partial P_t}{\partial V_{dc}}\bigg|_0 = a_4 akm_0 \cos\psi_0 \end{aligned} \tag{4.30}$$

From Eqs. (4.22) and (4.30), the damping torque contributed by the VSC-based reactive power stabilizer is obtained as

$$\Delta T_{rs-d} = a_4 ak \cos\psi_0 \left( V_{dc0} - m_0 K_{dm} \frac{K_9}{\omega_s^2 + K_9^2} \right) K_{rs} \Delta\omega \tag{4.31}$$

From Eqs. (4.24) and (4.30), the damping torque contributed by the VSC-based active power stabilizer is derived as

$$\Delta T_{as-d} = -a_4 akm_0 \left( V_{dc0} \sin\psi_0 + \cos\psi_0 K_{d\psi} \frac{K_9}{\omega_s^2 + K_9^2} \right) K_{as} \Delta\omega \tag{4.32}$$

From Fig. 4.3, it can be seen that $\delta_0$ increases when more active power is supplied by the generator. Hence, at a heavier load condition, $\psi_0$ is smaller, and more damping torque is provided by the VSC-based reactive power stabilizer as indicated by Eq. (4.31). This means that the VSC-based reactive power stabilizer is more effective in damping power oscillation when the single-machine infinite-bus power system operates at the heavier load condition. However, with the increase of load condition, $\cos\psi_0$ increases, but $\sin\psi_0$ decreases in Eq. (4.32). This means that with variations of power system load conditions, the damping torque provided by the VSC-based active power stabilizer changes less than that provided by the reactive power stabilizer. Hence, the VSC-based active power stabilizer is more robust to the variations of power system load conditions in damping power oscillations.

## 4.2 Damping Function of a Stabilizer Added on a Static Synchronous Series Compensator (SSSC) Installed in a Single-Machine Infinite-Bus Power System

### *4.2.1 Damping Torque Analysis of a SSSC Stabilizer Installed in a Single-Machine Infinite-Bus Power System*

#### 4.2.1.1 A Single-Machine Infinite-Bus Power System Installed with a Static Synchronous Series Compensator

A static synchronous series compensator (SSSC) is a "series STATCOM", a solid-state voltage source inverter generating a controllable AC voltage source, and connected in series to power transmission lines in a power system. The injected voltage is in quadrature with the line current and emulates an inductive or capacitive reactance so as to influence the power flow along the transmission lines. With the capability to change its reactance characteristic from capacitive to inductive, the SSSC is very effective in controlling power flow in the power system. An auxiliary stabilizing signal can be superimposed on the power flow control function of the SSSC so as to improve the damping of power system oscillations.

Figure 4.7 shows a single-machine infinite-bus power system installed with a SSSC, which consists of a series coupling transformer with a leakage reactance, $x_{sct}$, and a VSC connected to a DC capacitor. The exchange of reactive power between the SSSC and rest of the power system is achieved by controlling the magnitude and phase of the inserted voltage which is kept in quadrature with the line current (the inverter losses are ignored). Phasor diagrams in Fig. 4.8 show the basic operation principle of the SSSC, from which it can be seen that the compensation level can be controlled dynamically by regulating the magnitude of $\overline{V}_{inv}$. Mathematical description of the SSSC is

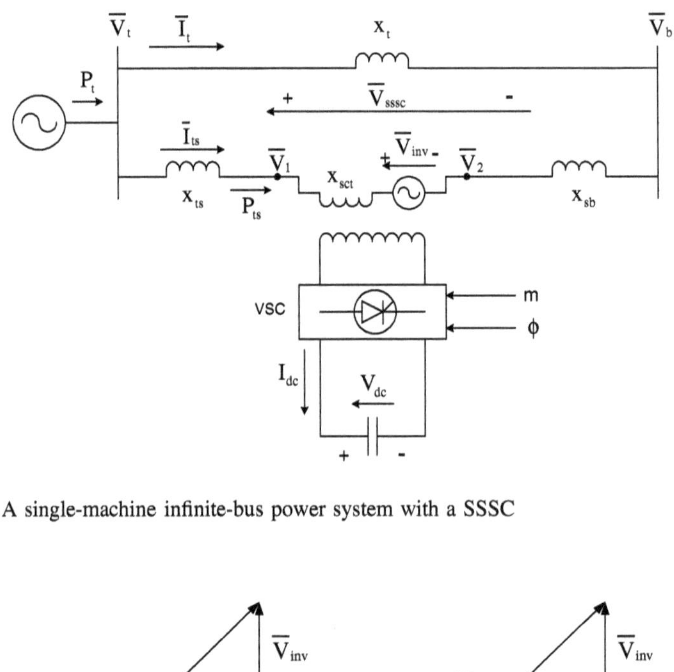


**Fig. 4.7** A single-machine infinite-bus power system with a SSSC

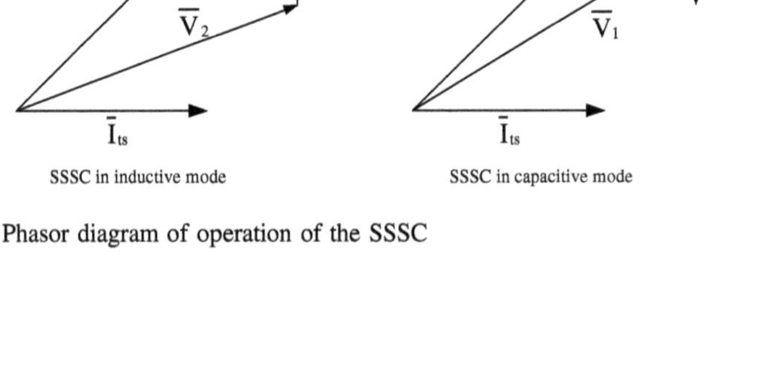


**Fig. 4.8** Phasor diagram of operation of the SSSC

$$
\begin{aligned}
&\bar{I}_{ts} = i_{tsd} + j i_{tsq} = I_{ts}\angle\varphi \\
&\bar{V}_{inv} = mkV_{dc}(\cos\phi + j\sin\phi) = mkV_{dc}\angle\phi, \quad \varphi + \phi = \pm 90^\circ \\
&\frac{dV_{dc}}{dt} = \frac{I_{dc}}{C_{dc}} = \frac{mk}{C_{dc}}(i_{tsd}\cos\phi + i_{tsq}\sin\phi)
\end{aligned} \tag{4.33}
$$

where k is the fixed ratio between the inverter's AC and DC voltages and m is the modulation ratio of the PWM algorithm implemented by the series VSC.

According to Fig. 4.8, a variable of equivalent reactance, $x_{dc}$, can be introduced to represent the AC voltage, $\bar{V}_{inv}$, to be

$$\bar{V}_{inv} = jx_{dc}\bar{I}_{ts} \tag{4.34}$$

From the circuit equation of power system of Fig. 4.7, it can have

$$\bar{V}_t - \bar{V}_b = j(x_{ts} + x_{sb} + x_{sct})\bar{I}_{ts} + \bar{V}_{inv} = jx_t\bar{I}_t \tag{4.35}$$

In the d–q coordinate, the above equation becomes

$$
\begin{aligned}
&-(x_{ts} + x_{sb} + x_{sct})i_{tsq} + j(x_{ts} + x_{sb} + x_{sct})i_{tsd} - x_{dc}i_{tsq} + jx_{dc}i_{tsd} \\
&\quad = -x_t i_{tq} + jx_t i_{td}
\end{aligned} \tag{4.36}
$$

Hence,

$$
\begin{aligned}
i_{tq} &= \frac{x_{ts} + x_{sb} + x_{sct} + x_{dc}}{x_t} i_{tsq} \\
i_{td} &= \frac{x_{ts} + x_{sb} + x_{sct} + x_{dc}}{x_t} i_{tsd}
\end{aligned} \tag{4.37}
$$

Considering the voltage equation of generator of Eq. (4.35), it can have

$$
\begin{aligned}
&\bar{V}_t - j(x_{ts} + x_{sb} + x_{sct})\bar{I}_{ts} = x_q(i_{tsq} + i_{tq}) + j[E'_q - x'_d(i_{tsd} + i_{td})] - j(x_{ts} + x_{sb} + x_{sct})\bar{I}_{ts} \\
&= x_q\left(1 + \frac{x_{ts} + x_{sb} + x_{sct} + x_{dc}}{x_t}\right)i_{tsq} + j\left[E'_q - x'_d\left(1 + \frac{x_{ts} + x_{sb} + x_{sct} + x_{dc}}{x_t}\right)i_{tsd}\right] \\
&\quad + (x_{ts} + x_{sb} + x_{sct})i_{tsq} - j(x_{ts} + x_{sb} + x_{sct})i_{tsd} \\
&= \bar{V}_{inv} + \bar{V}_b = -x_{dc}i_{tsq} + jx_{dc}i_{tsd} + V_b\sin\delta + jV_b\cos\delta
\end{aligned} \tag{4.38}
$$

From the above equation, it can be obtained that

$$
\begin{aligned}
i_{tsq} &= \frac{V_b\sin\delta}{x_{ts} + x_{sb} + x_{sct} + c_s x_q + x_{dc}} \\
i_{tsd} &= \frac{E'_q - V_b\cos\delta}{x_{ts} + x_{sb} + x_{sct} + c_s x'_d + x_{dc}}
\end{aligned} \tag{4.39}
$$

where

$$c_s = 1 + \frac{x_{ts} + x_{sb} + x_{sct} + x_{dc}}{x_t}$$

Thus,

$$\begin{aligned} P_t &= E'_q(i_{tq} + i_{tsq}) + (x_q - x'_d)(i_{td} + i_{tsd})(i_{tq} + i_{tsq}) \\ &= E'_q c_s i_{tsq} + (x_q - x'_d)c_s^2 i_{tsd} i_{tsq} \\ E_q &= E'_q + (x_d - x'_d)(i_{td} + i_{tsd}) = E'_q + (x_d - x'_d)c_s i_{tsd} \\ V_t &= \sqrt{[E'_q - x'_d(i_{td} + i_{tsd})]^2 + [x_q(i_{tq} + i_{tsq})]^2} \\ &= \sqrt{(E'_q - x'_d c_s i_{tsd})^2 + (x_q c_s i_{tsq})^2} \end{aligned} \tag{4.40}$$

From Eqs. (4.33) and (4.34), it can have

$$\overline{V}_{inv} = jx_{dc}\overline{I}_{ts} = -x_{dc}i_{tsq} + jx_{dc}i_{tsd} = mkV_{dc}(\cos\phi + j\sin\phi) \tag{4.41}$$

The above equation gives

$$\begin{aligned} \cos\phi &= \frac{-x_{dc}i_{tsq}}{mkV_{dc}} \\ \sin\phi &= \frac{x_{dc}i_{tsd}}{mkV_{dc}} \end{aligned} \tag{4.42}$$

By using Eq. (4.42), dynamic equation in Eq. (4.33) is converted to the following equivalent form

$$\frac{dV_{dc}}{dt} = 0 \tag{4.43}$$

Equation (4.43) shows that indeed, there is no exchange of active power between the SSSC and rest of the power system.

#### 4.2.1.2 Extended Heffron–Phillips Model of the Single-Machine Infinite-Bus Power System Installed with the SSSC and Damping Torque Analysis [4]

Linearization of Eq. (4.40) is

$$\begin{aligned} \Delta P_t &= K_1\Delta\delta + K_2\Delta E'_q + K_{px}\Delta x_{dc} \\ \Delta E_q &= K_4\Delta\delta + K_3\Delta E'_q + K_{qx}\Delta x_{dc} \\ \Delta V_t &= K_5\Delta\delta + K_6\Delta E'_q + K_{vx}\Delta x_{dc} \end{aligned} \tag{4.44}$$

From Eq. (4.41), it can have

$$x_{dc}I_{ts} = mkV_{dc} \tag{4.45}$$

Hence,

$$I_{ts0}\Delta x_{dc} + x_{dc0}\Delta I_{ts} = kV_{dc0}\Delta m \tag{4.46}$$

Because only the damping control function will be considered, it can be assumed that $x_{dc0} = 0$ as far as the damping control function at the steady-state operation is concerned. Hence, Eq. (4.44) becomes

$$\begin{aligned} \Delta P_t &= K_1\Delta\delta + K_2\Delta E'_q + K_{pm}\Delta m \\ \Delta E_q &= K_4\Delta\delta + K_3\Delta E'_q + K_{qm}\Delta m \\ \Delta V_t &= K_5\Delta\delta + K_6\Delta E'_q + K_{vm}\Delta m \end{aligned} \tag{4.47}$$

where

$$\begin{aligned} K_{pm} &= \frac{K_{px}kV_{dc0}}{I_{ts0}} \\ K_{qm} &= \frac{K_{qx}kV_{dc0}}{I_{ts0}} \\ K_{vm} &= \frac{K_{vx}kV_{dc0}}{I_{ts0}} \end{aligned}$$

By substituting Eq. (4.47) into Eq. (2.39) (without the PSS being considered), the extended Heffron–Phillips model of the single-machine infinite-bus power system with the SSSC installed can be obtained as

$$\begin{aligned} \Delta\dot{\delta} &= \omega_o\Delta\omega \\ \Delta\dot{\omega} &= \frac{1}{M}[-K_1\Delta\delta - D\Delta\omega - K_2\Delta E'_q - K_{pm}\Delta m] \\ \Delta\dot{E}'_q &= \frac{1}{T'_{d0}}(-K_4\Delta\delta - K_3\Delta E'_q + \Delta E'_{fd} - K_{qm}\Delta m) \\ \Delta\dot{E}'_{fd} &= -\frac{1}{T_A}\Delta E'_{fd} - \frac{K_A}{T_A}(K_5\Delta\delta + K_6\Delta E'_q + K_{vm}\Delta m) \end{aligned} \tag{4.48}$$

The model is shown in Fig. 4.9.

Active power supplied by the generator can be expressed as

$$\begin{aligned} P_t &= E'_q(i_{tsq} + i_{tq}) + (x_q - x'_d)(i_{tsd} + i_{td})(i_{tsq} + i_{tq}) \\ &= E'_q c_s i_{tsq} + (x_q - x'_d)c_s^2 i_{tsd} i_{tsq} \end{aligned} \tag{4.49}$$

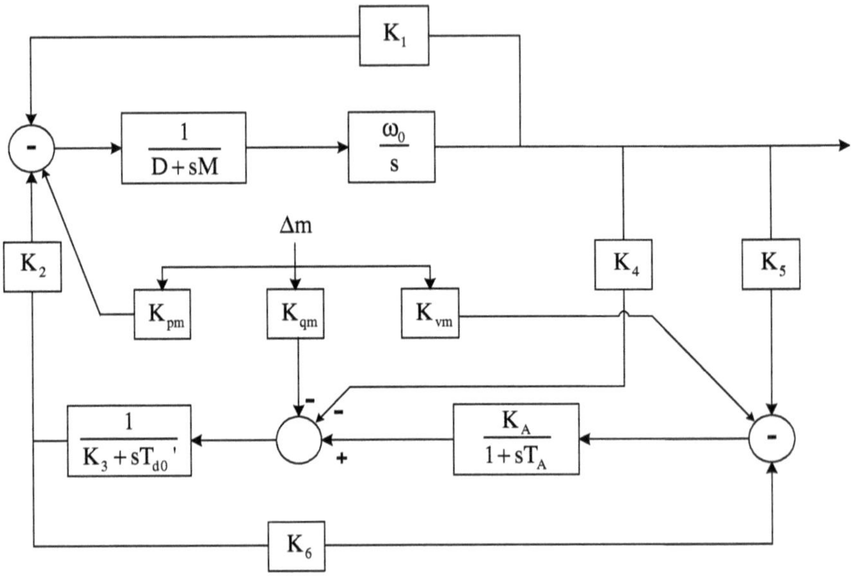


**Fig. 4.9** Extended Heffron–Phillips model of single-machine infinite-bus power system installed with a SSSC stabilizer

By using Eq. (4.39), it can have

$$P_t = \frac{c_s E_q' V_b}{x_{d\Sigma}'} \sin\delta - \frac{c_s^2 V_b^2}{2} \frac{(x_q - x_d')}{x_{d\Sigma}' x_{q\Sigma}} \sin 2\delta \tag{4.50}$$

where

$$\begin{aligned} x_{d\Sigma}' &= c_s x_d' + x_{ts} + x_{sb} + x_{sct} + x_{dc} \\ x_{q\Sigma} &= c_s x_q + x_{ts} + x_{sb} + x_{sct} + x_{dc} \end{aligned}$$

From Eq. (4.50), it can be defined that

$$\begin{aligned} P_{t0} &= \frac{c_s E_{q0}' V_b}{x_{d\Sigma}'} \sin\delta_0 - \frac{c_s^2 V_b^2}{2} \frac{(x_q - x_d')}{x_{d\Sigma}' x_{q\Sigma}} \sin 2\delta_0 = P_{t10} - P_{t20} \\ P_{t10} &= \frac{c_s E_{q0}' V_b}{x_{d\Sigma}'} \sin\delta_0 \\ P_{t20} &= \frac{c_s^2 V_b^2}{2} \frac{(x_q - x_d')}{x_{d\Sigma}' x_{q\Sigma}} \sin 2\delta_0 \end{aligned} \tag{4.51}$$

According to Eqs. (4.41) and (4.51), from Eq. (4.50), it can have

$$
\begin{aligned}
K_{px} = \left.\frac{\partial P_t}{\partial x_{dc}}\right|_0 &= \frac{1}{x_t c_s} P_{t10} - \frac{\left(\frac{X'_d}{X_t}+1\right)}{x'_{d\Sigma}} P_{t10} - \frac{2}{x_t c_s} P_{t20} + \frac{\left(\frac{X_q}{X_t}+1\right)}{x_{q\Sigma}} P_{t20} + \frac{\left(\frac{X'_d}{X_t}+1\right)}{x'_{d\Sigma}} P_{t20} \\
&= \frac{1}{x_t c_s}(P_{t10} - P_{t20}) - \frac{1}{x_t c_s} P_{t20} - \frac{\left(\frac{X'_d}{X_L}+1\right)}{x'_{d\Sigma}}(P_{t10} - P_{t20}) + \frac{\left(\frac{X_q}{X_L}+1\right)}{x_{q\Sigma}} P_{t20} \\
&= \left(\frac{1}{x_t c_s} - \frac{x'_d + x_t}{x_t x'_{d\Sigma}}\right) P_{t0} - \left(\frac{1}{x_t c_s} - \frac{x_q + x_t}{x_t x_{q\Sigma}}\right) P_{t20} \\
&= \frac{x'_{d\Sigma} - c_s x'_d - c_s x_t}{x_t c_s x'_{d\Sigma}} P_{t0} - \frac{x_{q\Sigma} - c_s x_q - x_t c_s}{x_t c_s x_{q\Sigma}} P_{t20} \\
&= \frac{-1}{c_s x'_{d\Sigma}} P_{t0} + \frac{1}{c_s x_{q\Sigma}} P_{t20}
\end{aligned}
\tag{4.52}
$$

From Eqs. (4.47) and (4.52), it can be obtained that

$$
K_{pm} = -\frac{kV_{dc0}}{c_s I_{ts0}}\left(\frac{1}{x'_{d\Sigma}} P_{t0} - \frac{1}{x_{q\Sigma}} P_{t20}\right) \tag{4.53}
$$

Taking the similar procedure of analysis to that from Eqs. (3.58) to (3.59), from Eq. (4.53), it can be concluded that

1. When the load conditions change, the SSSC stabilizer can always provide positive damping to power system oscillations if it is designed properly;
2. The heavier the load condition of the power system is, the more effective the SSSC stabilizer is;
3. When the load condition is unchanged, the SSSC stabilizer is less effective when the equivalent reactance of transmission line is bigger.

### 4.2.2 *Design of a SSSC Stabilizer*

Using a SSSC stabilizer as an example, this section demonstrates how a thyristor-based or VSC-based stabilizer can be designed effectively to suppress power system oscillations in a single-machine infinite-bus power system using the phase compensation method.

From Fig. 4.9, the forward path of the SSSC stabilizer is shown in Fig. 4.10. From Fig. 4.10, the transfer function of the forward path can be obtained as

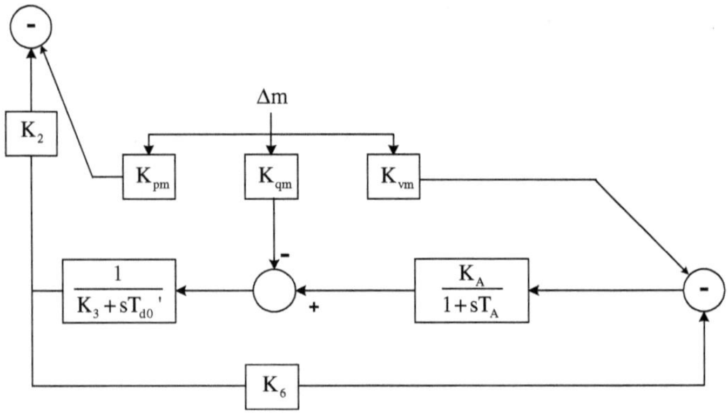


**Fig. 4.10** Forward path of stabilizing signal of SSSC stabilizer installed in a single-machine infinite-bus power system

$$
\begin{aligned}
F_{sssc}(s) &= K_{pm} - K_2K_{qm}\frac{\frac{1}{K_3+sT'_{d0}}}{1+K_6\frac{1}{K_3+sT'_{d0}}\frac{K_A}{1+sT_A}} \\
&\quad - K_2K_{vm}\frac{\frac{1}{K_3+sT'_{d0}}\frac{K_A}{1+sT_A}}{1+K_6\frac{1}{K_3+sT'_{d0}}\frac{K_A}{1+sT_A}} \\
&= K_{pm} - \frac{K_2[K_{qm}(1+sT_A)+K_{vm}K_A]}{(K_3+sT'_{d0})(1+sT_A)+K_6K_A}
\end{aligned}
\tag{4.54}
$$

Consider the case that the feedback signal and transfer function of the SSSC stabilizer are $\Delta\omega$ and $T_{sssc}(s)$, respectively, that is

$$
\Delta m = T_{sssc}(s)\Delta\omega \tag{4.55}
$$

If the amount of damping torque contribution required from the SSSC stabilizer is $D_{sssc}\Delta\omega$, the transfer function of SSSC stabilizer should be set by using the phase compensation method at the angular oscillation frequency, $\omega_s$, to satisfy

$$
F_{sssc}(j\omega_s)T_{sssc}(j\omega_s) = D_{sssc} \tag{4.56}
$$

In a more general case, the state-space representation of linearized model of the single-machine infinite-bus power system with the SSSC stabilizer installed can always be arranged in the following form:

$$
\begin{aligned}
\begin{bmatrix} \Delta\dot{\delta} \\ \Delta\dot{\omega} \\ \dot{\mathbf{x}} \end{bmatrix} &= \begin{bmatrix} 0 & \omega_0 & \mathbf{0} \\ -k & -d & -\mathbf{A}_{23} \\ \mathbf{A}_{31} & \mathbf{A}_{32} & \mathbf{A}_{33} \end{bmatrix} \begin{bmatrix} \Delta\delta \\ \Delta\omega \\ \mathbf{x} \end{bmatrix} + \begin{bmatrix} 0 \\ -B_2 \\ \mathbf{B}_3 \end{bmatrix} \Delta m \\
\Delta y &= [C_1 \quad C_2 \quad \mathbf{C}_3] \begin{bmatrix} \Delta\delta \\ \Delta\omega \\ \mathbf{x} \end{bmatrix} + d_m\Delta m
\end{aligned}
\tag{4.57}
$$

Equation (4.57) can be written as

$$\begin{aligned}\Delta\dot{\delta} &= \omega_0\Delta\omega \\ \Delta\dot{\omega} &= -k\Delta\delta - d\Delta\omega - \mathbf{A}_{23}\mathbf{x} - B_2\Delta m \\ \dot{\mathbf{x}} &= \mathbf{A}_{31}\Delta\delta + \mathbf{A}_{32}\Delta\omega + \mathbf{A}_{33}\mathbf{x} + \mathbf{B}_3\Delta m \\ \Delta y &= C_1\Delta\delta + C_2\Delta\omega + \mathbf{C}_3^T\mathbf{x} + d_m\Delta m\end{aligned} \tag{4.58}$$

From Eq. (4.58), it can have

$$\begin{aligned}s\Delta\delta &= \omega_0\Delta\omega \\ s\Delta\omega &= -k\Delta\delta - d\Delta\omega - \mathbf{A}_{23}\mathbf{x} - B_2\Delta m \\ s\mathbf{x} &= \mathbf{A}_{31}\Delta\delta + \mathbf{A}_{32}\Delta\omega + \mathbf{A}_{33}\mathbf{x} + \mathbf{B}_3\Delta m \\ \Delta y &= C_1\Delta\delta + C_2\Delta\omega + \mathbf{C}_3^T\mathbf{x} + d_m\Delta m\end{aligned} \tag{4.59}$$

Hence,

$$\mathbf{x} = (s\mathbf{I} - \mathbf{A}_{33})^{-1}\left(\mathbf{A}_{31} + \frac{s}{\omega_0}\mathbf{A}_{32}\right)\Delta\delta + (s\mathbf{I} - \mathbf{A}_{33})^{-1}\mathbf{B}_3\Delta m \tag{4.60}$$

From Eqs. (4.59) and (4.60), it can be obtained that

$$\begin{aligned}s\Delta\delta &= \omega_0\Delta\omega \\ s\Delta\omega &= -\left[k + d\frac{s}{\omega_0} + \mathbf{A}_{23}(s\mathbf{I} - \mathbf{A}_{33})^{-1}\left(\mathbf{A}_{31} + \mathbf{A}_{32}\frac{s}{\omega_0}\right)\right]\Delta\delta \\ &\quad - [\mathbf{A}_{23}(s\mathbf{I} - \mathbf{A}_{33})^{-1}\mathbf{B}_3 + B_2]\Delta m = -K(s)\Delta\delta - K_c(s)\Delta m \\ \Delta y &= C_1\Delta\delta + C_2\Delta\omega + \mathbf{C}_3^T\left[(s\mathbf{I} - \mathbf{A}_{33})^{-1}\left(\mathbf{A}_{31} + \frac{s}{\omega_0}\mathbf{A}_{32}\right)\Delta\delta + (s\mathbf{I} - \mathbf{A}_{33})^{-1}\mathbf{B}_3\Delta m\right] \\ &= \left[C_1\frac{\omega_0}{s} + C_2 + \mathbf{C}_3^T(s\mathbf{I} - \mathbf{A}_{33})^{-1}\left(\frac{\omega_0}{s}\mathbf{A}_{31} + \mathbf{A}_{32}\right)\right]\Delta\omega + [d_m + \mathbf{C}_3^T(s\mathbf{I} - \mathbf{A}_{33})^{-1}\mathbf{B}_3]\Delta m \\ &= K_o(s)\Delta\omega + K_{il}(s)\Delta m\end{aligned} \tag{4.61}$$

where

$$\begin{aligned}K(s) &= k + \frac{s}{\omega_0}d + \mathbf{A}_{23}^T(s\mathbf{I} - \mathbf{A}_{33})^{-1}\left(\mathbf{A}_{31} + \frac{s}{\omega_0}\mathbf{A}_{32}\right) \\ K_o(s) &= \left(\frac{\omega_0}{s}C_1 + C_2\right) + \mathbf{C}_3^T(s\mathbf{I} - \mathbf{A}_{33})^{-1}\left(\frac{\omega_0}{s}\mathbf{A}_{31} + \mathbf{A}_{32}\right) \\ K_c(s) &= \mathbf{A}_{23}^T(s\mathbf{I} - \mathbf{A}_{33})^{-1}\mathbf{B}_3 + B_2 \\ K_{il}(s) &= \mathbf{C}_3^T(s\mathbf{I} - \mathbf{A}_{33})^{-1}\mathbf{B}_3 + d_m\end{aligned}$$

Hence, according to Eq. (4.61), the system is shown in Fig. 4.11.

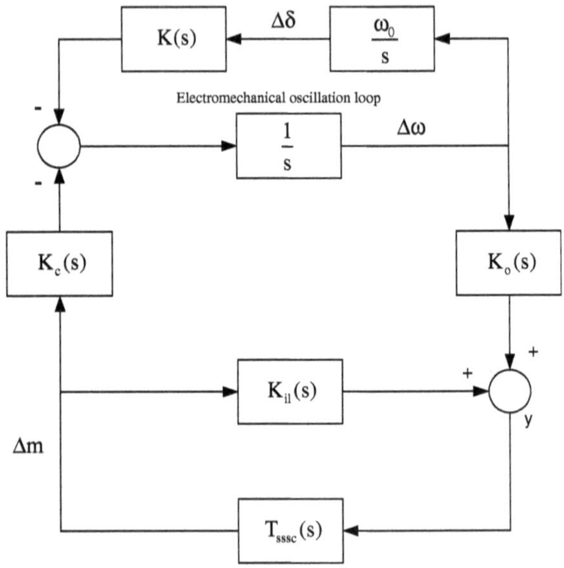


**Fig. 4.11** Closed-loop system installed with the SSSC stabilizer

From Fig. 4.11, it can be seen that with the SSSC stabilizer installed, the electric torque provided by the SSSC stabilizer to the electromechanical oscillation loop of generator is

$$\Delta T_{et} = \frac{K_c(j\omega_s)K_o(j\omega_s)T_{sssc}(j\omega_s)}{1 - K_{il}(j\omega_s)T_{sssc}(j\omega_s)}\Delta\omega \tag{4.62}$$

where $\omega_s$ is the angular frequency of power system oscillation. An ideal SSSC stabilizer should contribute a pure positive damping torque to the electromechanical oscillation loop, $\Delta T_{dt} = D_{sssc}\Delta\omega$, that is

$$\frac{K_c(j\omega_s)K_o(j\omega_s)T_{sssc}(j\omega_s)}{1 - K_{il}(j\omega_s)T_{sssc}(j\omega_s)} = D_{sssc} \tag{4.63}$$

Hence,

$$\begin{aligned} D_{sssc} &= [K_c(j\omega_s)K_o(j\omega_s) + D_{sssc}K_{il}(j\omega_s)]T_{sssc}(j\omega_s) \\ &= F_{sssc}(j\omega_s)T_{sssc}(j\omega_s) \end{aligned} \tag{4.64}$$

According to Eq. (4.64), the phase compensation method can be used to design the SSSC stabilizer by setting

$$\theta = -\varphi, \quad T_{sssc} = \frac{D_{sssc}}{F_{sssc}} \tag{4.65}$$

where $T_{sssc}(j\omega_s) = T_{sssc}\angle\theta F_{sssc}(j\omega_s) = F_{sssc}\angle\varphi$.

## 4.3 Damping Function of a Unified Power Flow Controller (UPFC) Installed in a Single-Machine Infinite-Bus Power System

### 4.3.1 *Mathematical Model of a Single-Machine Infinite-Bus Power System Installed with a UPFC*

#### 4.3.1.1 Dynamic Model of a UPFC

Figure 4.12 shows the configuration of a single-machine infinite-bus power system installed with a unified power flow controller (UPFC). The UPFC consists of an excitation transformer (ET), a boosting transformer (BT), two three-phase GTO-based voltage source converters (VSCs), and a DC link capacitor. In Fig. 4.12, $m_e, m_b$ and $\delta_e, \delta_b$ are the amplitude modulation ratio and phase angle of the pulse width modulation (PWM) implemented by the shunt and series VSC, respectively[5].

In Fig. 4.12, the shunt and series synchronous voltage source (SVS) can be denoted as

$$\begin{aligned} \overline{V}_{es} &= m_e k_e V_{dc} \angle \psi_{es} \\ \overline{V}_{bs} &= m_b k_b V_{dc} \angle \psi_{bs} \end{aligned} \tag{4.66}$$

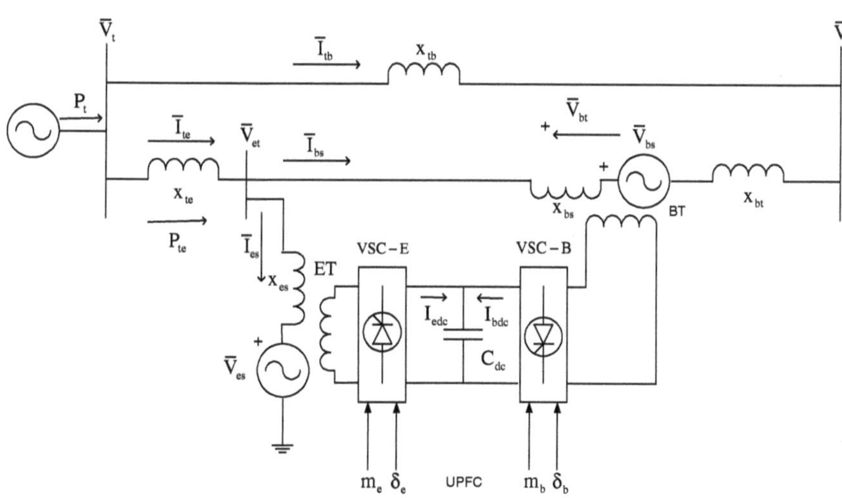


**Fig. 4.12** A single-machine infinite-bus power system installed with a UPFC

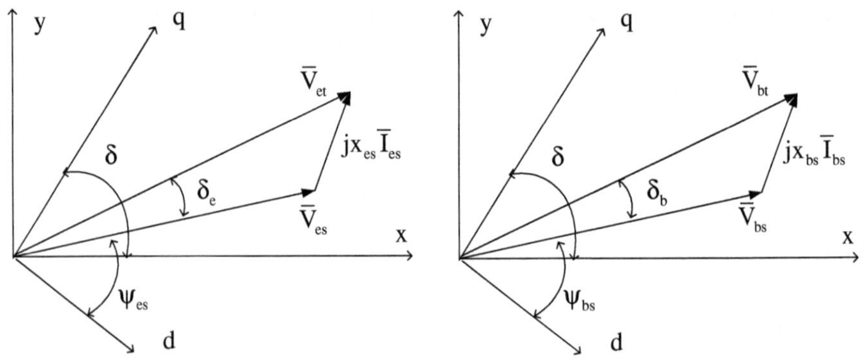


**Fig. 4.13** Phase diagram of shunt and series SVS

where $\psi_{es}$ and $\psi_{bs}$ are the angle between d-axis of d–q coordinate of the synchronous generator and voltage $\overline{V}_{es}$ and $\overline{V}_{bs}$, respectively, as shown in Fig. 4.13, which gives

$$\begin{aligned}\overline{V}_{et} &= jx_{es}\overline{I}_{es} + \overline{V}_{es}\\ \overline{V}_{bt} &= jx_{bs}\overline{I}_{bs} + \overline{V}_{bs}\end{aligned} \tag{4.67}$$

The active power injected into the DC side of the UPFC from the shunt and series transformer is the active power "consumed" by the shunt and series SVS, respectively, $P_{es} = \mathrm{Re}(\overline{V}_{es}\overline{I}^*_{es})$ and $P_{bs} = \mathrm{Re}(\overline{V}_{bs}\overline{I}^*_{bs})$, where Re() and * denote the real part and conjugate of a complex variable, respectively. From Eqs. (4.66) and (4.67), it can be obtained that

$$\begin{aligned}P_{es} &= \frac{V_{es}V_{et}}{x_{es}}\sin\delta_e = \frac{m_e k_e V_{dc} V_{et}}{x_e}\sin\delta_e\\ P_{bs} &= \frac{V_{bs}V_{bt}}{x_{bs}}\sin\delta_b = \frac{m_b k_b V_{dc} V_{bt}}{x_{bs}}\sin\delta_b\end{aligned} \tag{4.68}$$

The shunt and series active power injection should be equal to that "received" at the DC capacitor from the shunt and series VSC, respectively, that is

$$\begin{aligned}P_{es} &= \frac{m_e k_e V_{dc} V_{et}}{x_{es}}\sin\delta_e = I_{edc}V_{dc}\\ P_{bs} &= \frac{m_b k_b V_{dc} V_{bt}}{x_{bs}}\sin\delta_b = I_{bdc}V_{dc}\end{aligned} \tag{4.69}$$

From Eq. (4.69), it can have

$$
\begin{aligned}
I_{edc} &= \frac{m_e k_e V_{et}}{x_{es}} \sin \delta_e \\
I_{bdc} &= \frac{m_b k_b V_{bt}}{x_{bs}} \sin \delta_b
\end{aligned}
\tag{4.70}
$$

Alternatively, in the d–q coordinate as shown in Fig. 4.13,

$$
\begin{aligned}
P_{es} &= v_{esd} i_{esd} + v_{esq} i_{esq} = i_{esd} m_e k_e V_{dc} \cos \psi_{es} + i_{esq} m_e k_e V_{dc} \sin \psi_{es} = I_{edc} V_{dc} \\
P_{bs} &= v_{bsd} i_{bsd} + v_{bsq} i_{bsq} = i_{bsd} m_b k_b V_{dc} \cos \psi_{bs} + i_{bsq} m_b k_b V_{dc} \sin \psi_{bs} = I_{bdc} V_{dc}
\end{aligned}
\tag{4.71}
$$

From Eq. (4.71), it can have

$$
\begin{aligned}
I_{edc} &= i_{esd} m_e k_e \cos \psi_{es} + i_{esq} m_e k_e \sin \psi_{es} \\
I_{bdc} &= i_{bsd} m_b k_b \cos \psi_{bs} + i_{bsq} m_b k_b \sin \psi_{bs}
\end{aligned}
\tag{4.72}
$$

Hence, from Eq. (4.70), dynamic equation of the UPFC can be obtained as

$$
\dot{V}_{dc} = \frac{I_{edc} + I_{bdc}}{C_{dc}} = \frac{1}{C_{dc}} \left( \frac{m_e k_e V_{et}}{x_{es}} \sin \delta_e + \frac{m_b k_b V_{bt}}{x_{bs}} \sin \delta_b \right) \tag{4.73}
$$

Alternatively, from Eq. (4.72), it can have

$$
\dot{V}_{dc} = \frac{1}{C_{dc}} (i_{esd} m_e k_e \cos \psi_{es} + i_{esq} m_e k_e \sin \psi_{es} + i_{bsd} m_b k_b \cos \psi_{bs} + i_{bsq} m_b k_b \sin \psi_{bs}) \tag{4.74}
$$

where

$$
\begin{aligned}
\psi_{es} &= \tan^{-1} \frac{v_{etq}}{v_{etd}} - \delta_e \\
\psi_{bs} &= \tan^{-1} \frac{v_{btq}}{v_{btd}} - \delta_b
\end{aligned}
\tag{4.75}
$$

#### 4.3.1.2 Nonlinear Model of the Single-Machine Infinite-Bus Power System Installed with the UPFC

From Fig. 4.12, it can have

$$\begin{aligned}\overline{V}_t &= jx_{tb}\overline{I}_{tb} + \overline{V}_b \\ \overline{V}_t &= jx_{te}\overline{I}_{te} + \overline{V}_{et} \\ \overline{V}_{et} &= \overline{V}_{bt} + jx_{bt}\overline{I}_{bs} + \overline{V}_b\end{aligned} \tag{4.76}$$

which can be expressed in the d–q coordinate as

$$\begin{aligned}v_{td} + jv_{tq} &= jx_{tb}(i_{tbd} + ji_{tbq}) + V_b \sin\delta + jV_b \cos\delta \\ &= -x_{tb}i_{tbq} + V_b \sin\delta + jx_{tb}i_{tbd} + jV_b \cos\delta \\ v_{td} + jv_{tq} &= jx_{te}(i_{esd} + i_{bsd} + ji_{esq} + ji_{bsq}) + v_{etd} + jv_{etq} \\ v_{etd} + jv_{etq} &= v_{btd} + jv_{btq} + jx_{bt}i_{bsd} - x_{bt}i_{bsq} + V_b \sin\delta + jV_b \cos\delta\end{aligned} \tag{4.77}$$

From the first equation in Eq. (4.77), it can be obtained that

$$\begin{aligned}i_{tbd} &= \frac{v_{tq} - V_b \cos\delta}{x_{tb}} \\ i_{tbq} &= \frac{V_b \sin\delta - v_{td}}{x_{tb}}\end{aligned} \tag{4.78}$$

Because

$$\begin{aligned}v_{td} &= x_q(i_{teq} + i_{tbq}) \\ v_{tq} &= E'_q - x'_d(i_{ted} + i_{tbd})\end{aligned} \tag{4.79}$$

From Eqs. (4.78) and (4.79), it can have

$$\begin{aligned}v_{td} &= \frac{x_q x_{tb} i_{teq} + x_q V_b \sin\delta}{x_q + x_{tb}} \\ v_{tq} &= \frac{x_{tb}E'_q - x'_d x_{tb} i_{ted} + x'_d V_b \cos\delta}{x'_d + x_{tb}}\end{aligned} \tag{4.80}$$

The second and third equations in Eq. (4.77) give

$$
\begin{aligned}
v_{td} + jv_{tq} &= jx_{te}(i_{esd} + i_{bsd} + ji_{esq} + ji_{bsq}) - x_{es}i_{esq} + v_{esd} + jx_{es}i_{esd} + jv_{esq} \\
&\quad - x_{es}i_{esq} + v_{esd} + jx_{es}i_{esd} + jv_{esq} \\
&= -x_{bs}i_{bsq} + v_{bsd} + jx_{bs}i_{bsd} + jv_{bsq} + jx_{bt}i_{bsd} - x_{bt}i_{bsq} + V_b \sin\delta + jV_b \cos\delta
\end{aligned}
$$

That is

$$
\begin{aligned}
v_{td} &= -x_{te}(i_{esq} + i_{bsq}) - x_{es}i_{esq} + v_{esd} \\
v_{tq} &= x_{te}(i_{esd} + i_{bsd}) + x_{es}i_{esd} + v_{esq} \\
-x_{es}i_{esq} + v_{esd} &= -x_{bs}i_{bsq} + v_{bsd} - x_{bt}i_{bsq} + V_b \sin\delta \\
x_{es}i_{esd} + v_{esq} &= x_{bs}i_{bsd} + v_{bsq} + x_{bt}i_{bsd} + V_b \cos\delta
\end{aligned} \tag{4.81}
$$

From Eqs. (4.80) and (4.81), it can be obtained that

$$
\begin{aligned}
&[x_q x_{tb} + x_{te}(x_q + x_{tb}) + x_{es}(x_q + x_{tb})]i_{esq} \\
&\quad + [x_q x_{tb} + x_{te}(x_q + x_{tb})]i_{bsq} = (x_q + x_{tb})v_{esd} - x_q V_b \sin\delta \\
&[x_{te}(x'_d + x_{tb}) + x_{es}(x'_d + x_{tb}) + x'_d x_{te}]i_{esd} \\
&\quad + [x_{te}(x'_d + x_{tb}) + x'_d x_{tb}]i_{bsd} = x_{tb}E'_q + x'_d V_b \cos\delta - (x'_d + x_{tb})v_{esq} \\
&\quad - x_{es}i_{esq} + (x_{bs} + x_{bt})i_{bsq} = v_{bsd} - v_{esd} + V_b \sin\delta \\
&x_{es}i_{esd} - (x_{bs} + x_{bt})i_{bsd} = v_{bsq} - v_{esq} + V_b \cos\delta
\end{aligned}
$$

which in the matrix form is

$$
\begin{aligned}
\begin{bmatrix} i_{esq} \\ i_{bsq} \end{bmatrix} &= \begin{bmatrix} x_q x_{tb} + (x_{te} + x_{es})(x_q + x_{tb}) & x_q x_{tb} + x_{te}(x_q + x_{tb}) \\ -x_{es} & x_{es} + x_{bt} \end{bmatrix}^{-1} \\
&\quad \times \begin{bmatrix} (x_q + x_{tb})m_e k_e V_{dc} \cos\psi_{es} - x_q V_b \sin\delta \\ m_b k_b V_{dc} \cos\psi_{bs} - m_e k_e V_{dc} \cos\psi_{es} + V_b \sin\delta \end{bmatrix} \\
\begin{bmatrix} i_{esd} \\ i_{bsd} \end{bmatrix} &= \begin{bmatrix} (x_{te} + x_{es})(x'_d + x_{tb}) + x'_d x_{te} & x_{te}(x'_d + x_{tb}) + x'_d x_{tb} \\ x_{es} & -(x_{bs} + x_{bt}) \end{bmatrix}^{-1} \\
&\quad \times \begin{bmatrix} x_{tb}E'_q + x'_d V_b \cos\delta - (x'_d + x_{tb})m_e k_e V_{dc} \sin\psi_{es} \\ m_b k_b V_{dc} \sin\psi_{bs} - m_e k_e V_{dc} \sin\psi_{es} + V_b \cos\delta \end{bmatrix}
\end{aligned} \tag{4.82}
$$

For the single-machine infinite-bus power system with the UPFC installed, it can have

$$\begin{aligned} i_{ted} &= i_{esd} + i_{bsd} \\ i_{teq} &= i_{esq} + i_{bsq} \\ P_t &= v_{tq}(i_{teq} + i_{tbq}) + v_{td}(i_{ted} + i_{tbd}) \\ E_q &= E'_q + (x_d - x'_d)(i_{ted} + i_{tbd}) \\ V_t &= \sqrt{v_{td}^2 + v_{tq}^2} \end{aligned} \tag{4.83}$$

Equations (4.74), (4.75), (4.78), (4.80), (4.82), and (4.83), together with the differential equations of synchronous generator of Eq. (2.35), form the nonlinear mathematical model of the single-machine infinite-bus power system with the UPFC installed.

#### 4.3.1.3 Linearized Model of the Single-Machine Infinite-Bus Power System Installed with the UPFC

Linearization of Eq. (4.82) is

$$\begin{bmatrix} \Delta i_{esq} \\ \Delta i_{bsq} \end{bmatrix} = \mathbf{F}'_{\mathbf{q}} \begin{bmatrix} \Delta\delta \\ \Delta E'_q \\ \Delta V_{dc} \end{bmatrix} + \mathbf{G}'_{\mathbf{q}} \begin{bmatrix} \Delta m_e \\ \Delta\psi_{es} \\ \Delta m_b \\ \Delta\psi_{bs} \end{bmatrix}, \quad \begin{bmatrix} \Delta i_{esd} \\ \Delta i_{bsd} \end{bmatrix} = \mathbf{F}'_{\mathbf{d}} \begin{bmatrix} \Delta\delta \\ \Delta E'_q \\ \Delta V_{dc} \end{bmatrix} + \mathbf{G}'_{\mathbf{d}} \begin{bmatrix} \Delta m_e \\ \Delta\psi_{es} \\ \Delta m_b \\ \Delta\psi_{bs} \end{bmatrix} \tag{4.84}$$

where

$$\begin{aligned}
\mathbf{F}'_{\mathbf{q}} &= \frac{1}{x_{\Sigma 1}} \begin{bmatrix} x_{bs} + x_{bt} & -x_{te}(2x_q + x_{tb}) \\ x_{es} & x_q x_{te} + (x_{te} + x_{es})(x_q + x_{tb}) \end{bmatrix} \begin{bmatrix} f_{q11} & 0 & f_{q13} \\ f_{q21} & 0 & f_{q23} \end{bmatrix} \\
\mathbf{F}'_{\mathbf{d}} &= \frac{1}{x_{\Sigma 2}} \begin{bmatrix} -x_{bs} - x_{bt} & -x_{te} x_{tb} \\ -x_{es} & (x_{te} + x_{es})(x'_d + x_{tb}) - x'_d x_{te} \end{bmatrix} \begin{bmatrix} f_{d11} & x_L & f_{d13} \\ f_{d21} & 0 & f_{d23} \end{bmatrix} \\
\mathbf{G}'_{\mathbf{q}} &= \frac{1}{x_{\Sigma 1}} \begin{bmatrix} x_{bs} + x_{bt} & -x_{te}(2x_q + x_{tb}) \\ x_{es} & x_q x_{te} + (x_{te} + x_{es})(x_q + x_{tb}) \end{bmatrix} \begin{bmatrix} g_{q11} & g_{q12} & 0 & 0 \\ g_{q21} & g_{q22} & g_{q23} & g_{q24} \end{bmatrix} \\
\mathbf{G}'_{\mathbf{d}} &= \frac{1}{x_{\Sigma 2}} \begin{bmatrix} -x_{bs} - x_{bt} & -x_{te} x_{tb} \\ -x_{es} & (x_{te} + x_{es})(x'_d + x_{tb}) - x'_d x_{te} \end{bmatrix} \begin{bmatrix} g_{d11} & g_{d12} & 0 & 0 \\ g_{d21} & g_{d22} & g_{d23} & g_{d24} \end{bmatrix} \\
x_{\Sigma 1} &= [x_q x_{te} + (x_{te} + x_{es})(x_q + x_{tb})](x_{bs} + x_{bt}) + x_{es}^2(2x_q + x_{tb}) \\
x_{\Sigma 2} &= -[(x_{te} + x_{es})(x'_d + x_{tb}) - x'_d x_{te}](x_{bs} + x_{bt}) - x_{te}^2 x_{tb}
\end{aligned}$$

$$
\begin{aligned}
f_{q11} &= (x_q + x_{tb})m_{e0}k_e V_{dc0} \sin \psi_{es0} - x_q V_b \cos \delta_0 \\
f_{q13} &= (x_q + x_{tb})m_{e0}k_e \cos \psi_{es0} \\
f_{q21} &= m_{b0}k_b V_{dc0} \sin \psi_{bs0} - m_{e0}k_e V_{dc0} \sin \psi_{es0} + V_b \cos \delta_0 \\
f_{q23} &= m_{b0}k_b \cos \psi_{bs0} - m_{e0}k_e \cos \psi_{es0} \\
f_{d11} &= (x_d' + x_{tb})m_{e0}k_e V_{dc0} \cos \psi_{es0} - x_d' V_b \sin \delta_0 \\
f_{d13} &= -(x_d' + x_{tb})m_{e0}k_e \sin \psi_{es0} \\
f_{d21} &= m_{e0}k_e V_{dc} \cos \psi_{es0} - m_{b0}k_b V_{dc0} \cos \psi_{bs0} - V_b \sin \delta_0 \\
f_{d23} &= m_{b0}k_b \sin \psi_{bs0} - m_{e0}k_e \sin \psi_{es0} \\
g_{q11} &= (x_q + x_{tb})k_e V_{dc0} \cos \psi_{es0} \\
g_{d12} &= -(x_q + x_{tb})m_{e0}k_E V_{dc0} \sin \psi_{es0} \\
g_{q21} &= -k_e V_{dc0} \cos \psi_{es0} \\
g_{q22} &= m_{e0}k_e V_{dc0} \sin \psi_{es0} \\
g_{q23} &= k_b V_{dc0} \cos \psi_{bs0} \\
g_{q24} &= -m_{b0}k_b V_{dc0} \sin \psi_{bs0} \\
g_{d11} &= -(x_d' + x_{tb})k_e V_{dc0} \sin \psi_{es0} \\
g_{d12} &= -(x_d' + x_{tb})m_{e0}k_e V_{dc0} \cos \psi_{es0} \\
g_{d21} &= -k_e V_{dc0} \sin \psi_{e0} \\
g_{d22} &= -m_{e0}k_e V_{dc} \cos \psi_{es0} \\
g_{d23} &= k_b V_{dc0} \sin \psi_{bs0} \\
g_{d24} &= m_{b0}k_b V_{dc0} \cos \psi_{bs0}
\end{aligned}
$$

From Eqs. (4.66) and (4.67), it can have

$$
\begin{aligned}
v_{etd} + jv_{etq} &= jx_{es}(i_{esd} + ji_{esq}) + m_e k_e V_{dc} \cos \psi_{es} + jm_e k_e V_{dc} \sin \psi_{es} \\
v_{btd} + jv_{btq} &= jx_{bs}(i_{bsd} + ji_{bsq}) + m_b k_b V_{dc} \cos \psi_{bs} + jm_b k_b V_{dc} \sin \psi_{bs}
\end{aligned}
$$

The above equations give

$$
\begin{aligned}
v_{esd} &= -x_{es} i_{esq} + m_e k_e V_{dc} \cos \psi_{es} \\
v_{esq} &= x_{es} i_{esd} + m_e k_e V_{dc} \sin \psi_{es} \\
v_{bsd} &= -x_{bs} i_{bsq} + m_b k_b V_{dc} \cos \psi_{bs} \\
v_{bsq} &= x_{bs} i_{bsd} + m_b k_b V_{dc} \sin \psi_{bs}
\end{aligned}
$$

By using Eq. (4.84), linearization of the above equation can be obtained as

$$
\begin{aligned}
\Delta v_{esd} &= -x_{es}f'_{q11}\Delta\delta - x_E f'_{q12}\Delta E'_q - (x_{es}f'_{q13} - m_{e0}k_e\cos\psi_{es0})\Delta V_{dc} \\
&\quad - (x_{es}g'_{q11} - k_e V_{dc0}\cos\psi_{es0})\Delta m_e - (x_{es}g'_{q12} + m_{e0}k_e V_{dc0}\sin\psi_{es0})\Delta\psi_{es} \\
&\quad - x_{es}g'_{q13}\Delta m_b - x_{es}g'_{q14}\Delta\psi_{bs} \\
&= a_{ed1}\Delta\delta + a_{ed2}\Delta E'_q + a_{ed3}\Delta V_{dc} + a_{ed4}\Delta m_e + a_{ed5}\Delta\psi_{es} + a_{ed6}\Delta m_b + a_{ed7}\Delta\psi_{bs} \\
\Delta v_{esq} &= x_{es}f'_{d11}\Delta\delta + x_{es}f'_{d12}\Delta E'_q + (x_{es}f'_{d13} + m_{e0}k_e\sin\psi_{es0})\Delta V_{dc} \\
&\quad + (x_{es}g'_{d11} + k_e V_{dc0}\sin\psi_{es0})\Delta m_e \\
&\quad + (x_{es}g'_{d12} + m_{e0}k_e V_{dc0}\cos\psi_{es0})\Delta\psi_{es} + x_{es}g'_{d13}\Delta m_b + x_{es}g'_{d14}\Delta\psi_{bs} \\
&= a_{eq1}\Delta\delta + a_{eq2}\Delta E'_q + a_{eq3}\Delta V_{dc} + a_{eq4}\Delta m_e + a_{eq5}\Delta\psi_{es} + a_{eq6}\Delta m_b + a_{eq7}\Delta\psi_{bs} \\
\Delta v_{bsd} &= -x_{bs}f'_{q21}\Delta\delta - f'_{q22}\Delta E'_q - (x_{bs}f'_{q23} - m_{b0}k_b\cos\psi_{bs0})\Delta V_{dc} \\
&\quad - (x_{bs}g'_{q23} - k_b V_{dc0}\cos\psi_{bs0})\Delta m_b - (x_{bs}g'_{q24} + m_{b0}k_b V_{dc0}\sin\psi_{bs0})\Delta\psi_{bs} \\
&\quad - x_{bs}g'_{q21}\Delta m_e - x_{bs}g'_{q22}\Delta\psi_{es} \\
&= a_{bd1}\Delta\delta + a_{bd2}\Delta E'_q + a_{bd3}\Delta V_{dc} + a_{bd4}\Delta m_e + a_{bd5}\Delta\psi_{es} + a_{bd6}\Delta m_b + a_{bd7}\Delta\psi_{bs} \\
\Delta v_{bsq} &= x_{bs}f'_{d21}\Delta\delta + x_{bs}f'_{d22}\Delta E'_q + (x_{bs}f'_{d23} + m_{b0}k_b\sin\psi_{bs0})\Delta V_{dc} \\
&\quad + (x_{bs}g'_{d23} + k_b V_{dc0}\sin\psi_{bs0})\Delta m_b + (x_{bs}g'_{d24} + m_{b0}k_b V_{dc0}\cos\psi_{bs0})\Delta\psi_{bs} \\
&\quad + x_{bs}g'_{d21}\Delta m_e + x_{bs}g'_{d22}\Delta\psi_{es} \\
&= a_{bq1}\Delta\delta + a_{bq2}\Delta E'_q + a_{bq3}\Delta V_{dc} + a_{bq4}\Delta m_e + a_{bq5}\Delta\psi_{es} + a_{bq6}\Delta m_b + a_{bq7}\Delta\psi_{bs}
\end{aligned}
$$

By using the above equations, linearization of Eq. (4.75) can be obtained as

$$
\begin{aligned}
\Delta\psi_{es} &= a'_{e1}\Delta\delta + a'_{e2}\Delta E'_q + a'_{e3}\Delta V_{dc} + a'_{e4}\Delta m_e \\
&\quad + a'_{e5}\Delta\psi_{es} + a'_{e6}\Delta m_b + a'_{e7}\Delta\psi_{bs} - \Delta\delta_e \\
\Delta\psi_{bs} &= a'_{b1}\Delta\delta + a'_{b2}\Delta E'_q + a'_{b3}\Delta V_{dc} + a'_{b4}\Delta m_e \\
&\quad + a'_{b5}\Delta\psi_{es} + a'_{b6}\Delta m_b + a'_{b7}\Delta\psi_{bs} - \Delta\delta_b
\end{aligned}
$$

where

$$
\begin{aligned}
a'_{ei} &= \frac{1}{v^2_{esd0} + v^2_{esq0}}(v_{esd0}a_{eqi} - v_{esq0}a_{edi}) \\
a'_{bi} &= \frac{1}{v^2_{bsd0} + v^2_{bsq0}}(v_{bsd0}a_{bqi} - v_{bsq0}a_{bdi}) \\
i &= 1, 2, 3, 4, 5, 6, 7
\end{aligned}
$$

Rearrangement of above equations is

$$\begin{aligned}\begin{bmatrix}\Delta\psi_{es}\\ \Delta\psi_{bs}\end{bmatrix} &= \begin{bmatrix}1-a'_{e5} & -a'_{e7}\\ -a'_{b5} & 1-a'_{b7}\end{bmatrix}^{-1}\left(\begin{bmatrix}a'_{e1} & a'_{e2} & a'_{e3}\\ a'_{b1} & a'_{b2} & a'_{b3}\end{bmatrix}\begin{bmatrix}\Delta\delta\\ \Delta E'_q\\ \Delta V_{dc}\end{bmatrix}\right.\\ &\quad\left.+\begin{bmatrix}a'_{e4} & a'_{e6}\\ a'_{b4} & a'_{b6}\end{bmatrix}\begin{bmatrix}\Delta m_e\\ \Delta m_b\end{bmatrix}+\begin{bmatrix}-1 & 0\\ 0 & -1\end{bmatrix}\begin{bmatrix}\Delta\delta_e\\ \Delta\delta_b\end{bmatrix}\right)\\ &= \mathbf{A_{EB1}}\begin{bmatrix}\Delta\delta\\ \Delta E'_q\\ \Delta V_{dc}\end{bmatrix}+\mathbf{A_{EB2}}\begin{bmatrix}\Delta m_e\\ \Delta m_b\end{bmatrix}+\mathbf{A_{EB3}}\begin{bmatrix}\Delta\delta_e\\ \Delta\delta_b\end{bmatrix}\end{aligned} \tag{4.85}$$

Substituting Eq. (4.85) into Eq. (4.84), it can have

$$\begin{aligned}\begin{bmatrix}\Delta i_{esq}\\ \Delta i_{bsq}\end{bmatrix} &= \mathbf{F_q}\begin{bmatrix}\Delta\delta\\ \Delta E'_q\\ \Delta V_{dc}\end{bmatrix}+\mathbf{G_q}\begin{bmatrix}\Delta m_e\\ \Delta\delta_e\\ \Delta m_b\\ \Delta\delta_b\end{bmatrix}\\ \begin{bmatrix}\Delta i_{esd}\\ \Delta i_{bsd}\end{bmatrix} &= \mathbf{F_d}\begin{bmatrix}\Delta\delta\\ \Delta E'_q\\ \Delta V_{dc}\end{bmatrix}+\mathbf{G_d}\begin{bmatrix}\Delta m_e\\ \Delta\delta_e\\ \Delta m_b\\ \Delta\delta_b\end{bmatrix}\end{aligned} \tag{4.86}$$

Linearization of Eq. (4.80) is

$$\begin{aligned}\Delta v_{td} &= \frac{x_q V_b \sin\delta_0}{x_q + x_{tb}}\Delta\delta + \begin{bmatrix}\frac{x_q x_{tb}}{x_q + x_{tb}} & \frac{x_q x_{tb}}{x_q + x_{tb}}\end{bmatrix}\begin{bmatrix}\Delta i_{esq}\\ \Delta i_{bsq}\end{bmatrix}\\ \Delta v_{tq} &= \frac{-x'_d V_b \sin\delta_0}{x'_d + x_{tb}}\Delta\delta + \frac{x_{tb}}{x'_d + x_{tb}}\Delta E'_q + \begin{bmatrix}\frac{-x'_d x_{tb}}{x'_d + x_{tb}} & \frac{-x'_d x_{tb}}{x'_d + x_{tb}}\end{bmatrix}\begin{bmatrix}\Delta i_{esd}\\ \Delta i_{bsd}\end{bmatrix}\end{aligned} \tag{4.87}$$

Substituting Eq. (4.86) into (4.87) gives

$$\begin{aligned}\Delta v_{td} &= \mathbf{H_d}\begin{bmatrix}\Delta\delta\\ \Delta E'_q\\ \Delta V_{dc}\end{bmatrix}+\mathbf{L_d}\begin{bmatrix}\Delta m_e\\ \Delta\delta_e\\ \Delta m_b\\ \Delta\delta_b\end{bmatrix}\\ \Delta v_{tq} &= \mathbf{H_q}\begin{bmatrix}\Delta\delta\\ \Delta E'_q\\ \Delta V_{dc}\end{bmatrix}+\mathbf{L_q}\begin{bmatrix}\Delta m_e\\ \Delta\delta_e\\ \Delta m_b\\ \Delta\delta_b\end{bmatrix}\end{aligned} \tag{4.88}$$

where

$$\mathbf{H_d} = \begin{bmatrix} \frac{x_q V_b \sin\delta_0}{x_q + x_{tb}} & 0 & 0 \end{bmatrix} + \begin{bmatrix} \frac{x_q x_{tb}}{x_q + x_{tb}} & \frac{x_q x_{tb}}{x_q + x_{tb}} \end{bmatrix} \mathbf{F_q}$$
$$\mathbf{L_d} = \begin{bmatrix} \frac{x_q x_{tb}}{x_q + x_{tb}} & \frac{x_q x_{tb}}{x_q + x_{tb}} \end{bmatrix} \mathbf{G_q}$$
$$\mathbf{H_q} = \begin{bmatrix} \frac{-x'_d V_b \sin\delta_0}{x'_d + x_{tb}} & \frac{x_{tb}}{x'_d + x_{tb}} & 0 \end{bmatrix} + \begin{bmatrix} \frac{-x'_d x_{tb}}{x'_d + x_{tb}} & \frac{-x'_d x_{tb}}{x'_d + x_{tb}} \end{bmatrix} \mathbf{F_d}$$
$$\mathbf{L_q} = \begin{bmatrix} \frac{-x'_d x_{tb}}{x'_d + x_{tb}} & \frac{-x'_d x_{tb}}{x'_d + x_{tb}} \end{bmatrix} \mathbf{G_d}$$

Linearization of Eq. (4.78) is

$$\begin{aligned} \Delta i_{tbd} &= \frac{V_b \sin\delta_0}{x_{tb}} \Delta\delta + \frac{1}{x_{tb}} \Delta v_{tq} \\ \Delta i_{tbq} &= \frac{V_b \cos\delta_0}{x_{tb}} \Delta\delta - \frac{1}{x_{tb}} \Delta v_{td} \end{aligned} \tag{4.89}$$

Linearization of Eq. (4.83) is

$$\begin{aligned} \Delta i_{ted} &= \Delta i_{esd} + \Delta i_{bsd} \\ \Delta i_{teq} &= \Delta i_{esq} + \Delta i_{bsq} \\ \Delta P_t &= (i_{teq0} + i_{tbq0})\Delta v_{tq} + (i_{ted0} + i_{tbd0})\Delta v_{td} \\ &\quad + v_{tq0}(\Delta i_{teq} + \Delta i_{tbq}) + v_{td0}(\Delta i_{ted} + \Delta i_{tbd}) \\ \Delta E_q &= \Delta E'_q + (x_d - x'_d)(\Delta i_{ted} + \Delta i_{tbd}) \\ \Delta V_t &= \frac{v_{td0}}{V_{t0}} \Delta v_{td} + \frac{v_{tq0}}{V_{t0}} \Delta v_{tq} \end{aligned} \tag{4.90}$$

By using Eqs. (4.86), (4.88), and (4.89) and the first two equations in Eq. (4.90), the last three equations in Eq. (4.90) become

$$\begin{aligned} \Delta P_t &= (i_{teq0} + i_{tbq0})\Delta v_{tq} + (i_{ted0} + i_{tbd0})\Delta v_{td} \\ &\quad + v_{tq0}(\Delta i_{esq} + \Delta i_{bsq} + \Delta i_{tbq}) + v_{td0}(\Delta i_{esd} + \Delta i_{bsd} + \Delta i_{bsd}) \\ &= \left( v_{tq0} \frac{V_b \cos\delta_0}{x_{tb}} + v_{td0} \frac{V_b \sin\delta_0}{x_{tb}} \right) \Delta\delta \\ &\quad + \left( i_{ted0} + i_{tbd0} - \frac{v_{tq0}}{x_{tb}} \right) \Delta v_{td} + \left( i_{teq0} + i_{tbq0} + \frac{v_{td0}}{x_{tb}} \right) \Delta v_{tq} \\ &\quad + \begin{bmatrix} v_{tq0} & v_{tq0} \end{bmatrix} \begin{bmatrix} \Delta i_{esq} \\ \Delta i_{esq} \end{bmatrix} + \begin{bmatrix} v_{td0} & v_{td0} \end{bmatrix} \begin{bmatrix} \Delta i_{esd} \\ \Delta i_{esd} \end{bmatrix} \\ &= \begin{bmatrix} K_1 & K_2 & K_{pdc} \end{bmatrix} \begin{bmatrix} \Delta\delta \\ \Delta E'_q \\ \Delta V_{dc} \end{bmatrix} + \begin{bmatrix} K_{pme} & K_{p\delta e} & K_{pmb} & K_{p\delta b} \end{bmatrix} \begin{bmatrix} \Delta m_e \\ \Delta\delta_e \\ \Delta m_b \\ \Delta\delta_b \end{bmatrix} \end{aligned} \tag{4.91}$$

$$
\begin{aligned}
\Delta E_q &= \Delta E_q' + \left[\, (x_d - x_d') \quad (x_d - x_d') \,\right] \begin{bmatrix} \Delta i_{esd} \\ \Delta i_{bsd} \end{bmatrix} \\
&\quad + (x_d - x_d')\left( \frac{V_b \sin \delta_0}{x_{tb}} \Delta\delta + \frac{1}{x_{tb}} \Delta v_{tq} \right) \\
&= (x_d - x_d')\frac{V_b \sin \delta_0}{x_{tb}} \Delta\delta + \Delta E_q' + (x_d - x_d')\frac{1}{x_{tb}} \Delta v_{tq} \\
&\quad + \left[\, (x_d - x_d') \quad (x_d - x_d') \,\right] \begin{bmatrix} \Delta i_{esd} \\ \Delta i_{bsd} \end{bmatrix} \\
&= \left[\, K_4 \quad K_3 \quad K_{qdc} \,\right] \begin{bmatrix} \Delta\delta \\ \Delta E_q' \\ \Delta V_{dc} \end{bmatrix} + \left[\, K_{qme} \quad K_{q\delta e} \quad K_{qme} \quad K_{q\delta e} \,\right] \begin{bmatrix} \Delta m_e \\ \Delta\delta_e \\ \Delta m_b \\ \Delta\delta_b \end{bmatrix}
\end{aligned}
\tag{4.92}
$$

$$
\Delta V_t = \left[\, K_5 \quad K_6 \quad K_{vdc} \,\right] \begin{bmatrix} \Delta\delta \\ \Delta E_q' \\ \Delta V_{dc} \end{bmatrix} + \left[\, K_{vme} \quad K_{v\delta e} \quad K_{vmb} \quad K_{v\delta b} \,\right] \begin{bmatrix} \Delta m_e \\ \Delta\delta_e \\ \Delta m_b \\ \Delta\delta_b \end{bmatrix} \tag{4.93}
$$

where

$$
\begin{aligned}
\left[\, K_1 \quad K_2 \quad K_{pdc} \,\right] &= \left[ v_{tq0} \frac{V_b \cos \delta_2}{X_{tb}} + v_{ted0} \frac{V_b \sin \delta_0}{X_{tb}} \quad 0 \quad 0 \right] \\
&\quad + \left( i_{ted0} + i_{tbd0} - \frac{v_{teq0}}{x_{tb}} \right) \mathbf{H_d} + \left( i_{teq0} + i_{tbq0} + \frac{v_{ted0}}{x_{tb}} \right) \mathbf{H_q} \\
&\quad + \left[\, v_{tq0} \quad v_{tq0} \,\right] \mathbf{F_q} + \left[\, v_{td0} \quad v_{td0} \,\right] \mathbf{F_d} \\
\left[\, K_{pme} \quad K_{p\delta e} \quad K_{pmb} \quad K_{p\delta b} \,\right] &= \left(i_{ted0} + i_{tbd0} - \frac{v_{teq0}}{x_{tb}}\right) \mathbf{L_d} \\
&\quad + \left(i_{teq0} + i_{tbq0} + \frac{v_{ted0}}{x_{tb}}\right) \mathbf{L_q} + \left[\, v_{tq0} \quad v_{tq0} \,\right] \mathbf{G_q} + \left[\, v_{td0} \quad v_{td0} \,\right] \mathbf{G_d} \\
\left[\, K_4 \quad K_3 \quad K_{qdc} \,\right] &= \left[ (x_d - x_d') \frac{V_b \sin \delta_0}{X_{tb}} \quad 1 \quad 0 \right] \\
&\quad + \frac{(x_d - x_d')}{x_{tb}} \mathbf{H_q} + \left[\, (x_d - x_d') \quad (x_d - x_d') \,\right] \mathbf{F_d} \\
\left[\, K_{qme} \quad K_{q\delta e} \quad K_{qmb} \quad K_{q\delta b} \,\right] &= \frac{(x_d - x_d')}{x_{tb}} \mathbf{L_q} + \left[\, (x_d - x_d') \quad (x_d - x_d') \,\right] \mathbf{G_d} \\
\left[\, K_5 \quad K_6 \quad K_{vdc} \,\right] &= \frac{v_{td0}}{V_{t0}} \mathbf{H_d} + \frac{v_{tq0}}{E} \mathbf{H_q} \\
\left[\, K_{qme} \quad K_{q\delta e} \quad K_{qmb} \quad K_{q\delta b} \,\right] &= \frac{v_{td0}}{V_{t0}} \mathbf{L_d} + \frac{v_{tq0}}{V_{t0}} \mathbf{L_q}
\end{aligned}
$$

Linearization of Eq. (4.74) is

$$\begin{aligned}
C_{dc}\Delta\dot{V}_{dc} &= \left[\, m_{e0}k_e\cos\psi_{es0} \quad m_{b0}k_b\cos\psi_{bs0} \,\right]\begin{bmatrix}\Delta i_{esd}\\ \Delta i_{bsd}\end{bmatrix}\\
&\quad+\left[\, m_{e0}k_e\sin\psi_{es0} \quad m_{b0}k_b\sin\psi_{bs0} \,\right]\begin{bmatrix}\Delta i_{esq}\\ \Delta i_{bsq}\end{bmatrix}\\
&\quad+k_e i_{esd0}\cos\psi_{es0}\Delta m_e+k_e i_{esq0}\sin\psi_{es0}\Delta m_e+k_b i_{bsd0}\cos\psi_{bs0}\Delta m_b\\
&\quad+k_b i_{bsq0}\sin\psi_{bs0}\Delta m_b-m_{e0}k_e i_{esd0}\sin\psi_{es0}(-\Delta\delta+\Delta\delta_e)\\
&\quad+m_{e0}k_e i_{esq0}\cos\psi_{es0}(-\Delta\delta+\Delta\delta_e)-m_{b0}k_b i_{bsd0}\sin\psi_{bs0}(-\Delta\delta+\Delta\delta_b)\\
&\quad+m_{b0}k_b i_{bsq0}\cos\psi_{bs0}(-\Delta\delta+\Delta\delta_b)\\
&=(m_{e0}k_e i_{esd0}\sin\psi_{es0}-m_{e0}k_e i_{esq0}\cos\psi_{es0}+m_{b0}k_b i_{bsd0}\sin\psi_{bs0}\\
&\quad-m_{b0}k_b i_{bsq0}\cos\psi_{bs0})\Delta\delta+\left[\, m_{e0}k_e\cos\psi_{es0} \quad m_{b0}k_b\cos\psi_{bs0} \,\right]\begin{bmatrix}\Delta i_{esd}\\ \Delta i_{Bbsd}\end{bmatrix}\\
&\quad+\left[\, m_{e0}k_e\sin\psi_{es0} \quad m_{b0}k_b\sin\psi_{bs0} \,\right]\begin{bmatrix}\Delta i_{esq}\\ \Delta i_{bsq}\end{bmatrix}\\
&\quad+(k_e i_{esd0}\cos\psi_{es0}+k_e i_{esq0}\sin\psi_{es0})\Delta m_e\\
&\quad+(k_b i_{bsd0}\cos\psi_{bs0}+k_b i_{bsq0}\sin\psi_{bs0})\Delta m_b\\
&\quad+(m_{e0}k_e i_{esq0}\cos\psi_{es0}-m_{e0}k_e i_{esd0}\sin\psi_{es0})\Delta\delta_e\\
&\quad+(m_{b0}k_b i_{bsq0}\cos\psi_{bs0}-m_{b0}k_b i_{bsd0}\sin\psi_{bs0}\Delta\delta_b)\Delta\delta_b\\
&=\left[\, K_7 \quad K_8 \quad K_9 \,\right]\begin{bmatrix}\Delta\delta\\ \Delta E'_q\\ \Delta V_{dc}\end{bmatrix}+\left[\, K_{dme} \quad K_{d\delta e} \quad K_{dmb} \quad K_{d\delta b} \,\right]\begin{bmatrix}\Delta m_e\\ \Delta\delta_e\\ \Delta m_b\\ \Delta\delta_b\end{bmatrix}
\end{aligned} \tag{4.94}$$

where

$$\begin{aligned}
&\left[\, K_7 \quad K_8 \quad K_9 \,\right]\\
&=\begin{bmatrix} m_{e0}k_e i_{esd0}\sin\psi_{es0}-m_{e0}k_e i_{esq0}\cos\psi_{es0}+m_{b0}k_b i_{bsd0}\sin\psi_{bs0}-m_{b0}k_b i_{bsq0}\cos\psi_{bs0}\\ 0\\ 0\end{bmatrix}^T\\
&\quad+\left[\, m_{e0}k_e\cos\psi_{es0} \quad m_{b0}k_b\cos\psi_{B0} \,\right]\mathbf{F_d}+\left[\, m_{e0}k_e\sin\psi_{E0} \quad m_{b0}k_b\sin\psi_{B0} \,\right]\mathbf{F_q}
\end{aligned}$$

$$\begin{aligned}
\left[\, K_{dme} \quad K_{d\delta e} \quad K_{dmb} \quad K_{d\delta b} \,\right] &= \begin{bmatrix} k_e i_{esd0}\cos\psi_{es0}+k_e i_{esq0}\sin\psi_{es0}\\ m_{e0}k_e i_{esq0}\cos\psi_{es0}-m_{e0}k_e i_{esd0}\sin\psi_{es0}\\ k_b i_{bsd0}\cos\psi_{bs0}+k_b i_{bsq0}\sin\psi_{bs0}\\ m_{b0}k_b i_{bsq0}\cos\psi_{bs0}-m_{b0}k_b i_{bsd0}\sin\psi_{bs0}\end{bmatrix}^T\\
&\quad+\left[\, m_{e0}k_e\cos\psi_{es0} \quad m_{b0}k_b\cos\psi_{bs0} \,\right]\mathbf{G_d}+\left[\, m_{e0}k_e\sin\psi_{es0} \quad m_{b0}k_b\sin\psi_{bs0} \,\right]\mathbf{G_q}
\end{aligned}$$

By substituting Eqs. (4.91)–(4.93) into Eq. (2.37) (without the PSS control) and arranging the equations obtained together with Eq. (4.94) in matrix form, the extended Heffron–Phillips model of the single-machine infinite-bus power system with the UPFC installed can be obtained as

$$\begin{bmatrix} \Delta\dot{\delta} \\ \Delta\dot{\omega} \\ \Delta\dot{E}'_q \\ \Delta\dot{E}'_{fd} \\ \Delta\dot{V}_{dc} \end{bmatrix} = \begin{bmatrix} 0 & \omega_0 & 0 & 0 & 0 \\ -\frac{K_1}{M} & -\frac{D}{M} & -\frac{K_2}{M} & 0 & -\frac{K_{pdc}}{M} \\ -\frac{K_4}{T'_{d0}} & 0 & -\frac{K_3}{T'_{d0}} & \frac{1}{T'_{d0}} & -\frac{K_{qdc}}{T'_{d0}} \\ -\frac{K_A K_5}{T_A} & 0 & -\frac{K_A K_6}{T_A} & -\frac{1}{T_A} & -\frac{K_A K_{vdc}}{T_A} \\ \frac{K_7}{C_{dc}} & 0 & \frac{K_8}{C_{dc}} & 0 & \frac{K_9}{C_{dc}} \end{bmatrix} \begin{bmatrix} \Delta\delta \\ \Delta\omega \\ \Delta E'_q \\ \Delta E'_{fd} \\ \Delta V_{dc} \end{bmatrix}$$
$$\times \begin{bmatrix} 0 & 0 & 0 & 0 \\ -\frac{K_{pme}}{M} & -\frac{K_{p\delta e}}{M} & -\frac{K_{pmb}}{M} & -\frac{K_{p\delta b}}{M} \\ -\frac{K_{qme}}{T'_{d0}} & -\frac{K_{q\delta e}}{T'_{d0}} & -\frac{K_{qmB}}{T'_{d0}} & -\frac{K_{q\delta B}}{T'_{d0}} \\ -\frac{K_A K_{vme}}{T_A} & -\frac{K_A K_{v\delta e}}{T_A} & -\frac{K_A K_{vmb}}{T_A} & -\frac{K_A K_{v\delta b}}{T_A} \\ \frac{K_{dme}}{C_{dc}} & \frac{K_{d\delta e}}{C_{dc}} & \frac{K_{dmb}}{C_{dc}} & \frac{K_{d\delta b}}{C_{dc}} \end{bmatrix} \begin{bmatrix} \Delta m_e \\ \Delta\delta_e \\ \Delta m_b \\ \Delta\delta_b \end{bmatrix} \tag{4.95}$$

The linearized model is shown in Figs. 4.14 and 4.15, where

$$\begin{aligned} \mathbf{k_p} &= [K_{pdc} \quad K_{pme} \quad K_{p\delta e} \quad K_{pmb} \quad K_{p\delta b}] \\ \mathbf{k_q} &= [K_{qdc} \quad K_{qme} \quad K_{q\delta e} \quad K_{qmb} \quad K_{q\delta b}] \\ \mathbf{k_v} &= [K_{vdc} \quad K_{vme} \quad K_{v\delta e} \quad K_{vmb} \quad K_{v\delta b}] \\ \mathbf{k_d} &= [K_{dme} \quad K_{d\delta e} \quad K_{dmb} \quad K_{d\delta b}] \\ \mathbf{u_{upfc1}} &= [\Delta m_e \quad \Delta\delta_e \quad \Delta m_b \quad \Delta\delta_b]^T \\ \mathbf{u_{upfc}} &= [\Delta V_{dc} \quad \mathbf{u}^T_{\mathbf{upfc1}}]^T \end{aligned} \tag{4.96}$$

## 4.3.2 *Design of a UPFC Stabilizer Installed in a Single-Machine Infinite-Bus Power System*

### 4.3.2.1 Selection of Modulation Signal to Add the Damping Control Signal of the UPFC Stabilizer by Using the Controllability and Observability Index [6, 7]

Damping control signal of a UPFC stabilizer can be superimposed on a modulation signal of the PWM control, $\Delta m_e$, $\Delta\delta_e$, $\Delta m_b$, or $\Delta\delta_b$. Selection of the modulation signal to add the damping control signal ought to ensure the most effectiveness of

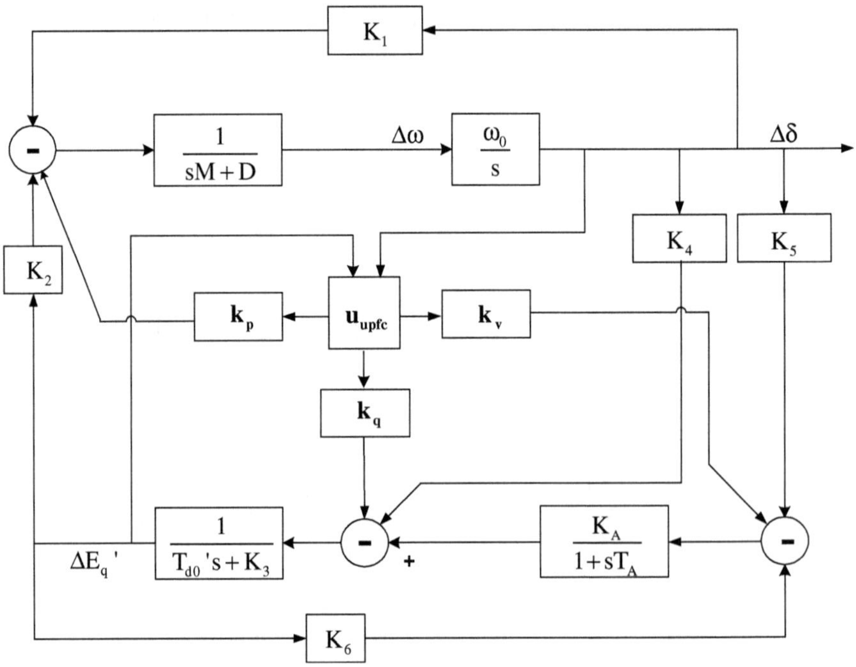


**Fig. 4.14** Extended Heffron–Phillips model of power system with a UPFC installed—part of the power system

**Fig. 4.15** Extended Heffron–Phillips model of power system with a UPFC installed—part of the UPFC

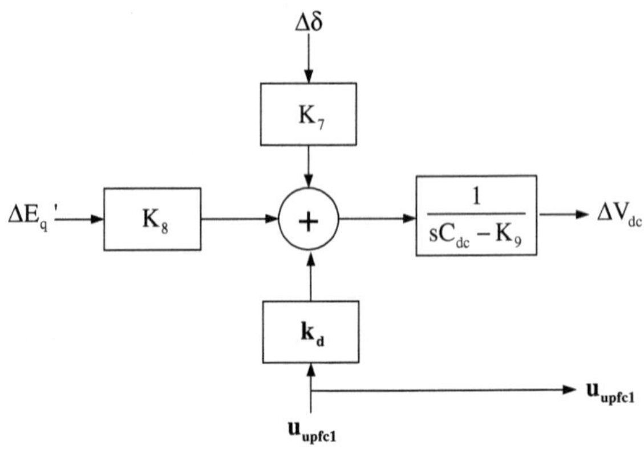


UPFC stabilizer. It can be carried out by use of the modal controllability and observability index introduced in Sect. 2.2.1.2.

Linearized model of the single-machine infinite-bus power system installed with the UPFC of Eq. (4.95) can be written as

$$\dot{\mathbf{X}} = \mathbf{A}\mathbf{X} + \sum_{k=1}^{4} \mathbf{b_k}\Delta u_k \tag{4.97}$$

where

$$\mathbf{X} = \begin{bmatrix} \Delta\delta \\ \Delta\omega \\ \Delta E'_q \\ \Delta E'_{fd} \\ \Delta V_{dc} \end{bmatrix}, \quad \mathbf{A} = \begin{bmatrix} 0 & \omega_o & 0 & 0 & 0 \\ -\frac{K_1}{M} & -\frac{D}{M} & -\frac{K_2}{M} & 0 & -\frac{K_{pdc}}{M} \\ -\frac{K_4}{T'_{d0}} & 0 & -\frac{K_3}{T'_{d0}} & \frac{1}{T'_{d0}} & -\frac{K_{qdc}}{T'_{d0}} \\ -\frac{K_A K_5}{T_A} & 0 & -\frac{K_A K_6}{T_A} & -\frac{1}{T_A} & -\frac{K_A K_{vdc}}{T_A} \\ \frac{K_7}{C_{dc}} & 0 & \frac{K_8}{C_{dc}} & 0 & \frac{K_9}{C_{dc}} \end{bmatrix},$$

$$\mathbf{b_1} = \begin{bmatrix} 0 \\ -\frac{K_{pme}}{M} \\ -\frac{K_{qme}}{T'_{d0}} \\ -\frac{K_A K_{vme}}{T_A} \\ \frac{K_{dme}}{C_{dc}} \end{bmatrix}, \mathbf{b_2} = \begin{bmatrix} 0 \\ -\frac{K_{p\delta e}}{M} \\ -\frac{K_{q\delta e}}{T'_{d0}} \\ -\frac{K_A K_{v\delta e}}{T_A} \\ \frac{K_{d\delta e}}{C_{dc}} \end{bmatrix}, \mathbf{b_3} = \begin{bmatrix} 0 \\ -\frac{K_{pmb}}{M} \\ -\frac{K_{qmb}}{T'_{d0}} \\ -\frac{K_A K_{vmb}}{T_A} \\ \frac{K_{dmb}}{C_{dc}} \end{bmatrix}, \mathbf{b_4} = \begin{bmatrix} 0 \\ -\frac{K_{p\delta b}}{M} \\ -\frac{K_{q\delta b}}{T'_{d0}} \\ -\frac{K_A K_{v\delta b}}{T_A} \\ \frac{K_{d\delta b}}{C_{dc}} \end{bmatrix},$$

$$\Delta u_1 = \Delta m_e, \quad \Delta u_2 = \Delta\delta_e, \quad \Delta u_3 = \Delta m_b, \quad \Delta u_4 = \Delta\delta_b$$

If the transfer function and feedback signal of the UPFC stabilizer are $T_{upfck}(s)$ and $y_k$, respectively, state-space representation of the power system with the UPFC stabilizer is

$$\begin{aligned} \dot{\mathbf{X}} &= \mathbf{A}\mathbf{X} + \sum_{k=1}^{4} \mathbf{b_k}\Delta u_k \\ y_k &= \mathbf{c}_k^T \mathbf{X} \\ \Delta u_k &= T_{upfck}(s) y_k \end{aligned} \tag{4.98}$$

If $\lambda_i$ is the electromechanical oscillation mode of the power system, from Eqs. (2.50 ) and (2.54), it can have

$$\begin{aligned} \mathbf{A}\mathbf{v_i} &= \lambda_i \mathbf{v_i} \\ \mathbf{w_i^T}\mathbf{A} &= \mathbf{w_i^T}\lambda_i \end{aligned} \tag{4.99}$$

where $\mathbf{v_i}$ and $\mathbf{w_i^T}$ are the right eigenvector and left eigenvector of state matrix, **A,** with respect to mode $\lambda_i$, respectively. From the state transformation of Eqs. (2.53), (2.55), (2.56), and (4.98), it can have

$$\begin{aligned}\dot{\mathbf{Z}} &= \mathbf{V}^{-1}\mathbf{AVZ} + \sum_{k=1}^{4}\mathbf{V}^{-1}\mathbf{b_k}\Delta u_k = \mathbf{\Lambda Z} + \sum_{k=1}^{4}\mathbf{V}^{-1}\mathbf{b_k}\Delta u_k \\ y_k &= \mathbf{c}_k^T\mathbf{VZ} \\ \Delta u_k &= T_{upfck}(s)y_k\end{aligned} \tag{4.100}$$

where

$$\mathbf{V} = [\mathbf{v_1} \quad \mathbf{v_2} \quad \ldots \quad \mathbf{v_n}], \quad \mathbf{\Lambda} = \begin{bmatrix}\lambda_1 & 0 & \mathbf{0} & \mathbf{0} \\ 0 & \lambda_2 & \mathbf{0} & \mathbf{0} \\ \mathbf{0} & \mathbf{0} & \ddots & \mathbf{0} \\ \mathbf{0} & \mathbf{0} & \mathbf{0} & \lambda_n\end{bmatrix}$$

According to Eq. (2.53),

$$\mathbf{V}^{-1} = [\mathbf{w_1} \quad \mathbf{w_2} \quad \ldots \quad \mathbf{w_M}]^T = \mathbf{W}^T \tag{4.101}$$

Equation (4.100) can be written as

$$\begin{aligned}z_j &= \frac{1}{s-\lambda_j} + \sum_{k=1}^{4}\mathbf{w}_j^T\mathbf{b_k}\Delta u_k \\ y_k &= \sum_j \mathbf{c}_k^T\mathbf{v_j}\mathbf{z_j}, \quad j = 1, 2, \ldots \\ \Delta u_k &= T_{upfck}(s)y_k\end{aligned} \tag{4.102}$$

The system represented by Eq. (4.102) is shown in Fig. 4.16.

From Fig. 4.16, it can be seen that $\mathbf{w}_i^T\mathbf{b_k}$ measures the influence of the UPFC stabilizer on the oscillation mode and $\mathbf{c}_k^T\mathbf{v_i}$ measures how much the oscillation mode is observed in the feedback signal of the stabilizer. Hence, it is defined that as

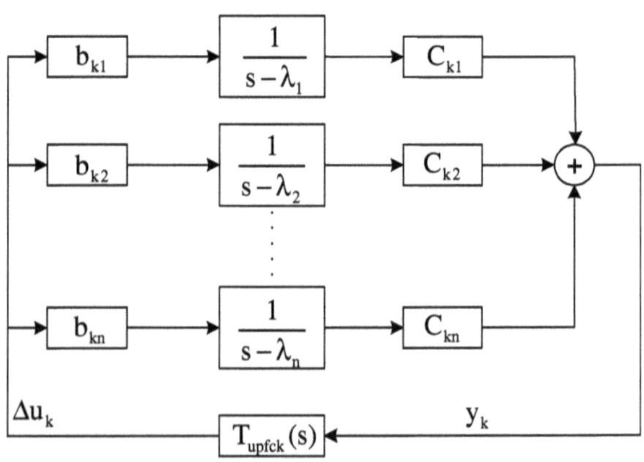


**Fig. 4.16** Modal decomposition of the system to be installed with UPFC stabilizer

far as the oscillation mode $\lambda_i$ is concerned, the controllability index and observability index of the UPFC stabilizer are, respectively,

$$b_{ik} = \mathbf{w_i}^T \mathbf{b_k}, \quad c_{ik} = \mathbf{c}_k^T \mathbf{v_i} \tag{4.103}$$

Their product is the residue

$$R_{ik} = b_{ik} c_{ik} \tag{4.104}$$

which measures the overall influence of the stabilizer on the oscillation mode.

Obviously, the controllability index, observability index, and residue are calculated from the open-loop system model as far as the UPFC stabilizer is concerned. Hence, it can be used to predict the effectiveness of the stabilizer to be installed. The controllability index can be used to select the most effective modulation signal to be superimposed with the damping control signal to enhance the damping of system electromechanical oscillation mode.

#### 4.3.2.2 Selection by Damping Torque Calculation

Electric torque contribution from a UPFC stabilizer can be calculated from the forward path of UPFC damping control signal. The forward path can be obtained from Figs. 4.14 and 4.15 with the blocks associated with $K_4$, $K_5$, and $K_7$ ignored as shown in Fig. 4.17,

$$\begin{aligned}
&\Delta u_1 = \Delta m_e, \quad \Delta u_2 = \Delta\delta_e, \quad \Delta u_3 = \Delta m_b, \quad \Delta u_4 = \Delta\delta_b \\
&\qquad \text{when } k = 1, k_{puk} = K_{dme}, \\
&\mathbf{k_{puk}} = \begin{bmatrix} K_{pdc} & K_{pme} \end{bmatrix}, \quad \mathbf{k_{quk}} = \begin{bmatrix} K_{qdc} & K_{qme} \end{bmatrix}, \quad \mathbf{k_{vuk}} = \begin{bmatrix} K_{vdc} & K_{vme} \end{bmatrix} \\
&\qquad \text{when } k = 2, k_{puk} = K_{d\delta e}, \\
&\mathbf{k_{puk}} = \begin{bmatrix} K_{pdc} & K_{p\delta e} \end{bmatrix}, \quad \mathbf{k_{quk}} = \begin{bmatrix} K_{qdc} & K_{q\delta e} \end{bmatrix}, \quad \mathbf{k_{vuk}} = \begin{bmatrix} K_{vdc} & K_{v\delta e} \end{bmatrix} \\
&\qquad \text{when } k = 3, k_{puk} = K_{dmb}, \\
&\mathbf{k_{puk}} = \begin{bmatrix} K_{pdc} & K_{pmb} \end{bmatrix}, \quad \mathbf{k_{quk}} = \begin{bmatrix} K_{qdc} & K_{qmb} \end{bmatrix}, \quad \mathbf{k_{vuk}} = \begin{bmatrix} K_{vdc} & K_{vmb} \end{bmatrix} \\
&\qquad \text{when } k = 4, k_{puk} = K_{d\delta b}, \\
&\mathbf{k_{puk}} = \begin{bmatrix} K_{pdc} & K_{p\delta b} \end{bmatrix}, \quad \mathbf{k_{quk}} = \begin{bmatrix} K_{qdc} & K_{q\delta b} \end{bmatrix}, \quad \mathbf{k_{vuk}} = \begin{bmatrix} K_{vdc} & K_{v\delta b} \end{bmatrix}
\end{aligned}$$

The forward path is described by the following state-space representation

$$\begin{aligned}
\dot{\mathbf{X}}_1 &= \mathbf{A}_1 \mathbf{X}_1 + \mathbf{b}_{1k} \Delta u_k \\
\Delta T_{uk} &= \mathbf{c}_{upfc1}^T \mathbf{X}_1 + k_{puk} \Delta u_k, \quad k = 1, 2, 3, 4
\end{aligned} \tag{4.105}$$

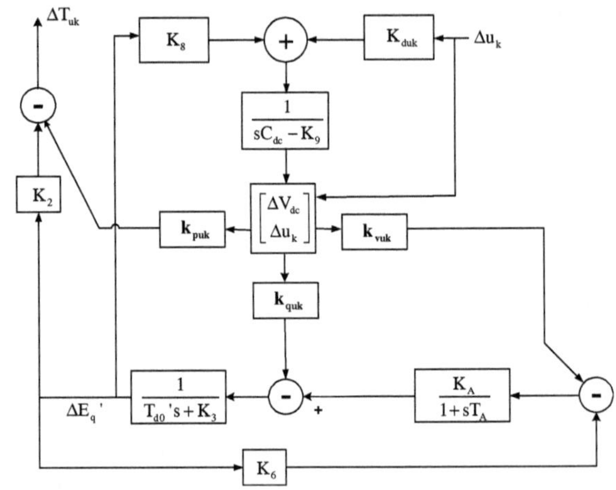


**Fig. 4.17** Forward path of UPFC stabilizer

where

$$\Delta u_1 = \Delta m_e, \quad \Delta u_2 = \Delta \delta_e, \quad \Delta u_3 = \Delta m_b, \quad \Delta u_4 = \Delta \delta_b$$

$$\mathbf{X_1} = \begin{bmatrix} \Delta E'_q \\ \Delta E'_{fd} \\ \Delta V_{dc} \end{bmatrix}, \quad \mathbf{A_1} = \begin{bmatrix} -\frac{K_3}{T'_{d0}} & \frac{1}{T'_{d0}} & -\frac{K_{qdc}}{T'_{d0}} \\ -\frac{K_A K_6}{T_A} & -\frac{1}{T_A} & -\frac{K_A K_{vdc}}{T_A} \\ \frac{K_8}{C_{dc}} & 0 & \frac{K_9}{C_{dc}} \end{bmatrix}$$

$$\mathbf{b_{11}} = \begin{bmatrix} K_{qme} & K_{vme} & K_{dme} \end{bmatrix}^T, \quad \mathbf{b_{12}} = \begin{bmatrix} K_{q\delta e} & K_{v\delta e} & K_{d\delta e} \end{bmatrix}^T,$$

$$\mathbf{b_{13}} = \begin{bmatrix} K_{qmb} & K_{vmb} & K_{dmb} \end{bmatrix}^T, \quad \mathbf{b_{14}} = \begin{bmatrix} K_{q\delta b} & K_{v\delta b} & K_{d\delta b} \end{bmatrix}^T,$$

$$\mathbf{c_{upfc1}} = \begin{bmatrix} K_2 & 0 & K_{pdc} \end{bmatrix}^T$$

From Eq. (4.105), transfer function of the forward path can be obtained as

$$F_{upfck}(s) = \mathbf{c}^T_{\mathbf{upfc1}}(s\mathbf{I} - \mathbf{A_1})^{-1}\mathbf{b_{1k}}, \quad k = 1, 2, 3, 4 \tag{4.106}$$

Hence, the electric torque contributed from the UPFC stabilizer is

$$\Delta T_{uk} = F_{upfck}(j\omega_s) T_{upfck}(j\omega_s) y_k, \quad k = 1, 2, 3, 4 \tag{4.107}$$

where $\omega_s$ is the angular frequency of power oscillation. From Eq. (4.107), it can be seen that the following index can also be used to predict the effectiveness of the UPFC stabilizer to be installed. The index can be used to select the modulation signal to add the damping control signal of the UPFC stabilizer

$$\left| F_{upfck}(j\omega_s) \right|, \quad k = 1, 2, 3, 4 \tag{4.108}$$

#### 4.3.2.3 Robustness of the Damping Control to the Variations of Power System Operating Conditions

Discussions so far have demonstrated that the effectiveness of a stabilizer may change with variations of power system operating conditions (load conditions and network configurations). Design of the stabilizer is carried out based on the linearized model which is obtained at a selected operating condition. Hence, the effectiveness of stabilizer is only guaranteed at the selected operating condition.

However, operating conditions of a power system are variable. In order to ensure that the stabilizer is effective over a certain range of power system operating conditions, when the stabilizer is designed its robustness to the variations of system operating conditions should be considered. This is the topic to be discussed in this section.

It is assumed that the set of system known operating conditions is $\Omega_0$ and the stabilizer is to be designed to ensure its robustness over $\Omega_0$, that is, to guarantee the effectiveness of the stabilizer in damping power system oscillations when system operating conditions change within $\Omega_0$. If it is known that at an operating condition $\sigma_r \in \Omega_0$, the stabilizer is least effective, $\sigma_r \in \Omega_0$ can be selected as the operating condition at which the stabilizer is designed. By doing so, the design can ensure the effectiveness of stabilizer over the set of system operating conditions and hence guarantees the robustness of the stabilizer to the variations of system operating conditions within set $\Omega_0$. Based on this principle, two examples to design a robust stabilizer are introduced as follows.

1. *Robust selection of modulation signal to add UPFC stabilizer*

Selection of a modulation signal to add the damping control signal of a UPFC stabilizer is in order to maximize its effectiveness. When the controllability index is used, the criterion of selection is

$$\Delta u_{selected} = \underset{\Delta u_k}{\mathrm{Max}}(b_{ik} = \mathbf{u}_i^T \mathbf{b}_k), \quad \Delta u_k \in \{\Delta m_e, \Delta m_b, \Delta\delta_e, \Delta\delta_b\} \tag{4.109}$$

When the damping torque index is used, according to Eq. (4.108), the criterion of selection is

$$\Delta u_{selected} = \underset{\Delta u_k}{\mathrm{Max}}\left[\left|F_{upfck}(j\omega_s)\right|, k = 1,2,3,4\right], \Delta u_k \in \{\Delta m_e, \Delta m_b, \Delta\delta_e, \Delta\delta_b\} \tag{4.110}$$

To consider the robustness of stabilizer to the variations of system operating conditions, the above criteria of Eqs. (4.109) and (4.110) should be calculated at a selected operating condition of the power system when

$$\sigma_{selected} = \underset{\sigma}{\mathrm{Min}}(b_{ik} = \mathbf{u}_i^T \mathbf{b}_k), \quad \sigma \in \Omega_0 \tag{4.111}$$

or

$$\sigma_{selected} = \underset{\sigma}{\mathrm{Min}}\left(\left|F_{upfck}(j\omega_s)\right|, k = 1, 2, 3, 4\right), \quad \sigma \in \Omega_0 \tag{4.112}$$

Equations (4.109)–(4.112) mean that the selection is made at an operating condition of power system where the stabilizer is predicted to be least effective and the most effective modulation signal is selected to add the UPFC stabilizer at the selected operating condition.

2. *Design of robust stabilizer by the phase compensation method*

Denote the transfer function of a stabilizer installed in a single-machine infinite-bus power system to be $T_{stab}(s) = K_{stab}T_{stabk}(s)$, where $K_{stab}$ is the gain of the stabilizer. The feedback signal of the stabilizer is $y_{stab}$, and its decomposition into the damping torque and synchronizing torque is

$$y_{stab} = K_{d-stab}\Delta\omega + K_{s-stab}\Delta\delta = \left(K_{d-stab} + \frac{\omega_0 K_{s-stab}}{j\omega_s}\right)\Delta\omega = K_y(j\omega_s)\Delta\omega \tag{4.113}$$

If the forward path of stabilizer to the electromechanical oscillation loop of generator is $F_{stab}(s)$, the damping torque provided by the stabilizer is

$$\Delta T_{D-stab} = \mathrm{Re}[F_{stab}(j\omega_s)K_y(j\omega_s)T_{stab}(j\omega_s)]\Delta\omega \tag{4.114}$$

and the synchronizing torque is

$$\Delta T_{S-stab} = -\mathrm{Im}[F_{stab}(j\omega_s)K_y(j\omega_s)T_{stab}(j\omega_s)]\frac{\omega_s}{\omega_0}\Delta\delta \tag{4.115}$$

From Eq. (4.114), it can be seen that with the variations of system operating conditions, change of the damping torque contributed by the stabilizer is mainly determined by variations of $F_{stab}(j\omega_s)K_y(j\omega_s)$. At an operating condition $\sigma_j \in \Omega_0$, if it is denoted that

$$F_{stab}(j\omega_s)K_y(j\omega_s) = H_j\angle\varphi_j, \quad \text{at } \sigma_j \in \Omega_0 \tag{4.116}$$

$$T_{stab}(j\omega_s) = H\angle - \phi \tag{4.117}$$

From Eqs. (4.114) and (4.115), the damping torque and synchronizing torque provided by the stabilizer at $\sigma_j \in \Omega_0$ can be obtained as

$$\Delta T_{d-stab} = H_j H\cos(\varphi_j - \phi)\Delta\omega \tag{4.118}$$

$$\Delta T_{s-stab} = -H_j H\sin(\varphi_j - \phi)\frac{\omega_s}{\omega_0}\Delta\delta \tag{4.119}$$

If it has

$$\varphi_{max} = \max(\varphi_j), \quad \sigma_j \in \Omega_0 \tag{4.120}$$

in order for the stabilizer to provide the power system with positive synchronizing torque at all known operating conditions $\sigma_j \in \Omega_0$, design of the stabilizer must ensure

$$\phi \geq \varphi_j, \sigma_j \in \Omega_0 \tag{4.120a}$$

Hence, simply it can choose

$$\phi = \varphi_{max} \tag{4.121}$$

Therefore, from Eqs. (4.118) and (4.121), the damping torque contributed by the stabilizer can be obtained as

$$\Delta T_{d-stab} = H_j H \cos(\varphi_j - \varphi_{max}) \Delta\omega \tag{4.122}$$

Equation (4.122) indicates that the effectiveness of stabilizer to be designed is measured by $H_j \cos(\varphi_j - \varphi_{max})$ at $\sigma_j \in \Omega_0$. Hence, the operating condition at which the stabilizer will be least effective can be selected by using the following criterion:

$$H_{selected} \cos(\varphi_{selected} - \varphi_{max}) = \underset{j}{\mathrm{Min}}[H_j \cos(\varphi_j - \varphi_{max}), \sigma_j \in \Omega_0] \tag{4.123}$$

At $\sigma_{selected} \in \Omega_0$ determined by using the criterion of Eq. (4.123), the stabilizer can be designed. The design will ensure that more damping torque will be provided by the stabilizer when the power system operates at other operating conditions. Therefore, the design guarantees the robustness of the stabilizer to the variations of power system operating conditions.

## 4.4 Examples

### *4.4.1 An Example Single-Machine Infinite-Bus Power System Installed with a BESS Stabilizer*

#### 4.4.1.1 Extended Heffron–Phillips Model of the Example Power System Installed with a BESS

Configuration of an example single-machine infinite-bus power system installed with a shunt-connected BESS is as same as that shown in Fig. 4.2. The model of BESS is shown in Fig. 4.18.

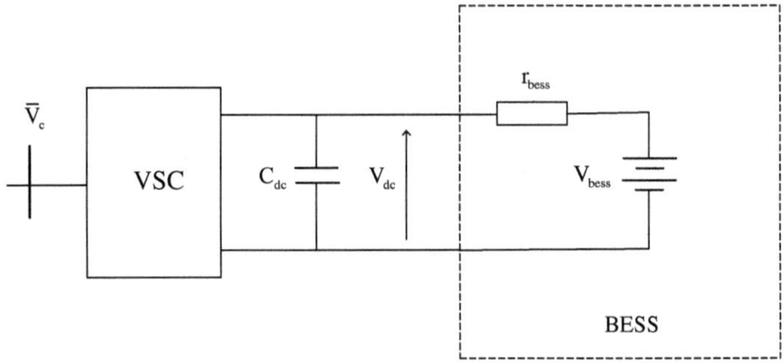


**Fig. 4.18** Model of BESS

$$I_{dc2} = \frac{V_{bess} - V_{dc}}{r_{bess}} \tag{4.124}$$

Parameters of the example system are as follows:

| | |
|---|---|
| Generator | $x_d = 1.0,\ x_q = 0.6,\ x'_d = 0.3,\ M = 8s.,\ D = 0,\ T'_{d0} = 5.044s.$ |
| The AVR | $K_A = 100,\ T_A = 0.01$ |
| Transmission line | $x_{ts} = 0.3,\ x_{sb} = 0.3,\ x_s = 0.15$ |
| The BESS | $k = 3,\ C_{dc} = 1.0,\ V_{dc0} = 1.0,\ V_{bess} = 1.0,\ r_{bess} = 0.01$ |
| Steady-state operating point | $P_{t0} = 0.5,\ V_{t0} = 1.0,\ V_b = 1.0,\ V_{s0} = 1.0$ |

Computation of initial values of various variables at the steady-state operation of example power system is as same as that given in Sect. 3.3.1.2 as follows:

$$Q_{b0} = \frac{V_b}{X_{sb}}\left[\sqrt{V_{s0}^2 - \left(\frac{x_{sb}P_{t0}}{V_b}\right)^2} - V_b\right] = -0.0337$$

$$\bar{I}_{sb0} = \frac{P_{t0} - jQ_{b0}}{V_b} = 0.5 + j0.0377$$

$$\bar{V}_{s0} = jx_{sb}\bar{I}_{sb} + V_b = 0.9887 + j0.15 = 1.0\angle 8.6^\circ$$

$$Q_{s0} = \frac{\sqrt{V_{t0}^2V_{s0}^2 - (P_{t0}x_{ts})^2} - V_{s0}^2}{x_{ts}} = -0.0377$$

$$\bar{I}_{ts0} = \frac{P_{t0} - jQ_{s0}}{\bar{V}_{s0}} = 0.4887 + j0.1123$$

$$\bar{I}_{s0} = \bar{I}_{ts0} - \bar{I}_{sb0} = -0.0113 + j0.0746$$

$$\bar{V}_{t0} = jx_{ts}\bar{I}_{ts0} + \bar{V}_{s0} = 0.955 + j0.2966$$

$$\bar{E}_Q = \bar{V}_{t0} + jx_q\bar{I}_{ts0} = 0.8876 + j0.5898 = 1.07\angle 33.6^\circ$$

Denote the real and imaginary part of a complex number by Re( ) and Im( ), respectively, from the phasor diagram of Fig. 4.3 it can have

$$
\begin{aligned}
i_{tsd0} &= \mathrm{Re}(\bar{I}_{ts0})\sin\delta_0 - \mathrm{Im}(\bar{I}_{ts0})\cos\delta_0 = 0.1769 \\
i_{tsq0} &= \mathrm{Re}(\bar{I}_{ts0})\cos\delta_0 + \mathrm{Im}(\bar{I}_{ts0})\sin\delta_0 = 0.4692 \\
v_{td0} &= \mathrm{Re}(\overline{V}_{t0})\sin\delta_0 - \mathrm{Im}(\overline{V}_{t0})\cos\delta_0 = 0.2815 \\
v_{tq0} &= \mathrm{Re}(\overline{V}_{t0})\cos\delta_0 + \mathrm{Im}(\overline{V}_{t0})\sin\delta_0 = 0.9596 \\
i_{sd0} &= \mathrm{Re}(\bar{I}_{s0})\sin\delta_0 - \mathrm{Im}(\bar{I}_{s0})\cos\delta_0 = -0.0684 \\
i_{sq0} &= \mathrm{Re}(\bar{I}_{s0})\cos\delta_0 + \mathrm{Im}(\bar{I}_{s0})\sin\delta_0 = 0.0318
\end{aligned}
$$

$$
\begin{aligned}
E'_{q0} &= E_Q - (x_d - x'_d)i_{tsd0} = 1.0126 \\
\overline{V}_{c0} &= \overline{V}_{s0} - jx_s\bar{I}_{s0} = 0.9999 + j0.1517 = 1.011\angle 8.6^\circ \\
m_0 &= \frac{V_{c0}}{kV_{dc0}} = 0.3371
\end{aligned}
$$

From Fig. 4.3, it can be seen that $\psi_0$ and $\psi_{x0} = 8.6^\circ$ is the phase of $\overline{V}_{c0}$ in the d–q coordinate and x–y coordinate, respectively. Hence,

$$
\psi_0 = \psi_{x0} + \frac{\pi}{2} - \delta_0 = 65.0250^\circ
$$

From Eqs. (4.7) and (4.15), it is obtained that

$$
\begin{bmatrix} c_{11} & c_{12} \\ c_{21} & c_{22} \end{bmatrix} = \begin{bmatrix} -0.6667 & 1.0000 \\ -2.6667 & 0.6667 \end{bmatrix}, \quad \begin{bmatrix} d_{11} & d_{12} \\ d_{21} & d_{22} \end{bmatrix} = \begin{bmatrix} -0.9524 & 1.4286 \\ -2.8571 & 0.9524 \end{bmatrix}
$$

By using Eqs. (4.16)–(4.19), following coefficients in the Heffron–Phillips model of Figs. 4.4 and 4.5 are obtained

$$
\begin{aligned}
&K_1 = 0.3330,\ K_2 = 0.6702,\ K_3 = 2,\ K_4 = 0.1845,\ K_5 = -0.0290, \\
&K_6 = 0.5483,\ K_7 = -1.9772,\ K_8 = 0.4067,\ K_9 = -0.0746 \\
&K_{dm} = -0.2212,\ K_{pm} = 0.5355,\ K_{qm} = -1.8130,\ K_{p\psi} = -0.7086, \\
&K_{q\psi} = -0.2847,\ K_{d\psi} = -2.6858,\ K_{pdc} = 0.1805,\ K_{qdc} = -0.6112, \\
&K_{vm} = 0.8882,\ K_{v\psi} = 0.0138,\ K_{vdc} = 0.2994
\end{aligned}
$$

From Eq. (4.124), it can have

$$\Delta I_{dc2} = -\frac{1}{r_{bess}}\Delta V_{dc} \tag{4.125}$$

From Eqs. (4.19) and (4.125), state equation of the example power system is obtained as

$$\dot{\mathbf{X}} = \mathbf{AX} + \mathbf{Bu} \tag{4.126}$$

where

$$\mathbf{X} = \begin{bmatrix} \Delta\delta \\ \Delta\omega \\ \Delta E'_q \\ \Delta E'_{fd} \\ \Delta V_{dc} \end{bmatrix}, \quad \mathbf{u} = \begin{bmatrix} \Delta\phi \\ \Delta m \end{bmatrix},$$

$$\mathbf{A} = \begin{bmatrix} 0 & \omega_0 & 0 & 0 & 0 \\ -\frac{K_1}{M} & -\frac{D}{M} & -\frac{K_2}{M} & 0 & -\frac{K_{pdc}}{M} \\ -\frac{K_4}{T'_{d0}} & 0 & -\frac{K_3}{T'_{d0}} & \frac{1}{T'_{d0}} & \frac{K_{qdc}}{T'_{d0}} \\ -\frac{K_A K_5}{T_A} & 0 & -\frac{K_A K_6}{T_A} & -\frac{1}{T_A} & -\frac{K_{vdc}}{T_A} \\ K_7 & 0 & K_8 & 0 & K_9 - \frac{1}{C_{dc} r_{bess}} \end{bmatrix} = \begin{bmatrix} 0 & 314.16 & 0 & 0 & 0 \\ -0.1302 & 0 & -0.0838 & 0 & -0.0226 \\ -0.0930 & 0 & -0.3965 & 0.1983 & 0.1212 \\ 428.242 & 0 & -5483.2 & -100 & -2994.2 \\ 0.7086 & 0 & 0.4067 & 0 & -100.0746 \end{bmatrix},$$

$$\mathbf{B} = \begin{bmatrix} 0 & 0 \\ -\frac{K_{p\psi}}{M} & -\frac{K_{pm}}{M} \\ -\frac{K_{q\psi}}{T'_{d0}} & -\frac{K_{qm}}{T'_{d0}} \\ -\frac{K_A K_{v\psi}}{T_A} & -\frac{K_A K_{vm}}{T_A} \\ K_{d\psi} & K_{dm} \end{bmatrix} = \begin{bmatrix} 0 & 0 \\ -0.0886 & -0.0669 \\ -0.0564 & 0.3594 \\ 138.495 & -8882 \\ 2.6858 & -0.2212 \end{bmatrix}$$

Eigenvalues of state matrix are

$$\lambda_1, \lambda_2 = 0.0591 \pm j6.4978$$
$$\lambda_3 = -13.0310$$
$$\lambda_4 = -100.2902$$
$$\lambda_5 = -87.2679$$

$\lambda_1$ and $\lambda_2$ are the pair of electromechanical oscillation mode of the example power system.

#### 4.4.1.2 Design of BESS Stabilizers by Use of the Phase Compensation Method

To work out the transfer function of the forward path from a BESS stabilizer to the electromechanical oscillation loop, the path related to $\Delta\delta$ in the Heffron–Phillips model of Figs. 4.4 and 4.5 can be ignored. Hence, from Fig. 4.4 with $\Delta\delta$ being ignored, it can have

$$\Delta V_{dc} = (K_8\Delta E'_q + K_{dm}\Delta m + K_{d\psi}\Delta\phi)\frac{1}{s - K_9 + \frac{1}{C_{dc}r_{bess}}} \tag{4.127}$$

Figure 4.19 shows the forward path from the BESS to the electromechanical oscillation loop of generator. From Fig. 4.19, it can be obtained

$$\begin{aligned}\Delta E'_q &= \frac{1}{sT'_{d0} + K_3}\left[\frac{K_A}{1+sT_A}(-K_{vdc}\Delta V_{dc} - K_{v\psi}\Delta\psi - K_{vm}\Delta m - K_6\Delta E'_q)\right.\\ &\quad\left. - K_{qdc}\Delta V_{dc} - K_{q\psi}\Delta\phi - K_{qm}\Delta m\right] \\ \Delta T_{et} &= K_2\Delta E'_q + K_{pdc}\Delta V_{dc} + K_{p\psi}\Delta\phi + K_{pm}\Delta m\end{aligned} \tag{4.128}$$

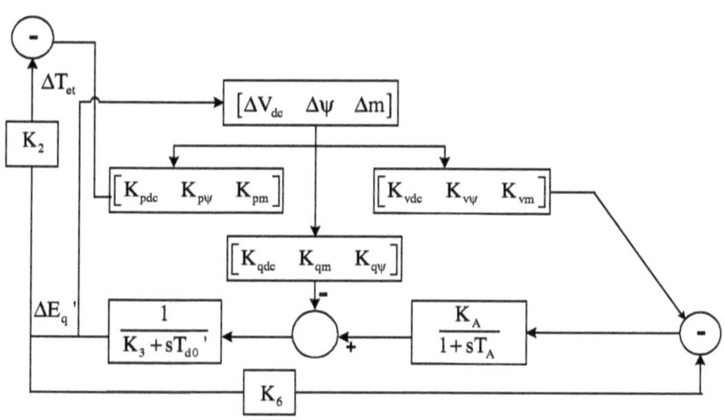


**Fig. 4.19** Forward path of BESS

Deleting variable $\Delta V_{dc}$ and $\Delta E'_q$ in Eqs. (4.126) and (4.127), it can be obtained that

$$\Delta T_{es} = F_{act}(s)\Delta\phi + F_{rct}(s)\Delta m \tag{4.129}$$

With $\omega_s = \mathrm{Im}(\lambda_1 \text{ or } \lambda_2) = 6.4978$, it can be computed that

$$\begin{aligned} F_{act}(j\omega_s) &= 0.7164 - j0.0017 = 0.7164\angle -0.136^\circ \\ F_{rct}(j\omega_s) &= -0.2467 + j0.4706 = 0.5313\angle 117.661^\circ \end{aligned} \tag{4.130}$$

A simple way to calculate the transfer function of the forward path with $s = j\omega_s$ is to obtain the following state-space representation of the system of Eqs. (4.127) and (4.128)

$$\begin{aligned} \dot{\mathbf{X}}_\mathbf{r} &= \mathbf{A_r X_r} + \mathbf{B_r u} \\ \Delta T_{et} &= \mathbf{C_r X_r} + \mathbf{D_r u} \end{aligned} \tag{4.131}$$

where $\mathbf{X_r} = \begin{bmatrix} \Delta E'_q \\ \Delta E'_{fd} \\ \Delta V_{dc} \end{bmatrix}$, $\mathbf{u} = \begin{bmatrix} \Delta\phi \\ \Delta m \end{bmatrix}$, $\mathbf{A_r}$ and $\mathbf{B_r}$ are obtained directly from $\mathbf{A}$ and $\mathbf{B}$ from Eq. (4.126) by deleting two state variables $\Delta\delta$ and $\Delta\omega$, and the output equation is obtained from the second equation in Eq. (4.128). That is

$$\mathbf{A_r} = \begin{bmatrix} -\frac{K_3}{T_{d0}'} & \frac{1}{T_{d0}'} & \frac{K_{qdc}}{T_{d0}'} \\ -\frac{K_A K_6}{T_A} & -\frac{1}{T_A} & -\frac{K_{vdc}}{T_A} \\ K_8 & 0 & K_9 - \frac{1}{C_{dc} r_{bess}} \end{bmatrix} = \begin{bmatrix} -0.3965 & 0.1983 & 0.1212 \\ -5483.2 & -100 & -2994.2 \\ 0.4067 & 0 & -100.0746 \end{bmatrix},$$

$$\mathbf{B_r} = \begin{bmatrix} -\frac{K_{p\psi}}{T_{d0}'} & -\frac{K_{pm}}{T_{d0}'} \\ -\frac{K_A K_{v\psi}}{T_A} & -\frac{K_A K_{vm}}{T_A} \\ K_{d\psi} & K_{dm} \end{bmatrix} = \begin{bmatrix} -0.0564 & 0.3594 \\ 138.495 & -8882 \\ 2.6858 & -0.2212 \end{bmatrix},$$

$$\mathbf{C_r} = \begin{bmatrix} K_2 \\ 0 \\ K_{pdc} \end{bmatrix} = \begin{bmatrix} 0.6702 \\ 0 \\ 0.1805 \end{bmatrix}, \quad \mathbf{D_r} = \begin{bmatrix} K_{p\psi} & K_{pm} \end{bmatrix} = \begin{bmatrix} 0.7086 & 0.5355 \end{bmatrix}$$

It can have

$$\Delta T_{et} = [\mathbf{C_r^T}(s\mathbf{I} - \mathbf{A_r})^{-1}\mathbf{B_r} + \mathbf{D_r}]\mathbf{u} = F_{act}(s)\Delta\phi + F_{rct}(s)\Delta m \tag{4.132}$$

Hence,

$$[F_{act}(j\omega_s) \quad F_{rct}(j\omega_s)] = \mathbf{C}_r^T(j\omega_s I - \mathbf{A}_r)^{-1}\mathbf{B}_r + \mathbf{D}_r$$
$$= [0.7164\angle -0.136^\circ \quad 0.5313\angle 117.661^\circ]$$

The above results are as same as those given in Eq. (4.130). Take the transfer function of the BESS stabilizer to be

$$T_{bess}(s) = K_{bess}\frac{(1+sT_2)}{(1+sT_1)} \quad \text{with } T_1 = 0.38s.$$

Deviation of active power delivered along the transmission line, $\Delta P_t$, is taken as the feedback signal of the stabilizer. Hence, the electric torque supplied by a BESS active power stabilizer is

$$\begin{aligned}\Delta T_{et-active} &= F_{act}(j\omega_s)\Delta\phi = F_{act}(j\omega_s)T_{bess}(j\omega_s)\Delta P_t \\ &= F_{act}(j\omega_s)T_{bess}(j\omega_s)(D + j\omega_s M)\Delta\omega\end{aligned} \tag{4.133}$$

The electric torque supplied by a BESS reactive power stabilizer is

$$\begin{aligned}\Delta T_{et-reactive} &= F_{rct}(j\omega_s)\Delta m = F_{rct}(j\omega_s)T_{bess}(j\omega_s)\Delta P_t \\ &= F_{rct}(j\omega_s)T_{bess}(j\omega_s)(D + j\omega_s M)\Delta\omega\end{aligned} \tag{4.134}$$

The BESS active and reactive power stabilizer can be designed by using the phase compensation method to supply a pure damping torque $\Delta T_{et-active} = 9\Delta\omega$ and $\Delta T_{et-reactive} = 20\Delta\omega$, respectively. The parameters of stabilizers obtained from the design are as follows:

1. The BESS active power stabilizers: $K_{bess} = 15.9567$, $T_{2-bess} = 0.3826s.$;
2. The BESS reactive power stabilizers: $K_{bess} = 99.7980$, $T_{2-bess} = 0.0151s.$

Table 4.1 gives the computational results of oscillation mode of example power system without and with the BESS stabilizers installed. They confirm the effectiveness of the BESS stabilizer in damping the power oscillations.

Figure 4.20 shows the simulation results of the example power system without and with the BESS stabilizers installed. In the simulation, a three-phase to-earth

**Table 4.1** Oscillation mode of example power system when $P_{t0} = 0.5$ p.u.

| Without any stabilizer | With active power stabilizer | With reactive power stabilizer |
|---|---|---|
| $0.0591 \pm j6.4978$ | $-0.6541 \pm j6.4621$ | $-0.5315 \pm j5.1291$ |

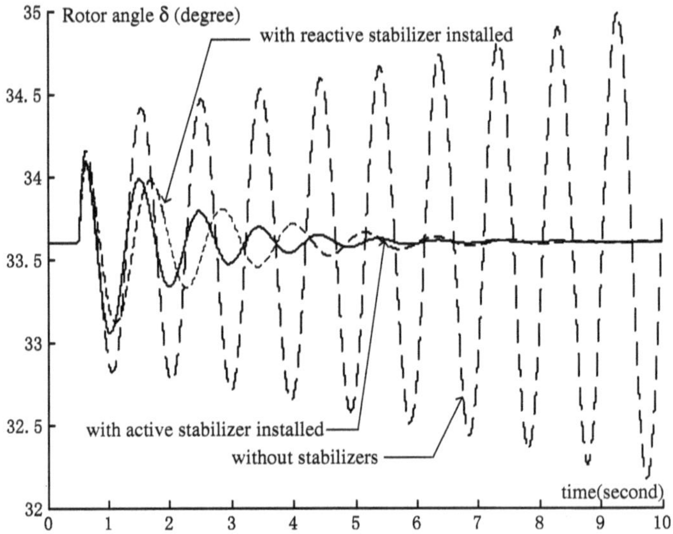


**Fig. 4.20** Simulation results of example power system when $P_{t0} = 0.5$ p.u.

short circuit occurred at 0.5 s of simulation for 100 ms. The results confirm the effectiveness of the BESS stabilizers designed by use of the phase compensation method to damp the power oscillation.

#### 4.4.1.3 Robustness of the BESS Stabilizers to the Variations of Power System Loading Conditions

Effectiveness of the BESS stabilizers is examined when $P_{t0}$ varies from 0.1 p.u. to 1.0 p.u. with $V_{t0} = 1.0$, $V_b = 1.0$, $V_{s0} = 1.0$ being fixed. Figure 4.21 shows the damping torque provided by the BESS stabilizers with variations of system load conditions. From Fig. 4.21, it can be seen that provision of the damping torque by the BESS active power stabilizer changes much less than that by the reactive power stabilizer. Hence, the active power stabilizer is more robust to the variations of system load conditions. Table 4.2 gives the computational results of electromechanical oscillation mode. Figures 4.22 and 4.23 are the results of simulation of example power system at different load conditions. Those results obviously confirm that the BESS active power stabilizer is more robust than the reactive power stabilizer to the variations of power system loading conditions as concluded by the discussion in Sect. 4.1.2.2.

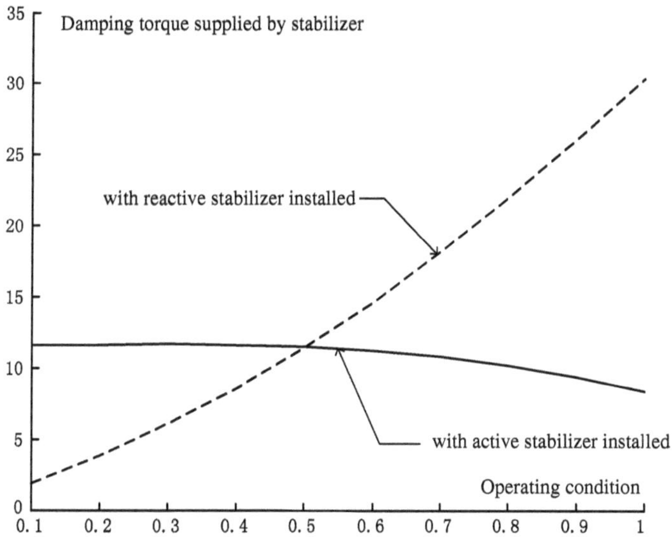


**Fig. 4.21** Damping torque provided by BESS stabilizers with variations of system loading conditions

**Table 4.2** Oscillation mode of example power system with variable loading conditions

| | $P_{t0} = 0.2$ p.u. | $P_{t0} = 0.9$ p.u. |
|---|---|---|
| Without any stabilizer | $0.0070 \pm j6.3064$ | $0.2505 \pm j6.8670$ |
| With active power stabilizer | $-0.6685 \pm j6.2676$ | $-0.5427 \pm j6.8615$ |
| With reactive power stabilizer | $-0.2050 \pm j5.7870$ | $-0.9353 \pm j4.6027$ |

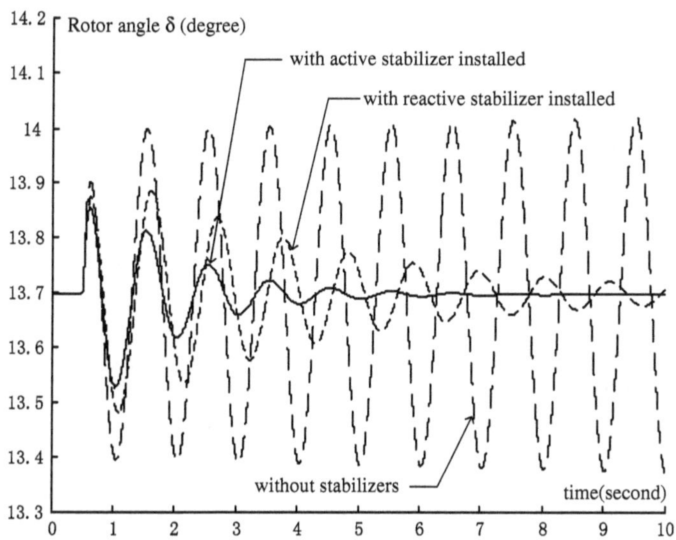


**Fig. 4.22** Simulation results of example power system when $P_{t0} = 0.2$ p.u.

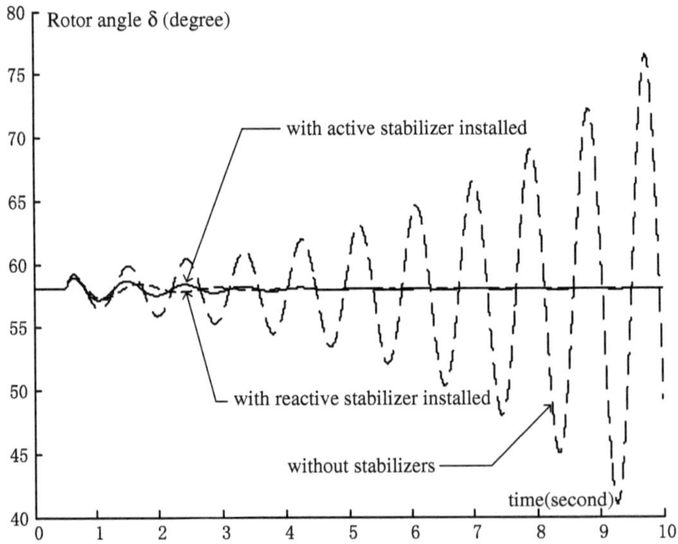


**Fig. 4.23** Simulation results of example power system when $P_{t0} = 0.9$ p.u.

### 4.4.2 *An Example Single-Machine Infinite-Bus Power System Installed with a UPFC Stabilizer*

#### 4.4.2.1 Extended Heffron–Phillips Model of Example Power System Installed with a UPFC

Configuration of an example single-machine infinite-bus power system installed with a UPFC is as same as that shown in Fig. 4.12. Parameters of example system are as follows:

| | |
|---|---|
| Generator | $x_d = 1.0$, $x_q = 0.6$, $x_d' = 0.3$, $M = 8s.$, $D = 0$, $T_{d0}' = 5.044s.$ |
| The AVR | $K_A = 100$, $T_A = 0.01s.$ |
| Transmission line | $x_{tb} = 0.3$, $x_{te} = 0.15$, $x_{bt} = 0.15.$ |
| The UPFC | $k_e = 3$, $k_b = 3$, $C_{dc} = 1.0$, $V_{dc0} = 1.0$, $x_{es} = 0.02$, $x_{bs} = 0.02.$ |
| Steady-state operating point | $P_{t0} = 0.1$, $V_{t0} = 1.05$, $V_b = 1.0.$ |

This example is used to demonstrate the design and effectiveness of the UPFC stabilizers. For the simplicity of presentation, normal control functions of the UPFC are not included in the following demonstration. The procedure will be exactly as same as what is presented below when the UPFC normal control functions are considered. Hence, it is assumed that $\bar{I}_{es} = 0, \bar{V}_{bt} = 0$ which can give

$$\bar{V}_{bs} = -jx_{bs}\bar{I}_{bs}, \quad \bar{V}_{es} = \bar{V}_{et} \tag{4.135}$$

Total reactance of the transmission lines is $x_t = \frac{x_{tb}(x_{te}+x_{bt})}{x_{tb}+x_{te}+x_{bt}} = 0.15$. From $\bar{V}_b\left(\frac{\bar{V}_{t0}-\bar{V}_b}{jx_t}\right)^* = P_{t0}+jQ_{b0}$, the reactive power received at the infinite bus bar $Q_{b0}$ can be obtained as

$$Q_{b0} = \frac{V_b}{x_t}\left[\sqrt{V_{t0}^2 - \left(\frac{x_t P_{t0}}{V_b}\right)^2} - V_b\right] = 0.3326 \tag{4.136}$$

Thus,

$$\begin{aligned}
\bar{I}_{tb0}+\bar{I}_{te0} &= \frac{P_{t0}-jQ_{b0}}{V_b} = 0.1 - j0.3326\\
\bar{V}_{t0} &= jx_t(\bar{I}_{tb0}+\bar{I}_{te0}) + \bar{V}_b = 1.0499 + j0.0150 = 1.05\angle 0.82^\circ\\
\bar{I}_{te0} &= \bar{I}_{bs0} = (\bar{I}_{tb0}+\bar{I}_{te0})\frac{x_{tb}}{x_{tb}+x_{te}+x_{bt}} = 0.05 - j0.1663 = 0.1737\angle -73.27^\circ\\
\bar{E}_Q &= \bar{V}_{t0} + jx_q(\bar{I}_{tb0}+\bar{I}_{te0}) = 1.2495 + j0.0750 = 1.252\angle 3.435^\circ\\
\bar{V}_{es0} &= \bar{V}_{et0} = \bar{V}_{t0} - jx_{te}\bar{I}_{te0} = 1.0249 + j0.0075 = 1.025\angle 0.419^\circ = m_{e0}k_eV_{dc0}\angle 86.99^\circ\\
\bar{V}_{bs0} &= -jx_{bs}\bar{I}_{bs} = -0.0033 - j0.001 = 0.0034\angle 16.86^\circ = m_{b0}k_bV_{dc0}\angle 86.56^\circ
\end{aligned}$$

Hence, $\delta_0 = 3.435^\circ$, $m_{e0} = 0.3417$, $m_{b0} = 0.0012$, $\delta_{e0} = 0$, $\delta_{b0} = 2.8459^\circ$,

$$\begin{aligned}
v_{td0} &= \mathrm{Re}(\bar{V}_{t0})\sin\delta_0 - \mathrm{Im}(\bar{V}_{t0})\cos\delta_0 = 0.0479,\\
v_{tq0} &= \mathrm{Re}(\bar{V}_{t0})\cos\delta_0 + \mathrm{Im}(\bar{V}_{t0})\sin\delta_0 = 1.0489,\\
i_{ted0} &= \mathrm{Re}(\bar{I}_{te0})\sin\delta_0 - \mathrm{Im}(\bar{I}_{te0})\cos\delta_0 = 0.1690,\\
i_{teq0} &= \mathrm{Re}(\bar{I}_{te0})\cos\delta_0 + \mathrm{Im}(\bar{I}_{te0})\sin\delta_0 = 0.0399,\\
i_{tbd0} &= \frac{v_{tq0} - V_b\cos\delta_0}{x_{tb}} = 0.1690,\\
i_{tbq0} &= \frac{V_b\sin\delta_0 - v_{td0}}{x_{tb}} = 0.0399,\\
v_{etd0} &= \mathrm{Re}(\bar{V}_{et0})\sin\delta_0 - \mathrm{Im}(\bar{V}_{et0})\cos\delta_0 = 0.0539,\\
v_{etq0} &= \mathrm{Re}(\bar{V}_{et0})\cos\delta_0 + \mathrm{Im}(\bar{V}_{et0})\sin\delta_0 = 1.0236,\\
i_{esd0} &= 0, i_{esq0} = 0,\\
i_{bsd0} &= \mathrm{Re}(\bar{I}_{bs0})\sin\delta_0 - \mathrm{Im}(\bar{I}_{bs0})\cos\delta_0 = 0.1690,\\
i_{bsq0} &= \mathrm{Re}(\bar{I}_{bs0})\cos\delta_0 + \mathrm{Im}(\bar{I}_{bs0})\sin\delta_0 = 0.0399,\\
E'_{q0} &= E_Q - (x_d - x'_d)(i_{ted0} + i_{tbd0}) = 1.1503
\end{aligned}$$

Various coefficient matrices in Eqs. (4.86) and (4.88) are computed to be

$$\mathbf{F_q} = \begin{bmatrix} 0.9805 & 0 & 0.4001 \\ -0.0134 & 0 & -0.2689 \end{bmatrix}; \quad \mathbf{F_d} = \begin{bmatrix} 0.0367 & 1.4073 & -7.9475 \\ 0.0408 & 0.1656 & 5.0655 \end{bmatrix}$$

$$\mathbf{G_q} = \begin{bmatrix} 1.1742 & 7.6144 & -0.9001 & -0.0174 \\ -0.7903 & -5.1251 & 0.9515 & 0.0184 \end{bmatrix}$$

$$\mathbf{G_d} = \begin{bmatrix} -23.3120 & 0.4196 & 14.8739 & -0.001 \\ 14.8800 & -0.2678 & -15.8655 & 0.0011 \end{bmatrix}$$

$$\mathbf{H_d} = \begin{bmatrix} 0.8589 & 0 & 0.0262 \end{bmatrix}$$

$$\mathbf{L_d} = \begin{bmatrix} 0.0768 & 0.4979 & 0.0103 & 0.0002 \end{bmatrix}$$

$$\mathbf{H_q} = \begin{bmatrix} -0.0416 & 0.2641 & 0.4323 \end{bmatrix}$$

$$\mathbf{L_q} = \begin{bmatrix} 1.2648 & -0.0228 & 0.1487 & 0 \end{bmatrix}$$

Following state equation of extended Heffron–Phillips model of Eq. (4.95) is obtained

$$\begin{bmatrix} \Delta\dot{\delta} \\ \Delta\dot{\omega} \\ \Delta\dot{E}'_q \\ \Delta\dot{E}'_{fd} \\ \Delta\dot{V}_{dc} \end{bmatrix} = \begin{bmatrix} 0 & 314.1593 & 0 & 0 & 0 \\ -0.2244 & 0 & -0.0173 & 0 & -0.0025 \\ -0.0192 & 0 & -0.5387 & 0.1983 & 0.2000 \\ 23.2301 & 0 & -2638.0 & -100 & -4330.5 \\ 1.0061 & 0 & 0.0759 & 0 & -0.0189 \end{bmatrix} \begin{bmatrix} \Delta\delta \\ \Delta\omega \\ \Delta E'_q \\ \Delta E'_{fd} \\ \Delta V_{dc} \end{bmatrix}$$
$$\begin{bmatrix} 0 & 0 & 0 & 0 \\ -0.0074 & -0.1301 & -0.0012 & -0.00005 \\ 0.5851 & -0.0105 & 0.0688 & -0.000004 \\ -12670 & 0.1463 & -1490.5 & 0.0127 \\ -0.0549 & 7.7987 & 0.0308 & -0.0172 \end{bmatrix} \begin{bmatrix} \Delta m_E \\ \Delta\delta_E \\ \Delta m_B \\ \Delta\delta_B \end{bmatrix} \tag{4.137}$$

Eigenvalues of state matrix are

$$\lambda_1 = -94.4375$$
$$\overline{\lambda}_2, \overline{\lambda}_3 = -0.0106 \pm j8.4249$$
$$\lambda_4 = -6.0714$$
$$\lambda_5 = -0.0276$$

Obviously, $\lambda_2, \lambda_3 = -0.0106 \pm j8.4249$ are the pair of electromechanical oscillation mode of example power system.

#### 4.4.2.2 Robust Selection of a Modulation Signal to Add the Damping Control Signal of the UPFC Stabilizer by Using the Damping Torque Analysis and Residue Index

With $V_{t0} = 1.05$, $V_b = 1.0$, $P_{t0}$ changes from 0.1 p.u. to 1.2 p.u. Deviation of modulation signal to add a UPFC stabilizer is $\Delta u_k$, $k = 1, 2, 3, 4$ where

$$\Delta u_1 = \Delta m_e, \quad \Delta u_2 = \Delta\delta_e, \quad \Delta u_3 = \Delta m_b, \quad \Delta u_4 = \Delta\delta_b \tag{4.138}$$

According to the definition of Eq. (4.108) and from Fig. 4.17, $|F_{upfck}(j\omega_s)|$, $k = 1, 2, 3, 4$ is calculated as shown in Fig. 4.24 ($j\omega_s = j8.4249$).

From Fig. 4.24, it can be seen that

$$\sigma_{selected} = \{P_{t0} = 0.1\} = \underset{\sigma}{\mathrm{Min}}(|F_{upfck}(j\omega_s)|,\ k = 1, 2, 3, 4,\ \sigma \in \{P_{t0} = 0.1 - 1.2\})$$

It means that the UPFC stabilizer should be designed at load condition $P_{t0} = 0.1$ when the stabilizer is least effective to damp the electromechanical oscillation mode. It is also shown in Fig. 4.23 that the stabilizer can provide the most amount of damping torque when it is added on the modulation signal $\Delta\delta_e$. That is

$$\Delta u_{selected} = \Delta\delta_e = \underset{\Delta u_k}{\mathrm{Max}}[|F_{upfck}(j\omega_s)|, k = 1, 2, 3, 4, \Delta u_k \in \{\Delta m_e, \Delta m_b, \Delta\delta_e, \Delta\delta_b\}]$$

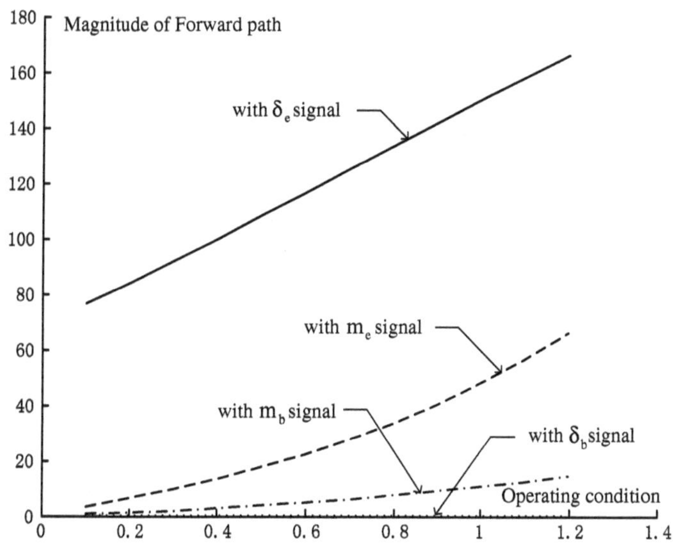


**Fig. 4.24** Computational result of $|F_{upfck}(j\omega_s)|, k = 1, 2, 3, 4$

Hence, to have a robust UPFC stabilizer, the stabilizer should be added on $\Delta\delta_e$ and the design carried out at operating condition $P_{t0} = 0.1$, $V_{t0} = 1.05$, $V_b = 1.0$.

The robust modulation signal to add a UPFC stabilizer can also be selected by using the residue index defined by Eq. (4.104). Take the feedback signal of stabilizer to be the locally available signal, the deviation of line active power $P_{te}$. Similar to Eq. (4.83), it has

$$P_{te} = v_{tq}i_{teq} + v_{td}i_{ted} \tag{4.139}$$

By using Eqs. (4.86) and (4.88), similar to the derivation of Eq. (4.91), linearization of Eq. (4.139) can be obtained to be

$$\begin{aligned}\Delta P_{te} &= i_{teq0}\Delta v_{tq} + i_{ted0}\Delta v_{td} + v_{tq0}(\Delta i_{esq} + \Delta i_{bsq}) + v_{td0}(\Delta i_{esd} + \Delta i_{bsd}) \\ &= \begin{bmatrix} c_1 & 0 & c_3 & 0 & c_5 \end{bmatrix}\begin{bmatrix}\Delta\delta \\ \Delta\omega \\ \Delta E'_q \\ \Delta E'_{fd} \\ \Delta V_{dc}\end{bmatrix} + \begin{bmatrix} d_1 & d_2 & d_3 & d_4 \end{bmatrix}\begin{bmatrix}\Delta m_e \\ \Delta\delta_e \\ \Delta m_b \\ \Delta\delta_b\end{bmatrix}\end{aligned} \tag{4.140}$$

where

$$\begin{aligned}\begin{bmatrix} c_1 & 0 & c_3 & 0 & c_5 \end{bmatrix} &= I_{tq0}\mathbf{H_q} + I_{td0}\mathbf{H_d} + V_{tq0}\mathbf{F_{q-1}} + V_{tq0}\mathbf{F_{q-2}} \\ &\quad + V_{td0}\mathbf{F_{d-1}} + V_{td0}\mathbf{F_{d-2}} = \begin{bmatrix} 1.1616 & 0 & 0.0859 & 0 & 0.0212 \end{bmatrix} \\ \begin{bmatrix} d_1 & d_2 & d_3 & d_4 \end{bmatrix} &= I_{tq0}\mathbf{L_q} + I_{td0}\mathbf{L_d} + V_{tq0}\mathbf{G_{q-1}} + V_{tq0}\mathbf{G_{q-2}} \\ &\quad + V_{td0}\mathbf{G_{d-1}} + V_{td0}\mathbf{G_{d-2}} = \begin{bmatrix} 0.0620 & 2.7016 & 0.0141 & 0.0011 \end{bmatrix}\end{aligned}$$

$\mathbf{F_{q-1}}$ and $\mathbf{F_{q-2}}$, $\mathbf{F_{d-1}}$ and $\mathbf{F_{d-2}}$, $\mathbf{G_{q-1}}$ and $\mathbf{G_{q-2}}$ and $\mathbf{G_{d-1}}$ and $\mathbf{G_{d-2}}$ are the first and second rows of $\mathbf{F_q}$, $\mathbf{F_d}$, $\mathbf{G_q}$, and $\mathbf{G_d}$, respectively.

At the selected operating condition $\sigma_{selected} = \{P_{t0} = 0.1, V_{t0} = 1.05, V_{b0} = 1.0\}$, for the oscillation mode $\bar{\lambda}_{2,3} = -0.0106 \pm j8.4249$, the right and left eigenvectors are obtained as

$$\bar{\mathbf{v}}_2 = \begin{bmatrix} -0.0091 + j0.0206 \\ 0.7702 + j0.3380 \\ -0.0013 + j0.0008 \\ -0.0000 + j0.0000 \\ -0.0010 - j0.0010 \end{bmatrix} \quad \text{and} \quad \bar{\mathbf{v}}_3 = \begin{bmatrix} -0.0091 - j0.0206 \\ 0.7702 - j0.3380 \\ -0.0013 - j0.0008 \\ -0.0000 - j0.0000 \\ -0.0010 + j0.0010 \end{bmatrix}$$

$$\bar{\mathbf{w}}_2 = \begin{bmatrix} -0.0899 - j0.2015 \\ 0.0054 - j0.0024 \\ 0.0021 - j0.0224 \\ 0.9747 \\ -0.0243 + j0.0107 \end{bmatrix} \quad \text{and} \quad \bar{\mathbf{w}}_3 = \begin{bmatrix} -0.0899 + j0.2015 \\ 0.0054 + j0.0024 \\ 0.0021 + j0.0224 \\ 0.9747 \\ -0.0243 - j0.0107 \end{bmatrix}$$

From Eqs. (4.102) and (4.103), the controllability of four input signals as far as the electromechanical oscillation mode is concerned is calculated as

$$k = 1, \quad \Delta u_1 = \Delta m_E,$$
$$\bar{b}_{21} = \bar{\mathbf{w}}_2^T \mathbf{b_1} = 2.4108 - j2.3637, \quad \bar{b}_{31} = \bar{\mathbf{w}}_3^T \mathbf{b_1} = 2.4108 + j2.3637$$
$$k = 2, \quad \Delta u_2 = \Delta \delta_E,$$
$$\bar{b}_{22} = \bar{\mathbf{w}}_2^T \mathbf{b_2} = -10.7641 - j5.1591, \quad \bar{b}_{32} = \bar{\mathbf{w}}_3^T \mathbf{b_2} = -10.7641 + j5.1591$$
$$k = 3, \quad \Delta u_3 = \Delta m_b,$$
$$\bar{b}_{23} = \bar{\mathbf{w}}_2^T \mathbf{b_3} = 0.2548 - j0.2928, \quad \bar{b}_{33} = \bar{\mathbf{w}}_3^T \mathbf{b_3} = 0.2548 + j0.2928$$
$$k = 4, \quad \Delta u_4 = \Delta \delta_b,$$
$$\bar{b}_{24} = \bar{\mathbf{w}}_2^T \mathbf{b_4} = -0.0023 - j0.0001, \quad \bar{b}_{34} = \bar{\mathbf{w}}_3^T \mathbf{b_4} = -0.0023 + j0.0001$$

where

$$\mathbf{b_1} = \begin{bmatrix} 0 \\ -0.0074 \\ 0.5851 \\ -12670 \\ -0.0549 \end{bmatrix}, \ \mathbf{b_2} = \begin{bmatrix} 0 \\ -0.1301 \\ -0.0105 \\ 0.1463 \\ 7.7986 \end{bmatrix}, \ \mathbf{b_3} = \begin{bmatrix} 0 \\ -0.0012 \\ 0.0688 \\ -1490.5 \\ 0.0308 \end{bmatrix}, \ \mathbf{b_4} = \begin{bmatrix} 0 \\ -0.0000 \\ -0.0000 \\ 0.0127 \\ -0.0172 \end{bmatrix},$$

The observability is given by

$$\bar{c}_{2k} = \mathbf{c}_\mathbf{k}^T \bar{\mathbf{v}}_\mathbf{2} = -0.0002 - j0.0029 \quad \text{and} \quad \bar{c}_{3k} = \mathbf{c}_\mathbf{k}^T \bar{\mathbf{v}}_\mathbf{3} = -0.0002 + j0.0029$$

where

$$\mathbf{c}_\mathbf{k}^T = [1.1616 \quad 0 \quad 0.1387 \quad 0 \quad 0.0202] \quad (k = 1, 2, 3, 4)$$

The residue index for four input signals $\{\Delta m_E, \Delta \delta_E, \Delta m_B, \Delta \delta_B\}$ of the UPFC stabilizer, thus, is obtained as

$$k = 1, \quad \Delta u_1 = \Delta m_E,$$
$$\overline{R}_{21} = \bar{b}_{21}\bar{c}_{21} = -0.0073 - j0.0065; \quad \overline{R}_{31} = \bar{b}_{31}\bar{c}_{31} = -0.0073 + j0.0065;$$
$$k = 2, \quad \Delta u_2 = \Delta \delta_E,$$
$$\overline{R}_{22} = \bar{b}_{22}\bar{c}_{22} = -0.0128 + j0.0322; \quad \overline{R}_{32} = \bar{b}_{32}\bar{c}_{32} = -0.0128 - j0.0322;$$
$$k = 3, \quad \Delta u_3 = \Delta m_b,$$
$$\overline{R}_{23} = \bar{b}_{23}\bar{c}_{23} = -0.0009 - j0.0007; \quad \overline{R}_{33} = \bar{b}_{33}\bar{c}_{33} = -0.0009 + j0.0007;$$
$$k = 4, \quad \Delta u_4 = \Delta \delta_b,$$
$$\overline{R}_{24} = \bar{b}_{24}\bar{c}_{24} = 0.0000 + j0.0000; \quad \overline{R}_{34} = \bar{b}_{34}\bar{c}_{34} = 0.0000 - j0.0000$$

Because $\overline{R}_{2k}$ and $\overline{R}_{3k}$ (k = 1, 2, 3, 4) are pairs of conjugate vectors, and the amplitude of residue measures the overall influence of the UPFC stabilizer on the oscillation mode. It can be obtained that

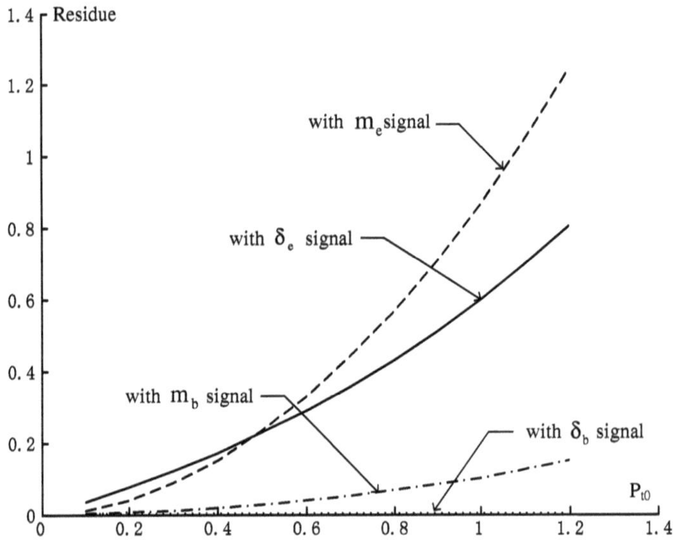


**Fig. 4.25** Computational result of $|R_{ki}|, k = 1, 2, 3, 4$

$$\begin{aligned}
|\overline{R}_{21}| &= |\overline{R}_{31}| = 0.0097 \\
|\overline{R}_{22}| &= |\overline{R}_{32}| = 0.035 \\
|\overline{R}_{23}| &= |\overline{R}_{33}| = 0.0011 \\
|\overline{R}_{24}| &= |\overline{R}_{34}| = 0
\end{aligned}$$

The largest residue indicates the most effective feedback signal to design the UPFC stabilizer. Hence,

$$u_{selected} = \delta_E \quad \text{where} \quad |\overline{R}_{ik}|_{max} = |\overline{R}_{i2}| = 0.035(i = 2, 3)$$

The selection is as same as that obtained previously by using the damping torque analysis. Variation of the residue with system load condition from 0.1 to 1.2 p.u. is shown in Fig. 4.25. Obviously, $P_{t0} = 0.1$ should be selected, which again is as same as that obtained previously.

#### 4.4.2.3 Design of the UPFC Stabilizer by Using the Phase Compensation Method

State Eq. (4.137) and output Eq. (4.140) are obtained at the selected operating condition $P_{t0} = 0.1$, $V_{t0} = 1.05$, $V_b = 1.0$ of the example power system.

With $\Delta\delta_e$ being selected to add the damping control signal of the UPFC stabilizer, it can be obtained that

$$k = 0.2244,\ d = 0,\ C_1 = 1.1616,\ C_2 = 0,\ \mathbf{C_3} = [0.0859 \quad 0 \quad 0.0212],$$
$$d_m = 2.7016$$
$$\mathbf{A_{23}} = [0.0173 \quad 0 \quad 0.0025]$$
$$\mathbf{A_{31}} = \begin{bmatrix} -0.0192 \\ 23.2301 \\ 1.0061 \end{bmatrix}, \quad \mathbf{A_{32}} = \begin{bmatrix} 0 \\ 0 \\ 0 \end{bmatrix}, \quad \mathbf{A_{33}} = \begin{bmatrix} -0.5387 & 0.1983 & 0.2 \\ -2638.0 & -100 & -4330.5 \\ 0.0759 & 0 & -0.0189 \end{bmatrix}$$
$$B_2 = 0.1301$$
$$\mathbf{B_3} = \begin{bmatrix} -0.0105 \\ 0.1463 \\ 7.7986 \end{bmatrix}$$

Results of computation according to Eq. (4.61) are ($j\omega_s = j8.4249$)

$$\overline{K}(j\omega_s) = k + \frac{j\omega_s}{\omega_0} d + \mathbf{A}_{23}^T (j\omega_s \mathbf{I} - \mathbf{A_{33}})^{-1} \left( \mathbf{A_{31}} + \frac{j\omega_s}{\omega_0} \mathbf{A_{32}} \right) = 0.2259 + j0.0006$$
$$\overline{K}_o(j\omega_s) = \left( \frac{\omega_0}{j\omega_s} C_1 + C_2 \right) + \mathbf{C}_3^T (j\omega_s \mathbf{I} - \mathbf{A_{33}})^{-1} \left( \frac{\omega_0}{j\omega_s} \mathbf{A_{31}} + \mathbf{A_{32}} \right) = 0.0663 - j43.5996$$
$$\overline{K}_c(j\omega_s) = \mathbf{A}_{23}^T (j\omega_s \mathbf{I} - \mathbf{A_{33}})^{-1} \mathbf{B_3} + B_2 = 0.1418 + j0.0047$$
$$\overline{K}_{il}(j\omega_s) = \mathbf{C}_3^T (j\omega_s \mathbf{I} - \mathbf{A_{33}})^{-1} \mathbf{B_3} + d_m = 2.7599 + j0.0155$$

Let the design of the UPFC stabilizer provide a pure damping torque $D_{upfc}\Delta\omega = 1\Delta\omega$. Take $\Delta P_{te}$ as the feedback signal. Thus,

$$\Delta\delta_e = T_{upfc}(s)\Delta P_{te} = K_{upfc} \frac{(1 + sT_2)}{(1 + sT_1)} \Delta P_{te} \tag{4.141}$$

where $T_1 = 0.3$ s.

From Eq. (4.64), the forward path of the stabilizer is calculated as

$$\overline{F}_{upfc}(j\omega_s) = \overline{K}_c(j\omega_s)\overline{K}_o(j\omega_s) + D_{upfc}\overline{K}_{il}(j\omega_s) = 2.9764 - j6.1674$$

Thus, parameters of the UPFC stabilizer are obtained as

$$K_{upfc} = 0.2689, \quad T_2 = 0.1288.$$

Table 4.3 gives the computational results of the electromechanical oscillation mode of example power system at three different load conditions. Figures 4.26, 4.27, and 4.28 are the results of nonlinear simulation. Those results confirm that the UPFC stabilizer is effectively designed in damping the power oscillation and robust to the variations of system load conditions.

**Table 4.3** Oscillation mode of example power system with variable load conditions

| | $P_{t0} = 0.1$ p.u. | $P_{t0} = 0.8$ p.u. | $P_{t0} = 1.2$ p.u. |
|---|---|---|---|
| Without stabilizer | $-0.0106 \pm j8.4249$ | $0.0133 \pm j8.9982$ | $0.1457 \pm j9.4250$ |
| With stabilizer | $-0.6880 \pm j7.0257$ | $-1.2283 \pm j7.5835$ | $-1.2930 \pm j8.1842$ |

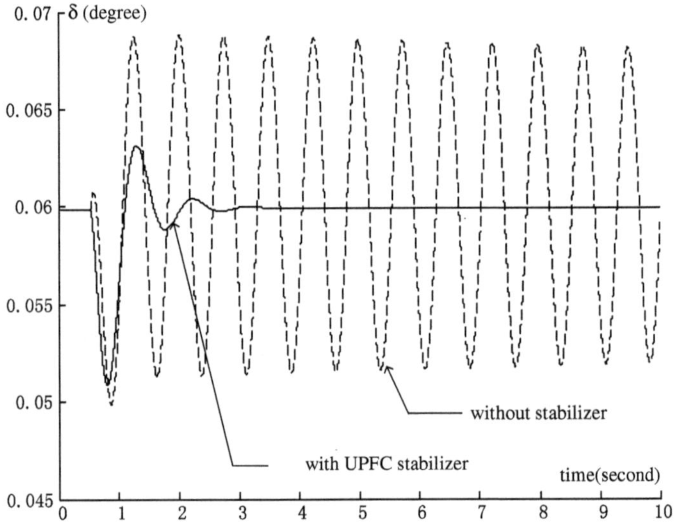


**Fig. 4.26** Simulation results of example power system when $P_{t0} = 0.1$ p.u.

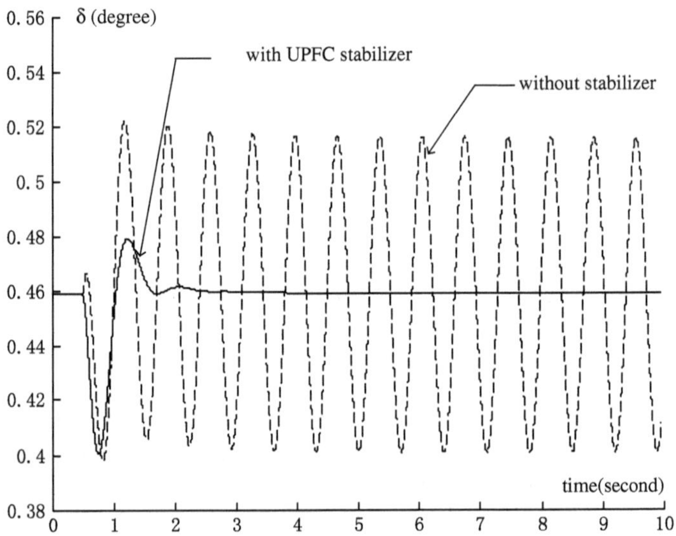


**Fig. 4.27** Simulation results of example power system when $P_{t0} = 0.8$ p.u.

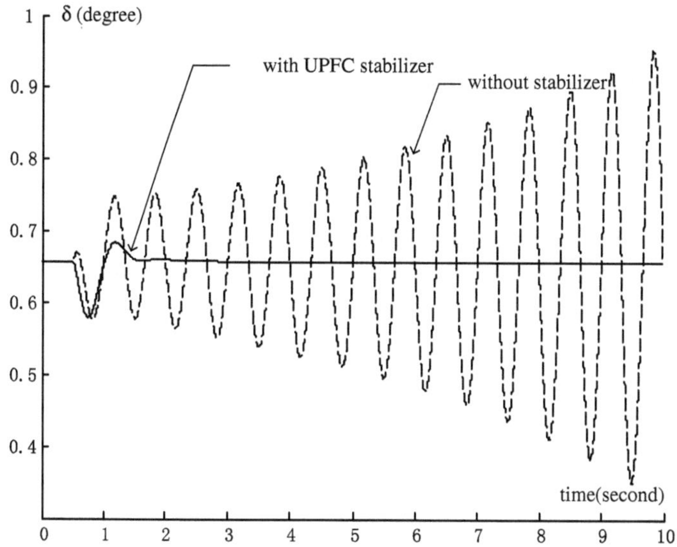


**Fig. 4.28** Simulation results of example power system when $P_{t0} = 1.2$ p.u.

In order to demonstrate that the robustness of the UPFC stabilizer is due to the selection of the load condition at $P_{t0} = 0.1$, $V_{t0} = 1.05$, $V_b = 1.0$ where the stabilizer is designed, the stabilizer is designed at another load condition $P_{t0} = 0.8$, $V_{t0} = 1.05$, $V_b = 1.0$. Then, its effectiveness is checked at different load conditions by modal computation presented in Table 4.4. From the results, it can be seen that though the UPFC stabilizer is designed properly at the selected load condition $P_{t0} = 0.8$, $V_{t0} = 1.05$, $V_b = 1.0$, where the power oscillation is effectively suppressed (see the second column of Table 4.4), its effectiveness is not ensured at $P_{t0} = 0.1$, $V_{t0} = 1.05$, $V_b = 1.0$ (see the first column of Table 4.4).

At the selected operating condition $P_{t0} = 0.1$, $V_{t0} = 1.05$, $V_b = 1.0$, the UPFC stabilizer can also be added to other modulation signals. Table 4.5 gives the results of designing the UPFC stabilizers added on other two modulation signals and the oscillation mode of example power system when stabilizers are installed. From Table 4.5, it can be seen that the gain value of the UPFC stabilizers added on $\Delta m_e$ and $\Delta m_b$ is much greater than that of the stabilizer added on $\Delta\delta_e$ in order to

**Table 4.4** Oscillation mode of example power system with the stabilizer designed at $P_{t0} = 0.8, V_{t0} = 1.05, V_b = 1.0$

| | $P_{t0} = 0.1$ p.u. | $P_{t0} = 0.8$ p.u. | $P_{t0} = 1.2$ p.u. |
|---|---|---|---|
| Without stabilizer | $-0.0106 \pm j8.4249$ | $0.0133 \pm j8.9982$ | $0.0133 \pm j8.9982$ |
| With stabilizer | $-0.4291 \pm j6.8578$ | $-0.7517 \pm j7.3203$ | $-0.7565 \pm j7.5281$ |

**Table 4.5** Results of designing the UPFC stabilizers added on other two modulation signals

| Modulation signal added with a UPFC stabilizer | Parameters of the UPFC stabilizer | Oscillation mode with the stabilizer installed |
|---|---|---|
| $\Delta m_e$ | $K_p = 1.6654; T_p = 0.0926$ | $-0.7030 \pm j6.9400$ |
| $\Delta m_b$ | $K_p = 15.3864; T_p = 0.1069$ | $-0.7078 \pm j7.0774$ |

suppress the power oscillation effectively. This means that more control cost is required. Hence, the UPFC stabilizer added on $\Delta\delta_e$ is the most efficient damping controller.

## References


1. CIGRE TF 30-01-08 Report, Modelling of power electronics equipment (FACTS) in load flow and stability programs (1999)
2. Wang HF (1999) Phillips-Heffron model of power systems installed with STATCOM and applications. IEE Proc Part C 146(5):521–527
3. Du W (2009) Power system small signal oscillation stability as affected by static synchronous compensator (STATCOM) and energy storage system (ESS). PhD thesis, University of Bath, UK
4. Wang HF (2000) Static synchronous series compensation to damp power system oscillations. Int J Power Syst Res 54(2)
5. Wang HF (1999) Applications of modelling UPFC into multi-machine power systems. IEE Proc Part C (3)
6. Wang HF (1999) Damping function of unified power flow controller. IEE Proc Part C. (1)
7. Wang HF (2000) A unified model for the analysis of FACTS devices in damping power system oscillations part III: unified power flow controller. IEEE Trans Power Delivery (3)

# Chapter 5
# A Multi-machine Power System Installed with Power System Stabilizers

## 5.1 Mathematical Model of a Multi-machine Power System Installed with Power System Stabilizers

### *5.1.1 A Two-Machine Power System Installed with Power System Stabilizers*

Figure 5.1 shows the configuration of a two-machine power system where two synchronous generators are connected by a transmission line. This simple system is used in this section to demonstrate the procedure of establishing the mathematical model of a multi-machine power system. Later, it will be used for introducing basic concepts and methods in the analysis of power system oscillations and design of PSSs in the multi-machine power system. For the purpose of demonstration, a load is connected at the terminal of two generators and a node along the transmission line, respectively. The load is represented by equivalent impedance $r_{Li} + jx_{Li}, i = 1, 2, 3$.

#### 5.1.1.1 Network Equations

Connections of two synchronous generators in the two-machine power system are given by following network equations expressed in a common x–y coordinate

$$
\begin{aligned}
\bar{I}_{g1} &= \frac{1}{r_{L1} + jx_{L1}}\bar{V}_{g1} + \frac{1}{jx_{13}}(\bar{V}_{g1} - \bar{V}_3) = (g_{L1} + jb_{L1} + jb_{13})\bar{V}_{g1} - jb_{13}\bar{V}_3 \\
\bar{I}_{g2} &= \frac{1}{r_{L2} + jx_{L2}}\bar{V}_{g2} + \frac{1}{jx_{23}}(\bar{V}_{g2} - \bar{V}_3) = (g_{L2} + jb_{L2} + jb_{23})\bar{V}_{g2} - jb_{23}\bar{V}_3
\end{aligned}
\quad (5.1)
$$



H. Wang and W. Du, *Analysis and Damping Control of Power System Low-frequency Oscillations*, Power Electronics and Power Systems,
DOI 10.1007/978-1-4899-7696-3_5

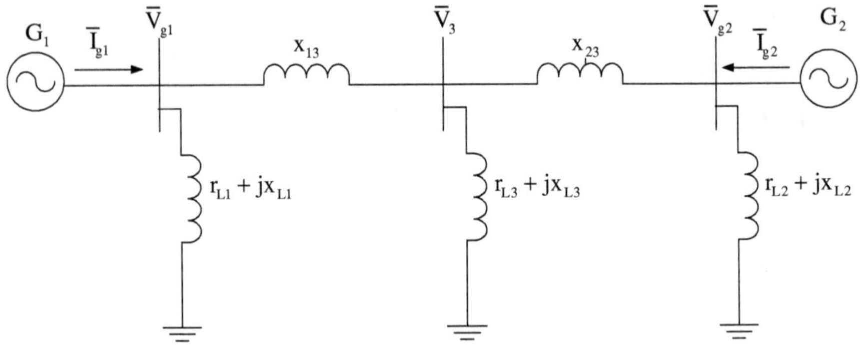


**Fig. 5.1** A two-machine power system

$$\begin{aligned}0 &= \frac{1}{r_{L3}+jx_{L3}}\bar{V}_3 + \frac{1}{jx_{13}}(\bar{V}_3-\bar{V}_{g1}) + \frac{1}{jx_{23}}(\bar{V}_3-\bar{V}_{g2}) \\ &= (g_{L3}+jb_{L3}+jb_{13}+jb_{23})\bar{V}_3 - jb_{13}\bar{V}_{g1} - jb_{23}\bar{V}_{g2}\end{aligned} \tag{5.2}$$

where $jb_{13}=\frac{1}{jx_{13}}, jb_{23}=\frac{1}{jx_{23}}, g_{Li}+jb_{Li}=\frac{1}{r_{Li}+jx_{Li}}, i=1,2,3.$

Matrix form of above equations is

$$\begin{bmatrix}\bar{I}_{g1}\\ \bar{I}_{g2}\\ 0\end{bmatrix} = \begin{bmatrix} g_{L1}+jb_{L1}+jb_{13} & 0 & -jb_{13} \\ 0 & g_{L2}+jb_{L2}+jb_{23} & -jb_{23} \\ -jb_{13} & -jb_{23} & g_{L3}+jb_{L3}+jb_{13}+jb_{23}\end{bmatrix}\begin{bmatrix}\bar{V}_{g1}\\ \bar{V}_{g2}\\ \bar{V}_3\end{bmatrix} \tag{5.3}$$

From Eq. (5.2), it can have

$$\bar{V}_3 = -\frac{jb_{13}}{g_{L3}+jb_{L3}+jb_{13}+jb_{23}}\bar{V}_{g1} - \frac{jb_{23}}{g_{L3}+jb_{L3}+jb_{13}+jb_{23}}\bar{V}_{g2} \tag{5.4}$$

Substituting above equation into Eq. (5.1), it can be obtained that

$$\begin{aligned}\begin{bmatrix}\bar{I}_{g1}\\ \bar{I}_{g2}\end{bmatrix} &= \begin{bmatrix} g_{L1}+jb_{L1}+jb_{13}+\frac{jb_{13}\times jb_{13}}{g_{L3}+jb_{\Sigma}} & \frac{jb_{13}\times jb_{23}}{g_{L3}+jb_{\Sigma}} \\ \frac{jb_{23}\times jb_{13}}{g_{L3}+jb_{\Sigma}} & g_{L2}+jb_{L2}+jb_{23}+\frac{jb_{23}\times jb_{23}}{g_{L3}+jb_{\Sigma}}\end{bmatrix}\begin{bmatrix}\bar{V}_{g1}\\ \bar{V}_{g2}\end{bmatrix} \\ &= \begin{bmatrix}\bar{y}_{N11} & \bar{y}_{N12}\\ \bar{y}_{N21} & \bar{y}_{N22}\end{bmatrix}\begin{bmatrix}\bar{V}_{g1}\\ \bar{V}_{g2}\end{bmatrix} = \bar{\mathbf{Y}}_{\mathbf{N}}\begin{bmatrix}\bar{V}_{g1}\\ \bar{V}_{g2}\end{bmatrix}\end{aligned} \tag{5.5}$$

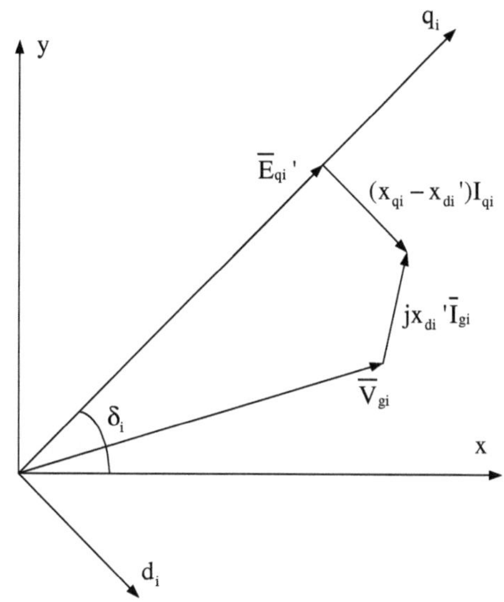


**Fig. 5.2** Phasor diagram in x–y coordinate and $d_i - q_i$ coordinate

where $b_{L\Sigma} = b_{L3} + b_{13} + b_{23}$ and $\bar{\mathbf{Y}}_{\mathbf{N}}$ is the network admittance matrix with only nodes of generator terminal left. Equation (5.2) can be rearranged as

$$\begin{bmatrix} i_{gx1} \\ i_{gy1} \\ i_{gx2} \\ i_{gy2} \end{bmatrix} = \begin{bmatrix} g_{N11} & -b_{N11} & g_{N12} & -b_{N12} \\ b_{N11} & g_{N11} & b_{N12} & g_{N12} \\ g_{N21} & -b_{N21} & g_{N22} & -b_{N22} \\ b_{N21} & g_{N21} & b_{N22} & g_{N22} \end{bmatrix} \begin{bmatrix} v_{gx1} \\ v_{gy1} \\ v_{gx2} \\ v_{gy2} \end{bmatrix} \tag{5.6}$$

where subscript x and y denotes x and y component of a variable in common x–y coordinate, respectively. According to the phasor diagram of Fig. 2.9, in d–q coordinate of the ith generator, $d_i$-$q_i$, it can have $(i = 1, 2)$

$$\bar{V}_{gi} = \bar{E}'_{qi} - jx'_{di}\bar{I}_{gi} + (x_{qi} - x'_{di})i_{qi} \tag{5.7}$$

From Fig. 5.2, it can be seen that in the common x–y coordinate, phase of $\bar{E}'_{qi}$ is $\delta_i$ and that of $(x_{iq} - x'_{id})i_{qi}$ is $\delta_i - 90°$. Hence, in x–y coordinate, Eq. (5.7) is $(i = 1, 2)$

$$\bar{V}_{gi} = E'_{qi}e^{j\delta_i} - jx'_{di}\bar{I}_{gi} + (x_{qi} - x'_{di})i_{qi}e^{j(\delta_i - 90°)} \tag{5.8}$$

In matrix form, Eq. (5.8) is

$$\begin{bmatrix} \bar{V}_{g1} \\ \bar{V}_{g2} \end{bmatrix} = \begin{bmatrix} e^{j\delta_1} & 0 \\ 0 & e^{j\delta_2} \end{bmatrix} \begin{bmatrix} E'_{q1} \\ E'_{q2} \end{bmatrix} - \begin{bmatrix} jx'_{d1} & 0 \\ 0 & jx'_{d2} \end{bmatrix} \begin{bmatrix} \bar{I}_{g1} \\ \bar{I}_{g2} \end{bmatrix} + \begin{bmatrix} (x_{q1} - x'_{d1}) i_{q1} e^{j(\delta_1 - 90^\circ)} & 0 \\ 0 & (x_{q2} - x'_{d2}) i_{q2} e^{j(\delta_2 - 90^\circ)} \end{bmatrix} \begin{bmatrix} i_{q1} \\ i_{q2} \end{bmatrix} \tag{5.9}$$

Denote

$$\bar{E}_{gi} = E'_{qi} e^{j\delta_i} + (x_{qi} - x'_{di}) i_{qi} e^{j(\delta_i - 90^\circ)}, \quad i = 1, 2 \tag{5.10}$$

as an "internal voltage" of generator. Equation (5.9) becomes

$$\begin{bmatrix} \bar{E}_{g1} \\ \bar{E}_{g2} \end{bmatrix} = \begin{bmatrix} \bar{V}_{g1} \\ \bar{V}_{g2} \end{bmatrix} + \begin{bmatrix} jx'_{d1} & 0 \\ 0 & jx'_{d2} \end{bmatrix} \begin{bmatrix} \bar{I}_{g1} \\ \bar{I}_{g2} \end{bmatrix} \tag{5.11}$$

Equation (5.11) indicates that introduction of internal voltage $\bar{E}_{gi}, i = 1, 2$ can be seen as addition of an extra node behind the node of generator terminal as shown in Fig. 5.3. The added extra "internal voltage" node is connected to the generator terminal through $x'_{di}, i = 1, 2$.

Substituting Eq. (5.5) into Eq. (5.11), it can be obtained that

$$\begin{bmatrix} \bar{I}_{g1} \\ \bar{I}_{g2} \end{bmatrix} = \left( \bar{\mathbf{Y}}_{\mathbf{N}}^{-1} + \begin{bmatrix} jx'_{d1} & 0 \\ 0 & jx'_{d2} \end{bmatrix} \right)^{-1} \begin{bmatrix} \bar{E}_{g1} \\ \bar{E}_{g2} \end{bmatrix} = \bar{\mathbf{Y}} \begin{bmatrix} \bar{E}_{g1} \\ \bar{E}_{g2} \end{bmatrix} \tag{5.12}$$

where matrix $\bar{\mathbf{Y}}$ is the network admittance matrix with only internal voltage nodes of generators left.

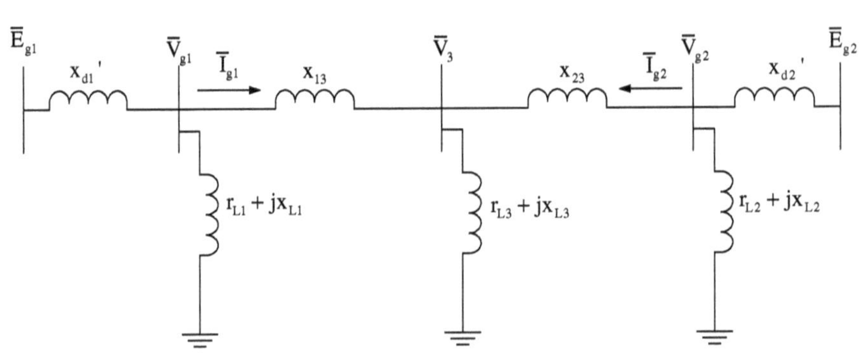


**Fig. 5.3** Introduction of internal voltage nodes of generators

That is $(i = 1, 2)$

$$\bar{I}_{gi} = i_{gxi} + ji_{gyi} = \sum_{j=1}^{2} \bar{y}_{ij}\bar{E}_{gj} = \sum_{j=1}^{2} \bar{y}_{ij}[E'_{qj}e^{j\delta_j} + (x_{qj} - x'_{dj})i_{qj}e^{j(\delta_j - 90^\circ)}] \tag{5.13}$$

where $\bar{y}_{ij} = y_{ij}e^{j\alpha_{ij}}$ is the element of $\bar{\mathbf{Y}}$. According to Fig. 5.2, in $d_i$-$q_i$ coordinate Eq. (5.13) is $(i = 1, 2)$

$$\begin{aligned}
\bar{I}_{gi} &= i_{di} + ji_{qi} = (i_{xi} + ji_{yi})e^{j(90^\circ - \delta_i)} = \sum_{j=1}^{2} \bar{y}_{ij}[E'_{qj}e^{j\delta_j} + (x_{qj} - x'_{dj})i_{qj}e^{j(\delta_j - 90^\circ)}]e^{j(90^\circ - \delta_i)} \\
&= \sum_{j=1}^{2} \bar{y}_{ij}[E'_{qj}\, e^{j(90^\circ + \delta_j - \delta_i)} + (x_{qj} - x'_{dj})i_{qj}e^{j(\delta_j - \delta_i)}] \\
&= \sum_{j=1}^{2} y_{ij}[E'_{qj}\, e^{j(90^\circ + \delta_j - \delta_i + \alpha_{ij})} + (x_{qj} - x'_{dj})i_{qj}e^{j(\delta_j - \delta_i + \alpha_{ij})}] \\
&= \sum_{j=1}^{2} y_{ij}[E'_{qj}\cos(90^\circ + \delta_j - \delta_i + \alpha_{ij}) + jE'_{qj}\sin(90^\circ + \delta_j - \delta_i + \alpha_{ij}) \\
&\quad + (x_{qj} - x'_{dj})i_{qj}\cos(\delta_j - \delta_i + \alpha_{ij}) + j(x_{qj} - x'_{dj})i_{qj}\sin(\delta_j - \delta_i + \alpha_{ij})] \\
&= \sum_{j=1}^{2} y_{ij}[-E'_{qj}\sin(\delta_j - \delta_i + \alpha_{ij}) + jE'_{qj}\cos(\delta_j - \delta_i + \alpha_{ij}) \\
&\quad + (x_{qj} - x'_{dj})i_{qj}\cos(\delta_j - \delta_i + \alpha_{ij}) + j(x_{qj} - x'_{dj})i_{qj}\sin(\delta_j - \delta_i + \alpha_{ij})]
\end{aligned} \tag{5.14}$$

From above equation, it can have $(i = 1, 2)$

$$\begin{aligned}
i_{di} &= \sum_{j=1}^{2} y_{ij}[-E'_{qj}\sin(\delta_j - \delta_i + \alpha_{ij}) + (x_{qj} - x'_{dj})i_{qj}\cos(\delta_j - \delta_i + \alpha_{ij})] \\
i_{qi} &= \sum_{j=1}^{2} y_{ij}[E'_{qj}\cos(\delta_j - \delta_i + \alpha_{ij}) + (x_{qj} - x'_{dj})i_{qj}\sin(\delta_j - \delta_i + \alpha_{ij})]
\end{aligned} \tag{5.15}$$

#### 5.1.1.2 Linearized Model When the Classical Model of Synchronous Generators Is Used

For the simplicity of discussion, firstly the simplest case is considered when a synchronous generator is represented by the so-called classic model, i.e., the rotor motion equation of generator. That is the model given by Eq. (2.41) with function of the AVR ignored and $E'_q$ being constant. In $d_i$-$q_i$ coordinate, model of generator is $(i = 1, 2)$

$$\begin{aligned}
\dot{\delta}_i &= \omega_o(\omega_i - 1)\\
\dot{\omega}_i &= \frac{1}{M_i}[P_{mi} - P_i - D_i(\omega_i - 1)]\\
P_i &= v_{di}i_{di} + v_{qi}i_{qi}\\
v_{di} &= x_{qi}i_{qi}\\
v_{qi} &= E'_{qi} - x'_{di}i_{di}
\end{aligned} \tag{5.16}$$

Hence, linearized classical model of generator is $(i = 1, 2)$

$$\begin{aligned}
\Delta\dot{\delta}_i &= \omega_o\Delta\omega_i\\
\Delta\dot{\omega}_i &= \frac{1}{M_i}(-\Delta P_i - D_i\Delta\omega_i)\\
\Delta P_i &= (v_{di0} - i_{qi0}x'_{di})\Delta i_{di} + (v_{qi0} + i_{di0}x_{qi})\Delta i_{qi}
\end{aligned} \tag{5.17}$$

Linearization of Eq. (5.15) is ($E'_{qi}$ is a constant, $i, j = 1, 2$)

$$\begin{aligned}
\Delta i_{di} &= h_{di}(\Delta\delta_j - \Delta\delta_i)\\
\Delta i_{qi} &= h_{qi}(\Delta\delta_j - \Delta\delta_i)\\
i &\neq j
\end{aligned} \tag{5.18}$$

where $h_{di}$ and $h_{di}$ $i = 1, 2$ are constants. Hence, from Eqs. (5.17) and (5.18), it can have $(i, j = 1, 2)$

$$\Delta P_i = k_i(\Delta\delta_j - \Delta\delta_i) \tag{5.19}$$

where $k_i$ $i = 1,\ 2$ are constants. State equation of the two-machine power system is

$$\begin{bmatrix}\Delta\dot{\delta}_1\\ \Delta\dot{\delta}_2\\ \Delta\dot{\omega}_1\\ \Delta\dot{\omega}_2\end{bmatrix} = \begin{bmatrix}0 & 0 & \omega_o & 0\\ 0 & 0 & 0 & \omega_o\\ \frac{k_1}{M_1} & -\frac{k_1}{M_1} & -\frac{D_1}{M_1} & 0\\ -\frac{k_2}{M_2} & \frac{k_2}{M_2} & 0 & -\frac{D_2}{M_2}\end{bmatrix}\begin{bmatrix}\Delta\delta_1\\ \Delta\delta_2\\ \Delta\omega_1\\ \Delta\omega_2\end{bmatrix} = \mathbf{A_4}\begin{bmatrix}\Delta\delta_1\\ \Delta\delta_2\\ \Delta\omega_1\\ \Delta\omega_2\end{bmatrix} \tag{5.20}$$

Equation (5.20) is equivalent to the following equation

$$\left(s^2\mathbf{I} + s\omega_o\begin{bmatrix}\frac{D_1}{M_1} & 0\\ 0 & \frac{D_2}{M_2}\end{bmatrix} + \omega_o\begin{bmatrix}\frac{k_1}{M_1} & -\frac{k_1}{M_1}\\ -\frac{k_2}{M_2} & \frac{k_2}{M_2}\end{bmatrix}\right)\begin{bmatrix}\Delta\delta_1\\ \Delta\delta_2\end{bmatrix} = \mathbf{0} \tag{5.21}$$

Characteristic equation of the system is

$$\begin{vmatrix} \lambda^2 + \lambda\omega_o \frac{D_1}{M_1} + \omega_o \frac{k_1}{M_1} & -\omega_o \frac{k_1}{M_1} \\ -\omega_o \frac{k_2}{M_2} & \lambda^2 + \lambda\omega_o \frac{D_2}{M_2} + \omega_o \frac{k_2}{M_2} \end{vmatrix} = 0 \tag{5.22}$$

Obviously, $\lambda = 0$ is one of the eigenvalues of the system. Because state matrix $\mathbf{A}_4$ is a real matrix, if there is a power oscillation, other three eigenvalues must be a pair of conjugate complex eigenvalues and a real eigenvalue. The pair of conjugate complex eigenvalues is the electromechanical oscillation mode, because the only dynamic of system is the rotor motion of generators. Therefore, in the two-machine power system, there is only one electromechanical oscillation mode. In fact, if it is denoted that $\Delta\delta_{12} = \Delta\delta_1 - \Delta\delta_2$, the state equation of system of Eq. (5.20) becomes

$$\begin{bmatrix} \Delta\dot{\delta}_{12} \\ \Delta\dot{\omega}_1 \\ \Delta\dot{\omega}_2 \end{bmatrix} = \begin{bmatrix} 0 & \omega_o & 0 \\ -\frac{k_1}{M_1} & -\frac{D_1}{M_1} & 0 \\ \frac{k_2}{M_2} & 0 & -\frac{D_2}{M_2} \end{bmatrix} \begin{bmatrix} \Delta\delta_{12} \\ \Delta\omega_1 \\ \Delta\omega_2 \end{bmatrix} = \mathbf{A}_3 \begin{bmatrix} \Delta\delta_{12} \\ \Delta\omega_1 \\ \Delta\omega_2 \end{bmatrix} \tag{5.23}$$

Obviously, the system can only have one pair of conjugate complex eigenvalues and hence one electromechanical oscillation mode. The reduced-order state matrix $\mathbf{A}_3$ does not have to have one zero eigenvalue when the relative angular position of generators is used as the state variable.

From Eq. (5.19), it can be seen that variations of the active power delivered along the transmission line connecting two generators are proportional to the relative angular position. For example, if $\Delta\delta_1 = \Delta\delta_2$, $\Delta P_i = 0$ $(i = 1, 2)$, there will be no variations of active power even if the absolute angular positions of generators vary. This means that the power oscillations are related only to the relative angular positions of generators.

#### 5.1.1.3 Heffron–Phillips Model

Heffron–Phillips model of the two-machine power system installed with the PSS can be established based on the following 4th-order model of synchronous generators $(i = 1, 2)$

$$\begin{aligned} \dot{\delta}_i &= \omega_o(\omega_i - 1) \\ \dot{\omega}_i &= \frac{1}{M_i}[P_{mi} - P_i - D_i(\omega_i - 1)] \\ \dot{E}'_{qi} &= \frac{1}{T'_{doi}}(-E_{qi} + E_{fd0i} + E'_{fdi}) \\ \dot{E}'_{fdi} &= -\frac{1}{T_{Ai}}E'_{fdi} + \frac{K_{Ai}}{T_{Ai}}(V_{refgi} - V_{gi} + u_{pssi}) \end{aligned} \tag{5.24}$$

where

$$\begin{aligned}
v_{id} &= x_{qi} i_{qi} \\
v_{qi} &= E'_{qi} - x'_{di} i_{di} \\
P_i &= v_{di} i_{di} + v_{qi} i_{qi} \\
E_{qi} &= E'_{qi} + (x_{di} - x'_{di}) i_{di} \\
V_{gi} &= \sqrt{v_{di}^2 + v_{qi}^2}
\end{aligned} \tag{5.25}$$

Linearization of Eq. (5.15) is $(i = 1, 2)$

$$\begin{aligned}
\Delta i_{di} &= \sum_{j=1}^{2} h_{dij} \Delta E'_{qj} + h_{dj} (\Delta \delta_j - \Delta \delta_i) \\
\Delta i_{qi} &= \sum_{j=1}^{2} h_{qij} \Delta E'_{qj} + h_{qj} (\Delta \delta_j - \Delta \delta_i) \\
& i \neq j
\end{aligned} \tag{5.26}$$

By using Eq. (5.26), linearization of Eq. (5.25) can be obtained as

$$\begin{aligned}
\Delta P_1 &= k_{1-1}(\Delta\delta_2 - \Delta\delta_1) + k_{2-11}\Delta E'_{q1} + k_{2-12}\Delta E'_{q2} \\
\Delta P_2 &= k_{1-2}(\Delta\delta_1 - \Delta\delta_2) + k_{2-21}\Delta E'_{q1} + k_{2-22}\Delta E'_{q2} \\
\Delta E_{q1} &= k_{4-1}(\Delta\delta_2 - \Delta\delta_1) + k_{3-11}\Delta E''_{q1} + k_{3-12}\Delta E'_{q2} \\
\Delta E_{q2} &= k_{4-2}(\Delta\delta_1 - \Delta\delta_2) + k_{3-21}\Delta E'_{q1} + k_{3-22}\Delta E'_{q2} \\
\Delta V_{g1} &= k_{5-1}(\Delta\delta_2 - \Delta\delta_1) + k_{6-11}\Delta E''_{q1} + k_{6-12}\Delta E'_{q2} \\
\Delta V_{g2} &= k_{5-2}(\Delta\delta_1 - \Delta\delta_2) + k_{6-21}\Delta E''_{q1} + k_{6-22}\Delta E'_{q2}
\end{aligned} \tag{5.27}$$

By substituting Eq. (5.27) into the linearization of Eq. (5.24), it can have for G1

$$\begin{aligned}
\Delta\dot{\delta}_1 &= \omega_o \Delta\omega_1 \\
\Delta\dot{\omega}_1 &= \frac{1}{M_1}[-k_{1-1}(\Delta\delta_2 - \Delta\delta_1) - k_{2-11}\Delta E'_{q1} - k_{2-12}\Delta E'_{q2} - D_1\Delta\omega_1] \\
\Delta\dot{E}'_{q1} &= \frac{1}{T'_{do1}}[-k_{4-1}(\Delta\delta_2 - \Delta\delta_1) - k_{3-11}\Delta E'_{q1} - k_{3-12}\Delta E'_{q2} + \Delta E'_{fd1}] \\
\Delta\dot{E}'_{fd1} &= -\frac{1}{T_{A1}}\Delta E'_{fd1} + \frac{K_{A1}}{T_{A1}}[-k_{5-1}(\Delta\delta_2 - \Delta\delta_1) - k_{6-11}\Delta E'_{q1} - k_{6-12}\Delta E'_{q2} + \Delta u_{pss1}]
\end{aligned} \tag{5.28}$$

and for G2

$$
\begin{aligned}
\Delta\dot{\delta}_2 &= \omega_o \Delta\omega_2 \\
\Delta\dot{\omega}_2 &= \frac{1}{M_2}[-k_{1-2}(\Delta\delta_1 - \Delta\delta_2) - k_{2-21}\Delta E'_{q1} - k_{2-22}\Delta E'_{q2} - D_2\Delta\omega_2] \\
\Delta\dot{E}'_{q2} &= \frac{1}{T'_{do2}}[-k_{4-2}(\Delta\delta_1 - \Delta\delta_2) - k_{3-21}\Delta E'_{q1} - k_{3-22}\Delta E'_{q2} + \Delta E'_{fd2}] \\
\Delta\dot{E}'_{fd2} &= -\frac{1}{T_{A2}}\Delta E'_{fd2} + \frac{K_{A2}}{T_{A2}}[-k_{5-2}(\Delta\delta_1 - \Delta\delta_2) - k_{6-21}\Delta E'_{q1} - k_{6-22}\Delta E'_{q2} + \Delta u_{pss2}]
\end{aligned} \tag{5.29}
$$

Equations (5.28) and (5.29) are the Heffron–Phillips model of the two-machine power system installed with PSS as shown in Fig. 5.4.

State-space representation of the Heffron–Phillips model is

$$
\begin{bmatrix} \Delta\dot{\boldsymbol{\delta}} \\ \Delta\dot{\boldsymbol{\omega}} \\ \Delta\dot{\mathbf{E}}'_{\mathbf{q}} \\ \Delta\dot{\mathbf{E}}'_{\mathbf{fd}} \end{bmatrix} = \begin{bmatrix} \mathbf{0} & \omega_o\mathbf{I} & \mathbf{0} & \mathbf{0} \\ -\mathbf{M}^{-1}\mathbf{K}_1 & -\mathbf{M}^{-1}\mathbf{D} & -\mathbf{M}^{-1}\mathbf{K}_2 & \mathbf{0} \\ -\mathbf{T}_{d0}^{-1}\mathbf{K}_4 & \mathbf{0} & -\mathbf{T}_{d0}^{-1}\mathbf{K}_3 & \mathbf{T}_{d0}^{-1} \\ -\mathbf{T}_A^{-1}\mathbf{K}_5\mathbf{K}_A & \mathbf{0} & -\mathbf{T}_A^{-1}\mathbf{K}_5\mathbf{K}_A & -\mathbf{T}_A^{-1} \end{bmatrix} \begin{bmatrix} \Delta\boldsymbol{\delta} \\ \Delta\boldsymbol{\omega} \\ \Delta\mathbf{E}'_{\mathbf{q}} \\ \Delta\mathbf{E}'_{\mathbf{fd}} \end{bmatrix} + \begin{bmatrix} \mathbf{0} \\ \mathbf{0} \\ \mathbf{0} \\ \mathbf{T}_A^{-1}\mathbf{K}_A \end{bmatrix} \Delta\mathbf{u}_{pss} \tag{5.30}
$$

where

$$
\Delta\boldsymbol{\delta} = \begin{bmatrix} \Delta\delta_1 \\ \Delta\delta_2 \end{bmatrix},\ \Delta\boldsymbol{\omega} = \begin{bmatrix} \Delta\omega_1 \\ \Delta\omega_2 \end{bmatrix},\ \Delta\mathbf{E}'_{\mathbf{q}} = \begin{bmatrix} \Delta E'_{q1} \\ \Delta E'_{q2} \end{bmatrix},\ \Delta\mathbf{E}'_{\mathbf{fd}} = \begin{bmatrix} \Delta E'_{fd1} \\ \Delta E'_{fd2} \end{bmatrix},
$$

$$
\mathbf{K}_1 = \begin{bmatrix} -k_{1-1} & k_{1-1} \\ k_{1-2} & -k_{1-2} \end{bmatrix},\ \mathbf{K}_2 = \begin{bmatrix} k_{2-11} & k_{2-12} \\ k_{2-21} & k_{2-22} \end{bmatrix},\ \mathbf{K}_3 = \begin{bmatrix} k_{3-11} & k_{3-12} \\ k_{3-21} & k_{3-22} \end{bmatrix},
$$

$$
\mathbf{K}_4 = \begin{bmatrix} -k_{4-1} & k_{4-1} \\ k_{4-2} & -k_{4-2} \end{bmatrix},\ \mathbf{K}_5 = \begin{bmatrix} -k_{5-1} & k_{5-1} \\ k_{5-2} & -k_{5-22} \end{bmatrix},\ \mathbf{K}_6 = \begin{bmatrix} k_{6-11} & k_{6-12} \\ k_{6-21} & k_{6-22} \end{bmatrix},
$$

$$
\mathbf{M} = \begin{bmatrix} M_1 & 0 \\ 0 & M_2 \end{bmatrix},\ \mathbf{D} = \begin{bmatrix} D_1 & 0 \\ 0 & D_2 \end{bmatrix},\ \mathbf{T}_{d0} = \begin{bmatrix} T'_{do1} & 0 \\ 0 & T'_{do2} \end{bmatrix},
$$

$$
\mathbf{K}_A = \begin{bmatrix} K_{A1} & 0 \\ 0 & K_{A2} \end{bmatrix},\ \mathbf{T}_A = \begin{bmatrix} T_{A1} & 0 \\ 0 & T_{A2} \end{bmatrix},\ \Delta\mathbf{u}_{pss} = \begin{bmatrix} \Delta u_{pss1} \\ \Delta u_{pss2} \end{bmatrix}
$$

The model can be shown in Fig. 5.5 which is the matrix form of Fig. 5.4.

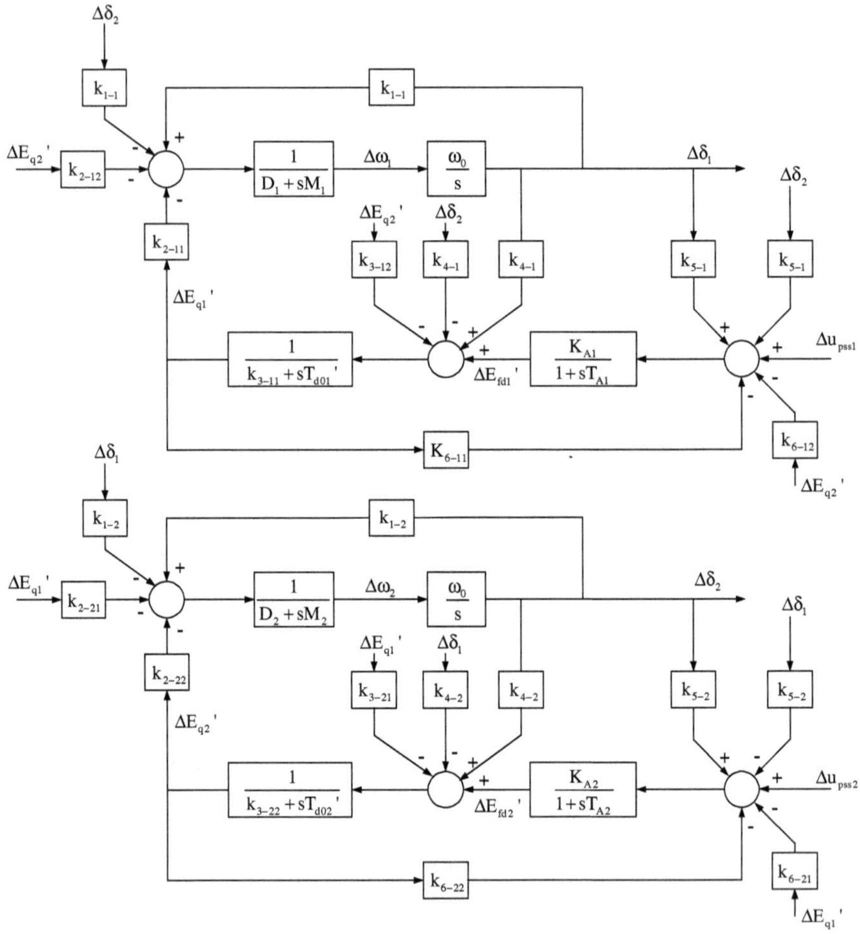


**Fig. 5.4** Heffron–Phillips model of two-machine power system installed with PSS

## 5.1.2 *A Multi-machine Power System Installed with Power System Stabilizers*

### 5.1.2.1 Heffron–Phillips Model of an N-Machine Power System Installed with Power System Stabilizers

In an N-machine power system installed with PSSs, the 4th-order model of synchronous generator is given by Eqs. (5.24) and (5.25) for $i = 1, 2, \ldots, N$. Equation (5.8) for $i = 1, 2, \ldots, N$ can be written in the following matrix form

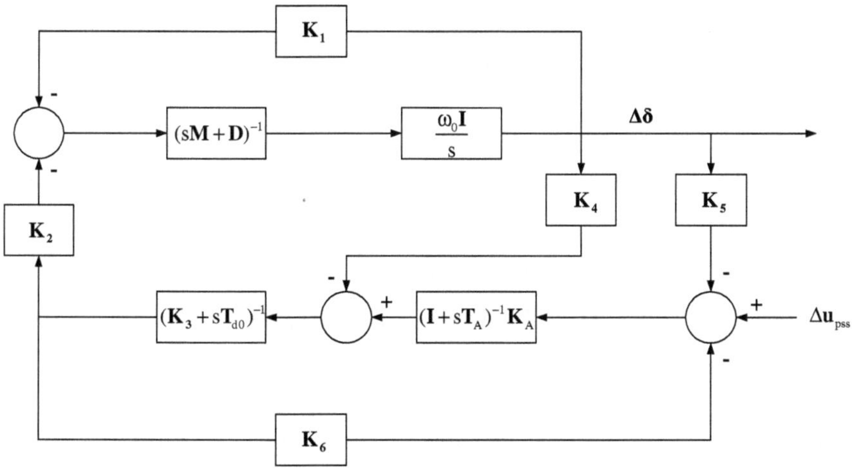


**Fig. 5.5** Heffron–Phillips model of two-machine power system in matrix form

$$\bar{\mathbf{V}}_{\mathbf{g}} = \mathbf{e}^{\mathbf{j}\delta}\mathbf{E}'_{\mathbf{q}} - \mathrm{j}\mathbf{x}'_{\mathbf{d}}\bar{\mathbf{I}}_{\mathbf{g}} + (\mathbf{x}_{\mathbf{q}} - \mathbf{x}'_{\mathbf{d}})\mathbf{e}^{\mathbf{j}(\delta-90^\circ)}\mathbf{i}_{\mathbf{q}} \tag{5.31}$$

where

$$\begin{aligned}
&\bar{\mathbf{V}}_{\mathbf{g}} = \begin{bmatrix}\bar{V}_{g1} & \bar{V}_{g2} & \cdots & \bar{V}_{gN}\end{bmatrix}^{T}, \bar{\mathbf{I}}_{\mathbf{g}} = \begin{bmatrix}\bar{I}_{g1} & \bar{I}_{g2} & \cdots & \bar{I}_{gN}\end{bmatrix}^{T}, \\
&\mathbf{E}'_{\mathbf{q}} = \begin{bmatrix}E'_{q1} & E'_{q2} & \cdots & E'_{q1}\end{bmatrix}^{T}, \mathbf{i}_{\mathbf{q}} = \begin{bmatrix}i_{q1} & i_{q2} & \cdots & i_{q1}\end{bmatrix}^{T}, \\
&\mathbf{e}^{\mathbf{j}\delta} = \mathrm{diag}(e^{j\delta_i}), \mathbf{x}'_{\mathbf{d}} = \mathrm{diag}(x'_{di}), (\mathbf{x}_{\mathbf{q}} - \mathbf{x}'_{\mathbf{d}})\mathbf{e}^{\mathbf{j}(\delta-90^\circ)} = \mathrm{diag}\left[(x_{qi} - x'_{di})e^{j(\delta_i-90^\circ)}\right]
\end{aligned} \tag{5.32}$$

In Eq. (5.32), $\mathrm{diag}(c_i)$ denotes an Nth-order diagonal matrix with $c_i$ being the ith diagonal element.

Let the network equation be

$$\bar{\mathbf{I}}_{\mathbf{g}} = \bar{\mathbf{Y}}_{\mathbf{N}}\bar{\mathbf{V}}_{\mathbf{g}} \tag{5.33}$$

where $\bar{\mathbf{Y}}_{\mathbf{N}}$ is the network admittance matrix with only the nodes of generator terminal left. From Eqs. (5.31) and (5.33), it can have

$$\bar{\mathbf{I}}_{\mathbf{g}} = \bar{\mathbf{Y}}\left[\mathbf{e}^{\mathbf{j}\delta}\mathbf{E}'_{\mathbf{q}} + (\mathbf{x}_{\mathbf{q}} - \mathbf{x}'_{\mathbf{d}})\mathbf{e}^{\mathbf{j}(\delta-90^\circ)}\mathbf{i}_{\mathbf{q}}\right] = \bar{\mathbf{Y}}\bar{\mathbf{E}}_{\mathbf{g}} \tag{5.34}$$

where $\bar{\mathbf{Y}} = (\bar{\mathbf{Y}}_{\mathbf{N}}^{-1} + \mathrm{j}\mathbf{x}'_{\mathbf{d}})^{-1}$. The above equation gives the generator current in x–y coordinate as $(i = 1, 2, \ldots, N)$

$$\bar{I}_{gi} = \sum_{j=1}^{N} \bar{y}_{ij}[E'_{qj}e^{j\delta_j} + (x_{qj} - x'_{dj})i_{qj}e^{j(\delta_j - 90^\circ)}] \tag{5.35}$$

where $\bar{y}_{ij} = y_{ij}e^{j\alpha_{ij}}$ is the elements of matrix $\bar{\mathbf{Y}}$. Hence, Eq. (5.15) can be obtained for $i = 1, 2, \ldots, N$. Their linearization can be expressed in the following matrix form

$$\begin{aligned} \mathbf{\Delta I_d} &= \mathbf{F_{dd}\Delta\delta} + \mathbf{G_{dd}\Delta E'_q} + \mathbf{H_{dd}\Delta I_q} \\ \mathbf{\Delta I_q} &= \mathbf{F_{qq}\Delta\delta} + \mathbf{G_{qq}\Delta E'_q} + \mathbf{H_{qq}\Delta I_q} \end{aligned} \tag{5.36}$$

where

$$\mathbf{\Delta\delta} = [\Delta\delta_1 \quad \Delta\delta_2 \quad \ldots \quad \Delta\delta_N]^T,\ \mathbf{\Delta E'_q} = [\Delta E'_{q1} \quad \Delta E'_{q2} \quad \cdots \quad \Delta E'_{qN}]^T,$$

$$\mathbf{\Delta I_d} = [\Delta i_{d1} \quad \Delta i_{d2} \quad \ldots \quad \Delta i_{dN}]^T,\ \mathbf{\Delta I_q} = [\Delta i_{q1} \quad \Delta i_{q2} \quad \ldots \quad \Delta i_{qN}]^T$$

$$f_{ddij} = -y_{ij}[E'_{qj0}\cos(\delta_{j0} - \delta_{i0} + \alpha_{ij}) + (x_{qj} - x'_{dj})i_{qj0}\sin(\delta_{j0} - \delta_{i0} + \alpha_{ij})],\ i \neq j,$$

$$f_{qqij} = -y_{ij}[E'_{qj0}\sin(\delta_{j0} - \delta_{i0} + \alpha_{ij}) - (x_{qj} - x'_{dj})i_{qj0}\cos(\delta_{j0} - \delta_{i0} + \alpha_{ij})],\ i \neq j,$$

$$f_{ddii} = -\sum_{j=1, j\neq i}^{N} f_{ddij},\ f_{qqii} = -\sum_{j=1, j\neq i}^{N} f_{qqij},$$

$$g_{ddij} = -y_{ij}\sin(\delta_{j0} - \delta_{i0} + \alpha_{ij}),\ g_{qqij} = y_{ij}\cos(\delta_{j0} - \delta_{i0} + \alpha_{ij}),$$

$$h_{ddij} = y_{ij}(x_{qj} - x'_{dj})\cos(\delta_{j0} - \delta_{i0} + \alpha_{ij}),\ h_{qqij} = y_{ij}(x_{qj} - x'_{dj})\sin(\delta_{j0} - \delta_{i0} + \alpha_{ij})$$

and $f_{ddij}$, $f_{qqij}$, $g_{ddij}$, $g_{qqij}$, $h_{ddij}$ and $h_{qqij}$ are the elements of matrix $\mathbf{F_{dd}}$, $\mathbf{F_{qq}}$, $\mathbf{G_{dd}}$, $\mathbf{G_{qq}}$, $\mathbf{H_{dd}}$ and $\mathbf{H_{qq}}$, respectively.

From Eq. (5.45), it can be obtained that

$$\begin{aligned} \mathbf{\Delta I_d} &= \mathbf{F_d\Delta\delta} + \mathbf{G_d\Delta E'_q} \\ \mathbf{\Delta I_q} &= \mathbf{F_q\Delta\delta} + \mathbf{G_q\Delta E'_q} \end{aligned} \tag{5.37}$$

where

$$\mathbf{F_d} = \mathbf{F_{dd}} + \mathbf{H_{dd}F_q},\quad \mathbf{G_d} = \mathbf{G_{dd}} + \mathbf{H_{dd}G_q},$$

$$\mathbf{F_q} = (\mathbf{I} - \mathbf{H_{qq}})^{-1}\mathbf{F_{qq}},\quad \mathbf{G_q} = (\mathbf{I} - \mathbf{H_{qq}})^{-1}\mathbf{G_{qq}}$$

Linearization of Eqs. (5.24) and (5.25) is

$$
\begin{aligned}
\Delta\dot{\delta}_i &= \omega_o \Delta\omega_i \\
\Delta\dot{\omega}_i &= \frac{1}{M_i}(-\Delta P_i - D_i \Delta\omega_i) \\
\Delta\dot{E}'_{qi} &= \frac{1}{T'_{doi}}(-\Delta E_{qi} + \Delta E'_{fdi}) \\
\Delta\dot{E}'_{fdi} &= -\frac{1}{T_{Ai}}\Delta E'_{fdi} + \frac{K_{Ai}}{T_{Ai}}(-\Delta V_{gi} + \Delta u_{pssi})
\end{aligned}
\tag{5.38}
$$

$$
\begin{aligned}
\Delta v_{di} &= x_{qi} \Delta i_{qi} \\
\Delta v_{qi} &= \Delta E'_{qi} - x'_{di} \Delta i_{di} \\
\Delta P_i &= v_{di0} \Delta i_{id} + v_{qi0} \Delta i_{qi} + i_{di0} \Delta v_{di} + i_{qi0} \Delta v_{qi} \\
\Delta E_{qi} &= \Delta E'_{qi} + (x_{di} - x'_{di}) \Delta i_{di} \\
\Delta V_{gi} &= \frac{v_{gdi0}}{V_{gi0}} \Delta v_{gdi} + \frac{v_{qi0}}{V_{gi0}} \Delta v_{gqi}
\end{aligned}
\tag{5.39}
$$

In matrix form, the above equations are

$$
\begin{aligned}
\mathbf{\Delta\dot{\delta}} &= \omega_o \mathbf{I} \mathbf{\Delta\omega} \\
\mathbf{\Delta\dot{\omega}} &= \mathbf{M}^{-1}(-\mathbf{\Delta P} - \mathbf{D}\mathbf{\Delta\omega}) \\
\mathbf{\Delta\dot{E}'_q} &= \mathbf{T'^{-1}_{d0}}(-\mathbf{\Delta E_q} + \mathbf{\Delta E'_{fd}}) \\
\mathbf{\Delta\dot{E}'_{fd}} &= -\mathbf{T_A^{-1}} \mathbf{\Delta E'_{fd}} + \mathbf{T_A^{-1}} \mathbf{K_A}(-\mathbf{\Delta V_g} + \mathbf{\Delta u_{pss}})
\end{aligned}
\tag{5.40}
$$

$$
\begin{aligned}
\mathbf{\Delta P} &= \mathbf{I_{q0}} \mathbf{\Delta E'_q} + (\mathbf{V_{d0}} - \mathbf{I_{q0}} \mathbf{X'_d}) \mathbf{\Delta I_d} + (\mathbf{V_{q0}} + \mathbf{I_{d0}} \mathbf{X_q}) \mathbf{\Delta I_q} \\
\mathbf{\Delta E_q} &= \mathbf{\Delta E'_q} + (\mathbf{X_d} - \mathbf{X'_d}) \mathbf{\Delta I_d} \\
\mathbf{\Delta V_g} &= \mathbf{V'^{-1}_{g0}} \mathbf{V_{gq0}} \mathbf{\Delta E'_q} - \mathbf{V'^{-1}_{g0}} \mathbf{V_{gq0}} \mathbf{X'_d} \mathbf{\Delta I_d} + \mathbf{V'^{-1}_{g0}} \mathbf{V_{gd0}} \mathbf{X_q} \mathbf{\Delta I_q}
\end{aligned}
\tag{5.41}
$$

where

$$
\begin{aligned}
&\mathbf{\Delta\omega} = [\Delta\omega_1 \quad \Delta\omega_2 \quad \dots \quad \Delta\omega_N]^T,\ \mathbf{\Delta E'_{fd}} = [\Delta E'_{fd1} \quad \Delta E'_{fd2} \quad \dots \quad \Delta E'_{fdN}]^T, \\
&\mathbf{\Delta P} = [\Delta P_1 \quad \Delta P_2 \quad \dots \quad \Delta P_N]^T,\ \mathbf{\Delta E_q} = [\Delta E_{q1} \quad \Delta E_{q2} \quad \dots \quad \Delta E_{qN}]^T, \\
&\mathbf{\Delta V_g} = [\Delta V_{g1} \quad \Delta V_{g2} \quad \dots \quad \Delta V_{gN}]^T,\ \mathbf{\Delta u_{pss}} = [\Delta u_{pss1} \quad \Delta u_{pss2} \quad \dots \quad \Delta u_{pssN}]^T, \\
&\mathbf{M} = \mathrm{diag}(M_i),\ \mathbf{D} = \mathrm{diag}(D_i),\ \mathbf{T'_{d0}} = \mathrm{diag}(T'_{d0i}),\ \mathbf{T_A} = \mathrm{diag}(T_{Ai}),\ \mathbf{K_A} = \mathrm{diag}(K_{Ai}), \\
&\mathbf{I_{d0}} = \mathrm{diag}(i_{di0}),\ \mathbf{I_{q0}} = \mathrm{diag}(i_{qi0}),\ \mathbf{V_{d0}} = \mathrm{diag}(v_{di0}),\ \mathbf{V_{q0}} = \mathrm{diag}(v_{qi0}),\ \mathbf{V_{g0}} = \mathrm{diag}(V_{gi0}), \\
&\mathbf{X'_d} = \mathrm{diag}(x'_{di}),\ \mathbf{X_q} = \mathrm{diag}(x_{qi}),\ \mathbf{X_d} = \mathrm{diag}(x_{di})
\end{aligned}
$$

Substituting Eq. (5.37) into Eq. (5.41), it can have

$$\begin{aligned}\Delta \mathbf{P} &= \mathbf{K_1}\Delta\boldsymbol{\delta} + \mathbf{K_2}\Delta\mathbf{E}'_{\mathbf{q}}\\ \Delta\mathbf{E_q} &= \mathbf{K_3}\Delta\mathbf{E}'_{\mathbf{q}} + \mathbf{K_4}\Delta\boldsymbol{\delta}\\ \Delta\mathbf{V_g} &= \mathbf{K_5}\Delta\boldsymbol{\delta} + \mathbf{K_6}\Delta\mathbf{E}'_{\mathbf{q}}\end{aligned} \tag{5.42}$$

By substituting Eq. (5.42) into Eq. (5.40), Heffron–Phillips model of the N-machine power system with PSS installed can be obtained as

$$\begin{bmatrix}\Delta\dot{\boldsymbol{\delta}}\\ \Delta\dot{\boldsymbol{\omega}}\\ \Delta\dot{\mathbf{E}}'_{\mathbf{q}}\\ \Delta\dot{\mathbf{E}}'_{\mathbf{fd}}\end{bmatrix} = \begin{bmatrix}\mathbf{0} & \boldsymbol{\omega_o}\mathbf{I} & \mathbf{0} & \mathbf{0}\\ -\mathbf{M}^{-1}\mathbf{K_1} & -\mathbf{M}^{-1}\mathbf{D} & -\mathbf{M}^{-1}\mathbf{K_2} & \mathbf{0}\\ -\mathbf{T}_{\mathbf{d0}}^{-1}\mathbf{K_4} & \mathbf{0} & -\mathbf{T}_{\mathbf{d0}}^{-1}\mathbf{K_3} & \mathbf{T}_{\mathbf{d0}}^{-1}\\ -\mathbf{T}_{\mathbf{A}}^{-1}\mathbf{K_5}\mathbf{K_A} & \mathbf{0} & -\mathbf{T}_{\mathbf{A}}^{-1}\mathbf{K_5}\mathbf{K_A} & -\mathbf{T}_{\mathbf{A}}^{-1}\end{bmatrix}\begin{bmatrix}\Delta\boldsymbol{\delta}\\ \Delta\boldsymbol{\omega}\\ \Delta\mathbf{E}'_{\mathbf{q}}\\ \Delta\mathbf{E}'_{\mathbf{fd}}\end{bmatrix} + \begin{bmatrix}\mathbf{0}\\ \mathbf{0}\\ \mathbf{0}\\ \mathbf{T}_{\mathbf{A}}^{-1}\mathbf{K_A}\end{bmatrix}\Delta\mathbf{u}_{\mathbf{pss}} \tag{5.43}$$

It is in the exactly same format to that of two-machine power system given by Eq. (5.30). Hence, the model can also be shown in Fig. 5.5.

#### 5.1.2.2 Linearized Model When Full Mathematical Model of Synchronous Generators Is Used

Linearized full dynamic mathematical model of the ith synchronous generator in the N-machine power system is $(i = 1, 2, \ldots, N)$

$$\begin{aligned}\Delta\dot{\psi}_{di} &= \Delta v_{di} + r_{ai}\Delta i_{di} + \omega_0\Delta\psi_{qi} + \psi_{qi0}\Delta\omega_i\\ \Delta\dot{\psi}_{qi} &= \Delta v_{qi} + r_{ai}\Delta i_{qi} - \omega_0\Delta\psi_{di} - \psi_{di0}\Delta\omega_i\\ \Delta\dot{\psi}_{fi} &= \Delta v_{fi} - r_{fi}\Delta i_{fi}\\ \Delta\dot{\psi}_{Di} &= -r_{Di}\Delta i_{Di}\\ \Delta\dot{\psi}_{Qi} &= -r_{Qi}\Delta i_{Qi}\end{aligned} \tag{5.44}$$

$$\begin{aligned}\begin{bmatrix}\Delta i_{di}\\ \Delta i_{fi}\\ \Delta i_{Di}\end{bmatrix} &= \begin{bmatrix}x_{di} & x_{adi} & x_{adi}\\ x_{adi} & x_{fi} & x_{adi}\\ x_{adi} & x_{adi} & x_{Di}\end{bmatrix}^{-1}\begin{bmatrix}\Delta\psi_{di}\\ \Delta\psi_{fi}\\ \Delta\psi_{Di}\end{bmatrix}\\ \begin{bmatrix}\Delta i_{qi}\\ \Delta i_{Qi}\end{bmatrix} &= \begin{bmatrix}x_{qi} & x_{aqi}\\ x_{aqi} & x_{Qi}\end{bmatrix}^{-1}\begin{bmatrix}\Delta\psi_{qi}\\ \Delta\psi_{Qi}\end{bmatrix}\end{aligned} \tag{5.45}$$

$$
\begin{aligned}
\Delta\dot{\delta}_i &= \omega_0 \Delta\omega_i \\
\Delta\dot{\omega}_i &= -\frac{1}{M_i}(\Delta T_{ei} + D_i \Delta\omega_i) \\
\Delta T_{ei} &= \psi_{di0}\Delta i_{qi} - \psi_{qi0}\Delta i_{di} + i_{qi0}\Delta\psi_{di} - i_{di0}\Delta\psi_{qi}
\end{aligned}
\tag{5.46}
$$

$$
\begin{aligned}
\Delta\dot{v}'_{fi} &= -\frac{1}{T_{Ai}}\Delta v'_{fi} - \frac{K_{Ai}}{T_{Ai}}(\Delta V_{gi} + \Delta u_{pssi}) \\
\Delta V_{gi} &= \frac{v_{gdi0}}{V_{gi0}}\Delta v_{gdi} + \frac{v_{gqi0}}{V_{gi0}}\Delta v_{gqi}
\end{aligned}
\tag{5.47}
$$

In matrix form, Eqs. (5.44) to (5.47) can be arranged as

$$
\begin{aligned}
\dot{\mathbf{X}}_{\mathbf{g}} &= \mathbf{A_g X_g} + \mathbf{B_g \Delta u_{pss}} + \mathbf{B_{gv} \Delta V_{dq}} \\
\mathbf{\Delta I_{dq}} &= \mathbf{C_g X_g}
\end{aligned}
\tag{5.48}
$$

where

$$
\begin{aligned}
&\boldsymbol{\Delta\psi_d} = \begin{bmatrix} \Delta\psi_{d1} & \Delta\psi_{d2} & \cdots & \Delta\psi_{dN} \end{bmatrix}^T, \boldsymbol{\Delta\psi_q} = \begin{bmatrix} \Delta\psi_{q1} & \Delta\psi_{q2} & \cdots & \Delta\psi_{qN} \end{bmatrix}^T, \\
&\boldsymbol{\Delta\psi_f} = \begin{bmatrix} \Delta\psi_{f1} & \Delta\psi_{f2} & \cdots & \Delta\psi_{fN} \end{bmatrix}^T, \boldsymbol{\Delta\psi_D} = \begin{bmatrix} \Delta\psi_{D1} & \Delta\psi_{D2} & \cdots & \Delta\psi_{DN} \end{bmatrix}^T, \\
&\boldsymbol{\Delta\psi_Q} = \begin{bmatrix} \Delta\psi_{Q1} & \Delta\psi_{Q2} & \cdots & \Delta\psi_{QN} \end{bmatrix}^T, \mathbf{\Delta v'_f} = \begin{bmatrix} \Delta v'_{f1} & \Delta v'_{f2} & \cdots & \Delta v'_{fN} \end{bmatrix}^T, \\
&\mathbf{X_g} = \begin{bmatrix} \boldsymbol{\Delta\delta}^T & \boldsymbol{\Delta\omega}^T & \boldsymbol{\Delta\psi_d^T} & \boldsymbol{\Delta\psi_q^T} & \boldsymbol{\Delta\psi_f^T} & \boldsymbol{\Delta\psi_D^T} & \boldsymbol{\Delta\psi_Q^T} & \mathbf{\Delta v_f'^T} \end{bmatrix}^T \\
&\mathbf{\Delta u_{pss}} = \begin{bmatrix} \Delta u_{pss1} & \Delta u_{pss2} & \cdots & \Delta u_{pssN} \end{bmatrix}^T \\
&\mathbf{\Delta V_{dq}} = \begin{bmatrix} \Delta v_{gd1} & \Delta v_{gq1} & \Delta v_{gd2} & \Delta v_{gq2} & \cdots & \cdots & \Delta v_{gdN} & \Delta v_{gqN} \end{bmatrix}^T \\
&\mathbf{\Delta I_{dq}} = \begin{bmatrix} \Delta i_{d1} & \Delta i_{q1} & \Delta i_{d2} & \Delta i_{q2} & \cdots & \cdots & \Delta i_{dN} & \Delta i_{qN} \end{bmatrix}^T
\end{aligned}
$$

Conversion equations between x–y and $d_i - q_i$ coordinate systems as shown in Fig. 5.2 are $(i = 1, 2, \ldots, N)$

$$
\begin{bmatrix} f_{di} \\ f_{qi} \end{bmatrix} = \begin{bmatrix} \sin\delta_i & -\cos\delta_i \\ \cos\delta_i & \sin\delta_i \end{bmatrix} \begin{bmatrix} f_{xi} \\ f_{yi} \end{bmatrix}, \begin{bmatrix} f_{xi} \\ f_{yi} \end{bmatrix} = \begin{bmatrix} \sin\delta_i & \cos\delta_i \\ -\cos\delta_i & \sin\delta_i \end{bmatrix} \begin{bmatrix} f_{di} \\ f_{qi} \end{bmatrix} \tag{5.49}
$$

where $\bar{f}_i$ is a phasor $(i = 1, 2, \ldots, N)$. Linearization of above equations is

$$
\begin{aligned}
\begin{bmatrix} \Delta i_{di} \\ \Delta i_{qi} \end{bmatrix} &= \begin{bmatrix} \sin\delta_{i0} & -\cos\delta_{i0} \\ \cos\delta_{i0} & \sin\delta_{i0} \end{bmatrix} \begin{bmatrix} \Delta i_{xi} \\ \Delta i_{yi} \end{bmatrix} + \begin{bmatrix} i_{xi0}\cos\delta_{i0} + i_{yi0}\sin\delta_{i0} \\ -i_{xi0}\sin\delta_{i0} + i_{yi0}\cos\delta_{i0} \end{bmatrix} \Delta\delta_i \\
\begin{bmatrix} \Delta v_{gdi} \\ \Delta v_{gqi} \end{bmatrix} &= \begin{bmatrix} \sin\delta_{i0} & -\cos\delta_{i0} \\ \cos\delta_{i0} & \sin\delta_{i0} \end{bmatrix} \begin{bmatrix} \Delta v_{gxi} \\ \Delta v_{gyi} \end{bmatrix} + \begin{bmatrix} v_{gxi0}\cos\delta_{i0} + v_{gyi0}\sin\delta_{i0} \\ -v_{gxi0}\sin\delta_{i0} + v_{gyi0}\cos\delta_{i0} \end{bmatrix} \Delta\delta_i
\end{aligned}
\tag{5.50}
$$

In matrix form, above equation can be written as

$$\begin{aligned}\mathbf{\Delta I_{dq}} &= \mathbf{T_{g0}\Delta I_{xy}} + \mathbf{B_{gI}X_g}\\ \mathbf{\Delta V_{dq}} &= \mathbf{T_{g0}\Delta V_{xy}} + \mathbf{B_{gV}X_g}\end{aligned} \tag{5.51}$$

where

$$\begin{aligned}\mathbf{\Delta V_{xy}} &= \begin{bmatrix}\Delta v_{gx1} & \Delta v_{gy1} & \Delta v_{gx2} & \Delta v_{gy2} & \cdots & \cdots & \Delta v_{gxN} & \Delta v_{gyN}\end{bmatrix}^T\\ \mathbf{\Delta I_{xy}} &= \begin{bmatrix}\Delta i_{x1} & \Delta i_{y1} & \Delta i_{x2} & \Delta i_{y2} & \cdots & \cdots & \Delta i_{xN} & \Delta i_{yN}\end{bmatrix}^T\end{aligned}$$

In the x–y coordinate, network equation of Eq. (5.33) can be written as

$$\begin{bmatrix} i_{x1}\\ i_{y1}\\ i_{x2}\\ i_{y2}\\ \vdots\\ i_{xN}\\ i_{yN}\end{bmatrix} = \begin{bmatrix} \begin{bmatrix} g_{N11} & -b_{N11}\\ b_{N11} & g_{N11}\end{bmatrix} & \begin{bmatrix} g_{N12} & -b_{N12}\\ b_{N12} & g_{N12}\end{bmatrix} & \cdots & \begin{bmatrix} g_{N1N} & -b_{N1N}\\ b_{N1N} & g_{N1N}\end{bmatrix}\\ \begin{bmatrix} g_{N21} & -b_{N21}\\ b_{N21} & g_{N21}\end{bmatrix} & \begin{bmatrix} g_{N22} & -b_{N22}\\ b_{N22} & g_{N22}\end{bmatrix} & \cdots & \begin{bmatrix} g_{N2N} & -b_{N2N}\\ b_{N2N} & g_{N2N}\end{bmatrix}\\ \vdots & \vdots & \vdots & \vdots\\ \begin{bmatrix} g_{NN1} & -b_{NN1}\\ b_{NN1} & g_{NN1}\end{bmatrix} & \begin{bmatrix} g_{NN2} & -b_{NN2}\\ b_{NN2} & g_{NN2}\end{bmatrix} & \cdots & \begin{bmatrix} g_{NNN} & -b_{NNN}\\ b_{NNN} & g_{NNN}\end{bmatrix}\end{bmatrix} \begin{bmatrix} v_{gx1}\\ v_{gy1}\\ v_{gx2}\\ v_{gy2}\\ \vdots\\ v_{gxN}\\ v_{gyN}\end{bmatrix} \tag{5.52}$$

where $\bar{y}_{Nij} = g_{Nij} + jb_{Nij}$ is the elements of network admittance matrix $\mathbf{\bar{Y}_N}$. Linearization of Eq. (5.52) is

$$\mathbf{\Delta I_{xy}} = \mathbf{Y_G \Delta V_{xy}} \tag{5.53}$$

where

$$\mathbf{Y_G} = \begin{bmatrix} \begin{bmatrix} g_{N11} & -b_{N11}\\ b_{N11} & g_{N11}\end{bmatrix} & \begin{bmatrix} g_{N12} & -b_{N12}\\ b_{N12} & g_{N12}\end{bmatrix} & \cdots & \begin{bmatrix} g_{N1N} & -b_{N1N}\\ b_{N1N} & g_{N1N}\end{bmatrix}\\ \begin{bmatrix} g_{N21} & -b_{N21}\\ b_{N21} & g_{N21}\end{bmatrix} & \begin{bmatrix} g_{N22} & -b_{N22}\\ b_{N22} & g_{N22}\end{bmatrix} & \cdots & \begin{bmatrix} g_{N2N} & -b_{N2N}\\ b_{N2N} & g_{N2N}\end{bmatrix}\\ \vdots & \vdots & \vdots & \vdots\\ \begin{bmatrix} g_{NN1} & -b_{NN1}\\ b_{NN1} & g_{NN1}\end{bmatrix} & \begin{bmatrix} g_{NN2} & -b_{NN2}\\ b_{NN2} & g_{NN2}\end{bmatrix} & \cdots & \begin{bmatrix} g_{NNN} & -b_{NNN}\\ b_{NNN} & g_{NNN}\end{bmatrix}\end{bmatrix}$$

From Eqs. (5.48), (5.51), and (5.53), it can have

$$\begin{aligned}\mathbf{\dot{X}_g} &= \mathbf{A_gX_g} + \mathbf{B_g\Delta u_{pss}} + \mathbf{B_{gv}}(\mathbf{T_{g0}\Delta V_{xy}} + \mathbf{B_{gV}X_g})\\ \mathbf{C_gX_g} &= \mathbf{T_{g0}Y_G\Delta V_{xy}} + \mathbf{B_{gI}X_g}\end{aligned} \tag{5.54}$$

Hence, state equation of the power system is

$$\dot{\mathbf{X}}_{\mathbf{g}} = \mathbf{A}\mathbf{X}_{\mathbf{g}} + \mathbf{B}_{\mathbf{g}}\Delta\mathbf{u}_{\mathbf{pss}} \tag{5.55}$$

where

$$\mathbf{A} = \mathbf{A}_{\mathbf{g}} + \mathbf{B}_{\mathbf{gv}}[\mathbf{T}_{\mathbf{g0}}(\mathbf{T}_{\mathbf{g0}}\mathbf{Y}_{\mathbf{G}})^{-1}(\mathbf{C}_{\mathbf{g}} - \mathbf{B}_{\mathbf{gI}}) + \mathbf{B}_{\mathbf{gV}}\mathbf{X}_{\mathbf{g}}]$$

## 5.2 Modal Analysis and Control of Power System Oscillations in a Multi-machine Power System Installed with Power System Stabilizers

### *5.2.1 Eigensolution for the Analysis of Power System Oscillations*

#### 5.2.1.1 Participation Factor, Correlation Ratio of Electromechanical Loop, Modal Shape, and Eigensolution

Modal analysis is a method to study power system small-signal angular stability by calculating various eigenparameters of state matrix of state-space representation of power system linearized model of Eqs. (5.43) and (5.55).

In Sect. 2.2.1.1, following definition on eigenvalues $\lambda_i$, corresponding left and right eigenvectors, $\mathbf{v_i}$ and $\mathbf{w}_i^T$, of state matrix $\mathbf{A}$ is given

$$\begin{aligned} &\mathbf{A}\mathbf{v_i} = \lambda_i \mathbf{v_i}, \\ &\mathbf{w}_i^T\mathbf{A} = \mathbf{w}_i^T\lambda_i, \quad i = 1, 2, \ldots, M \end{aligned} \tag{5.56}$$

Also following two matrices are constructed by eigenvectors as

$$\mathbf{V} = [\mathbf{v_1} \quad \mathbf{v_2} \quad \cdots \quad \mathbf{v_n}], \mathbf{V}^{-1} = \begin{bmatrix} \mathbf{w}_1^T \\ \mathbf{w}_2^T \\ \vdots \\ \mathbf{w}_n^T \end{bmatrix} = \mathbf{W}^T \tag{5.57}$$

Equations (2.61) and (2.66) establish the following relationship between the state variables $x_i(t)$, $i = 1, 2, \ldots, M$ and modes of the system $z_i(t) = z_i(0)e^{\lambda_i t}$, $i = 1, 2, \ldots, M$ to be

$$x_k(t) = v_{k1}z_1(0)e^{\lambda_1 t} + v_{k2}z_2(0)e^{\lambda_2 t} + \cdots + v_{kn}z_n(0)e^{\lambda_M t} \tag{5.58}$$

$$z_i(t) = w_{1i}x_1(t) + w_{2i}x_2(t) + \cdots + w_{Mi}x_M(t) \tag{5.59}$$

where $v_{ki}$ is the kth-row ith-column element of matrix $\mathbf{V}$ and $w_{ki}$ the ith-row kth-column element of matrix $\mathbf{W}$.

From Eq. (5.58), it can be seen that the magnitude of $v_{ki}$ measures how much the ith mode $z_i(t) = z_i(0)e^{\lambda_i t}$, $i = 1, 2, \ldots, M$, contributes to the kth state variable $x_k(t)$. Thus, $|v_{ki}|$ is a kind of measurement of "observability" of the ith mode in the kth state variable. Equation (5.59) indicates that the magnitude of $|w_{ki}|$ measures the influence of the kth state variable $x_k(t)$ on the ith mode $z_i(t) = z_i(0)e^{\lambda_i t}$, $i = 1, 2, \ldots, M$ of the system. It is a kind of measurement of "controllability" of the kth state variable on the ith mode. Hence, $|v_{ki}w_{ki}|$ measures how much the ith mode and the kth state variable are connected. Its normalized value is defined to be the participation factor

$$p_{ki} = \frac{|v_{ki}w_{ki}|}{\sum_{i=1}^{M} |v_{ki}w_{ki}|} \tag{5.60}$$

If it is assumed that $x_k(0) = 1$, $x_j(0) = 0$ $(j \neq k, j = 1, 2, \ldots, n)$, from Eq. (5.59), it can have

$$z_i(0) = w_{ki}, \ i = 1, 2, \ldots, M \tag{5.61}$$

By substituting Eq. (5.61) into Eq. (5.58) it can have

$$x_k(t) = \sum_{i=1}^{n} v_{ki}w_{ki}e^{\lambda_i t} \tag{5.62}$$

Equation (5.62) indicates that the participation factor $|v_{ki}w_{ki}|$ measures how much the ith mode excited by the kth state variable participates the time response of the kth state variable. That is why the index defined by Eq. (5.60) is named as the participation factor.

As it is pointed out previously in Chap. 2, the electromechanical power oscillations are closely related to the rotor motion equation of generators where state variables are the deviation of rotor position $\Delta\delta_i$ and speed $\Delta\omega_i$, $i = 1, 2, \ldots, M$ in the state Eqs. (5.43) or (5.55). Hence for a pair of conjugate complex eigenvalue, $\bar{\lambda}_i$, of state matrix, if the associated participation factors of state variables $\Delta\delta_i$ and $\Delta\omega_i$, $i = 1, 2, \ldots, M$ are much higher than those of other state variables, this means $\bar{\lambda}_i$ is the dominant oscillation mode responsible for the electromechanical oscillation. Hence, $\bar{\lambda}_i$ is an electromechanical oscillation mode. That is, if

$$\sigma_i = \frac{p_{ki}(x_k = \Delta\delta) + p_{ki}(x_k = \Delta\omega)}{\sum_k p_{ki}(x_k \neq \Delta\delta \text{ and } x_k \neq \Delta\omega)} \gg 1; \left(\sum_k p_{ki} = 1\right) \tag{5.63}$$

$\bar{\lambda}_i$ is an electromechanical oscillation mode. $\sigma_i$ defined by Eq. (5.63) is called the correlation ratio of electromechanical loop, which measure how much closely the eigenvalue $\bar{\lambda}_i$ is related to the rotor motion of generators (hence the electromechanical oscillation). $\sigma_i$ is often used to identify whether an eigenvalue $\bar{\lambda}_i$ is an oscillation mode or not.

For the oscillation mode of interests, $\bar{\lambda}_i$, let its corresponding (right) eigenvector be $\bar{\mathbf{v}}_i = [\bar{v}_{1i} \quad \bar{v}_{2i} \quad \ldots \quad \bar{v}_{Mi}]^T$. Assume the rotor speed of generators, $\Delta\omega_j$ $(j = 1, 2, \ldots, N)$, to be the (N + 1)th to (2N)th state variables of the system of Eq. (4.1), i.e.,

$$\Delta\omega_j = \Delta x_{j+N}, \quad j = 1, 2, \ldots, N \tag{5.64}$$

Equation (5.58) can be written as

$$\begin{bmatrix} x_1(t) \\ \vdots \\ \Delta\omega_1 \\ \Delta\omega_2 \\ \vdots \\ \Delta\omega_N \\ \vdots \\ x_M(t) \end{bmatrix} = v_1 z_1(0) e^{\lambda_1 t} + \cdots + \begin{bmatrix} \bar{v}_{1i} \\ \vdots \\ \bar{v}_{N+1i} \\ \bar{v}_{N+2i} \\ \vdots \\ \bar{v}_{2Ni} \\ \vdots \\ \bar{v}_{Mi} \end{bmatrix} z_i(0) e^{\bar{\lambda}_i t} + \cdots \tag{5.65}$$

The (N + 1)th to (2N)th elements of $\bar{\mathbf{v}}_i$, $\bar{v}_{ki} = v_{ki}\angle\varphi_{ki}$, $k = N + 1,\ N + 2, \ldots, 2N$ describe the mode shape of oscillation mode $\bar{\lambda}_i$, because $\bar{v}_{ki} = v_{ki}\angle\varphi_{ki}$ weighs the contribution of the oscillation mode to the rotor motion of involved generators. By sketching the phasor diagram of $\bar{v}_{ki} = v_{ki}\angle\varphi_{ki}$, relative positions of rotor motion of involved generators can be found to determine how generators participate the power oscillation. For example, let

$$\bar{\lambda}_i = -\xi_i + j\omega_i, \bar{v}_{N+1i} = |v_{N+1i}|\angle 0^\circ, \bar{v}_{N+2i} = |v_{N+2i}|\angle 180^\circ \tag{5.66}$$

Equations (5.65) and (5.66) indicate that the rotor motion of the 1st and 2nd generator, $\Delta\omega_1$ and $\Delta\omega_2$, must have component $|v_{N+1i}|z_i(0)e^{-\xi_i}\cos\omega_i t$ and $|v_{N+2i}|z_i(0)e^{-\xi_i}\cos(\omega_i t + 180^\circ)$, respectively. Hence, as far as oscillation mode $\bar{\lambda}_i = -\xi_i + j\omega_i$ is concerned, the 1st and 2nd generators oscillate against each other. Thus, by looking at $\bar{v}_{ki} = v_{ki}\angle\varphi_{ki}$, $k = N + 1, N + 2, \ldots, 2N$, how groups of generators take part in the oscillation associated with the mode can be determined.

Procedure of modal analysis from the eigensolution of power system oscillations normally is

1. Calculation of eigenvalues of state matrix;
2. Identification of electromechanical oscillation modes by checking the imaginary part of eigenvalues within the range $2\pi f_O (f_O = 0.1 - 2\ \text{Hz})$ and computing the correlation ratio of electromechanical loop;
3. Examining how much each generator is involved in a particular oscillation mode by calculating the modal participation index;
4. Determining how groups of generators oscillate against each other by computing the modal shape.

#### 5.2.1.2 Selective Reduced-Order Calculation of Electromechanical Oscillation Modes

Commonly used numerical method of eigensolution is the QR decomposition. It is a method to compute all eigenvalues of system state matrix. From above introduction on the eigensolution for the analysis of power system oscillations, it can be seen that in fact the majority of computational results of the QR method is redundant, because only eigenvalues and associated eigenvectors related to the electromechanical oscillation modes are needed. In addition, for a large-scale power system with a high-dimensional state matrix, numerical difficulty of QR method may occur. Hence, a lot of effort has been spent by many researchers to propose and develop numerical methods for the efficient and effective eigensolution of large-scale power system. The reduced-order modal analysis is one of the methods to carry out the eigensolution from the reduced-order matrices rather than directly from the system high-dimensional state matrix so as to avoid the numerical difficulty. The selective modal analysis is another computational strategy to compute only the eigenvalues of interests from system state matrix to save computational time and resources. In this section, a selective reduced-order method of modal computation is introduced by using Heffron–Phillips model.

Heffron–Phillips model of N-machine power system of Eq. (5.43) can be written as (without PSS being considered)

$$\begin{aligned}
s\boldsymbol{\Delta\delta} &= \omega_0 \boldsymbol{\Delta\omega} \\
s\boldsymbol{\Delta\omega} &= \mathbf{M}^{-1}(-\mathbf{K}_1\boldsymbol{\Delta\delta} - \mathbf{K}_2\boldsymbol{\Delta E}'_q - \mathbf{D}\boldsymbol{\Delta\omega}) \\
s\boldsymbol{\Delta E}'_q &= \mathbf{T}'^{-1}_{d0}(-\mathbf{K}_3\boldsymbol{\Delta E}'_q - \mathbf{K}_4\boldsymbol{\Delta\delta} + \boldsymbol{\Delta E}'_{fd}) \\
s\boldsymbol{\Delta E}'_{fd} &= -\mathbf{T}_A^{-1}\boldsymbol{\Delta E}'_{fd} + \mathbf{T}_A^{-1}\mathbf{K}_A(-\mathbf{K}_5\boldsymbol{\Delta\delta} - \mathbf{K}_6\boldsymbol{\Delta E}'_q)
\end{aligned} \tag{5.67}$$

From Eq. (5.67), it can be obtained that

$$\begin{aligned}&\{s^2\mathbf{M}+s\mathbf{D}+\omega_0\mathbf{K}_1-\mathbf{K}_2[(\mathbf{I}+s\mathbf{T}_\mathbf{A})(\mathbf{K}_3+s\mathbf{T}'_{\mathbf{d0}})+\mathbf{K}_\mathbf{A}\mathbf{K}_6]^{-1}\\&[(\mathbf{I}+s\mathbf{T}_\mathbf{A})\mathbf{K}_4+\mathbf{K}_\mathbf{A}\mathbf{K}_5]\}\boldsymbol{\Delta\delta}=\mathbf{F}(s)\boldsymbol{\Delta\delta}=0\end{aligned} \tag{5.68}$$

Obviously, if $\bar{\lambda}_i$ is an electromechanical oscillation mode, it must have

$$\mathrm{DET}[\mathbf{F}(\bar{\lambda}_i)]=0 \tag{5.69}$$

where $\mathrm{DET}[\mathbf{M}]=0$ denotes the determinant of matrix $\mathbf{M}$ being equal to zero.

Usually, the oscillation modes of interests are poorly damped, close to the imaginary axis of the complex plan. The angular oscillation frequency of electromechanical oscillation modes is close to the natural oscillation frequency, which can be determined from the computation of eigenvalues of matrix $-\omega_o\mathbf{M}^{-1}\mathbf{K}_1$. In fact, in the N-machine power system, linearized 2N rotor motion equations of generators are (with $\mathbf{D}$ being ignored)

$$\begin{bmatrix}\Delta\dot{\boldsymbol{\delta}}\\ \Delta\dot{\boldsymbol{\omega}}\end{bmatrix}=\begin{bmatrix}0 & \omega_o\mathbf{I}\\ \mathbf{K}_1 & 0\end{bmatrix}\begin{bmatrix}\Delta\boldsymbol{\delta}\\ \Delta\boldsymbol{\omega}\end{bmatrix} \tag{5.70}$$

The characteristic equation of the system is

$$\left|\lambda\mathbf{I}-\begin{bmatrix}0 & \omega_o\mathbf{I}\\ \mathbf{K}_1 & 0\end{bmatrix}\right|=|\lambda\mathbf{I}-\mathbf{A}_r|=0 \tag{5.71}$$

The equivalent form of Eq. (5.71) is

$$(s^2\mathbf{M}+\omega_o\mathbf{K}_1)\boldsymbol{\Delta\delta}=(s^2\mathbf{I}+\omega_o\mathbf{M}^{-1}\mathbf{K}_1)\boldsymbol{\Delta\delta}=0 \tag{5.72}$$

Hence, 2N eigenvalues of system state matrix $\mathbf{A}_r$ are also solutions of the following equation

$$\left|s^2\mathbf{I}+\omega_o\mathbf{M}^{-1}\mathbf{K}_1\right|=0 \tag{5.73}$$

Therefore, if N eigenvalues of matrix $-\omega_o\mathbf{M}^{-1}\mathbf{K}_1$ are $\eta_j$, $j=1,2,\ldots,N$, 2N eigenvalues of matrix $\mathbf{A}_r$ are $\pm\sqrt{\eta_j}$, $j=1,2,\ldots,N$. It can be proved that all eigenvalues of $-\omega_o\mathbf{M}^{-1}\mathbf{K}_1$ are real negative numbers except one zero eigenvalue. Hence, $2N-2$ eigenvalues of $\mathbf{A}_r$ are $N-1$ pairs of the conjugate imaginary roots $\pm j\sqrt{\eta_j}$, $j=1,2,\ldots,N$. They are in fact the $N-1$ electromechanical oscillation modes of the power system when only the rotor motion of generators is considered. $\sqrt{\eta_j}\neq 0$, $j=1,2,\ldots,N-1$, are the $N-1$ angular natural oscillation frequency of electromechanical oscillation modes.

For a poorly damped oscillation mode of interests, $\bar{\lambda}_i=\xi_i\pm j\omega_i$, because it is close to the imaginary axis of the complex plan and the angular oscillation

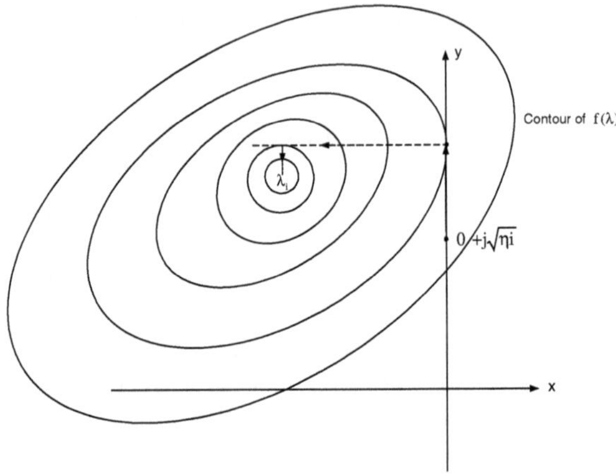


**Fig. 5.6** Optimal direct one-dimensional searching along the direction of imaginary and real axis on complex plane

frequency $\omega_i$ is close to the angular natural oscillation frequency, $\sqrt{\eta_i}$ , $\bar{\lambda}_i(0) = 0 + j\sqrt{\eta_i}$ should be a good initial guess of oscillation mode, $\bar{\lambda}_i = \xi_i \pm j\omega_i$. Therefore, following objective function can be established

$$f(\lambda) = |DET[\mathbf{F}(\lambda)]| \tag{5.74}$$

A direct searching method can be used to find the minimal optimum of above objective function on the complex plan, starting from $0 + j\sqrt{\eta_i}$. That will be the oscillation mode of interests, $\bar{\lambda}_i = \xi_i \pm j\omega_i$, which satisfies Eq. (5.74). The direct searching can be implemented in the direction of imaginary and real axis iteratively as shown in Fig. 5.6. If $\bar{\lambda}_i = \xi_i \pm j\omega_i$ is very close to $0 + j\sqrt{\eta_i}$ , $\bar{\lambda}_i = \xi_i \pm j\omega_i$ will be found by applying one-dimensional optimal searching just a couple of times in each of two directions.

## *5.2.2 Design of Power System Stabilizers in Multi-machine Power System*

### 5.2.2.1 Coordinated Design of Multiple Power System Stabilizers

In an N-machine power system, it is assumed that there are L oscillation modes of interests which need to be damped by installing PSSs in the power system. The first task of designing PSSs is to determine the effective installing locations of PSSs. Though nowadays it may be the case that every generator is equipped with a PSS, it is still necessary to determine how the PSS is designed to effectively damp a particular oscillation mode. This is an equivalent task to selecting the installing locations of the PSS.

By computing the participation factors and/or the correlation ratio of electromechanical loop, it can be decided which generators are most strongly related to a particular oscillation mode of interests. This is one of the most commonly used methods to select the installing locations of PSSs to damp the particular oscillation mode of interests. Without loss of generality, it can be assumed that L generators in the N-machine power system are selected as the most effective installing locations of PSSs to damp L oscillation modes of interests. The ith oscillation mode of interests $\bar{\lambda}_i = \xi_i \pm j\omega_i$ is selected to be damped by the PSS installed on the jth generator in the power system, denoted as the jth PSS. In total, L PSSs are to be designed to damp L oscillation modes. (It may be possible that more than L PSSs are needed to damp L oscillation modes in some cases.)

A simple way to design L PSSs is to set parameters of L PSSs one by one in a sequence. For example, method of pole assignment introduced in Sect. 2.2.2.3 can be used to assign the ith oscillation mode $\bar{\lambda}_i = \xi_i \pm j\omega_i$ to a target position $\bar{\lambda}_i^* = \xi_i^* \pm j\omega_i^*$ by setting the parameters of the jth PSS. Afterwards, the (i + 1)th oscillation mode is assigned by designing the (j + 1)th PSS, etc., until L oscillation modes are assigned to the target positions by setting parameters of L PSSs. However, this method of sequential design of L PSSs has a problem of "eigenvalue drifting", because when the (j + 1)th PSS is designed to assign the (i + 1)th oscillation mode, the ith oscillation mode which has been assigned may drift away from the target position. Hence at the end, L PSSs may not assign L oscillation modes to their target positions at all.

To solve the problem of "eigenvalue drift" associated with the method of sequential design of multiple PSSs, parameters of L PSSs can be set to move L oscillation modes to the target positions simultaneously. This strategy of simultaneous tuning of parameters of all PSSs is also referred to as the coordinated design of multiple PSSs. So far, there have been many methods of simultaneous tuning or coordinated design of PSSs being proposed and developed. The majority of those proposed methods adopt a certain objective function,

$$f(\mathbf{p}),\ \mathbf{p} \in P \tag{5.75}$$

where $\mathbf{p}$ is the vector of parameters of L PSSs and P is the parameter space of L PSSs. A method of optimization can be used to find the optimum of objective function, which is the solution of coordinated design of L PSSs.

So far, methods of optimization for the coordinated design of multiple PSSs can be classified into two categories: (1) mathematical methods and (2) heuristic methods. Mathematical methods include linear programming, nonlinear programming and other mathematical algorithms of optimization. Examples of heuristic methods are genetic algorithm, simulated annealing, and particle swarm algorithm. Those methods have all been tried for optimally tuning parameters of multiple PSSs to effectively damp power system oscillations. Application of mathematical methods must have a computable objective function and sometimes is constrained by the problem of local optimum, while heuristic methods can handle non-computable objective function and are more effective to find the global optima.

In the following section, a mathematical method is introduced to tune parameters of multiple PSSs installed in a multi-machine power system on the basis of Heffron–Phillips model. In Sect. 7.4.1, a heuristic method to tune parameters of multiple stabilizers added on a UPFC will be introduced.

#### 5.2.2.2 Parameter Tuning Algorithm for the Design of Multiple Power System Stabilizers

Let the objective function to simultaneously tune parameters of L PSSs to assign L oscillation modes, $\bar{\lambda}_i = \xi_i \pm j\omega_i$, to target positions $\bar{\lambda}_i^* = \xi_i^* \pm j\omega_i^*, i = 1, 2, \ldots, L$ be

$$f(\mathbf{p}) = \sum_{i=1}^{L} [\bar{\lambda}_i(\mathbf{p}) - \bar{\lambda}_i^*(\mathbf{p}^*)]^2 \tag{5.76}$$

The minimal optimum $\mathbf{p}^* \in P$ is when the objective function is equal to zero and L oscillation modes are moved to the target positions. If the deepest decent method in nonlinear programming is used, the optimal searching algorithm is

$$\mathbf{p}(k+1) = \mathbf{p}(k) - st \times \nabla f[\mathbf{p}(k)] \tag{5.77}$$

where st is the optimal one-dimensional searching length and $\nabla f[\mathbf{p}(k)]$ is the searching decent direction of objective function at the (k + 1)th step. A selective modal algorithm to implement the deepest decent searching can be developed as follows.

Heffron–Phillips model of N-machine power system of Eq. (5.43) can be written as (with PSSs being considered)

$$\begin{aligned}
s\boldsymbol{\Delta\delta} &= \omega_o \mathbf{I} \boldsymbol{\Delta\omega} \\
s\boldsymbol{\Delta\omega} &= \mathbf{M}^{-1}(-\mathbf{K}_1\boldsymbol{\Delta\delta} - \mathbf{K}_2\boldsymbol{\Delta}\mathbf{E}'_{\mathbf{q}} - \mathbf{D}\boldsymbol{\Delta\omega}) \\
s\boldsymbol{\Delta}\mathbf{E}'_{\mathbf{q}} &= \mathbf{T}'^{-1}_{\mathbf{d0}}(-\mathbf{K}_3\boldsymbol{\Delta}\mathbf{E}'_{\mathbf{q}} - \mathbf{K}_4\boldsymbol{\Delta\delta} + \boldsymbol{\Delta}\mathbf{E}'_{\mathbf{fd}}) \\
s\boldsymbol{\Delta}\mathbf{E}'_{\mathbf{fd}} &= -\mathbf{T}_{\mathbf{A}}^{-1}\boldsymbol{\Delta}\mathbf{E}'_{\mathbf{fd}} + \mathbf{T}_{\mathbf{A}}^{-1}\mathbf{K}_{\mathbf{A}}(-\mathbf{K}_5\boldsymbol{\Delta\delta} - \mathbf{K}_6\boldsymbol{\Delta}\mathbf{E}'_{\mathbf{q}} + \boldsymbol{\Delta}\mathbf{u}_{\mathbf{pss}})
\end{aligned} \tag{5.78}$$

From Eq. (5.78), it can be obtained that

$$\begin{aligned}
&\{s^2\mathbf{M} + s\mathbf{D} + \omega_o\mathbf{K}_1 - \mathbf{K}_2[(\mathbf{I} + s\mathbf{T}_{\mathbf{A}})(\mathbf{K}_3 + s\mathbf{T}'_{\mathbf{d0}}) + \mathbf{K}_{\mathbf{A}}\mathbf{K}_6]^{-1} \\
&[(\mathbf{I} + s\mathbf{T}_{\mathbf{A}})\mathbf{K}_4 + \mathbf{K}_{\mathbf{A}}\mathbf{K}_5]\}\boldsymbol{\Delta\delta} + \mathbf{K}_2[(\mathbf{I} + s\mathbf{T}_{\mathbf{A}})(\mathbf{K}_3 + s\mathbf{T}'_{\mathbf{d0}}) \\
&\quad + \mathbf{K}_{\mathbf{A}}\mathbf{K}_6]^{-1}\mathbf{K}_{\mathbf{A}}\boldsymbol{\Delta}\mathbf{u}_{\mathbf{pss}} = 0
\end{aligned} \tag{5.79}$$

Let the transfer function of the jth PSS be $G_j(s)$ and the feedback signal $\Delta\omega_j$. $\mathbf{\Delta u_{pss}}$ can be denoted as

$$\mathbf{\Delta u_{pss}} = \frac{s}{\omega_0}\mathbf{G_{pss}}(\mathbf{p}, s)\mathbf{\Delta\delta} \tag{5.80}$$

where $\mathbf{G_{pss}}(\mathbf{p}, s) = \mathrm{diag}(G_{pssj})$, $G_{pssj} = G_j(s)$ when the jth generator is installed with a PSS; otherwise, $G_{pssj} = 0$. From Eqs. (5.79) and (5.80), it can have

$$\begin{aligned}&\{s^2\mathbf{M} + s\mathbf{D} + \omega_o\mathbf{K_1} - \mathbf{K_2}[(\mathbf{I} + s\mathbf{T_A})(\mathbf{K_3} + s\mathbf{T}'_{\mathbf{d0}}) + \mathbf{K_A}\mathbf{K_6}]^{-1}\\&[(\mathbf{I} + s\mathbf{T_A})\mathbf{K_4} + \mathbf{K_A}\mathbf{K_5} - \mathbf{K_A}\frac{s}{\omega_0}\mathbf{G_{pss}}(\mathbf{p}, s)\mathbf{\Delta\delta}]\}\mathbf{\Delta\delta} = \mathbf{F}(\mathbf{p}, s)\mathbf{\Delta\delta} = 0\end{aligned} \tag{5.81}$$

Hence, it must have

$$\mathrm{DET}[\mathbf{F}(\mathbf{p}, \lambda_i)] = 0 \tag{5.82}$$

Equation (5.82) indicates that there should exist a nonzero vector $\bar{\mathbf{v}}_\mathbf{i}$ and $\bar{\mathbf{w}}_\mathbf{i}$ satisfying

$$\begin{aligned}\mathbf{F}(\mathbf{p}, \lambda_i)\bar{\mathbf{v}}_\mathbf{i} &= 0\\ \mathbf{w}_\mathbf{i}^T\mathbf{F}(\mathbf{p}, \lambda_i) &= 0\end{aligned} \tag{5.83}$$

Let $\mathbf{p} = [p_1 \quad p_2 \quad \ldots \quad p_s]^T$. A small increment of $p_k$, $\Delta p_k$, will cause that of $\lambda_i$ and $\bar{\mathbf{v}}_\mathbf{i}$, $\Delta\lambda_i$ and $\Delta\bar{\mathbf{v}}_\mathbf{i}$. It can have

$$\mathbf{F}[\mathbf{p}(p_k + \Delta p_k), \lambda_i + \Delta\lambda_i](\bar{\mathbf{v}}_\mathbf{i} + \Delta\bar{\mathbf{v}}_\mathbf{i}) = 0 \tag{5.84}$$

Because

$$\mathbf{F}[\mathbf{p}(p_k + \Delta p_k), \lambda_i + \Delta\lambda_i] = \mathbf{F}(\mathbf{p}, \lambda_i) + \frac{\partial\mathbf{F}(\mathbf{p}, \lambda_i)}{\partial\lambda_i}\Delta\lambda_i + \frac{\partial\mathbf{F}(\mathbf{p}, \lambda_i)}{\partial p_k}\Delta p_k \tag{5.85}$$

From Eqs. (5.83), (5.84), and (5.85), it can be obtained that

$$\frac{\partial\lambda_i}{\partial p_k} = -\frac{\bar{\mathbf{w}}_\mathbf{i}^T\frac{\partial\mathbf{F}(\mathbf{p},\lambda_i)}{\partial p_k}\bar{\mathbf{v}}_\mathbf{i}}{\bar{\mathbf{w}}_\mathbf{i}^T\frac{\partial\mathbf{F}(\mathbf{p},\lambda_i)}{\partial\lambda_i}\bar{\mathbf{v}}_\mathbf{i}} \tag{5.86}$$

From Eq. (5.76), it can be obtained that

$$\nabla f[\mathbf{p}(k)] = 2(\lambda_i - \lambda_i^*)\frac{\partial\lambda_i}{\partial\mathbf{p}(k)} \tag{5.87}$$

where $\frac{\partial \lambda_i}{\partial \mathbf{p}(k)} = \left[\frac{\partial \lambda_i}{\partial p_1} \ \frac{\partial \lambda_i}{\partial p_2} \ \cdots \ \frac{\partial \lambda_i}{\partial p_s}\right]^T$ and $\frac{\partial \lambda_i}{\partial p_k}$ $(k = 1, 2, \ldots, s)$ is calculated by using Eq. (5.86).

When the deepest decent algorithm of Eqs. (5.77) and (5.87) is implemented, it needs to compute the oscillation mode of interests, $\lambda_i$, at each step of optimal searching. Because it is an iterative algorithm, at the (k + 1) step, $\lambda_i(k+1)$ is very close to $\lambda_i(k)$ and it should have $\mathbf{DETF}[\mathbf{p}(k+1), \lambda_i(k+1)] = 0$. Hence, set up the following objective function,

$$f_{k+1}(\lambda_i) = |\mathbf{DETF}[\mathbf{p}, \lambda_i]| \tag{5.88}$$

The direct searching method introduced based on Eq. (5.74) can be used to find $\lambda_i(k+1)$ on the complex plane, starting from $\lambda_i(k)$.

In the multi-machine power system, objective of installation of multiple PSSs is to ensure sufficient damping to be provided to the oscillation modes of interests. Hence, pole assignment to the exact target positions, such as that to be achieved by the objective function of Eq. (5.76), is usually not necessary. A more relaxed objective function can be used for the coordinated design of multiple PSSs. For example, following objective function

$$f(\mathbf{p}) = \min(\zeta_i) \geq 0.05 \tag{5.89}$$

where

$$\lambda_i = \xi_i \pm j\omega_i \in \{\text{all oscillation modes to be damped by the PSS design}\},$$
$$\zeta_i = -\frac{\zeta_i}{\sqrt{\xi_i^2 + \omega_i^2}}$$

is for the coordinated design of multiple PSSs to ensure the damping of all oscillation modes to be greater than .05. For the objective function of Eq. (5.89), some heuristic methods of optimization, such as the genetic algorithm, can be used.

Strategies of sequential setting and simultaneous tuning for the design of multiple PSSs can also be jointly used to achieve more efficient and practical design. For example, it can be assumed that in the N-machine power system, the jth PSS provides a pure damping torque, $D_{pssj}\Delta\omega_j (j = 1, 2, \ldots, L)$, to the electromechanical oscillation loop of the jth generator. With this assumption, Eq. (5.81) becomes

$$\begin{aligned}&\{s^2\mathbf{M} + s(\mathbf{D} + \mathbf{D}_{pss}) + \omega_o\mathbf{K}_1 - \mathbf{K}_2[(\mathbf{I} + s\mathbf{T}_A)(\mathbf{K}_3 + s\mathbf{T}'_{d0}) + \mathbf{K}_A\mathbf{K}_6]^{-1} \\ &[(\mathbf{I} + s\mathbf{T}_A)\mathbf{K}_4 + \mathbf{K}_A\mathbf{K}_5\}\Delta\boldsymbol{\delta} = \mathbf{F}(\mathbf{p}, s)\Delta\boldsymbol{\delta} = 0\end{aligned} \tag{5.90}$$

where

$$\mathbf{D_{pss}} = \text{diag}(d_{pssj}),\ d_{pssj} = D_{pssj}$$

when the jth generator is installed with a PSS; otherwise, $d_{pssj} = 0\ (j = 1, 2, \ldots, L)$, $\mathbf{p} = \begin{bmatrix} D_{pssj} & D_{pss(j+1)} & \cdots & D_{pss(j+L)} \end{bmatrix}^T$.

The method of coordinated design of PSSs introduced from Eqs. (5.76) to (5.88) above can be used to set the damping torque provision from each PSS for the multiple PSSs to damp all the oscillation modes of interests. Afterwards, each PSS can be designed individually by using the phase compensation method to ensure that the right amount of damping torque, $D_{pssj}\Delta\omega_j (j = 1, 2, \ldots, L)$, is supplied to each generator.

### 5.2.3 *Fixed Modes Associated with PSS Control*

A PSS is a local controller, and hence, an N-machine power system with L PSSs installed forms a decentralized control system. Fixed modes are an important concept about the controllability and observability of a decentralized control system, a natural generalization of the well-known concept of uncontrollable and unobservable modes in centralized control systems. In this section, a mathematical proof is presented to show that in the decentralized control of N-machine power system by PSSs, any electromechanical oscillation mode is not a fixed mode. This means that in the N-machine power system, any oscillation mode can be damped by the appropriate design of multiple PSSs. In other words, in the parameter space of coordinated design of multiple PSSs, the optimum always exists if the objective function is appropriately set. Theoretical proof given below is based on the Heffron–Phillips model of N-machine power system installed with multiple PSSs.

Let the realization of the jth PSS, $\Delta u_{pssj} = G_j(s)\Delta\omega_j$, be

$$\begin{aligned} \Delta u_{pssj} &= \mathbf{H_j z_j}(t) + k_j\Delta\omega_j \\ \dot{\mathbf{z}}_{\mathbf{j}}(t) &= \mathbf{F_j z_j}(t) + \mathbf{S_j}\Delta\omega_j \end{aligned} \tag{5.91}$$

Hence, it can have

$$\begin{aligned} \mathbf{\Delta u_{pss}} &= \mathbf{Hz}(t) + \mathbf{Ky} \\ \dot{\mathbf{z}}(t) &= \mathbf{Fz}(t) + \mathbf{Sy} \end{aligned} \tag{5.92}$$

where

$$\mathbf{z}(t) = \begin{bmatrix} \mathbf{z}_1(\mathbf{t})^T & \mathbf{z}_2(\mathbf{t})^T & \cdots & \mathbf{z}_L(\mathbf{t})^T \end{bmatrix}^T, \ y = \begin{bmatrix} \Delta\omega_j & \Delta\omega_{j+1} & \cdots & \Delta\omega_{j+L} \end{bmatrix}^T,$$
$$\mathbf{H} = \text{blockdiag}(\mathbf{H_j}), \ \mathbf{F} = \text{blockdiag}(\mathbf{F_j}), \ \mathbf{S} = \text{blockdiag}(\mathbf{S_j}), \ \mathbf{K} = \text{diag}(k_j)$$

and $\text{blockdiag}(\mathbf{M_j})$ denotes a block diagonal matrix with the diagonal matrix to be $\mathbf{M_j}$.

From Eq. (5.92), state equation of Heffron–Phillips model of N-machine power system with L PSSs installed of Eq. (5.43) can be written as

$$\begin{aligned} \dot{\mathbf{x}}(t) &= \mathbf{A}\mathbf{x}(t) + \mathbf{B}\Delta\mathbf{u}_{\mathbf{pss}} \\ \mathbf{y} &= \mathbf{C}\mathbf{x}(t) \end{aligned} \tag{5.93}$$

where

$$\mathbf{A} = \begin{bmatrix} \mathbf{0} & \omega_0\mathbf{I} & \mathbf{0} & \mathbf{0} \\ -\mathbf{M}^{-1}\mathbf{K}_1 & -\mathbf{M}^{-1}\mathbf{D} & -\mathbf{M}^{-1}\mathbf{K}_2 & \mathbf{0} \\ -\mathbf{T}_{\mathbf{d0}}^{-1}\mathbf{K}_4 & \mathbf{0} & -\mathbf{T}_{\mathbf{d0}}^{-1}\mathbf{K}_3 & \mathbf{T}_{\mathbf{d0}}^{-1} \\ -\mathbf{T}_{\mathbf{A}}^{-1}\mathbf{K}_5\mathbf{K}_{\mathbf{A}} & \mathbf{0} & -\mathbf{T}_{\mathbf{A}}^{-1}\mathbf{K}_6\mathbf{K}_{\mathbf{A}} & -\mathbf{T}_{\mathbf{A}}^{-1} \end{bmatrix},$$

$$\mathbf{B} = \begin{bmatrix} \mathbf{0} \\ \mathbf{0} \\ \mathbf{0} \\ \mathbf{T}_{\mathbf{A}}^{-1}\mathbf{K}_{\mathbf{A}} \end{bmatrix}, \ \mathbf{C} = \begin{bmatrix} \mathbf{0} & \mathbf{I} & \mathbf{0} & \mathbf{0} \end{bmatrix}$$

The set of fixed modes of the decentralized control system of Eqs. (5.92) and (5.93), denoted as $\Lambda(\mathbf{A}, \mathbf{B}, \mathbf{C}, \mathbf{K})$, is defined as

$$\Lambda(\mathbf{A}, \mathbf{B}, \mathbf{C}, \mathbf{K}) = \bigcap_{\mathbf{K} \in \mathbf{K}'} \lambda[\mathbf{A} + \mathbf{BKC}] \tag{5.94}$$

where $\lambda[\mathbf{A} + \mathbf{BKC}]$ denotes the set of all eigenvalues of matrix $\mathbf{A} + \mathbf{BKC}$, and $\mathbf{K}'$ is the following set

$$\mathbf{K}' = \{\mathbf{K} | \mathbf{K} = \text{diag}(k_j), \mathbf{K} \in R^{N \times N}\} \tag{5.95}$$

Definition of the fixed modes given by Eqs. (5.94) and (5.95) in fact means that the fixed modes are not affected by variations of any $k_j$. This is the key in the following proof.

From Eq. (5.93), it can have

$$\mathbf{A}+\mathbf{BKC}=\begin{bmatrix} \mathbf{0} & \omega_o\mathbf{I} & \mathbf{0} & \mathbf{0} \\ -\mathbf{M}^{-1}\mathbf{K}_1 & -\mathbf{M}^{-1}\mathbf{D} & -\mathbf{M}^{-1}\mathbf{K}_2 & \mathbf{0} \\ -\mathbf{T}_{d0}^{-1}\mathbf{K}_4 & \mathbf{0} & -\mathbf{T}_{d0}^{-1}\mathbf{K}_3 & \mathbf{T}_{d0}^{-1} \\ -\mathbf{T}_A^{-1}\mathbf{K}_5\mathbf{K}_A & -\mathbf{T}_A^{-1}\mathbf{K}_A\mathbf{K} & -\mathbf{T}_A^{-1}\mathbf{K}_5\mathbf{K}_A & -\mathbf{T}_A^{-1} \end{bmatrix} \tag{5.96}$$

If $\lambda_o$ is a fixed mode, it must be the eigenvalue of matrix $\mathbf{A}+\mathbf{BKC}$. Hence,

$$\begin{bmatrix} \mathbf{0} & \omega_o\mathbf{I} & \mathbf{0} & \mathbf{0} \\ -\mathbf{M}^{-1}\mathbf{K}_1 & -\mathbf{M}^{-1}\mathbf{D} & -\mathbf{M}^{-1}\mathbf{K}_2 & \mathbf{0} \\ -\mathbf{T}_{d0}^{-1}\mathbf{K}_4 & \mathbf{0} & -\mathbf{T}_{d0}^{-1}\mathbf{K}_3 & \mathbf{T}_{d0}^{-1} \\ -\mathbf{T}_A^{-1}\mathbf{K}_5\mathbf{K}_A & -\mathbf{T}_A^{-1}\mathbf{K}_A\mathbf{K} & -\mathbf{T}_A^{-1}\mathbf{K}_5\mathbf{K}_A & -\mathbf{T}_A^{-1} \end{bmatrix}\begin{bmatrix} \mathbf{v}_1 \\ \mathbf{v}_2 \\ \mathbf{v}_3 \\ \mathbf{v}_4 \end{bmatrix}=\lambda_o\begin{bmatrix} \mathbf{v}_1 \\ \mathbf{v}_2 \\ \mathbf{v}_3 \\ \mathbf{v}_4 \end{bmatrix} \tag{5.97}$$

That is,

$$\begin{aligned} &\omega_o\mathbf{v}_2=\lambda_o\mathbf{v}_1 \\ &-\mathbf{K}_1\mathbf{v}_1-\mathbf{D}\mathbf{v}_2-\mathbf{K}_2\mathbf{v}_3=\lambda_o\mathbf{M}\mathbf{v}_2 \\ &-\mathbf{K}_4\mathbf{v}_1-\mathbf{K}_3\mathbf{v}_3+\mathbf{v}_4=\lambda_o\mathbf{T}_{d0}\mathbf{v}_3 \\ &-\mathbf{K}_5\mathbf{K}_A\mathbf{v}_1-\mathbf{K}_A\mathbf{K}\mathbf{v}_2-\mathbf{K}_6\mathbf{K}_A\mathbf{v}_3-\mathbf{v}_4=\lambda_o\mathbf{T}_A\mathbf{v}_4 \end{aligned} \tag{5.98}$$

From Eq. (5.98), it can be obtained that

$$\left[\mathbf{A}_4\lambda_o^4+\mathbf{A}_3\lambda_o^3+\mathbf{A}_2\lambda_o^2+\left(\mathbf{A}_1-\frac{\mathbf{K}_A\mathbf{K}}{\omega_o}\right)\lambda_o+\mathbf{A}_0\right]\mathbf{v}_1=\mathbf{0} \tag{5.99}$$

where

$$\begin{aligned} \mathbf{A}_0&=(\mathbf{K}_3+\mathbf{K}_A\mathbf{K}_6)\mathbf{K}_2^{-1}\mathbf{K}_1-\mathbf{K}_4-\mathbf{K}_A\mathbf{K}_5 \\ \mathbf{A}_1&=(\mathbf{T}_A\mathbf{K}_3+\mathbf{T}_{d0})\mathbf{K}_2^{-1}\mathbf{K}_1+\frac{1}{\omega_0}(\mathbf{K}_3+\mathbf{K}_A\mathbf{K}_6)\mathbf{K}_2^{-1}\mathbf{D}-\mathbf{T}_A\mathbf{K}_4 \\ \mathbf{A}_2&=\frac{1}{\omega_0}(\mathbf{T}_A\mathbf{K}_3+\mathbf{T}_{d0})\mathbf{K}_2^{-1}\mathbf{D}+\mathbf{T}_A\mathbf{T}_{d0}\mathbf{K}_2^{-1}\mathbf{K}_1+\frac{1}{\omega_0}(\mathbf{K}_3+\mathbf{K}_A\mathbf{K}_6)\mathbf{K}_2^{-1}\mathbf{M} \\ \mathbf{A}_3&=\frac{1}{\omega_0}\mathbf{T}_A\mathbf{T}_{d0}\mathbf{K}_2^{-1}\mathbf{D}+\frac{1}{\omega_0}(\mathbf{T}_A\mathbf{K}_3+\mathbf{T}_{d0})\mathbf{K}_2^{-1}\mathbf{M} \\ \mathbf{A}_4&=\frac{1}{\omega_0}\mathbf{T}_A\mathbf{T}_{d0}\mathbf{K}_2^{-1}\mathbf{M} \end{aligned}$$

$\mathbf{v}\neq\mathbf{0}$; hence, $\mathbf{v}_1\neq\mathbf{0}$. It should have

$$\mathrm{DET}\left[\mathbf{A}_4\lambda_o^4+\mathbf{A}_3\lambda_o^3+\mathbf{A}_2\lambda_o^2+\left(\mathbf{A}_1-\frac{\mathbf{K}_A\mathbf{K}}{\omega_o}\right)\lambda_o+\mathbf{A}_0\right]=0 \tag{5.100}$$

Denote

$$
\begin{aligned}
\mathbf{F}(\lambda_o, k_1, k_2, \ldots k_N) &= \mathbf{A_4}\lambda_o^4 + \mathbf{A_3}\lambda_o^3 + \mathbf{A_2}\lambda_o^2 + \left(\mathbf{A_1} - \frac{\mathbf{K_A K}}{\omega_o}\right)\lambda_o + \mathbf{A_0} \\
&= \begin{bmatrix} f_{11}(\lambda_o, k_1) & f_{12}(\lambda_o) & \ldots & f_{1N}(\lambda_o) \\ f_{21}(\lambda_o) & f_{22}(\lambda_o, k_2) & \ldots & f_{2N}(\lambda_o) \\ \vdots & \vdots & \vdots & \vdots \\ f_{N1}(\lambda_o) & f_{N2}(\lambda_o) & \ldots & f_{NN}(\lambda_o, k_N) \end{bmatrix}
\end{aligned} \tag{5.101}
$$

Also denote

$$
\mathbf{F}_k(\lambda_o, k_1, k_2, \ldots k_N) = \begin{bmatrix} f_{11}(\lambda_o, k_1) & f_{12}(\lambda_o) & \ldots & f_{1N}(\lambda_o) \\ f_{21}(\lambda_o) & f_{22}(\lambda_o, k_2) & \ldots & f_{2N}(\lambda_o) \\ \vdots & \vdots & \vdots & \vdots \\ f_{N1}(\lambda_o) & f_{N2}(\lambda_o) & \ldots & f_{NN}(\lambda_o, k_N) \end{bmatrix} \tag{5.102}
$$

Obviously, $\mathbf{F}_N(\lambda_o, k_1, k_2, \ldots k_N) = \mathbf{F}(\lambda_o, k_1, k_2, \ldots k_N)$. Let the companion matrix of $\mathbf{F}_k(\lambda_o, k_1, k_2, \ldots k_k)$ be $\mathbf{F}_k^*(\lambda_o, k_1, k_2, \ldots k_k)$ and denote

$$
\mathbf{F}_k^*(\lambda_o, k_1, k_2, \ldots k_k) = \begin{bmatrix} f_{k,11}^* & f_{k,12}^* & \cdots & f_{k,1k}^* \\ f_{k,21}^* & f_{k,22}^* & \cdots & f_{k,2k}^* \\ \vdots & \vdots & \vdots & \vdots \\ f_{k,k1}^* & f_{k,k2}^* & \cdots & f_{k,kk}^* \end{bmatrix} \tag{5.103}
$$

From Eq. (5.102), it can have

$$
f_{k,kk}^* = \mathrm{DET}[\mathbf{F}_{k-1}(\lambda_o, k_1, k_2, \ldots k_{k-1})] \tag{5.104}
$$

Denote

$$
g_k(\lambda_o, k_1, k_2, \cdots k_k) = \mathrm{DET}[\mathbf{F}_k(\lambda_o, k_1, k_2, \cdots k_k)] \tag{5.105}
$$

When $k = N$, from Eqs. (5.100) and (5.105) it can have

$$
\begin{aligned}
g_N(\lambda_o, k_1, k_2, \ldots k_N) &= \mathrm{DET}[\mathbf{F}_N(\lambda_o, k_1, k_2, \ldots k_N)] \\
&= \mathrm{DET}[\mathbf{F}(\lambda_o, k_1, k_2, \ldots k_N)] = 0
\end{aligned} \tag{5.106}
$$

Because $\lambda_o$ is a fixed mode,

$$
\frac{\partial \lambda_o}{\partial k_j} = 0 (j = 1, 2, \ldots, N) \tag{5.107}
$$

From Eq. (5.106), it can have

$$\frac{\partial g_N}{\partial \lambda_o}\frac{\partial \lambda_o}{\partial k_N}+\frac{\partial g_N}{\partial k_N}=0 \tag{5.108}$$

From Eqs. (5.107) and (5.108), it can be obtained that

$$\frac{\partial g_N}{\partial k_N}=0 \tag{5.109}$$

For any matrix $\mathbf{M}(x)$,

$$\frac{d\mathrm{DET}[\mathbf{M}(x)]}{dx}=\mathrm{Trace}[\mathbf{M}^*(x)\frac{d\mathbf{M}(x)}{dx}^T] \tag{5.110}$$

where $\mathbf{M}^*(x)$ is the companion matrix of $\mathbf{M}(x)$; thus,

$$\frac{\partial g_N}{\partial k_N}=\mathrm{Trace}[\mathbf{F}_N^*(\lambda_o,k_1,k_2,\ldots k_N)\frac{\partial \mathbf{F}_N(\lambda_o,k_1,k_2,\ldots k_N)^T}{\partial k_N}]=0 \tag{5.111}$$

From Eq. (5.101), it can be obtained that

$$\begin{aligned}\frac{\partial \mathbf{F}_N(\lambda_o,k_1,k_2,\cdots k_N)^T}{\partial k_N}&=\mathrm{diag}\begin{bmatrix}0 & 0 & \cdots & 0 & \frac{\partial f_{NN}(\lambda_o,k_N)}{\partial k_N}\end{bmatrix}\\&=\mathrm{diag}\begin{bmatrix}0 & 0 & \cdots & 0 & \frac{\lambda_o}{\omega_0}K_{NA}\end{bmatrix}\end{aligned} \tag{5.112}$$

Hence from Eqs. (5.103), (5.111), and (5.112), it can have

$$\frac{\partial g_N}{\partial k_N}=\frac{\lambda_o}{\omega_0}K_{NA}f^*_{N,NN}=0 \tag{5.113}$$

$K_{NA}\neq 0$ and if $\lambda_o$ is an electromechanical oscillation mode, $\lambda_o \neq 0$. Hence from Eqs. (5.113), (5.104), and (5.105), it can have

$$f^*_{N,NN}=\mathrm{DET}[\mathbf{F}_{N-1}(\lambda_o,k_1,k_2,\ldots k_{N-1})]=g_{N-1}(\lambda_o,k_1,k_2,\ldots k_{N-1})=0 \tag{5.114}$$

From Eq. (5.114), it should have

$$\frac{\partial g_{N-1}}{\partial \lambda_o}\frac{\partial \lambda_o}{\partial k_{N-1}}+\frac{\partial g_{N-1}}{\partial k_{N-1}}=0 \tag{5.115}$$

From Eqs. (5.101), (5.107), and (5.110), it can be obtained

$$
\begin{aligned}
\frac{\partial g_{N-1}}{\partial k_{N-1}} &= \text{Trace}\left[\mathbf{F}^*_{N-1}(\lambda_o, k_1, k_2, \ldots k_{N-1})\frac{\partial \mathbf{F}_N(\lambda_o, k_1, k_2, \ldots k_{N-1})^T}{\partial k_{N-1}}\right] \\
&= \text{Trace}\left\{\mathbf{F}^*_{N-1}(\lambda_o, k_1, k_2, \ldots k_{N-1})\text{diag}\begin{bmatrix}0 & 0 & \ldots & \frac{\lambda_o}{\lambda_0}\mathbf{K}_{(N-1)A}\end{bmatrix}\right\} \\
&= \frac{\partial_o}{\partial_0}\mathbf{K}_{(N-1)A}f^*{}_{(N-1),(N-1)(N-1)} = \frac{\partial_o}{\partial_0}\mathbf{K}_{(N-1)A}g_{N-2}(\partial_o, k_1, k_2, \ldots k_{N-2}) = 0
\end{aligned}
\tag{5.116}
$$

Again, $K_{(N-1)A} \neq 0$ and $\lambda_o \neq 0$, it can have

$$
g_{N-2}(\lambda_o, k_1, k_2, \ldots k_{N-2}) = \text{DET}[\mathbf{F}_{N-2}(\lambda_o, k_1, k_2, \ldots k_{N-2})] = 0 \tag{5.117}
$$

Above procedure from Eqs. (5.106) to (5.117) can be carried on iteratively until it is obtained that

$$
\frac{\partial g_1}{\partial k_1} = \frac{\lambda_o}{\omega_0}K_{1A} = 0 \tag{5.118}
$$

It contradicts to that $K_{1A} \neq 0$ and $\lambda_o \neq 0$. Hence, $\lambda_o$ cannot be a fixed mode and an electromechanical oscillation mode.

## 5.3 An Example Three-Machine Power System

### 5.3.1 *Example Power System and Its Linearized Heffron–Phillips Model*

#### 5.3.1.1 System Parameters and Operating Conditions

Configuration of a three-machine six-node example power system is shown in Fig. 5.7 [1]. Parameters of generators and the AVRs are given in Table 5.1. Parameters of transformers and lines are given in Table 5.2[1]. Initial conditions of example power system for the load flow computation are

$$
\begin{aligned}
V_{G1} &= 1.04\angle 0, V_{G2} = 1.025\angle 0, V_{G3} = 1.025\angle 0, \\
V_{1,2,3} &= 1.00\angle 0, V_{A,B,C} = 1.00\angle 0, \\
S_{G1} &= 0.716 + j0.27, S_{G2} = 1.63 + j0.067, S_{G3} = 0.85 - j0.109 \\
\text{Load}_A &= 1.25 + j0.5, \text{Load}_B = 0.9 + j0.3, \text{Load}_C = 1 + j0.35.
\end{aligned}
$$

Tables 5.3 and 5.4 are the results of the load flow computation.

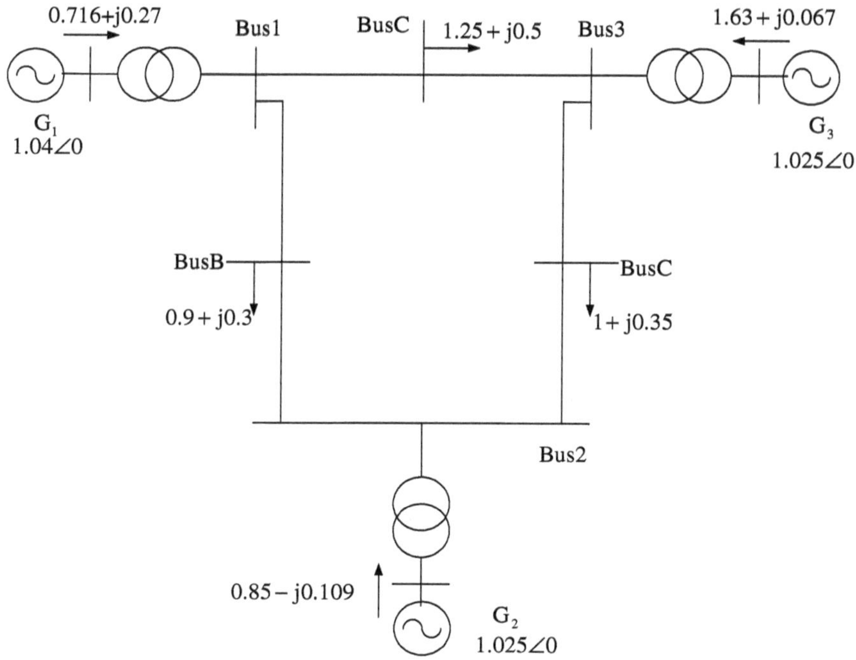


**Fig. 5.7** Configuration of a three-machine example power system [1]

**Table 5.1** Parameters of generators and the AVRs [1, 2]

| | G1 | G2 | G3 |
|---|---|---|---|
| $x_d$ | 0.146 | 0.8958 | 1.313 |
| $x'_d$ | 0.0608 | 0.1189 | 0.1813 |
| $T'_{d0}$ | 8.96 | 6 | 5.89 |
| $x_q$ | 0.0969 | 0.8645 | 1.258 |
| $x'_q$ | 0.0969 | 0.1969 | 0.25 |
| M | 47.2 | 12.8 | 6.02 |
| D | 0 | 0 | 0 |
| $K_A$ | 200 | 200 | 200 |
| $T_A$ | 0.02 | 0.02 | 0.02 |

#### 5.3.1.2 Calculation of Initial Values of State Variables

From active and reactive power output of each generator, $P_i$ and $Q_i, i = 1,2,3$, and its terminal voltage, $\bar{V}_{gi}, i = 1,2,3$, output current of each generator can be calculated as

**Table 5.2** Parameters of transformers and lines [1]

| Node | Node | Resistance | Reactance | Susceptance |
|---|---|---|---|---|
| 1 | A | 0.01 | 0.085 | 0.088 |
| 1 | B | 0.017 | 0.092 | 0.079 |
| A | 2 | 0.032 | 0.161 | 0.153 |
| B | 3 | 0.039 | 0.17 | 0.170 |
| 2 | C | 0.085 | 0.072 | 0.0745 |
| C | 3 | 0.0119 | 0.1008 | 0.1045 |
| G1 | 1 | 0 | 0.0576 | 0 |
| G2 | 2 | 0 | 0.0625 | 0 |
| G3 | 3 | 0 | 0.0586 | 0 |

**Table 5.3** Results of load flow computation—node voltage

| Node | Magnitude of voltage | Phase of voltage (deg.) |
|---|---|---|
| G1 | 1.04 | 0.000 |
| G2 | 1.025 | 0.163 |
| G3 | 1.025 | 0.081 |
| 1 | 1.018 | −0.039 |
| A | 0.977 | −0.070 |
| B | 1.007 | −0.065 |
| C | 1.012 | 0.012 |
| 2 | 1.021 | 0.065 |
| 3 | 1.030 | 0.034 |

**Table 5.4** Results of load flow computation—active and reactive power

| Starting node (SN) | Ending node (EN) | Active power at SN | Reactive power at SN | Active power at EN | Reactive power at EN |
|---|---|---|---|---|---|
| 4 | 5 | 0.412 | 0.411 | −0.409 | −0.468 |
| 4 | 6 | 0.306 | 0.036 | −0.304 | −0.108 |
| 5 | 8 | −0.841 | −0.117 | 0.865 | 0.084 |
| 6 | 9 | −0.596 | −0.061 | 0.609 | −0.065 |
| 8 | 7 | 0.765 | 0.018 | −0.760 | −0.054 |
| 7 | 9 | −0.240 | −0.204 | 0.241 | 0.103 |
| G1 | 4 | 0.718 | 0.396 | −0.718 | −0.360 |
| G2 | 8 | 1.630 | 0.142 | −1.630 | 0.017 |
| G3 | 9 | 0.850 | −0.072 | −0.850 | 0.112 |

$$\bar{I}_{gi} = \frac{P_i - jQ_i}{\bar{V}_{gi}^*}, \quad i = 1,2,3$$

The imaginary voltage of each generator, $\bar{E}_{Qi}$, $i = 1,2,3$, can be computed as

$$\bar{E}_{Qi} = V_{gi} + jx_{qi}\bar{I}_{gi}, \quad i = 1,2,3$$

Phase of $\bar{E}_{Qi}$, $i = 1,2,3$ is the angular position of each generator, $\delta_i$, $i = 1,2,3$. From $\delta_i$, $i = 1,2,3$, computational results of output current and terminal voltage of each generator can be transformed from the common x–y coordinate to the d–q coordinate as (Eq. (5.49))

$$\begin{bmatrix} i_{di} \\ i_{qi} \end{bmatrix} = \begin{bmatrix} \sin\delta_i & -\cos\delta_i \\ \cos\delta_i & \sin\delta_i \end{bmatrix} \begin{bmatrix} v_{gxi} \\ v_{gyi} \end{bmatrix} \begin{bmatrix} v_{di} \\ v_{qi} \end{bmatrix} = \begin{bmatrix} \sin\delta_i & -\cos\delta_i \\ \cos\delta_i & \sin\delta_i \end{bmatrix} \begin{bmatrix} v_{gxi} \\ v_{gyi} \end{bmatrix}$$

According to Eq. (5.25), it can have

$$E'_{qi} = v_{qi} + x'_{di} i_{di}$$
$$E_{qi} = E'_{qi} + (x_{di} - x'_{di}) i_{di}$$

Since $E'_{fdi0} = 0$, from the third equation in Eq. (5.24) it can have

$$E_{fd0i} = E_{qi0}$$

Computational results of all the above are presented in Table 5.5.

#### 5.3.1.3 Linearized Heffron–Phillips Model

From Table 5.2 and Fig. 5.7, following network admittance matrix can be established.

$$\bar{\mathbf{Y}}_L = \begin{bmatrix} \bar{\mathbf{Y}}_{11} & \bar{\mathbf{Y}}_{12} \\ \bar{\mathbf{Y}}_{21} & \bar{\mathbf{Y}}_{22} \end{bmatrix}$$

**Table 5.5** Initial value of state variables

| | G1 | G2 | G3 |
|---|---|---|---|
| $\delta_{i0}$ (rad.) | 0.0621 | 1.0388 | 0.9199 |
| $\omega_{i0}$ | 1 | 1 | 1 |
| $E'_{qi0}$ | 1.0637 | 0.8119 | 0.7882 |
| $E'_{fqi0}$ | 1.0998 | 1.8302 | 1.4334 |

where

$$\bar{\mathbf{Y}}_{11} = \begin{bmatrix} -\mathrm{j}17.3611 & 0 & 0 \\ 0 & -\mathrm{j}16 & 0 \\ 0 & 0 & -\mathrm{j}17.0648 \end{bmatrix},$$

$$\bar{\mathbf{Y}}_{12} = \begin{bmatrix} \mathrm{j}17.3611 & 0 & 0 & 0 & 0 & 0 \\ 0 & 0 & 0 & 0 & \mathrm{j}16 & 0 \\ 0 & 0 & 0 & 0 & 0 & \mathrm{j}17.0648 \end{bmatrix},$$

$$\bar{\mathbf{Y}}_{21} = \begin{bmatrix} \mathrm{j}17.3611 & 0 & 0 \\ 0 & 0 & 0 \\ 0 & 0 & 0 \\ 0 & 0 & 0 \\ 0 & \mathrm{j}16 & 0 \\ 0 & 0 & \mathrm{j}17.0648 \end{bmatrix},$$

$$\bar{\mathbf{Y}}_{22-1} = \begin{bmatrix} 3.3074 - \mathrm{j}39.3089 & -1.3652 + \mathrm{j}11.6041 & -1.9422 + \mathrm{j}10.5107 \\ -1.3652 + \mathrm{j}11.6041 & 3.8620 - \mathrm{j}18.0714 & 0 \\ -1.9422 + \mathrm{j}10.5107 & 0 & 4.1115 - \mathrm{j}16.1367 \\ 0 & 0 & 0 \\ 0 & -1.1876 + \mathrm{j}\,5.9751 & 0 \\ 0 & 0 & -1.2820 + \mathrm{j}5.5882 \end{bmatrix},$$

$$\bar{\mathbf{Y}}_{22-2} = \begin{bmatrix} 0 & 0 & 0 \\ 0 & -1.1876 + \mathrm{j}5.9751 & 0 \\ 0 & 0 & -1.2820 + \mathrm{j}5.5882 \\ 3.7482 - \mathrm{j}23.6449 & -1.6171 + \mathrm{j}13.6980 & -1.1551 + \mathrm{j}9.7843 \\ -1.6171 + \mathrm{j}13.6980 & 2.8047 - \mathrm{j}35.4456 & 0 \\ -1.1551 + \mathrm{j}9.7843 & 0 & 2.4371 - \mathrm{j}32.1539 \end{bmatrix},$$

$$\bar{\mathbf{Y}}_{22} = [\bar{\mathbf{Y}}_{22-1} \;\; \bar{\mathbf{Y}}_{22-2}]$$

By keeping three nodes of generator terminals via deleting all the other nodes in the above network admittance matrix, it can have

$$\bar{\mathbf{Y}}_{\mathbf{N}} = \bar{\mathbf{Y}}_{11} - \bar{\mathbf{Y}}_{12}\bar{\mathbf{Y}}_{22}^{-1}\bar{\mathbf{Y}}_{21}$$
$$= \begin{bmatrix} 1.0994 - \mathrm{j}4.7578 & 0.0954 + \mathrm{j}2.2197 & 0.0057 + \mathrm{j}2.2568 \\ 0.0954 + \mathrm{j}2.2197 & 0.7374 - \mathrm{j}5.1374 & 0.1249 + \mathrm{j}2.8146 \\ 0.0057 + \mathrm{j}2.2568 & 0.1249 + \mathrm{j}2.8146 & 0.7236 - \mathrm{j}5.0289 \end{bmatrix}$$

From Eq. (5.34), it can be obtained

$$\bar{\mathbf{Y}} = \left(\bar{\mathbf{Y}}_{\mathbf{N}}^{-1} + \mathrm{j}\mathbf{x}_{\mathbf{d}}'\right)$$

$$= \begin{bmatrix} 0.8312 - j3.0501 & 0.2806 + j1.4829 & 0.2054 + j1.2048 \\ 0.2806 + j1.4829 & 0.4192 - j2.7499 & 0.2119 + j1.0795 \\ 0.2054 + j1.2048 & 0.2119 + j1.0795 & 0.2759 - j2.3757 \end{bmatrix}$$

$$= \begin{bmatrix} 3.1614\angle{-1.3047} & 1.5092\angle 1.3838 & 1.2221\angle 1.4019 \\ 1.5092\angle 1.3838 & 2.7816\angle{-1.4195} & 1.1001\angle 1.3770 \\ 1.2221\angle 1.4019 & 1.1001\angle 1.3770 & 2.3917\angle{-1.4552} \end{bmatrix}$$

Thus according to Eq. (5.37), it can have

$$\mathbf{F_{dd}} = \begin{bmatrix} -0.1450 & 0.1484 & -0.0034 \\ -1.4886 & 2.4386 & -0.9500 \\ -1.1275 & -0.8120 & 1.9395 \end{bmatrix}, \mathbf{F_{qq}} = \begin{bmatrix} 2.8414 & -1.5907 & -1.2507 \\ -0.6023 & 1.2064 & -0.6042 \\ -0.6478 & -0.8347 & 1.4825 \end{bmatrix},$$

$$\mathbf{G_{dd}} = \begin{bmatrix} 3.0501 & -1.0626 & -0.9434 \\ -0.5975 & 2.7499 & -1.0467 \\ -0.6326 & -1.0970 & 2.3757 \end{bmatrix}, \mathbf{G_{qq}} = \begin{bmatrix} 0.8312 & -1.0718 & -0.7770 \\ 1.3859 & 0.4192 & 0.3384 \\ 1.0457 & 0.0823 & 0.2759 \end{bmatrix},$$

$$\mathbf{H_{dd}} = \begin{bmatrix} 0.0300 & -0.7991 & -0.8366 \\ 0.0500 & 0.3126 & 0.3643 \\ 0.0377 & 0.0614 & 0.2971 \end{bmatrix}, \mathbf{H_{qq}} = \begin{bmatrix} -0.1101 & 0.7923 & 1.0157 \\ 0.0216 & -2.0503 & 1.1270 \\ 0.0228 & 0.8179 & -2.5579 \end{bmatrix}.$$

From Eq. (5.38), following matrices are calculated

$$\mathbf{F_{d}} = \begin{bmatrix} 0.3252 & -0.0095 & -0.3157 \\ -1.5477 & 2.4103 & -0.8626 \\ -1.1306 & -0.8935 & 2.0241 \end{bmatrix}, \mathbf{F_{q}} = \begin{bmatrix} 2.1589 & -1.3566 & -0.8023 \\ -0.2670 & 0.3235 & -0.0565 \\ -0.2296 & -0.1689 & 0.3985 \end{bmatrix},$$

$$\mathbf{G_{d}} = \begin{bmatrix} 2.2175 & -1.2518 & -1.1666 \\ -0.1557 & 2.7758 & -0.9860 \\ -0.3995 & -1.1026 & 2.3982 \end{bmatrix}, \mathbf{G_{q}} = \begin{bmatrix} 1.6110 & -0.8094 & -0.4956 \\ 0.6318 & 0.1512 & 0.1474 \\ 0.4495 & 0.0527 & 0.1083 \end{bmatrix}.$$

Thus according to Eqs. (5.41) and (5.42), it can have

$$\mathbf{K}_1 = (\mathbf{V_d} - \mathbf{I_q} \cdot \mathbf{X}'_{\mathbf{d}}) \cdot \mathbf{F_d} + (\mathbf{V_q} + \mathbf{I_d} \cdot \mathbf{X_q}) \cdot \mathbf{F_q},$$
$$\mathbf{K}_2 = \mathbf{I_q} + (\mathbf{V_d} - \mathbf{I_q} \cdot \mathbf{X}'_{\mathbf{d}}) \cdot \mathbf{G_d} + (\mathbf{V_q} + \mathbf{I_d} \cdot \mathbf{X_q}) \cdot \mathbf{G_q},$$
$$\mathbf{K}_3 = \mathbf{I} + (\mathbf{X_d} \cdot \mathbf{X}'_{\mathbf{d}}) \cdot \mathbf{G_d},$$
$$\mathbf{K}_4 = (\mathbf{X_d} \cdot \mathbf{X}'_{\mathbf{d}}) \cdot \mathbf{F_d},$$
$$\mathbf{K}_5 = -\mathbf{V}_{\mathbf{gen}}^{-1} \cdot \mathbf{V_q} \cdot \mathbf{X}'_{\mathbf{d}} \cdot \mathbf{F_d} + \mathbf{V}_{\mathbf{gen}}^{-1} \cdot \mathbf{V_d} \cdot \mathbf{X_q} \cdot \mathbf{F_q},$$
$$\mathbf{K}_6 = \mathbf{V}_{\mathbf{gen}}^{-1} \cdot \mathbf{V_q} - \mathbf{V}_{\mathbf{gen}}^{-1} \cdot \mathbf{V_d} \cdot \mathbf{X}'_{\mathbf{d}} \cdot \mathbf{G_d} + \mathbf{V}_{\mathbf{gen}}^{-1} \cdot \mathbf{V_d} \cdot \mathbf{X_q} \cdot \mathbf{G_q}$$

Coefficient matrices of Heffron–Phillips model of example power system as given by Eq. (5.43) are obtained to be

$$\mathbf{K}_1 = \begin{bmatrix} 2.3373 & -1.4640 & -0.8733 \\ -1.5290 & 2.2160 & -0.6870 \\ -1.0599 & -0.8201 & 1.8800 \end{bmatrix}, \mathbf{K}_2 = \begin{bmatrix} 2.4574 & -0.9034 & -0.5628 \\ 1.0247 & 3.0671 & -0.4060 \\ 0.3694 & -0.6458 & 2.3234 \end{bmatrix},$$

$$\mathbf{K}_3 = \begin{bmatrix} 1.1889 & -0.1067 & -0.0994 \\ -0.1209 & 3.1565 & -0.7660 \\ -0.4521 & -1.2478 & 3.7141 \end{bmatrix}, \mathbf{K}_4 = \begin{bmatrix} 0.0277 & -0.0008 & -0.0269 \\ -1.2024 & 1.8725 & -0.6701 \\ -1.2795 & -1.0112 & 2.2907 \end{bmatrix},$$

$$\mathbf{K}_5 = \begin{bmatrix} -0.0068 & -0.0076 & 0.0143 \\ -0.0596 & 0.0315 & 0.0281 \\ -0.0780 & -0.0499 & 0.1278 \end{bmatrix}, \mathbf{K}_6 = \begin{bmatrix} 0.8732 & 0.0711 & 0.0678 \\ 0.4315 & 0.5292 & 0.1730 \\ 0.4691 & 0.1829 & 0.4790 \end{bmatrix}.$$

$$\mathbf{M} = \begin{bmatrix} 47.2 & 0 & 0 \\ 0 & 12.8 & 0 \\ 0 & 0 & 6.02 \end{bmatrix}, \mathbf{D} = \begin{bmatrix} 0 & 0 & 0 \\ 0 & 0 & 0 \\ 0 & 0 & 0 \end{bmatrix},$$

$$\mathbf{K_A} = \begin{bmatrix} 200 & 0 & 0 \\ 0 & 200 & 0 \\ 0 & 0 & 200 \end{bmatrix}, \mathbf{T_A} = \begin{bmatrix} 0.02 & 0 & 0 \\ 0 & 0.02 & 0 \\ 0 & 0 & 0.02 \end{bmatrix},$$

$$\mathbf{T}'_{\mathbf{d0}} = \begin{bmatrix} 8.96 & 0 & 0 \\ 0 & 6.00 & 0 \\ 0 & 0 & 5.89 \end{bmatrix}.$$

## 5.3.2 Modal Analysis of Example Power System

### 5.3.2.1 Electromechanical Oscillation Modes of Example Power System

Eigenvalues of state matrix of Heffron–Phillips model obtained in the above section are calculated as

$$\begin{aligned}
\bar{\lambda}_{1,2} &= -25.0316 \pm j28.2289 \\
\bar{\lambda}_{3} &= -34.2648 \\
\bar{\lambda}_{4,5} &= -25.0343 \pm j9.5156 \\
\bar{\lambda}_{6} &= -13.5354 \\
\bar{\lambda}_{7,8} &= -1.5051 \pm j9.1860 \\
\bar{\lambda}_{9,10} &= -0.1736 \pm j7.0590 \\
\bar{\lambda}_{11,12} &= 0.0000 \pm j0.0000
\end{aligned}$$

Frequency of $\bar{\lambda}_{4,5}$, $\bar{\lambda}_{7,8}$, and $\bar{\lambda}_{9,10}$ is $\frac{9.5156}{2\pi} = 1.5145$ Hz, $\frac{9.1860}{2\pi} = 1.4620$ Hz, and $\frac{7.0590}{2\pi} = 1.1235$ Hz, respectively, falling within the range of power system low-frequency oscillation 0.1–2 Hz. They can be identified to be the electromechanical oscillation modes of the example power system. To confirm the identification, participation factor and the correlation ratio of electromechanical loop can be calculated for $\bar{\lambda}_4 = -25.0343 + j9.5156$, $\bar{\lambda}_7 = -1.5051 + j9.1860$, and $\bar{\lambda}_9 = -0.1736 + j7.0590$ as follows.

Right and left eigenvectors corresponding to $\bar{\lambda}_4 = -25.0343 + j9.5156$ are calculated as

$$\bar{\mathbf{v}}_7 = \begin{bmatrix} -0.6852 - j17.3385 \\ -0.6595 + j5.9033 \\ 1.3447 + j11.4352 \\ -64.7503 + j192.9710 \\ 31.8356 - j61.9802 \\ 32.9147 - 130.9908 \\ -1.9362 - 396.4038 \\ -2.5524 + j88.6491 \\ 2.5028 + j50.7603 \\ -0.5973 - j1.5444 \\ 0.18208 + j0.5224 \\ 0.12978 + j0.2957 \end{bmatrix}, \bar{\mathbf{w}}_7 = \begin{bmatrix} -j0.0001 \\ 0.0002 + j0.0003 \\ 0.0001 + j0.0001 \\ 0 \\ 0 \\ 0 \\ 0.0020 + j0.0008 \\ -0.0044 - j0.0017 \\ -0.0024 - j0.0010 \\ -0.5202 - j0.0006 \\ 0.7505 + j0.0000 \\ 0.4076 + j0.0030 \end{bmatrix}$$

By use of Eq. (5.60), participation factors corresponding to each of state variables are calculated as

$$p_{41} = 0.0017, p_{42} = 0.0022, p_{43} = 0.0018.$$

Thus, the correlation ratio of electromechanical loop can be calculated by using Eq. (5.63) as $\sigma_4 = 0.0040$, which is smaller than 1. Hence, $\bar{\lambda}_4 = -25.0343 + j9.5156$ is not an electromechanical oscillation mode.

Right and left eigenvectors corresponding to $\bar{\lambda}_7 = -1.5051 + j9.1860$ are calculated as

$$\bar{\mathbf{v}}_7 = \begin{bmatrix} 1.0451 + j2.2184 \\ 2.2567 + j5.5000 \\ -3.3018 - j7.7184 \\ 68.1811 - j46.9144 \\ 170.8669 - j105.1733 \\ -239.0480 + j152.08774 \\ -0.1000 + j2.5611 \\ -2.449 + j6.2889 \\ 3.5529 - j8.2378 \\ 0.0009 + j0.0057 \\ -0.0042 + j0.0224 \\ 0.0067 - j0.0301 \end{bmatrix}, \bar{\mathbf{w}}_7 = \begin{bmatrix} 0.0007 - j0.0011 \\ 0.0073 - j0.0198 \\ -0.0252 + j0.0502 \\ 0 \\ 0.0005 + j0.0003 \\ -0.0013 - j0.0010 \\ 0.0005 - j0.0001 \\ 0.0039 + j0.0037 \\ -0.0042 - j0.0188 \\ 0.0017 + j0.0447 \\ -0.1917 + j0.1403 \\ 0.9683 \end{bmatrix}$$

By use of Eq. (5.60), participation factors corresponding to each of state variables are calculated as

$$p_{71} = 0.1427, p_{72} = 0.3021, p_{73} = 0.0618.$$

Thus, the correlation ratio of electromechanical loop can be calculated by using Eq. (5.63) as $\sigma_7 = 4.8776$, which is greater than 1. Hence, $\bar{\lambda}_7 = -1.5051 + j9.1860$ is an electromechanical oscillation mode.

Similarly, right and left eigenvectors corresponding to $\bar{\lambda}_9 = -0.1736 + j7.0590$ are calculated as

$$\bar{\mathbf{v}}_9 = \begin{bmatrix} 0.0830 + j2.4518 \\ -0.0192 - j1.8299 \\ -0.0638 - j0.62186 \\ 108.9599 - j6.3716 \\ -81.3708 + j2.8540 \\ -27.5890 + j3.5176 \\ -0.6337 + j0.5454 \\ 0.9340 - j0.2801 \\ 0.2936 - j0.0842 \\ -0.0012 + j0.0014 \\ 0.0029 - j0.0014 \\ 0.0009 - j0.0004 \end{bmatrix}, \bar{\mathbf{w}}_9 = \begin{bmatrix} 0.0018 - j0.0581 \\ -0.0085 + j0.1649 \\ -0.0082 + j0.0985 \\ 0.0013 + j0.0001 \\ -0.0037 - j0.0003 \\ -0.0022 - j0.0002 \\ 0.0015 + j0.0016 \\ -0.0066 - j0.0195 \\ -0.0030 - j0.0134 \\ -0.0973 + j0.0918 \\ 0.8031 \\ 0.5441 - j0.0031i \end{bmatrix}$$

By use of Eq. (5.60), participation factors corresponding to each of state variables are calculated as

$$p_{91} = 0.0031, p_{92} = 0.1254, p_{93} = 0.4717.$$

Thus, the correlation ratio of electromechanical loop can be calculated by using Eq. (5.63) as $\rho_9 = 34.3291$, which is greater than 1. Hence, $\bar{\lambda}_9 = -0.1736 + j7.0590$ is also an electromechanical oscillation mode.

According to Eq. (5.66), mode shape of electromechanical oscillation mode can be identified from the elements of right eigenvector corresponding to $\Delta\omega_j, j = 1, 2, 3$, which is the 4th, 5th, and 6th element of right eigenvector, respectively. For $\bar{\lambda}_7 = -1.5051 + j9.1860$, the elements of right eigenvector are

$$\bar{v}_{74} = 0.0000 + j0.0000 = 0\angle 42.2017^\circ$$
$$\bar{v}_{75} = 0.0005 + j0.0003 = 0.0006\angle 29.5912^\circ$$
$$\bar{v}_{76} = -0.0013 - j0.0010 = 0.0017\angle -144.0796^\circ$$

The mode shape is illustrated in Fig. 5.8. It shows that the electromechanical oscillation associated with this mode is that of G3 against G2.

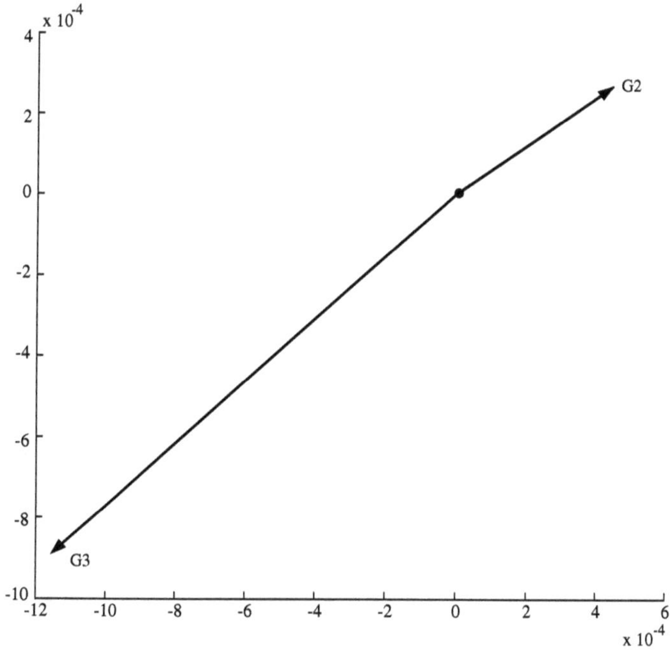


**Fig. 5.8** Mode shape of $\bar{\lambda}_7 = -1.5051 + j9.1860$

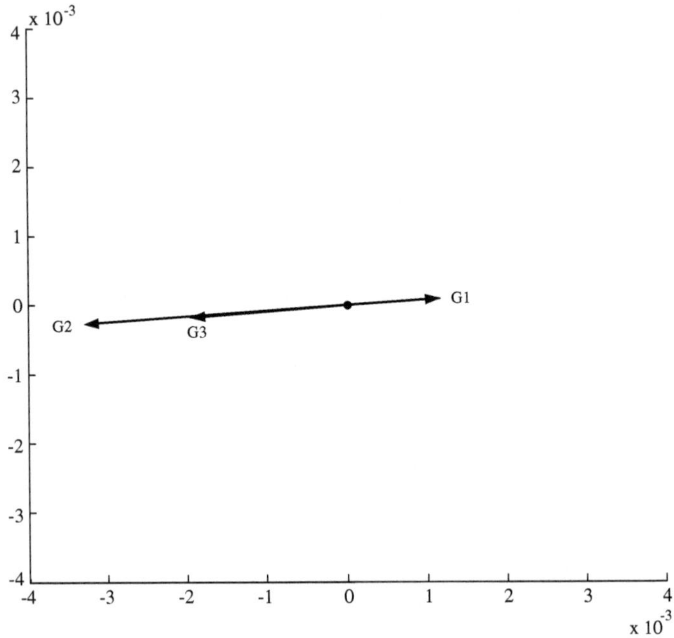


**Fig. 5.9** Mode shape of $\bar{\lambda}_9 = -0.1736 + j7.0590$

Similarly, for $\bar{\lambda}_9 = -0.1736 + j7.0590$, the elements of right eigenvector are

$$\begin{aligned} \bar{v}_{94} &= 0.0013 + j0.0001 = 0.0013\angle 3.2273 \\ \bar{v}_{95} &= -0.0037 - j0.0003 = 0.0037\angle -175.6366 \\ \bar{v}_{96} &= -0.0022 - j0.0002 = 0.0022\angle -173.8540 \end{aligned}$$

The mode shape is illustrated in Fig. 5.9. It indicates that the electromechanical oscillation associated with this mode is that of G1 against G2 and G3.

#### 5.3.2.2 Selective Reduced-Order Calculation of Electromechanical Oscillation Modes of Example Power System

From the Heffron–Phillips model of example power system, the compact model of system of Eq. (5.68) in the form of transfer function matrix is obtained as

$$\begin{aligned} \mathbf{F}(s) = s^2\mathbf{M} + s\mathbf{D} + \omega_0\mathbf{K}_1 - \mathbf{K}_2[(\mathbf{I} + s\mathbf{T}_A)(\mathbf{K}_3 + s\mathbf{T}'_{d0}) + \mathbf{K}_A\mathbf{K}_6]^{-1} \\ \times [(\mathbf{I} + s\mathbf{T}_A)\mathbf{K}_4 + \mathbf{K}_A\mathbf{K}_k] \end{aligned}$$

The Hooke-Jeeves optimum searching method in nonlinear programming can be used to find the solution of following objective function of Eq. (5.74) to calculate electromechanical oscillation modes as follows.

$$f(\bar{\lambda}) = \left|\mathrm{DEL}[\mathbf{F}(\bar{\lambda})]\right| \tag{5.74}$$

Step 1: Calculate eigenvalues of matrix $-\omega_0\mathbf{M}^{-1}\mathbf{K}_1$ as $\eta_i, i = 1, 2$. Thus, $\sqrt{\eta_i} \neq 0, j = 1, 2$ is the angular natural oscillation frequency of the electromechanical oscillation modes. Choose $\bar{\lambda}_i(0) = 0 + j\sqrt{\eta_i}$ on the complex plane as the starting point of optimum searching.
Step 2: Choose an initial searching scale $S_0 > 0$, a minimum searching scale $S_F > 0$, and an accelerating factor $\alpha > 0$. Search on the complex plane from $\bar{\lambda}_i(k)$ to $\bar{\lambda}_i(k+1)$. Three different patterns of searching on the complex plane from $\bar{\lambda}_i(k)$ to $\bar{\lambda}_i(k+1)$ are illustrated in Fig. 5.10 where S is the searching step and $\alpha = 2$. In Fig. 5.10 if $f(\bar{\lambda}_i \pm S) < f(\bar{\lambda}_i)$ or $f(\bar{\lambda}_i \pm jS) < f(\bar{\lambda}_i)$, the searching is successful in horizontal and vertical direction. Otherwise, it is a failure. From Fig. 5.10, it can be seen that (1) pattern 1—there is one successful searching in both horizontal and

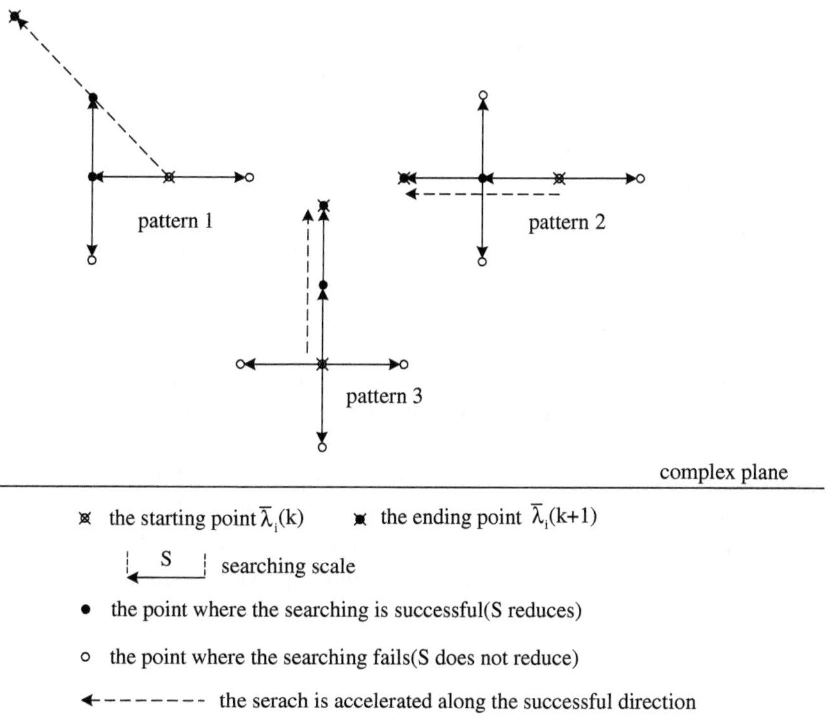


**Fig. 5.10** Illustration of searching patterns of the Hooke-Jeeves optimum searching method in one step

vertical direction; (2) pattern 2—only one searching in the horizontal direction is successful; (3) pattern 3—only one searching in the vertical direction is successful.
Step 3: If searching in Step 2 fails, reduce the searching scale S by half and then go back to Step 2;
Step 4: If at $\bar{\lambda}_i(k_F)$, the searching scale becomes less than $S_F$, stop the searching and take $\bar{\lambda}_i(k_F)$ as the solution of objective function of Eq. (5.74), i.e., the ith oscillation mode of the power system.

Figure 5.11 presents the flow chart of the Hooke-Jeeves optimum searching method as explained above for the selective reduced-order calculation of the electromechanical oscillation modes

The minimum searching scale $S_F > 0$ is the maximum distance of $\bar{\lambda}_i(k_F)$ to the position where the real solution $\bar{\lambda}_i$ locates as shown in Fig. 5.12, where the last searching scale is not greater than $2S_F$. Therefore, $S_F$ defines the accuracy of optimum searching which is fully controllable. Of course, theoretically, the algorithm can reach a solution which is as accurate as it is wished.

In the case that two or more starting points might converge to the same solution, preventing techniques, such as the addition of a penalty or barrier function into the objective function [3], can be applied in the direct searching.

The computational complexity (CC) [3] or computational cost (CC) [4] normally is used to estimate the computational efficiency of an algorithm where one operation of an addition or multiplication is defined as one CC. The CC of multiplication of two n-order full matrices (none of them is a diagonal matrix) is about $n^3$ and the inverse calculation of an n-order full matrix is $\frac{n^3}{3}$. Thus, the CC to form the polynomial matrix model of Eq. (5.74) (i.e. forming all coefficient matrices in $\mathbf{F}(s)$) is estimated as $2n^3/3 + 32n^3$. The CC in calculating the determinant of an n-order complex matrix is about $4n^3/3$. Thus, the CC of the Hooke-Jeeves optimum searching in one step is about $16n^3/3$. Therefore, the CC of the Hooke-Jeeves optimum searching method is $(k_{HS} + K_{HF})16n^3/3$, where $k_{HS}$ is the times of successful searching and $K_{HF}$ the times of searching failure. $K_{HF}$ can be estimated as $k_{HF} = \log(S_0/S_F)/\log R$, where R is the ratio of reducing the searching scale when the searching in one step fails. If R = 2 $S_0 = 0.01$, $S_F = 0.000001$, we can have $k_{HF} \approx 13$. Hence, the total CC of the reduced-order method is about $102n^3 + k_{HS}16n^3/3 + o(n^2)$.

Figures 5.13 and 5.14 show the trajectories of optimal searching starting from (0, 7.5128) and (0, 10.5647), respectively. Searching results of the electromechanical oscillation modes are $\bar{\lambda}_1 = -0.1736 + j7.0590$ and $\bar{\lambda}_2 = -1.5051 + j9.1860$, which are exactly as same as those obtained in the previous section where they are denoted as $\bar{\lambda}_9 = -0.1736 + j7.0590$ and $\bar{\lambda}_7 = -1.5051 + j9.1860$.

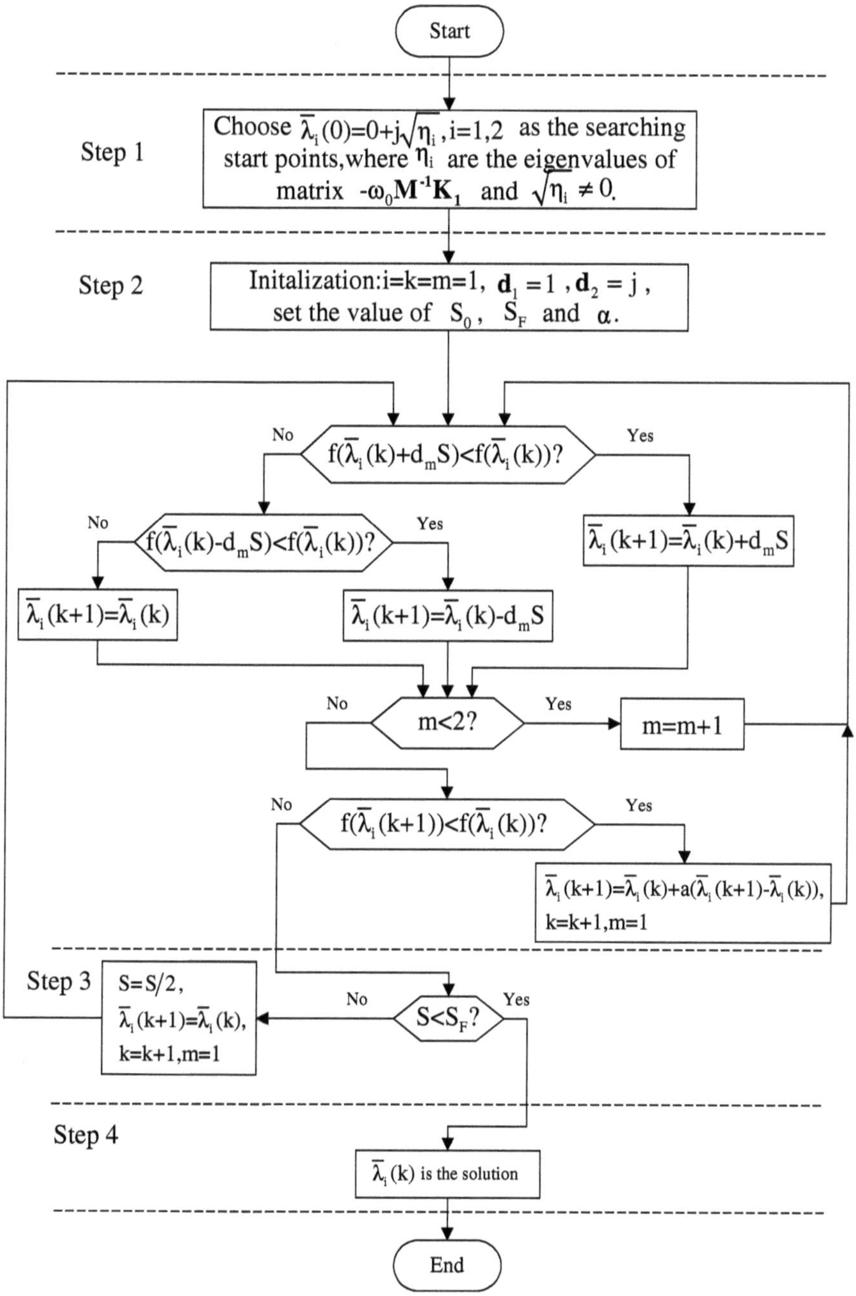


**Fig. 5.11** Flow chart of Hooke-Jeeves optimum searching method

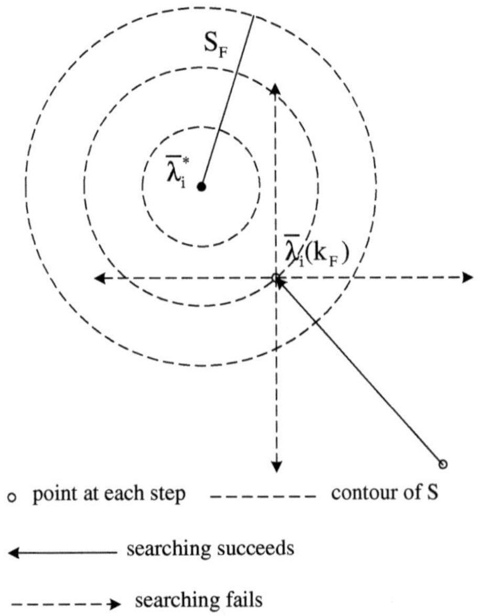


**Fig. 5.12** Illustration on the searching to arrive at its final point

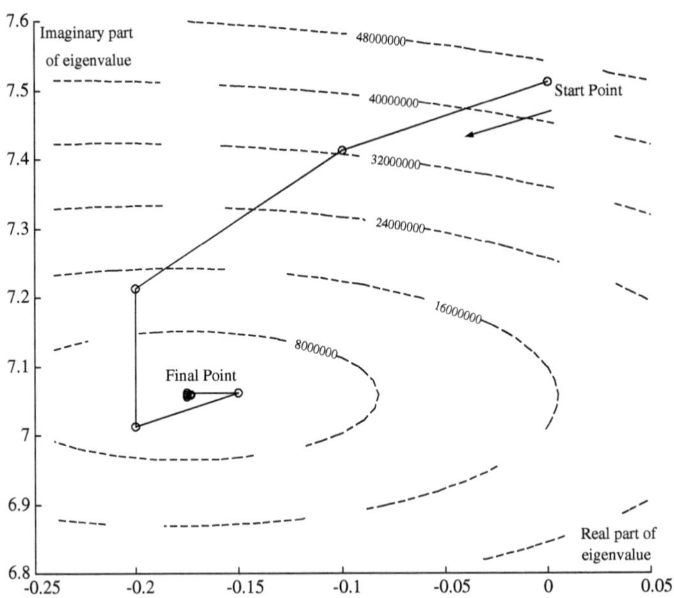


**Fig. 5.13** Path of $\bar{\lambda}_1$ calculation using direct searching algorithm

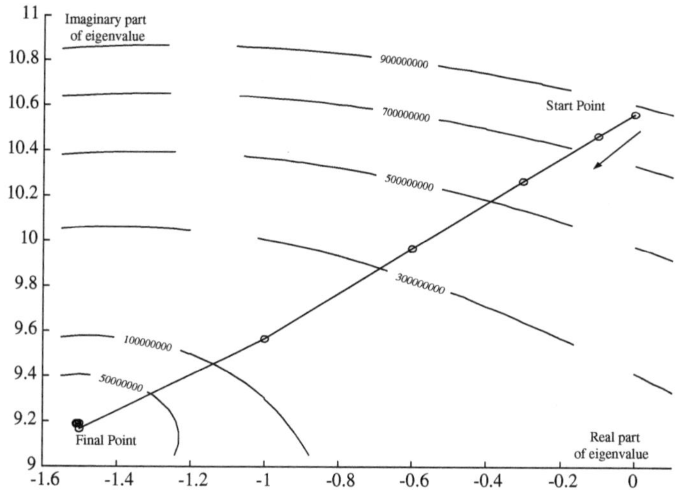


**Fig. 5.14** Path of $\bar{\lambda}_2$ calculation using direct searching algorithm

#### 5.3.2.3 Coordinated Design of PSSs

Damping ratio of $\bar{\lambda}_1$ is 0.0246 and that of $\bar{\lambda}_2$ is 0.1617. Now PSSs will be installed in three machines to enhance the damping ratio of $\bar{\lambda}_1$ to 0.1. Parameters of PSSs can be set by use of the method introduced previously in Sect. 5.2.2 to move $\bar{\lambda}_1$ to $\bar{\lambda}_1^* = -0.7094 + j7.0590$. Hence, following objective function is set up

$$f(\mathbf{p}) = [\mathrm{Re}(\bar{\lambda}_1(\mathbf{p})) - \mathrm{Re}(\bar{\lambda}_1^*(\mathbf{p}^*))]^2 + [\mathrm{Im}(\bar{\lambda}_1(\mathbf{p})) - \mathrm{Im}(\bar{\lambda}_1^*(\mathbf{p}^*))]^2$$

where $\mathbf{p} = [D_{pss1} \quad D_{pss2} \quad D_{pss3}]$ and $D_{pssi}$ is the coefficient of damping torque provided by the PSS installed in the ith generator.

Choose $\mathbf{p}_0 = [0 \quad 0 \quad 0]$ with $f(\mathbf{p}_0) = 0.2872$ to start the optimal searching of the solution of the above objective function by use of the method introduced in Sect. 5.2.2.2. In 52 iterations, the searching stops to have $\mathbf{p}_{52} = [21.73 \quad 11.84 \quad 21.94]$ with $f(\mathbf{p}_{52}) = 9.2687 \times 10^{-15}$, $\bar{\lambda}_1$ is moved to $\bar{\lambda}_{c1}$. Figure 5.15 is the trajectories of $D_{pssi}, i = 1, 2, 3$ in respective to the iterative searching. The optimal searching method used is the Hooke-Jeeves optimum searching. Figure 5.16 is the trajectory of the objective function during the optimum searching with respect to the movement of oscillation mode $\bar{\lambda}_1$.

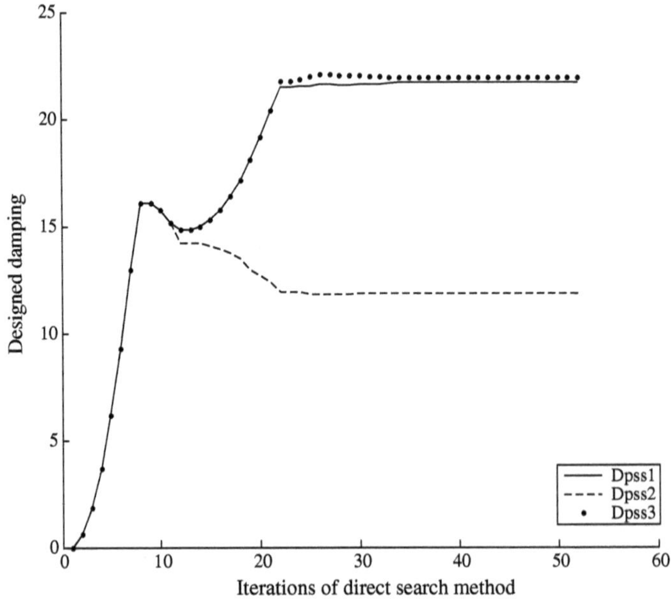


**Fig. 5.15** Trajectories of $D_{pssi}$, i = 1, 2, 3 in respective to the iterative searching

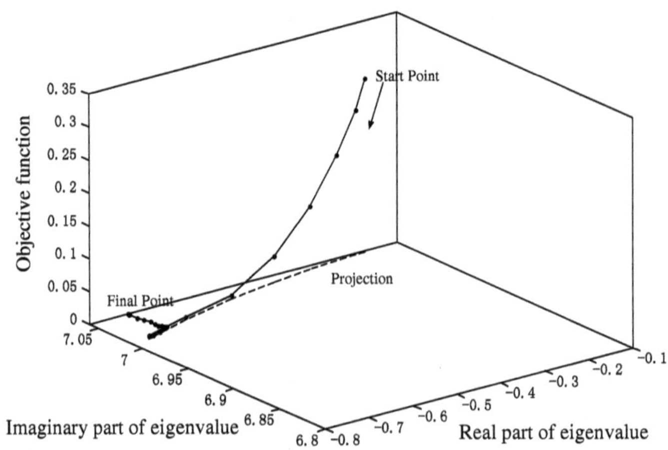


**Fig. 5.16** Trajectory of the objective function in respective to the movement of $\bar{\lambda}_1$

**Table 5.6** Parameter of the PSSs

| | $K_{pss}$ | $T_2$ | $T_4$ |
|---|---|---|---|
| G1 | 4.1272 | 0.096 | 0.096 |
| G2 | 1.4864 | 0.095 | 0.095 |
| G3 | 3.0945 | 0.1 | 0.1 |

From Fig. 5.5, the forward path matrix from the stabilizing signals of PSSs to the electromechanical oscillation loops of generators is obtained as

$$\mathbf{F}_{\mathbf{pss}}(\bar{\lambda}_1^*) = \mathbf{K_2K_A}\left((\mathbf{K_3} + \bar{\lambda}_1^*\mathbf{T'_{d0}})(\mathbf{I} + \bar{\lambda}_1^*\mathbf{T_A}) + \mathbf{K_6K_A}\right)^{-1}$$
$$= \begin{bmatrix} 4.0590\angle{-29.34^\circ} & 1.7719\angle 153.70^\circ & 1.0376\angle 155.74^\circ \\ 0.6433\angle 113.76^\circ & 6.1802\angle{-28.79^\circ} & 2.6963\angle 136.09^\circ \\ 1.1801\angle 125.70^\circ & 2.6184\angle 140.40^\circ & 5.3227\angle{-31.52^\circ} \end{bmatrix}$$

The forward path of the PSS installed in each generator as the diagonal element of the above matrix is

$$F_{pss11}(\bar{\lambda}_1^*) = 4.0590\angle{-29.34^\circ}$$
$$F_{pss22}(\bar{\lambda}_1^*) = 6.1802\angle{-28.79^\circ}$$
$$F_{pss33}(\bar{\lambda}_1^*) = 5.3227\angle{-31.52^\circ}$$

Let the transfer function of the PSS be

$$T_{pssi}(s) = K_{pssi}\frac{(1 + sT_{2i})}{(1 + sT_{1i})}\frac{(1 + sT_{4i})}{(1 + sT_{3i})}, \quad T_{1i} = T_{3i} = 0.05$$

By use of the phase compensation method for the PSS to compensate the forward path of the ith generator $F_{ii}(\bar{\lambda}_i^*), i = 1, 2, 3$ to provide a pure damping torque $D_{pssi}^*$, where $\begin{bmatrix} D_{pss1}^* & D_{pss2}^* & D_{pss3}^* \end{bmatrix} = \mathbf{p}_{52} = [21.73 \quad 11.84 \quad 21.94]$, parameters of PSSs are set and listed in Table 5.6.

With PSSs installed, the closed-loop state equation is

$$s\mathbf{\Delta X} = \mathbf{A_{pss}}\mathbf{\Delta X}$$

where

$$\mathbf{A_{pss}} = \begin{bmatrix} \mathbf{A_{pss11}} & \mathbf{A_{pss12}} & \mathbf{A_{pss13}} \\ \mathbf{A_{pss21}} & \mathbf{A_{pss22}} & \mathbf{A_{pss23}} \end{bmatrix}$$

$$\mathbf{A}_{\mathbf{pss11}} = \begin{bmatrix} 0 & 0 & 0 & 314.1592 & 0 & 0 \\ 0 & 0 & 0 & 0 & 314.1592 & 0 \\ 0 & 0 & 0 & 0 & 0 & 314.1592 \\ -0.0495 & 0.0310 & 0.0185 & 0 & 0 & 0 \\ 0.1195 & -0.1731 & 0.0537 & 0 & 0 & 0 \\ 0.1761 & 0.1362 & -0.3123 & 0 & 0 & 0 \\ -0.0031 & 0.0001 & 0.0030 & 0 & 0 & 0 \\ 0.2004 & -0.3121 & 0.1117 & 0 & 0 & 0 \\ 0.2172 & 0.1717 & -0.3889 & 0 & 0 & 0 \end{bmatrix}$$

$$\mathbf{A}_{\mathbf{pss12}} = \begin{bmatrix} 0 & 0 & 0 & 0 & 0 & 0 \\ 0 & 0 & 0 & 0 & 0 & 0 \\ 0 & 0 & 0 & 0 & 0 & 0 \\ -0.0521 & 0.0191 & 0.0119 & 0 & 0 & 0 \\ -0.0801 & -0.2396 & 0.0317 & 0 & 0 & 0 \\ -0.0614 & 0.1073 & -0.3860 & 0 & 0 & 0 \\ -0.1327 & 0.0119 & 0.0111 & 0.1116 & 0 & 0 \\ 0.0202 & -0.5261 & 0.1277 & 0 & 0.1667 & 0 \\ 0.0768 & 0.2119 & -0.6306 & 0 & 0 & 0.1698 \end{bmatrix}$$

$$\mathbf{A}_{\mathbf{pss13}} = \begin{bmatrix} 0 & 0 & 0 & 0 & 0 & 0 \\ 0 & 0 & 0 & 0 & 0 & 0 \\ 0 & 0 & 0 & 0 & 0 & 0 \\ 0 & 0 & 0 & 0 & 0 & 0 \\ 0 & 0 & 0 & 0 & 0 & 0 \\ 0 & 0 & 0 & 0 & 0 & 0 \\ 0 & 0 & 0 & 0 & 0 & 0 \\ 0 & 0 & 0 & 0 & 0 & 0 \\ 0 & 0 & 0 & 0 & 0 & 0 \end{bmatrix}$$

$$\mathbf{A}_{\mathbf{pss21}} = \begin{bmatrix} 67.5657 & 75.7691 & -143.3348 & 0 & 0 & 0 \\ 595.9666 & -314.8297 & -281.1369 & 0 & 0 & 0 \\ 779.5679 & 498.8968 & -1278.4646 & 0 & 0 & 0 \\ -0.0951 & 0.0596 & 0.0355 & 20 & 0 & 0 \\ -0.7703 & 0.4825 & 0.2878 & 162.048 & 0 & 0 \\ 0.2270 & -0.3289 & 0.1020 & 0 & 20 & 0 \\ 0.8193 & -1.1874 & 0.3681 & 0 & 72.2 & 0 \\ 0.3521 & 0.2724 & -0.6246 & 0 & 0 & 20 \\ 3.9227 & 3.0351 & -6.9579 & 0 & 0 & 222.8 \end{bmatrix}$$

$$
\mathbf{A_{pss22}} = \begin{bmatrix}
-8731.9338 & -710.95188 & -678.1403 & -50 & 0 & 0 \\
-4315.4400 & -5291.9761 & -1729.6500 & 0 & -50 & 0 \\
-4691.1745 & -1828.9008 & -4789.5838 & 0 & 0 & -50 \\
-0.1000 & 0.0367 & 0.0229 & 0 & 0 & 0 \\
-0.8099 & 0.2977 & 0.1855 & 0 & 0 & 0 \\
-0.1521 & -0.4553 & 0.0603 & 0 & 0 & 0 \\
-0.5491 & -1.6435 & 0.2176 & 0 & 0 & 0 \\
-0.1227 & 0.2146 & -0.7719 & 0 & 0 & 0 \\
-1.3673 & 2.3902 & -8.5990 & 0 & 0 & 0
\end{bmatrix}
$$

$$
\mathbf{A_{pss23}} = \begin{bmatrix}
0 & 10000 & 0 & 0 & 0 & 0 \\
0 & 0 & 0 & 10000 & 0 & 0 \\
0 & 0 & 0 & 0 & 0 & 10000 \\
-20 & 0 & 0 & 0 & 0 & 0 \\
-77.648 & -20 & 0 & 0 & 0 & 0 \\
0 & 0 & -20 & 0 & 0 & 0 \\
0 & 0 & -34.2 & -20 & 0 & 0 \\
0 & 0 & 0 & 0 & -20 & 0 \\
0 & 0 & 0 & 0 & -111.4 & -20
\end{bmatrix}
$$

Eigenvalues of state matrix are calculated as

$$
\begin{aligned}
&\bar{\lambda}_1 = -44.7474 + j0, \bar{\lambda}_{2,3} = -23.7299 \pm j27.9946, \bar{\lambda}_4 = -33.1534 + j0, \\
&\bar{\lambda}_{5,6} = -19.9770 \pm j11.9203, \bar{\lambda}_{7,8} = -12.9863 \pm j13.6705, \\
&\bar{\lambda}_{9,10} = \mathbf{-0.6094 \pm j7.0235}, \bar{\lambda}_{11,12} = -4.5207 \pm j8.0856, \\
&\bar{\lambda}_{13} = -0.2772 + j0, \bar{\lambda}_{14} = 0 + j0, \bar{\lambda}_{15} = -23.6389 + j0, \\
&\bar{\lambda}_{16} = -12.1456 + j0, \bar{\lambda}_{17} = -17.7485 + j0, \bar{\lambda}_{18} = -15.9316 + j0
\end{aligned}
$$

Obviously, $\bar{\lambda}_1$ is moved to $\bar{\lambda}_{c1} = -0.6094 + j7.0235$, slightly away from the target required position $\bar{\lambda}_1^* = -0.7094 + j7.0590$. In Sect. 6.4.1, it is examined why this happens.

Simulation results of example power system without and with PSSs installed are presented in Fig. 5.17. At 0.5 s of simulation, a 3-phase short-circuit occurs at Bus 1, and 0.1 s later the short-circuit fault is cleared. Fig. 5.17, confirms the effectiveness of PSSs designed by use of the introduced method.

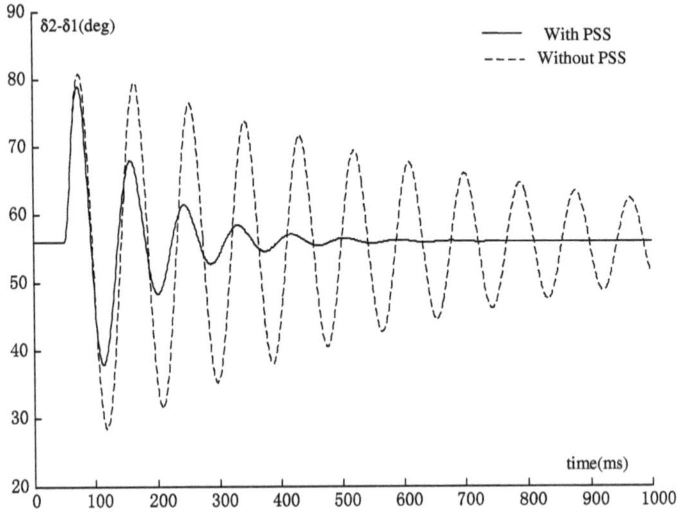


**Fig. 5.17** Simulation results with a 3-phase short-circuit fault at Bus 1

# References

1. Anderson PM, Fouad AA (1977) Power system control and stability. The Iowa State University Press
2. Wang XF, Song YH, Irving M (2011) Modern power systems analysis. Springer, Berlin
3. Bazarra MS et al (1993) Non-linear programming. Wiley, Hoboken
4. Marchuk GI (1994) Numerical methods and applications. CRC Press Inc, Boca Raton

# Chapter 6
# Multi-machine Power System Installed with Thyristor-Based FACTS Stabilizers

## 6.1 Mathematical Model of a Multi-machine Power System Installed with Thyristor-Based FACTS Stabilizers

### 6.1.1 *Heffron–Phillips Model [1]*

#### 6.1.1.1 Heffron–Phillips Model of an N-Machine Power System Installed with an SVC Stabilizer

Figure 6.1 shows an N-machine power system with nodes of internal voltage of generators added. From Fig. 6.1, the following network equations can be established.

$$\begin{bmatrix} \mathbf{0} \\ \bar{\mathbf{I}}_{\mathbf{g}} \end{bmatrix} = \begin{bmatrix} \bar{\mathbf{Y}}_{11} & \bar{\mathbf{Y}}_{12} \\ \bar{\mathbf{Y}}_{21} & \bar{\mathbf{Y}}_{22} \end{bmatrix} \begin{bmatrix} \bar{\mathbf{V}}_{\mathbf{m}} \\ \bar{\mathbf{E}}_{\mathbf{g}} \end{bmatrix} \tag{6.1}$$

where $\bar{\mathbf{I}}_{\mathbf{g}} = \begin{bmatrix} \bar{\mathrm{I}}_{\mathrm{g1}} & \bar{\mathrm{I}}_{\mathrm{g2}} & \cdots & \bar{\mathrm{I}}_{\mathrm{gN}} \end{bmatrix}^{\mathrm{T}}$, $\bar{\mathbf{E}}_{\mathbf{g}} = \begin{bmatrix} \bar{\mathrm{E}}_{\mathrm{g1}} & \bar{\mathrm{E}}_{\mathrm{g2}} & \cdots & \bar{\mathrm{E}}_{\mathrm{gN}} \end{bmatrix}^{\mathrm{T}}$, $\bar{\mathbf{V}}_{\mathbf{m}}$ is the vector of voltage at all non-internal nodes of generators in the power system and $\bar{\mathbf{E}}_{\mathbf{g}}$ is defined by Eq. (5.34). "Deleting" all non-internal nodes is to remove $\bar{\mathbf{V}}_{\mathbf{m}}$ in the network equations of Eq. (6.1) to have

$$\bar{\mathbf{I}}_{\mathbf{g}} = (\bar{\mathbf{Y}}_{21}\bar{\mathbf{Y}}_{11}^{-1}\bar{\mathbf{Y}}_{12} + \bar{\mathbf{Y}}_{22})\bar{\mathbf{E}}_{\mathbf{g}} \tag{6.2}$$

where only N internal nodes of generators are kept. Comparing Eqs. (6.2) and (5.34), it can be seen that in fact

$$\bar{\mathbf{Y}} = \bar{\mathbf{Y}}_{21}\bar{\mathbf{Y}}_{11}^{-1}\bar{\mathbf{Y}}_{12} + \bar{\mathbf{Y}}_{22} \tag{6.3}$$



H. Wang and W. Du, *Analysis and Damping Control of Power System Low-frequency Oscillations*, Power Electronics and Power Systems,
DOI 10.1007/978-1-4899-7696-3_6

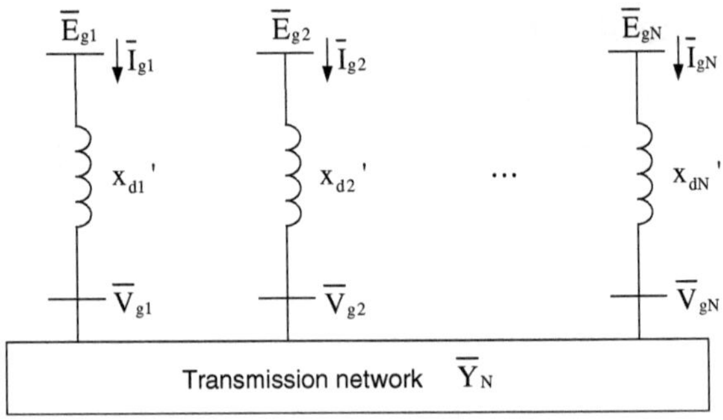


**Fig. 6.1** Network of an N-machine power system including internal nodes of generators

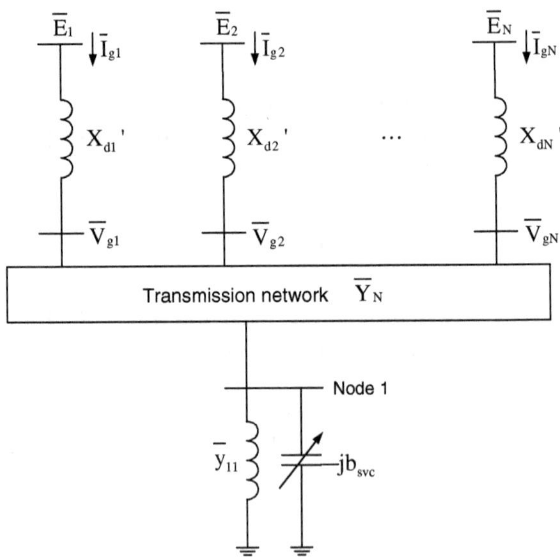


**Fig. 6.2** An N-machine power system installed with an SVC-based stabilizer

Now consider an N-machine power system installed with an SVC stabilizer as shown in Fig. 6.2. Without loss of generality of discussion, it can be assumed that (1) the SVC stabilizer is installed at node 1 in the transmission network, and (2) after other nodes are deleted in the network equations of the N-machine power system, there are only N internal nodes of generators and node 1 left. Hence, the network equations with those N + 1 nodes left are

$$\begin{bmatrix} 0 \\ \mathbf{I_g} \end{bmatrix} = \begin{bmatrix} \bar{Y}_{11} & \bar{\mathbf{Y}}_{12} \\ \bar{\mathbf{Y}}_{21} & \bar{\mathbf{Y}}_{22} \end{bmatrix} \begin{bmatrix} \bar{V}_{1n} \\ \mathbf{E_g} \end{bmatrix} \tag{6.4}$$

where $\bar{Y}_{11} = \bar{y}_{11} + jb_{svc}$ and $jb_{svc}$ is the equivalent admittance of the SVC.

From Eq. (6.4), it can be obtained that

$$\begin{aligned}\bar{\mathbf{I}}_{\mathbf{g}} &= \frac{1}{(\bar{y}_{11}+jb_{svc})}\bar{\mathbf{Y}}_{21}\bar{\mathbf{Y}}_{12}\bar{\mathbf{E}}_{\mathbf{g}}+\bar{\mathbf{Y}}_{22}\bar{\mathbf{E}}_{\mathbf{g}} \\ &= \left[\frac{1}{(\bar{y}_{11}+jb_{svc})}\bar{\mathbf{Y}}_{21}\bar{\mathbf{Y}}_{12}+\bar{\mathbf{Y}}_{22}\right]\bar{\mathbf{E}}_{\mathbf{g}} = \bar{\mathbf{Y}}\bar{\mathbf{E}}_{\mathbf{g}}\end{aligned} \tag{6.5}$$

From Eq. (6.5), it can be seen that the elements of network admittance matrix, $\bar{\mathbf{Y}}$, can be denoted as

$$\bar{y}_{ij} = \bar{y}'_{ij} + \bar{y}_{ij}(b_{svc}) \tag{6.6}$$

From Eqs. (5.35), (6.5), and (6.6), it can have

$$\begin{aligned}\bar{I}_{gi} = i_{di}+ji_{qi} &= \sum_{j=1}^{N}\bar{y}_{ij}[E'_{qj}e^{j(90^\circ+\delta_j-\delta_i)}+(x_{qj}-x'_{dj})i_{qj}e^{j(\delta_j-\delta_i)}] \\ &= \sum_{j=1}^{N}y'_{ij}[E'_{qj}e^{j(90^\circ+\alpha_{ij}+\delta_j-\delta_i)}+(x_{qj}-x'_{dj})i_{qj}e^{j(\alpha_{ij}+\delta_j-\delta_i)}] \\ &\quad+\sum_{j=1}^{N}\bar{y}_{ij}(b_{svc})[E'_{qj}e^{j(90^\circ+\delta_j-\delta_i)}+(x_{qj}-x'_{dj})i_{qj}e^{j(\delta_j-\delta_i)}]\end{aligned} \tag{6.7}$$

where $\bar{y}'_{ij} = y'_{ij}e^{j\alpha_{ij}}$. From Eq. (6.7), it can be obtained that

$$\begin{aligned}i_{di} &= \sum_{j=1}^{N}y'_{ij}[-E'_{qj}\sin(\alpha_{ij}-\delta_{ij})+(x_{qj}-x'_{dj})i_{qj}\cos(\alpha_{ij}-\delta_{ij})] \\ &\quad+\sum_{j=1}^{N}\mathrm{Re}[\bar{y}_{ij}(b_{svc})][E'_{qj}\sin\delta_{ij}+(x_{qj}-x'_{dj})i_{qj}\cos\delta_{ij}] \\ &\quad-\sum_{j=1}^{N}\mathrm{Im}[\bar{y}_{ij}(b_{svc})][E'_{qj}\cos\delta_{ij}-(x_{qj}-x'_{dj})i_{qj}\sin\delta_{ij}] \\ i_{qi} &= \sum_{j=1}^{N}y_{ij}[E'_{qj}\cos(\alpha_{ij}-\delta_{ij})+(x_{qj}-x'_{dj})i_{qj}\sin(\alpha_{ij}-\delta_{ij})] \\ &\quad+\sum_{j=1}^{N}\mathrm{Re}[\bar{y}_{ij}(b_{svc})][E'_{qj}\cos\delta_{ij}-(x_{qj}-x'_{dj})i_{qj}\sin\delta_{ij}] \\ &\quad+\sum_{j=1}^{N}\mathrm{Im}[\bar{y}_{ij}(b_{svc})][E'_{qj}\sin\delta_{ij}+(x_{qj}-x'_{dj})i_{qj}\cos\delta_{ij}]\end{aligned} \tag{6.8}$$

where $\mathrm{Re}[\bar{y}_{ij}(b_{svc})]$ and $\mathrm{Im}[\bar{y}_{ij}(b_{svc})]$ denote the real and imaginary parts of $\bar{y}_{ij}(b_{svc})$, respectively.

Linearization of Eq. (6.8) is

$$\begin{aligned}\Delta \mathbf{I_d} &= \mathbf{F_{dd}}\Delta\delta + \mathbf{G_{dd}}\Delta \mathbf{E}'_\mathbf{q} + \mathbf{H_{dd}}\Delta \mathbf{I_q} + \mathbf{L_{dd}}\Delta b_{svc} \\ \Delta \mathbf{I_q} &= \mathbf{F_{qq}}\Delta\delta + \mathbf{G_{qq}}\Delta \mathbf{E}'_\mathbf{q} + \mathbf{H_{qq}}\Delta \mathbf{I_q} + \mathbf{L_{qq}}\Delta b_{svc}\end{aligned} \tag{6.9}$$

The above equation can give

$$\begin{aligned}\Delta \mathbf{I_d} &= \mathbf{F_d}\Delta\delta + \mathbf{G_d}\Delta \mathbf{E}'_\mathbf{q} + \mathbf{L_d}\Delta b_{svc} \\ \Delta \mathbf{I_q} &= \mathbf{F_q}\Delta\delta + \mathbf{G_q}\Delta \mathbf{E}'_\mathbf{q} + \mathbf{L_q}\Delta b_{svc}\end{aligned} \tag{6.10}$$

Substituting Eq. (6.10) into Eqs. (5.40) and (5.41), the Heffron–Phillips model of the N-machine power system installed with the SVC stabilizer can be obtained as

$$\begin{aligned}\Delta\dot{\delta} &= \omega_0 \Delta\omega \\ \Delta\dot{\omega} &= \mathbf{M}^{-1}(-\mathbf{K_1}\Delta\delta - \mathbf{K_2}\Delta \mathbf{E}'_\mathbf{q} - \mathbf{D}\Delta\omega + \mathbf{K_P}\Delta b_{svc}) \\ \Delta\dot{\mathbf{E}}'_\mathbf{q} &= \mathbf{T}'^{-1}_\mathbf{d0}(-\mathbf{K_3}\Delta \mathbf{E}'_\mathbf{q} - \mathbf{K_4}\Delta\delta + \Delta \mathbf{E}'_\mathbf{fd} + \mathbf{K_q}\Delta b_{svc}) \\ \Delta\dot{\mathbf{E}}'_\mathbf{fd} &= -\mathbf{T}_\mathbf{A}^{-1}\Delta \mathbf{E}'_\mathbf{fd} + \mathbf{T}_\mathbf{A}^{-1}\mathbf{K_A}(-\mathbf{K_5}\Delta\delta - \mathbf{K_6}\Delta \mathbf{E}'_\mathbf{q} + \mathbf{K_V}\Delta b_{svc})\end{aligned} \tag{6.11}$$

#### 6.1.1.2 Heffron–Phillips Model of an N-Machine Power System Installed with a TCSC or TCPS Stabilizer

Without loss of generality of discussion, it can be assumed that a TCSC stabilizer is installed between nodes 1 and 2 in an N-machine power system as shown in Fig. 6.3. The following network admittance matrix can be formed with only the internal nodes of generators, nodes 1 and 2 left,

$$\begin{bmatrix} \mathbf{0} \\ \bar{\mathbf{I}}_\mathbf{g} \end{bmatrix} = \begin{bmatrix} \bar{\mathbf{Y}}_{11} & \bar{\mathbf{Y}}_{12} \\ \bar{\mathbf{Y}}_{21} & \bar{\mathbf{Y}}_{22} \end{bmatrix} \begin{bmatrix} \bar{\mathbf{V}}_{12} \\ \bar{\mathbf{E}}_\mathbf{g} \end{bmatrix} \tag{6.12}$$

where

$$\bar{\mathbf{Y}}_{11} = \begin{bmatrix} \bar{y}_{11} + \frac{1}{\bar{Z}_{12} - jX_{tcsc}} & \bar{y}_{12} - \frac{1}{\bar{Z}_{12} - jX_{tcsc}} \\ \bar{y}_{21} - \frac{1}{\bar{Z}_{12} - jX_{tcsc}} & \bar{y}_{22} + \frac{1}{\bar{Z}_{12} - jX_{tcsc}} \end{bmatrix}$$

$$\bar{\mathbf{V}}_{12} = \begin{bmatrix} \bar{V}_1 & \bar{V}_2 \end{bmatrix}^T$$

Hence, the network admittance matrix with only the internal nodes of generators left is

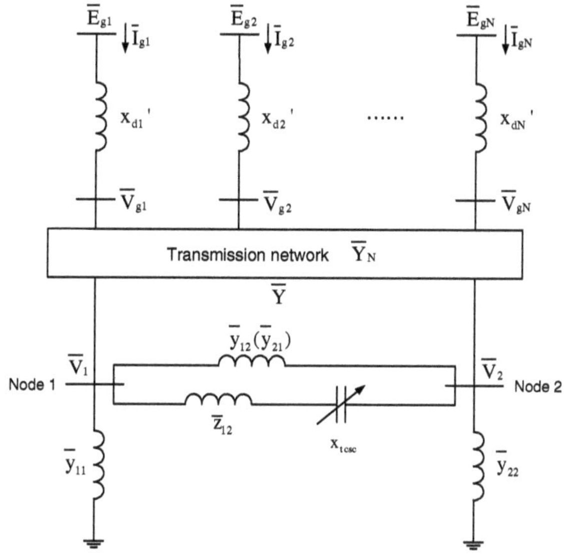


**Fig. 6.3** An N-machine power system installed with a TCSC-based stabilizer

$$\bar{\mathbf{Y}} = \bar{\mathbf{Y}}_{21}\bar{\mathbf{Y}}_{11}^{-1}\bar{\mathbf{Y}}_{12} + \bar{\mathbf{Y}}_{22} \tag{6.13}$$

Obviously, the element of the network admittance matrix $\bar{y}_{ij}$ can be denoted as

$$\bar{y}_{ij} = \bar{y}'_{ij} + \bar{y}_{ij}(x_{tcsc}) \tag{6.14}$$

From Eqs. (5.35), (6.12), and (6.14), it can have

$$\begin{aligned}
\bar{I}_{gi} = i_{di} + ji_{qi} &= \sum_{j=1}^{N} \bar{y}_{ij}[E'_{qj}e^{j(90^\circ + \delta_j - \delta_i)} + (x_{qj} - x'_{dj})i_{qj}e^{j(\delta_j - \delta_i)}] \\
&= \sum_{j=1}^{N} y'_{ij}[E'_{qj}e^{j(90^\circ + \alpha_{ij} + \delta_j - \delta_i)} + (x_{qj} - x'_{dj})i_{qj}e^{j(\alpha_{ij} + \delta_j - \delta_i)}] \\
&\quad + \sum_{j=1}^{N} \bar{y}_{ij}(x_{tcsc})[E'_{qj}e^{j(90^\circ + \delta_j - \delta_i)} + (x_{qj} - x'_{dj})i_{qj}e^{j(\delta_j - \delta_i)}]
\end{aligned} \tag{6.15}$$

Similar to Eqs. (6.7)–(6.10), by using Eq. (6.15), the Heffron–Phillips model of the N-machine power system installed with the TCSC stabilizer can be obtained as

$$\begin{aligned}
\Delta\dot{\boldsymbol{\delta}} &= \omega_o \Delta\boldsymbol{\omega} \\
\Delta\dot{\boldsymbol{\omega}} &= \mathbf{M}^{-1}(-\mathbf{K}_1\Delta\boldsymbol{\delta} - \mathbf{K}_2\Delta\mathbf{E}'_{\mathbf{q}} - \mathbf{D}\Delta\boldsymbol{\omega} + \mathbf{K}_{\mathbf{P}}\Delta x_{tcsc}) \\
\Delta\dot{\mathbf{E}}'_{\mathbf{q}} &= \mathbf{T}'^{-1}_{\mathbf{d0}}(-\mathbf{K}_3\Delta\mathbf{E}'_{\mathbf{q}} - \mathbf{K}_4\Delta\boldsymbol{\delta} + \Delta\mathbf{E}'_{\mathbf{fd}} + \mathbf{K}_{\mathbf{q}}\Delta x_{tcsc}) \\
\Delta\dot{\mathbf{E}}'_{\mathbf{fd}} &= -\mathbf{T}_{\mathbf{A}}^{-1}\Delta\mathbf{E}'_{\mathbf{fd}} + \mathbf{T}_{\mathbf{A}}^{-1}\mathbf{K}_{\mathbf{A}}(-\mathbf{K}_5\Delta\boldsymbol{\delta} - \mathbf{K}_6\Delta\mathbf{E}'_{\mathbf{q}} + \mathbf{K}_{\mathbf{V}}\Delta x_{tcsc})
\end{aligned} \tag{6.16}$$

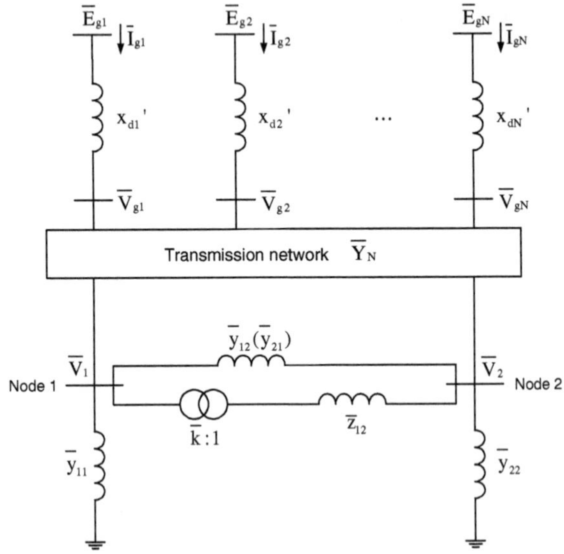


**Fig. 6.4** An N-machine power system installed with a TCPS-based stabilizer

Similarly, for the N-machine power system with a TCPS stabilizer installed between nodes 1 and 2 as shown in Fig. 6.4, where $\bar{k} = ke^{j\phi}$, it can have

$$\begin{bmatrix} \mathbf{0} \\ \bar{\mathbf{I}}_{\mathbf{g}} \end{bmatrix} = \begin{bmatrix} \bar{\mathbf{Y}}_{11} & \bar{\mathbf{Y}}_{12} \\ \bar{\mathbf{Y}}_{21} & \bar{\mathbf{Y}}_{22} \end{bmatrix} \begin{bmatrix} \bar{\mathbf{V}}_{12} \\ \bar{\mathbf{E}}_{\mathbf{g}} \end{bmatrix} \tag{6.17}$$

where

$$\bar{\mathbf{Y}}_{11} = \begin{bmatrix} \bar{y}_{11} + \frac{1}{k^2\bar{z}_{12}} & \bar{y}_{12} - \frac{e^{-j\phi}}{k\bar{z}_{12}} \\ \bar{y}_{21} - \frac{e^{j\phi}}{k\bar{z}_{12}} & \bar{y}_{22} + \frac{1}{\bar{z}_{12}} \end{bmatrix}$$

$$\bar{\mathbf{V}}_{\mathbf{12n}} = \begin{bmatrix} \bar{V}_1 & \bar{V}_2 \end{bmatrix}^T$$

Hence, the network admittance matrix with only the internal nodes of generators left can be obtained as

$$\bar{\mathbf{Y}} = \bar{\mathbf{Y}}_{\mathbf{21}} \bar{\mathbf{Y}}_{\mathbf{11}}^{-1} \bar{\mathbf{Y}}_{\mathbf{12}} + \bar{\mathbf{Y}}_{\mathbf{22}} \tag{6.18}$$

and the element of the network admittance matrix $\bar{y}_{ij}$ can be denoted as

$$\bar{y}_{ij} = \bar{y}'_{ij} + \bar{y}_{ij}(\phi) \tag{6.19}$$

Hence, for the system with the TCPS stabilizer installed, it can have

$$\begin{aligned}\bar{I}_{gi} = i_{di} + ji_{qi} &= \sum_{j=1}^{N} \bar{y}_{ij}[E'_{qj}e^{j(90^\circ + \delta_j - \delta_i)} + (x_{qj} - x'_{dj})i_{qj}e^{j(\delta_j - \delta_i)}] \\ &= \sum_{j=1}^{N} y'_{ij}[E'_{qj}e^{j(90^\circ + \alpha_{ij} + \delta_j - \delta_i)} + (x_{qj} - x'_{dj})i_{qj}e^{j(\alpha_{ij} + \delta_j - \delta_i)}] \\ &+ \sum_{j=1}^{N} \bar{y}_{ij}(\phi)[E'_{qj}e^{j(90^\circ + \delta_j - \delta_i)} + (x_{qj} - x'_{dj})i_{qj}e^{j(\delta_j - \delta_i)}]\end{aligned} \tag{6.20}$$

By taking the similar procedure from Eqs. (6.7) to (6.10), from Eq. (6.20), the Heffron–Phillips model of the N-machine power system installed with the TCPS stabilizer can be obtained as

$$\begin{aligned}\Delta\dot{\boldsymbol{\delta}} &= \omega_0\Delta\boldsymbol{\omega} \\ \Delta\dot{\boldsymbol{\omega}} &= \mathbf{M}^{-1}(-\mathbf{K}_1\Delta\boldsymbol{\delta} - \mathbf{K}_2\Delta\mathbf{E}'_\mathbf{q} - \mathbf{D}\Delta\boldsymbol{\omega} + \mathbf{K}_\mathbf{P}\Delta\phi) \\ \Delta\dot{\mathbf{E}}'_\mathbf{q} &= \mathbf{T}'^{-1}_{\mathbf{d0}}(-\mathbf{K}_3\Delta\mathbf{E}'_\mathbf{q} - \mathbf{K}_4\Delta\boldsymbol{\delta} + \Delta\mathbf{E}'_\mathbf{fd} + \mathbf{K}_\mathbf{q}\Delta\phi) \\ \Delta\dot{\mathbf{E}}'_\mathbf{fd} &= -\mathbf{T}_\mathbf{A}^{-1}\Delta\mathbf{E}'_\mathbf{fd} + \mathbf{T}_\mathbf{A}^{-1}\mathbf{K}_\mathbf{A}(-\mathbf{K}_5\Delta\boldsymbol{\delta} - \mathbf{K}_6\Delta\mathbf{E}'_\mathbf{q} + \mathbf{K}_\mathbf{V}\Delta\phi)\end{aligned} \tag{6.21}$$

The Heffron–Phillips model of the N-machine power system installed with an SVC stabilizer, TCSC stabilizer, or TCPS stabilizer can be expressed in a unified form as shown in Fig. 6.5. In Fig. 6.5, $\Delta u_{facts-s} = \Delta b_{svc}$ for the SVC stabilizer, $\Delta u_{facts-s} = \Delta x_{tcsc}$ for the TCSC stabilizer, and $\Delta u_{facts-s} = \Delta\phi$ for the TCPS stabilizer, respectively.

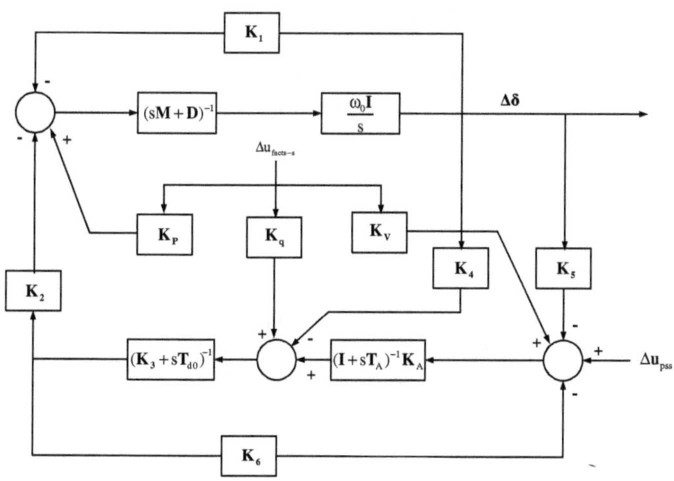


**Fig. 6.5** Unified Heffron–Phillips model of an N-machine power system installed with a thyristor-based FACTS stabilizer

### 6.1.2 *General Linearized Model of an N-Machine Power System Installed with Multiple Thyristor-Based FACTS Stabilizers*

In this section, a more general procedure to establish the linearized model of an N-machine power system installed with multiple thyristor-based FACTS stabilizers is introduced. In the established model, the generator can be described by either the reduced-order dynamic model which is used in the above section to establish the Heffron–Phillips model, or the full linearized model given by Eqs. (5.44)–(5.47). In addition, the normal control functions of thyristor-based FACTS controllers and the full dynamics of stabilizers are also included in the established model.

#### 6.1.2.1 Linearized Model of Generators

When full linearized mathematical model of generators is used, by substituting Eq. (5.51) into Eq. (5.48), it can be obtained that

$$\begin{aligned} \dot{\mathbf{X}}_{\mathbf{g}} &= \mathbf{A}_{\mathbf{gxy}}\mathbf{X}_{\mathbf{g}} + \mathbf{B}_{\mathbf{g}}\Delta\mathbf{u}_{\mathbf{pss}} + \mathbf{B}_{\mathbf{gxy}}\Delta\mathbf{V}_{\mathbf{xy}} \\ \Delta\mathbf{I}_{\mathbf{xy}} &= \mathbf{C}_{\mathbf{gxy}}\mathbf{X}_{\mathbf{g}} + \mathbf{D}_{\mathbf{gxy}}\Delta\mathbf{V}_{\mathbf{xy}} \end{aligned} \tag{6.22}$$

where

$$\begin{aligned} \mathbf{A}_{\mathbf{gxy}} &= \mathbf{A}_{\mathbf{g}} + \mathbf{B}_{\mathbf{gv}}\mathbf{B}_{\mathbf{gV}} \\ \mathbf{B}_{\mathbf{gxy}} &= \mathbf{B}_{\mathbf{gv}}\mathbf{T}_{\mathbf{g0}} \\ \mathbf{C}_{\mathbf{gxy}} &= \mathbf{T}_{\mathbf{g0}}^{-1}(\mathbf{C}_{\mathbf{g}} - \mathbf{B}_{\mathbf{gI}}) \\ \mathbf{D}_{\mathbf{gxy}} &= \mathbf{0} \end{aligned}$$

In the case that the reduced-order model of generators which is used for establishing the Heffron–Phillips mode, from Eq. (2.31), it can have

$$i_{qi} = \frac{v_{gdi}}{x_{qi}},\ i_{di} = \frac{E'_d - v_{gqi}}{x'_d} \tag{6.23}$$

Linearization of the above equation is

$$\Delta i_{qi} = \frac{\Delta v_{gdi}}{x_{qi}},\ \Delta i_{di} = \frac{\Delta E'_q - \Delta v_{gqi}}{x'_q} \tag{6.24}$$

From Eq. (6.24) and (5.39), it can be obtained that

$$
\begin{aligned}
\mathbf{\Delta I_{dq}} &= \mathbf{E_0 \Delta E'_{iq}} + \mathbf{X_{dq1} \Delta V_{dq}} \\
\mathbf{\Delta P} &= \mathbf{V_{dq0} \Delta I_{dq}} + \mathbf{I_{dq0} \Delta V_{dq}} \\
\mathbf{\Delta E_q} &= \mathbf{\Delta E'_q} + \mathbf{X'_{dd} \Delta I_{dq}} \\
\mathbf{\Delta V} &= \mathbf{V_{0dq} \Delta V_{dq}}
\end{aligned}
\tag{6.25}
$$

where

$$
\begin{aligned}
&\mathbf{E_0} = \{e_{ij}\},\ e_{ij} = \frac{1}{x'_{di}}, \\
&\text{for } i = 1, 3, \ldots, 2N-1,\ j = 1, 2, \ldots, N;\ \text{otherwise}\ \ e_{ij} = 0 \\
&\mathbf{X_{dq1}} = \operatorname{diag}(x_{dqi}),\ x_{dqi} = -\frac{1}{x'_{di}},\ \ \text{for}\ \ i = 1, 3, \ldots, 2N-1; \\
&x_{dqi} = -\frac{1}{x_{qi}} = 0,\ \ \text{for}\ \ i = 2, 4, \ldots 2N \\
&\mathbf{V_{dq0}} = \{V_{dq0ij}\},\ V_{dq0ij} = v_{gdi0},\ V_{dq0ij+1} = v_{gqi0}, \\
&\text{for } i\ = 1, 2, \ldots, N, j = 1, 3, \ldots, 2N-1; \text{otherwise}\ V_{dq0ij} = 0 \\
&\mathbf{I_{dq0}} = \{I_{dq0ij}\},\ I_{dq0ij} = i_{gdi0},\ I_{dq0ij+1} = i_{gqi0}, \\
&\text{for } i = 1, 2, \ldots, N,\ j = 1, 3, \ldots, 2N-1; \text{otherwise}\ I_{dq0ij} = 0 \\
&\mathbf{X'_{dd}} = \{x'_{ddij}\},\ x'_{ddij} = (x_{di} - x'_{di}), \\
&\text{for } i = 1, 2, \ldots, N,\ j\ = 1, 3, \ldots, 2N-1; \text{otherwise}\ x'_{ddij} = 0 \\
&\mathbf{V_{0dq}} = \{V_{0dqij}\},\ V_{0dqij} = \frac{v_{gdi0}}{V_{gi0}},\ V_{0dqij+1} = \frac{v_{gqi0}}{V_{gi0}}, \\
&\text{for } i\ = 1, 2, \ldots, N,\ j\ = 1, 3, \ldots, 2N-1; \text{otherwise}\ V_{dq0ij} = 0
\end{aligned}
$$

By substituting the above equation into (5.40), it can be obtained that

$$
\begin{aligned}
\mathbf{\Delta\dot{\delta}} &= \omega_0 \mathbf{\Delta\omega} \\
\mathbf{\Delta\dot{\omega}} &= \mathbf{M^{-1}}(-\mathbf{D\Delta\omega} - \mathbf{V_{dq0}E_0\Delta E'_{iq}}) - \mathbf{M^{-1}}(\mathbf{V_{dq0}X_{dq1}} + \mathbf{I_{dq0}})\mathbf{\Delta V_{dq}} \\
\mathbf{\Delta\dot{E}'_q} &= \mathbf{T'^{-1}_{d0}}(-\mathbf{\Delta E'_q} - \mathbf{X'_{dd}E_0\Delta E'_{iq}} + \mathbf{\Delta E'_{fd}}) - \mathbf{T'^{-1}_{d0}X'_{dd}X_{dq1}\Delta V_{dq}} \\
\mathbf{\Delta\dot{E}'_{fd}} &= -\mathbf{T_A^{-1}\Delta E'_{fd}} - \mathbf{T_A^{-1}K_A V_{0dq}\Delta V_{dq}} + \mathbf{T_A^{-1}K_A\Delta u_{pss}} \\
\mathbf{\Delta I_{dq}} &= \mathbf{E_0\Delta E'_{iq}} + \mathbf{X_{dq1}\Delta V_{dq}}
\end{aligned}
\tag{6.26}
$$

which can be written as

$$\begin{bmatrix} \Delta\dot{\delta} \\ \Delta\dot{\omega} \\ \Delta\dot{E}'_q \\ \Delta\dot{E}'_{fd} \end{bmatrix} = \begin{bmatrix} 0 & \omega_o I & 0 & 0 \\ 0 & -M^{-1}D & -M^{-1}V_{dq0}E_0 & 0 \\ 0 & 0 & -T_{d0}^{-1}(I + X'_{dd}E_0) & T_{d0}^{-1} \\ 0 & 0 & 0 & -T_A^{-1} \end{bmatrix} \begin{bmatrix} \Delta\delta \\ \Delta\omega \\ \Delta E'_q \\ \Delta E'_{fd} \end{bmatrix}$$

$$\begin{bmatrix} 0 \\ -M^{-1}(V_{dq0}X_{dq1} + I_{dq0}) \\ -T'^{-1}_{d0}X'_{dd}X_{dq1} \\ -T_A^{-1}K_A V_{0dq} \end{bmatrix} \Delta V_{dq} + \begin{bmatrix} 0 \\ 0 \\ 0 \\ T_A^{-1}K_A \end{bmatrix} \Delta u_{pss} \tag{6.27}$$

$$\Delta I_{dq} = \begin{bmatrix} 0 & 0 & E_0 & 0 \end{bmatrix} \begin{bmatrix} \Delta\delta \\ \Delta\omega \\ \Delta E'_q \\ \Delta E'_{fd} \end{bmatrix} + X_{dq1}\Delta V_{dq}$$

Obviously, the above equation can be written in the similar form to Eq. (5.48) as

$$\begin{aligned} \dot{X}_g &= A_g X_g + B_g \Delta u_{pss} + B_{gv}\Delta V_{dq} \\ \Delta I_{dq} &= C_g X_g + D_g \Delta V_{dq} \end{aligned} \tag{6.28}$$

where

$$X_g = \begin{bmatrix} \Delta\delta \\ \Delta\omega \\ \Delta E'_q \\ \Delta E'_{fd} \end{bmatrix}, A_g = \begin{bmatrix} 0 & \omega_o I & 0 & 0 \\ 0 & -M^{-1}D & -M^{-1}V_{dq0}E_0 & 0 \\ 0 & 0 & -T_{d0}^{-1}(I + X'_{dd}E_0) & T_{d0}^{-1} \\ 0 & 0 & 0 & -T_A^{-1} \end{bmatrix},$$

$$B_{gv} = \begin{bmatrix} 0 \\ -M^{-1}(V_{dq0}X_{dq1} + I_{dq0}) \\ -T'^{-1}_{d0}X'_{dd}X_{dq1} \\ -T_A^{-1}K_A V_{0dq} \end{bmatrix}, B_g = \begin{bmatrix} 0 \\ 0 \\ 0 \\ T_A^{-1}K_A \end{bmatrix},$$

$$C_g = \begin{bmatrix} 0 & 0 & E_0 & 0 \end{bmatrix}, D_g = X_{dq1}$$

By substituting Eq. (5.51) into the above equation, it can be obtained that

$$\begin{aligned} \dot{X}_g &= A_{gxy} X_g + B_g \Delta u_{pss} + B_{gxy}\Delta V_{xy} \\ \Delta I_{xy} &= C_{gxy} X_g + D_{gxy} \Delta V_{xy} \end{aligned} \tag{6.29}$$

where

$$\begin{aligned}\mathbf{A_{gxy}} &= \mathbf{A_g} + \mathbf{B_{gv}}\mathbf{B_{gV}}\\ \mathbf{B_{gxy}} &= \mathbf{B_{gv}}\mathbf{T_{g0}}\\ \mathbf{C_{gxy}} &= \mathbf{T_{g0}^{-1}}(\mathbf{C_g} - \mathbf{B_{gI}})\\ \mathbf{D_{gxy}} &= \mathbf{T_{g0}^{-1}}\mathbf{D_g}\end{aligned}$$

The format of Eqs. (6.22) and (6.29) is exactly the same, which will be used as the general linearized model of generators in the N-machine power system with no PSSs installed.

Let the transfer function of the ith PSS be $G_i(s)$ and the feedback signal $\Delta\omega_i$, that is

$$\Delta u_{pssi} = G_i(s)\Delta\omega_i \tag{6.30}$$

If the state-space realization of Eq. (6.30) is

$$\begin{aligned}\dot{\mathbf{X}}_{\mathbf{pssi}} &= \mathbf{A_{pssi}}\mathbf{X_{pssi}} + \mathbf{B_{pssi}}\Delta\omega_i\\ \Delta u_{pssi} &= \mathbf{C_{pssi}}\mathbf{X_{pssi}} + \mathbf{D_{pssi}}\Delta\omega_i\end{aligned} \tag{6.31}$$

State-space representation of L PSSs installed in the N-machine power system can be written as

$$\begin{aligned}\dot{\mathbf{X}}_{\mathbf{pss}} &= \mathbf{A_{pss}}\mathbf{X_{pss}} + \mathbf{B_{pss}}\boldsymbol{\Delta\omega}\\ \boldsymbol{\Delta}\mathbf{u_{pss}} &= \mathbf{C_{pss}}\mathbf{X_{pss}} + \mathbf{D_{pss}}\boldsymbol{\Delta\omega}\end{aligned} \tag{6.32}$$

where

$$\mathbf{X_{pss}} = \begin{bmatrix}\mathbf{X_{pss1}}\\ \mathbf{X_{pss2}}\\ \vdots\\ \mathbf{X_{pssL}}\end{bmatrix}, \quad \mathbf{B_{pss}} = \begin{bmatrix}\mathbf{B_{pss1}} & \mathbf{0} & \cdots & \mathbf{0}\\ \mathbf{0} & \mathbf{B_{pss2}} & \cdots & \mathbf{0}\\ \vdots & \vdots & \cdots & \vdots\\ \mathbf{0} & \mathbf{0} & \cdots & \mathbf{B_{pssL}}\end{bmatrix},$$

$$\mathbf{A_{pss}} = \begin{bmatrix}\mathbf{A_{pss1}} & \mathbf{0} & \cdots & \mathbf{0}\\ \mathbf{0} & \mathbf{A_{pss2}} & \cdots & \mathbf{0}\\ \vdots & \vdots & \cdots & \vdots\\ \mathbf{0} & \mathbf{0} & \cdots & \mathbf{A_{pssL}}\end{bmatrix}, \quad \mathbf{C_{pss}} = \begin{bmatrix}\mathbf{C_{pss1}} & \mathbf{0} & \cdots & \mathbf{0}\\ \mathbf{0} & \mathbf{C_{pss2}} & \cdots & \mathbf{0}\\ \vdots & \vdots & \cdots & \vdots\\ \mathbf{0} & \mathbf{0} & \cdots & \mathbf{C_{pssL}}\end{bmatrix},$$

$$\mathbf{D_{pss}} = \begin{bmatrix}\mathbf{D_{pss1}} & \mathbf{0} & \cdots & \mathbf{0}\\ \mathbf{0} & \mathbf{D_{pss2}} & \cdots & \mathbf{0}\\ \vdots & \vdots & \cdots & \vdots\\ \mathbf{0} & \mathbf{0} & \cdots & \mathbf{D_{pssL}}\end{bmatrix}$$

By substituting Eq. (6.32) into Eqs. (6.22) or (6.29), it can be obtained that

$$\begin{aligned}\dot{\mathbf{X}}_{\mathbf{gp}} &= \mathbf{A}_{\mathbf{gp}}\mathbf{X}_{\mathbf{gp}} + \mathbf{B}_{\mathbf{gp}}\Delta\mathbf{V}_{\mathbf{xy}} \\ \Delta\mathbf{I}_{\mathbf{xy}} &= \mathbf{C}_{\mathbf{gp}}\mathbf{X}_{\mathbf{gp}} + \mathbf{D}_{\mathbf{gp}}\Delta\mathbf{V}_{\mathbf{xy}}\end{aligned} \tag{6.33}$$

where

$$\mathbf{X}_{\mathbf{gp}} = \begin{bmatrix} \mathbf{X}_{\mathbf{g}} \\ \mathbf{X}_{\mathbf{pss}} \end{bmatrix}, \mathbf{A}_{\mathbf{gp}} = \begin{bmatrix} \mathbf{A}_{\mathbf{gxy}} + \mathbf{B}_{\mathbf{g}}\mathbf{D}_{\mathbf{Dpss}} & \mathbf{B}_{\mathbf{g}}\mathbf{C}_{\mathbf{pss}} \\ \mathbf{B}_{\mathbf{Dpss}} & \mathbf{A}_{\mathbf{pss}} \end{bmatrix},$$

$$\mathbf{B}_{\mathbf{gp}} = \begin{bmatrix} \mathbf{B}_{\mathbf{gxy}} \\ \mathbf{0} \end{bmatrix}, \mathbf{C}_{\mathbf{gp}} = \begin{bmatrix} \mathbf{C}_{\mathbf{gxy}} & \mathbf{0} \end{bmatrix}, \mathbf{D}_{\mathbf{gp}} = \begin{bmatrix} \mathbf{D}_{\mathbf{gp}} \\ \mathbf{0} \end{bmatrix},$$

$$\mathbf{D}_{\mathbf{Dpss}} = \begin{bmatrix} \mathbf{0} & \mathbf{0} & \mathbf{0} \\ \mathbf{0} & \mathbf{D}_{\mathbf{pss}} & \mathbf{0} \\ \mathbf{0} & \mathbf{0} & \mathbf{0} \end{bmatrix}, \mathbf{B}_{\mathbf{Dpss}} = \begin{bmatrix} \mathbf{0} & \mathbf{0} & \mathbf{0} \\ \mathbf{0} & \mathbf{B}_{\mathbf{pss}} & \mathbf{0} \\ \mathbf{0} & \mathbf{0} & \mathbf{0} \end{bmatrix}$$

Equation (6.33) is the linearized model of generators with the PSSs installed.

#### 6.1.2.2 Linearized Model of an SVC Installed in the N-Machine Power System

For a thyristor-controlled reactor and fixed capacitor (TCR-FC) type of SVC with an SVC stabilizer as shown in Fig. 3.1, from Eq. (3.1), it can be obtained that for the ith SVC installed in the N-machine power system

$$b_{svci} = \frac{1}{x_{svcci}} - \frac{2\alpha_i - \sin 2\alpha_i}{2\pi x_{svcli}} \tag{6.34}$$

Without loss of generality, denote the installing location of the SVC as the ith node in the power system. It can be assumed that the transfer function of voltage controller and stabilizer in Fig. 3.1 is $T_{vsvci}(s)$ and $T_{ssvci}(s)$, respectively, and the feedback signal of the SVC stabilizer is the integral signal of the active power, $P_{ij}$, delivered through the node where the SVC is installed as shown in Fig. 6.6. From Figs. 3.1 and 6.6, it can have

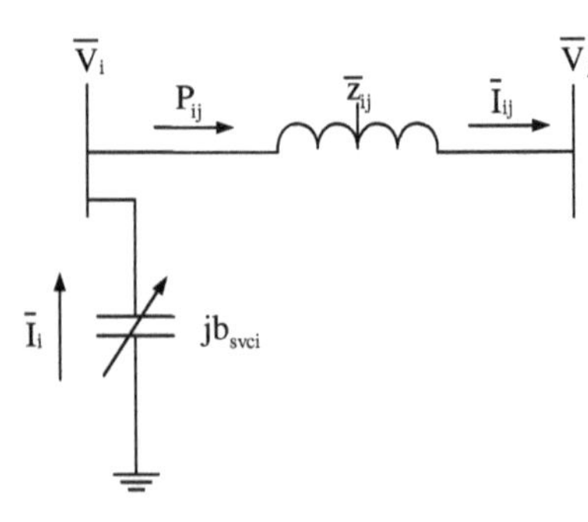


**Fig. 6.6** An SVC installed on the ith node in the N-machine power system

$$\alpha_i = \alpha_{vi} + \alpha_{si} \tag{6.35}$$

where

$$\begin{aligned} \alpha_{vi} &= T_{vsvci}(s)(V_i - V_{iref}) \\ \alpha_{si} &= T_{ssvci}(s)\frac{1}{s}(P_{ij} - P_{ijref}) \end{aligned} \tag{6.36}$$

where $V_{iref}$ and $P_{ijref}$ are the reference signal of the SVC voltage and stabilizing control, respectively.

Linearization of Eqs. (6.34) and (6.36) is

$$\Delta b_{svci} = -\frac{1 - \cos 2\alpha_{i0}}{\pi x_{svcli}}(\Delta\alpha_{vi} + \Delta\alpha_{si}) \tag{6.37}$$

$$\begin{aligned} \Delta\alpha_{vi} &= T_{vsvci}(s)\Delta V_i \\ \Delta\alpha_{si} &= T_{ssvci}(s)\frac{1}{s}\Delta P_{ij} \end{aligned} \tag{6.38}$$

where

$$\Delta V_i = \frac{v_{ix0}}{V_{i0}}\Delta v_{ix} + \frac{v_{iy0}}{V_{i0}}\Delta v_{iy} = \begin{bmatrix} \frac{v_{ix0}}{V_{i0}} & \frac{v_{iy0}}{V_{i0}} \end{bmatrix}\begin{bmatrix} \Delta v_{ix} \\ \Delta v_{iy} \end{bmatrix} = \begin{bmatrix} \frac{v_{ix0}}{V_{i0}} & \frac{v_{iy0}}{V_{i0}} \end{bmatrix}\mathbf{\Delta V_{ixy}}$$

From Fig. 6.6, it can have

$$\begin{aligned} P_{ij} &= \mathrm{Re}(\frac{\bar{V}_i - \bar{V}_j}{\bar{z}_{ij}}\bar{V}_i^*) = \mathrm{Re}[(g_{ij} + jb_{ij})(\bar{V}_i\bar{V}_i^* - \bar{V}_j\bar{V}_i^*)] \\ &= g_{ij}(v_{xi}^2 + v_{yi}^2) + b_{ij}(v_{xi}v_{yi} - v_{yi}v_{xj}) \end{aligned} \tag{6.39}$$

where * denotes the conjugate of a phasor. Linearization of the above equation gives

$$\begin{aligned} \Delta P_{ij} &= 2g_{ij}(v_{xi0}\Delta v_{xi} + v_{yi0}\Delta v_{yi}) + b_{ij}(v_{yj0}\Delta v_{xi} - v_{xj0}\Delta v_{yi}) \\ &\quad + b_{ij}(v_{xi0}\Delta v_{yj} - v_{yi0}\Delta v_{xj}) \\ &= (2g_{ij}v_{xi0} + b_{ij}v_{yj0})\Delta v_{xi} + (2g_{ij}v_{yi0} - b_{ij}v_{xj0})\Delta v_{yi} \\ &\quad + b_{ij}(v_{xi0}\Delta v_{yj} - v_{yj0}\Delta v_{xj}) \\ &= \begin{bmatrix} 2g_{ij}v_{xi0} + b_{ij}v_{yj0} & 2g_{ij}v_{yi0} - b_{ij}v_{xj0} \end{bmatrix}\begin{bmatrix} \Delta v_{xi} \\ \Delta v_{yi} \end{bmatrix} \\ &\quad + \begin{bmatrix} b_{ij}v_{xi0} & -b_{ij}v_{yi0} \end{bmatrix}\begin{bmatrix} \Delta v_{xj} \\ \Delta v_{yj} \end{bmatrix} \\ &= \begin{bmatrix} 2g_{ij}v_{xi0} + b_{ij}v_{yj0} & 2g_{ij}v_{yi0} - b_{ij}v_{xj0} \end{bmatrix}\mathbf{\Delta V_{xyi}} \\ &\quad + \begin{bmatrix} b_{ij}v_{xi0} & -b_{ij}v_{yi0} \end{bmatrix}\mathbf{\Delta V_{xyj}} \\ &= \mathbf{p_{svci}\Delta V_{xyi}} + \mathbf{p'_{svcj}\Delta V_{xyj}} \end{aligned} \tag{6.40}$$

Hence, from Eqs. (6.38) and (6.40), it can have

$$\begin{aligned}\Delta\alpha_{vi} &= T_{vsvci}(s)\left[\frac{v_{xi0}}{V_{i0}} \quad \frac{v_{yi0}}{V_{i0}}\right]\mathbf{\Delta V_{xyi}} \\ \Delta\alpha_{si} &= T_{ssvci}(s)\frac{1}{s}(\mathbf{p_{svci}\Delta V_{xyi}} + \mathbf{p'_{svcj}\Delta V_{xyj}})\end{aligned} \tag{6.41}$$

Let the state-space realization of the voltage controller and stabilizer of the SVC given by Eq. (6.41) be

$$\begin{aligned}\dot{\mathbf{X}}_{\mathbf{VSVCi}} &= \mathbf{A_{VSVCi}X_{VSVCi}} + \mathbf{B_{VSVCi}\Delta V_{xyi}} \\ \Delta\alpha_{vi} &= \mathbf{C_{VSVCi}X_{VSVCi}} + \mathbf{D_{VSVCi}\Delta V_{xyi}}\end{aligned} \tag{6.42}$$

$$\begin{aligned}\dot{\mathbf{X}}_{\mathbf{SSVCi}} &= \mathbf{A_{SSVCi}X_{SSVCi}} + \mathbf{B_{SSVCi}\Delta V_{xyi}} + \mathbf{B_{SSVCj}\Delta V_{xyj}} \\ \Delta\alpha_{si} &= \mathbf{C_{SSVCi}X_{SSVCi}} + \mathbf{D_{SSVCi}\Delta V_{xyi}} + \mathbf{D_{SSVCj}\Delta V_{xyj}}\end{aligned} \tag{6.43}$$

According to Fig. 6.6, the injected current from the SVC into the ith node is

$$\bar{I}_i = i_{xi} + ji_{yi} = -jb_{svci}\bar{V}_i = -jb_{svci}(v_{xi} + jv_{yi}) \tag{6.44}$$

The above equation gives

$$i_{xi} = b_{svci}v_{yi},\ i_{yi} = -b_{svci}v_{xi} \tag{6.45}$$

By using Eq. (6.37), linearization of Eq. (6.45) can be obtained as

$$\begin{aligned}\begin{bmatrix}\Delta i_{xi} \\ \Delta i_{yi}\end{bmatrix} &= \begin{bmatrix}v_{yi0} \\ -v_{xi0}\end{bmatrix}\Delta b_{svci} + \begin{bmatrix}b_{svci0} & 0 \\ 0 & -b_{svci0}\end{bmatrix}\begin{bmatrix}\Delta v_{xi} \\ \Delta v_{yi}\end{bmatrix} \\ &= -\frac{1-\cos 2\alpha_{i0}}{\pi X_{svcli}}\begin{bmatrix}v_{yi0} \\ -v_{xi0}\end{bmatrix}(\Delta\alpha_{vi} + \Delta\alpha_{si}) + \begin{bmatrix}b_{svci0} & 0 \\ 0 & -b_{svci0}\end{bmatrix}\mathbf{\Delta V_{xyi}}\end{aligned} \tag{6.46}$$

When only the SVC voltage control function is considered, by substituting Eq. (6.42) into (6.46), it can have

$$\begin{aligned}\begin{bmatrix}\Delta i_{xi} \\ \Delta i_{yi}\end{bmatrix} &= \begin{bmatrix}v_{yi0} \\ -v_{xi0}\end{bmatrix}\Delta b_{svci} + \begin{bmatrix}b_{svci0} & 0 \\ 0 & -b_{svci0}\end{bmatrix}\begin{bmatrix}\Delta v_{xi} \\ \Delta v_{yi}\end{bmatrix} \\ &= -\frac{1-\cos 2\alpha_{i0}}{\pi x_{svcli}}\begin{bmatrix}v_{yi0} \\ -v_{xi0}\end{bmatrix}(\mathbf{C_{VSVCi}X_{VSVCi}} + \mathbf{D_{VSVCi}\Delta V_{xyi}} + \Delta\alpha_{Si}) \\ &\quad + \begin{bmatrix}b_{svci0} & 0 \\ 0 & -b_{svci0}\end{bmatrix}\mathbf{\Delta V_{xyi}}\end{aligned} \tag{6.47}$$

That is

$$\mathbf{\Delta I_{SVCi}} = \mathbf{C_{SVCi}X_{VSVCi}} + \mathbf{D_{SVCi}\Delta V_{xyi}} + \mathbf{b_{SSVCi}}\Delta\alpha_{si} \tag{6.48}$$

where

$$\mathbf{\Delta I_{SVCi}} = \begin{bmatrix} \Delta i_{xi} \\ \Delta i_{yi} \end{bmatrix}, \mathbf{C_{SVCi}} = -\frac{1-\cos 2\alpha_{i0}}{\pi x_{svcli}} \begin{bmatrix} v_{yi0} \\ -v_{xi0} \end{bmatrix} \mathbf{C_{VSVCi}},$$
$$\mathbf{D_{SVCi}} = -\frac{1-\cos 2\alpha_{i0}}{\pi x_{svcli}} \begin{bmatrix} v_{yi0} \\ -v_{xi0} \end{bmatrix} \mathbf{D_{VSVCi}}, \mathbf{b_{SSVCi}} = -\frac{1-\cos 2\alpha_{i0}}{\pi x_{svcli}} \begin{bmatrix} v_{yi0} \\ -v_{xi0} \end{bmatrix}$$

Equations (6.42) and (6.48) are the linearized model of the SVC without the stabilizer. Feedback signal of the SVC stabilizer is $\Delta P_{ij}$. Hence, output equation is Eq. (6.40).

When the SVC stabilizer is considered, by substituting Eq. (6.43) into Eq. (6.47), it can be obtained that

$$\begin{aligned} \begin{bmatrix} \Delta i_{xi} \\ \Delta i_{yi} \end{bmatrix} = & -\frac{1-\cos 2\alpha_{i0}}{\pi x_{svcli}} \begin{bmatrix} v_{yi0} \\ -v_{xi0} \end{bmatrix} (\mathbf{C_{VSVCi}X_{VSVCi}} + \mathbf{C_{SSVCi}X_{SSVCi}}) \\ & -\frac{1-\cos 2\alpha_{i0}}{\pi x_{svcli}} \begin{bmatrix} v_{yi0} \\ -v_{xi0} \end{bmatrix} (\mathbf{D_{VSVCi}\Delta V_{xyi}} + \mathbf{D_{SSVCi}\Delta V_{xyi}} + \mathbf{D_{SSVCj}\Delta V_{xyj}}) \\ & + \begin{bmatrix} b_{svci0} & 0 \\ 0 & -b_{svci0} \end{bmatrix} \mathbf{\Delta V_{xyi}} \end{aligned} \tag{6.49}$$

Put the state equation of the SVC voltage controller of Eq. (6.42) and stabilizer of Eq. (6.43) together

$$\mathbf{\dot{X}_{SVCi}} = \mathbf{A_{SVCi}X_{SVCi}} + \mathbf{B_{SVCi}\Delta V_{xyi}} + \mathbf{B_{SVCj}\Delta V_{xyj}} \tag{6.50}$$

where

$$\mathbf{X_{SVCi}} = \begin{bmatrix} \mathbf{X_{VSVCi}} \\ \mathbf{X_{SSVCi}} \end{bmatrix}, \mathbf{A_{SVCi}} = \begin{bmatrix} \mathbf{A_{VSVCi}} & \mathbf{0} \\ \mathbf{0} & \mathbf{A_{SSVCi}} \end{bmatrix},$$
$$\mathbf{B_{SVCi}} = \begin{bmatrix} \mathbf{B_{VSVCi}} \\ \mathbf{B_{SSVCi}} \end{bmatrix}, \mathbf{B_{SVCj}} = \begin{bmatrix} \mathbf{0} \\ \mathbf{B_{SSVCj}} \end{bmatrix}$$

Equation (6.49) can be written as

$$\mathbf{\Delta I_{SVCi}} = \mathbf{C_{SVCi}X_{SVCi}} + \mathbf{D_{SVCi}\Delta V_{xyi}} + \mathbf{D_{SVCj}\Delta V_{xyj}} \tag{6.51}$$

where

$$\Delta\mathbf{I}_{\mathbf{SVCi}} = \begin{bmatrix} \Delta i_{xi} \\ \Delta i_{yi} \end{bmatrix}$$

$$\mathbf{C}_{\mathbf{SVCi}} = \begin{bmatrix} -\frac{1-\cos 2\alpha_{i0}}{\pi X_{svcli}} \begin{bmatrix} v_{yi0} \\ -v_{xi0} \end{bmatrix} \mathbf{C}_{\mathbf{VSVCi}} - \frac{1-\cos 2\alpha_{i0}}{\pi X_{svcli}} \begin{bmatrix} v_{yi0} \\ -v_{xi0} \end{bmatrix} \mathbf{C}_{\mathbf{SSVCi}} \end{bmatrix}$$

$$\mathbf{D}_{\mathbf{SVCi}} = -\frac{1-\cos 2\alpha_{i0}}{\pi x_{svcli}} \begin{bmatrix} v_{yi0} \\ -v_{xi0} \end{bmatrix} (\mathbf{D}_{\mathbf{VSVCi}} + \mathbf{D}_{\mathbf{SSVCi}}) + \begin{bmatrix} b_{svci0} & 0 \\ 0 & -b_{svci0} \end{bmatrix}$$

$$\mathbf{D}_{\mathbf{SVCj}} = -\frac{1-\cos 2\alpha_{i0}}{\pi x_{svcli}} \begin{bmatrix} v_{yi0} \\ -v_{xi0} \end{bmatrix} \mathbf{D}_{\mathbf{SSVCj}}$$

Equations (6.50) and (6.51) are the linearized model of the SVC, including voltage controller and stabilizer, installed at the ith node in the N-machine power system.

#### 6.1.2.3 Linearized Model of a TCSC Installed in the N-Machine Power System

At the location in the N-machine power system where a TCSC is installed, two extra nodes, denoted as i and j, at the terminals of the TCSC can be created as shown in Fig. 6.7. From Fig. 6.7, it can be seen that the function of the TCSC is electrically equivalent to the injection of current into two nodes,

$$\bar{I}_i = \frac{\bar{V}_j - \bar{V}_i}{jx_{tcsci}}, \ \bar{I}_j = \frac{\bar{V}_i - \bar{V}_j}{jx_{tcsci}} \tag{6.52}$$

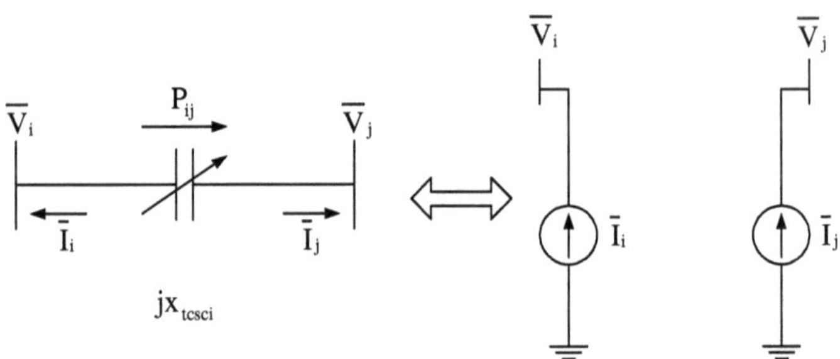


**Fig. 6.7** Creation of two extra nodes at the location where the TCSC is installed

That is

$$\begin{aligned}\bar{I}_i &= i_{xi} + ji_{yi} = -\frac{j}{x_{tcsci}}(v_{xj} + jv_{yj} - v_{xi} - jv_{yi}) \\ \bar{I}_j &= i_{xj} + ji_{yj} = -\frac{j}{x_{tcsci}}(v_{xi} + jv_{yi} - v_{xj} - jv_{yj})\end{aligned} \tag{6.53}$$

The above equation gives

$$\begin{aligned}i_{xi} &= \frac{1}{x_{tcsci}}(v_{yj} - v_{yi}),\ i_{yi} = \frac{1}{x_{tcsci}}(v_{xi} - v_{xj}), \\ i_{xj} &= \frac{1}{x_{tcsci}}(v_{yi} - v_{yj}),\ i_{yj} = \frac{1}{x_{tcsci}}(v_{xj} - v_{xi})\end{aligned} \tag{6.54}$$

The normal control function of the TCSC is the load flow regulation. Let the transfer function of the TCSC load flow regulator be $T_{ltcsci}(s)$. It can be assumed that the transfer function of the TCSC stabilizer and feedback signal is $T_{stcsci}(s)$ and the integral of deviation of the active power delivered along the transmission line, $P_{ij}$, where the TCSC is installed. Control signal of the TCSC stabilizer can be superimposed on the TCSC load flow controller, that is

$$x_{tcsci} = [T_{ltcsci}(s) + \frac{1}{s}T_{stcsci}(s)](P_{ij} - P_{ijref}) \tag{6.55}$$

where $P_{ijref}$ is the reference signal of the TCSC load flow and stabilizing control.

Linearization of Eq. (6.55) is

$$\Delta x_{tcsci} = [T_{ltcsci}(s) + \frac{1}{s}T_{stcsci}(s)]\Delta P_{ij} \tag{6.56}$$

From Fig. 6.7, it can have

$$\begin{aligned}P_{ij} &= \mathrm{Re}(\frac{\bar{V}_i - \bar{V}_j}{jx_{tcsci}}\bar{V}_i^*) = \mathrm{Re}[\frac{1}{jx_{tcsci}}(\bar{V}_i\bar{V}_i^* - \bar{V}_j\bar{V}_i^*)] \\ &= \frac{1}{x_{tcsci}}(v_{yi}v_{xj} - v_{xi}v_{yj})\end{aligned} \tag{6.57}$$

Linearization of the above equation is

$$\begin{aligned}\Delta P_{ij} &= -\frac{1}{x_{tcsci0}^2}(v_{yi0}v_{xj0} - v_{xi0}v_{yj0})\Delta v_{tcsci} \\ &\quad + \frac{1}{x_{tcsci0}}(v_{xj0}\Delta v_{yi} - v_{yj0}\Delta v_{xi}) + \frac{1}{x_{tcsci0}}(v_{yj0}\Delta v_{xj} - v_{xi0}\Delta v_{yj}) \\ &= -a_{tcsc0}\Delta x_{tcsci} + \mathbf{a_{tcsci}}\boldsymbol{\Delta}\mathbf{V_{xyi}} + \mathbf{a_{tcscj}}\boldsymbol{\Delta}\mathbf{V_{xyj}}\end{aligned} \tag{6.58}$$

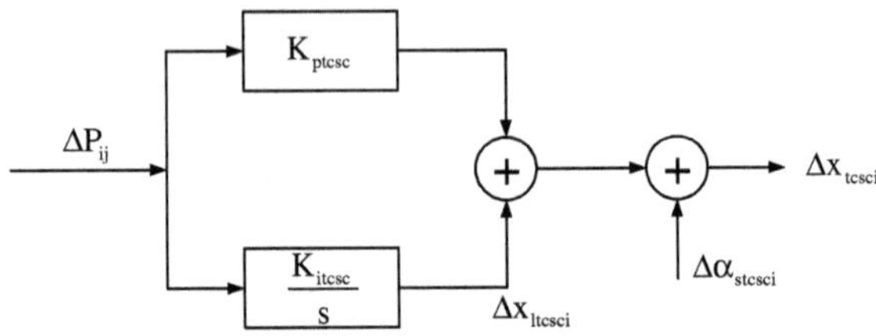


**Fig. 6.8** TCSC with a PI load flow controller superimposed by a stabilizing signal

where

$$
\begin{aligned}
a_{tcsc0} &= \frac{1}{x_{tcsci0}^2}\left(v_{yi0}v_{xj0} - v_{xi0}v_{yj0}\right) \\
\mathbf{a_{tcsci}} &= \begin{bmatrix} -\frac{1}{X_{tcsci0}}v_{yj0} & \frac{1}{X_{tcsci0}}v_{xj0} \end{bmatrix} \\
\mathbf{a_{tcscj}} &= \begin{bmatrix} \frac{1}{X_{tcsci0}}v_{yj0} & -\frac{1}{X_{tcsci0}}v_{xi0} \end{bmatrix}
\end{aligned}
$$

For the case that the TCSC stabilizer is not installed, Eq. (6.56) becomes

$$
\Delta x_{tcsci} = T_{ltcsci}(s)\Delta P_{ij} + \Delta\alpha_{stcsci} \tag{6.59}
$$

where $\Delta\alpha_{stcsci}$ is the stabilizing signal. Without loss of generality, it can be assumed that the TCSC load flow controller adopts a proportional and integral (PI) control law as shown in Fig. 6.8. From Fig. 6.8, it can have

$$
\begin{aligned}
\Delta x_{tcsci} &= \Delta\alpha_{stcsci} + K_{ptcsc}\Delta P_{ij} + \Delta x_{ltcsci} \\
\Delta \dot{x}_{ltcsci} &= K_{itcsc}\Delta P_{ij}
\end{aligned} \tag{6.60}
$$

From Eqs. (6.58) and (6.60), it can have

$$
\begin{aligned}
\Delta x_{tcsci} = {} & \frac{1}{1 + K_{ptcsc}a_{tcsc0}}\Delta\alpha_{stcsci} + \frac{1}{1 + K_{ptcsc}a_{tcsc0}}\Delta x_{ltcsci} \\
& + \frac{K_{ptcsc}}{1 + K_{ptcsc}a_{tcsc0}}\left(\mathbf{a_{tcsci}}\Delta\mathbf{V_{xyi}} + \mathbf{a_{tcscj}}\Delta\mathbf{V_{xyj}}\right)
\end{aligned} \tag{6.61}
$$

$$
\begin{aligned}
\Delta \dot{x}_{ltcsci} = {} & -\frac{K_{itcsc}a_{tcsc0}}{1 + K_{ptcsc}a_{tcsc0}}\Delta x_{ltcsci} - \frac{K_{itcsc}a_{tcsc0}}{1 + K_{ptcsc}a_{tcsc0}}\Delta\alpha_{stcsci} \\
& + \left(K_{itcsc} - \frac{K_{ptcsc}K_{itcsc}a_{tcsc0}}{1 + K_{ptcsc}a_{tcsc0}}\right)\left(\mathbf{a_{tcsci}}\Delta\mathbf{V_{xyi}} + \mathbf{a_{tcscj}}\Delta\mathbf{V_{xyj}}\right)
\end{aligned} \tag{6.62}
$$

Equation (6.62) can be written as

$$
\Delta \dot{x}_{ltcsci} = A_{ltcsci}\Delta x_{ltcsci} + \mathbf{B_{LTCSCi}}\Delta\mathbf{V_{xyi}} + \mathbf{B_{LTCSCj}}\Delta\mathbf{V_{xyj}} + b_{ltcsci}\Delta\alpha_{stcsci} \tag{6.63}
$$

Linearization of Eq. (6.54) is

$$\begin{aligned}
\Delta i_{xi} &= -\frac{1}{x_{tcsci0}^2}(v_{yj0} - v_{yi0})\Delta x_{tcsci} + \frac{1}{x_{tcsci0}}(\Delta v_{yj} - \Delta v_{yi}),\\
\Delta i_{yi} &= -\frac{1}{x_{tcsci0}^2}(v_{xi0} - v_{xj0})\Delta x_{tcsci} + \frac{1}{x_{tcsci0}}(\Delta v_{xi} - \Delta v_{xj}),\\
\Delta i_{xj} &= -\frac{1}{x_{tcsci0}^2}(v_{yi0} - v_{yj0})\Delta x_{tcsci} + \frac{1}{x_{tcsci0}}(\Delta v_{yi} - \Delta v_{yj}),\\
\Delta i_{yj} &= -\frac{1}{x_{tcsci0}^2}(v_{xj0} - v_{xi0})\Delta x_{tcsci} + \frac{1}{x_{tcsci0}}(\Delta v_{xj} - \Delta v_{xi})
\end{aligned} \tag{6.64}$$

That is

$$\begin{bmatrix}\Delta i_{xi}\\ \Delta i_{yi}\\ \Delta i_{xj}\\ \Delta i_{yj}\end{bmatrix} = \begin{bmatrix}-\frac{1}{x_{tcsci0}^2}(v_{yj0} - v_{yi0})\\ -\frac{1}{x_{tcsci0}^2}(v_{xi0} - v_{xj0})\\ -\frac{1}{x_{tcsci0}^2}(v_{yi0} - v_{yj0})\\ -\frac{1}{x_{tcsci0}^2}(v_{xj0} - v_{xi0})\end{bmatrix}\Delta x_{tcsci} + \frac{1}{x_{tcsci0}}\begin{bmatrix}0 & -1 & 0 & 1\\ 1 & 0 & -1 & 0\\ 0 & 1 & 0 & -1\\ -1 & 0 & 1 & 0\end{bmatrix}\begin{bmatrix}\Delta v_{xi}\\ \Delta v_{yi}\\ \Delta v_{xj}\\ \Delta v_{yj}\end{bmatrix} \tag{6.65}$$

By using Eq. (6.61), the above equation becomes

$$\mathbf{\Delta I_{TCSCij}} = \mathbf{C_{TCSCij}}\Delta x_{ltcsci} + \mathbf{b_{TCSCij}}\Delta\alpha_{stcsci} + \mathbf{D_{TCSCij}}\mathbf{\Delta V_{xyij}} \tag{6.66}$$

where

$$\mathbf{\Delta I_{TCSCij}} = \begin{bmatrix}\Delta i_{xi}\\ \Delta i_{yi}\\ \Delta i_{xj}\\ \Delta i_{yj}\end{bmatrix},\ \mathbf{\Delta V_{xyij}} = \begin{bmatrix}\Delta v_{xi}\\ \Delta v_{yi}\\ \Delta v_{xj}\\ \Delta v_{yj}\end{bmatrix},$$

$$\mathbf{b_{TCSCij}} = \mathbf{C_{TCSCij}} = \frac{1}{1 + K_{ptcsc}a_{tcsc0}}\begin{bmatrix}-\frac{1}{x_{tcsci0}^2}(v_{yj0} - v_{yi0})\\ -\frac{1}{x_{tcsci0}^2}(v_{xi0} - v_{xj0})\\ -\frac{1}{x_{tcsci0}^2}(v_{yi0} - v_{yj0})\\ -\frac{1}{x_{tcsci0}^2}(v_{xj0} - v_{xi0})\end{bmatrix},$$

$$\mathbf{D_{TCSCij}} = \frac{K_{ptcsc}}{1 + K_{ptcsc}a_{tcsc0}}\begin{bmatrix}\mathbf{a_{tcsci}} & \mathbf{0}\\ \mathbf{0} & \mathbf{a_{tcscj}}\end{bmatrix} + \frac{1}{x_{tcsci0}}\begin{bmatrix}0 & -1 & 0 & 1\\ 1 & 0 & -1 & 0\\ 0 & 1 & 0 & -1\\ -1 & 0 & 1 & 0\end{bmatrix} \tag{6.67}$$

Equations (6.63) and (6.66) are the linearized model of the TCSC without the TCSC stabilizer installed. Output equation can be obtained from Eqs. (6.58) and (6.61) as

$$\Delta P_{ij} = d_{tcsci}\Delta\alpha_{stcsci} + c_{tcsci}\Delta x_{ltcsci} + \mathbf{p}_{\mathbf{tcsci}}\Delta\mathbf{V}_{\mathbf{xyi}} + \mathbf{p}'_{\mathbf{tcscj}}\Delta\mathbf{V}_{\mathbf{xyj}} \tag{6.68}$$

where

$$d_{tcsci} = \frac{-a_{tcsc0}}{1 + K_{ptcsc}a_{tcsc0}},\ c_{tcsci} = \frac{-a_{tcsc0}}{1 + K_{ptcsc}a_{tcsc0}},$$

$$\mathbf{p}_{\mathbf{tcsci}} = \left(1 - \frac{a_{tcsc0}K_{ptcsc}}{1 + K_{ptcsc}a_{tcsc0}}\right)\mathbf{a}_{\mathbf{tcsci}},\ \mathbf{p}'_{\mathbf{tcscj}} = \left(1 - \frac{a_{tcsc0}K_{ptcsc}}{1 + K_{ptcsc}a_{tcsc0}}\right)\mathbf{a}_{\mathbf{tcscj}}$$

When the TCSC stabilizer is installed as shown in Eq. (6.55), denote $T_{tcsci}(s) = T_{ltcsci}(s) + \frac{1}{s}T_{stcsci}(s)$. From Eqs. (6.56) and (6.58), it can have

$$\begin{aligned} \Delta x_{tcsci} &= \frac{T_{tcsci}(s)}{1 + a_{tcsc0}T_{tcsci}(s)}(\mathbf{a}_{\mathbf{tcsci}}\Delta\mathbf{V}_{\mathbf{xyi}} + \mathbf{a}_{\mathbf{tcscj}}\Delta\mathbf{V}_{\mathbf{xyj}}) \\ &= T'_{tcsc\,i}(s)(\mathbf{a}_{\mathbf{tcsci}}\Delta\mathbf{V}_{\mathbf{xyi}} + \mathbf{a}_{\mathbf{tcscj}}\Delta\mathbf{V}_{\mathbf{xyj}}) \end{aligned} \tag{6.69}$$

Let the state-space realization of Eq. (6.69) be

$$\dot{\mathbf{X}}_{\mathbf{TCSCi}} = \mathbf{A}_{\mathbf{TCSCi}}\mathbf{X}_{\mathbf{TCSCi}} + \mathbf{B}_{\mathbf{TCSCi}}\Delta\mathbf{V}_{\mathbf{xyi}} + \mathbf{B}_{\mathbf{TCSCj}}\Delta\mathbf{V}_{\mathbf{xyj}} \tag{6.70}$$

$$\Delta x_{tcsci} = \mathbf{C}_{\mathbf{TCSCi}}\mathbf{X}_{\mathbf{TCSCi}} + \mathbf{D}_{\mathbf{TCSCi}}\Delta\mathbf{V}_{\mathbf{xyi}} + \mathbf{D}_{\mathbf{TCSCj}}\Delta\mathbf{V}_{\mathbf{xyj}} \tag{6.71}$$

By substituting Eq. (6.71) into Eq. (6.65), it can be obtained that

$$\Delta\mathbf{I}_{\mathbf{TCSCij}} = \mathbf{C}_{\mathbf{TCSCij}}\mathbf{X}_{\mathbf{TCSCi}} + \mathbf{D}_{\mathbf{TCSCij}}\Delta\mathbf{V}_{\mathbf{xyij}} \tag{6.72}$$

where

$$\Delta\mathbf{I}_{\mathbf{TCSCij}} = \begin{bmatrix} \Delta i_{xi} \\ \Delta i_{yi} \\ \Delta i_{xj} \\ \Delta i_{yj} \end{bmatrix},\ \mathbf{C}_{\mathbf{TCSCij}} = \begin{bmatrix} -\frac{1}{X^2_{tcsci0}}(v_{yj0} - v_{yi0}) \\ -\frac{1}{X^2_{tcsci0}}(v_{xi0} - v_{xj0}) \\ -\frac{1}{X^2_{tcsci0}}(v_{yi0} - v_{yj0}) \\ -\frac{1}{X^2_{tcsci0}}(v_{xj0} - v_{xi0}) \end{bmatrix}\mathbf{C}_{\mathbf{TCSCi}},$$

$$\Delta\mathbf{V}_{\mathbf{xyij}} = \begin{bmatrix} \Delta i_{xi} \\ \Delta i_{yi} \\ \Delta i_{xj} \\ \Delta i_{yj} \end{bmatrix},\ \mathbf{D}_{\mathbf{TCSCij}} = \begin{bmatrix} \mathbf{D}_{\mathbf{TCSCi}} & \mathbf{0} \\ \mathbf{0} & \mathbf{D}_{\mathbf{TCSCj}} \end{bmatrix} + \frac{1}{x_{tcsci0}}\begin{bmatrix} 0 & -1 & 0 & 1 \\ 1 & 0 & -1 & 0 \\ 0 & 1 & 0 & -1 \\ -1 & 0 & 1 & 0 \end{bmatrix}$$

Equations (6.70) and (6.72) are the linearized model of the TCSC with stabilizer installed between ith and jth nodes in the N-machine power system.

#### 6.1.2.4 Linearized Model of a TCPS Installed in the N-Machine Power System

Without loss of generality, it can be assumed that a TCPS is installed between nodes i and j in an N-machine power system as shown in Fig. 6.9. From Fig. 6.9, it can have

$$\frac{\bar{V}_i'}{\bar{V}_i} = \frac{\bar{I}_{ij}^*}{\bar{I}_{ij}'^*} = e^{j\phi_i} \tag{6.73}$$

$$\begin{aligned} \bar{I}_{ij}' &= \frac{\bar{V}_i' - \bar{V}_j}{\bar{Z}_{ij}} = \frac{\bar{V}_i e^{j\phi_i} - \bar{V}_i + \bar{V}_i - \bar{V}_j}{\bar{Z}_{ij}} \\ \bar{I}_{ij} &= \bar{I}_{ij}' e^{-j\phi_i} = \frac{\bar{V}_i - \bar{V}_j e^{-j\phi_i}}{\bar{Z}_{ij}} = \frac{\bar{V}_i - \bar{V}_j + \bar{V}_j - \bar{V}_j e^{-j\phi_i}}{\bar{Z}_{ij}} \end{aligned} \tag{6.74}$$

Hence, installation of the TCPS is electrically equivalent to the addition of a current source at nodes i and j, respectively, as shown in Fig. 6.9, where

$$\begin{aligned} \bar{I}_j &= \frac{\bar{V}_i e^{j\phi_i} - \bar{V}_i}{\bar{Z}_{ij}} \\ \bar{I}_i &= \frac{\bar{V}_j e^{-j\phi_i} - \bar{V}_j}{\bar{Z}_{ij}} \end{aligned} \tag{6.75}$$

Denote $\frac{1}{\bar{Z}_{ij}} = g_{ij} + jb_{ij}$. From Eq. (6.75), it can be obtained that

$$\begin{aligned} i_{xi} &= (g_{ij}v_{xj} + b_{ij}v_{yj})(\cos\phi_i - 1) + (g_{ij}v_{yj} + b_{ij}v_{xj})\sin\phi_i \\ i_{yi} &= -(g_{ij}v_{xj} + b_{ij}v_{yj})\sin\phi_i + (g_{ij}v_{yj} + b_{ij}v_{xj})(\cos\phi_i - 1) \\ i_{xj} &= (g_{ij}v_{xi} + b_{ij}v_{yi})(\cos\phi_i - 1) - (g_{ij}v_{yi} + b_{ij}v_{xi})\sin\phi_i \\ i_{yj} &= (g_{ij}v_{xi} + b_{ij}v_{yi})\sin\phi_i + (g_{ij}v_{yi} + b_{ij}v_{xi})(\cos\phi_i - 1) \end{aligned} \tag{6.76}$$

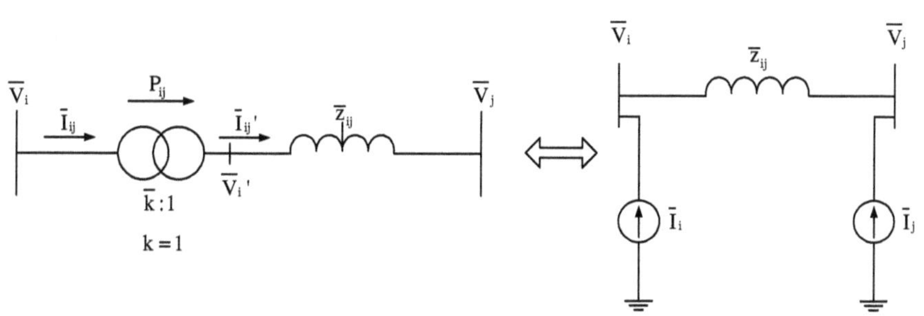


**Fig. 6.9** Installation of a TCPS between node i and j in an N-machine power system

Linearization of the above equation is

$$\begin{aligned}
\Delta i_{xi} &= [-(g_{ij}v_{xj0} + b_{ij}v_{yj0})\sin\phi_{i0} + (g_{ij}v_{yj0} + b_{ij}v_{xj0})\cos\phi_{i0}]\Delta\phi_i \\
&\quad + [g_{ij}(\cos\phi_{i0} - 1) + b_{ij}\sin\phi_{i0}]\Delta v_{xj} + [b_{ij}(\cos\phi_{i0} - 1) + g_{ij}\sin\phi_{i0}]\Delta v_{yj} \\
\Delta i_{yi} &= [-(g_{ij}v_{xj0} + b_{ij}v_{yj0})\cos\phi_{i0}] - (g_{ij}v_{yj0} + b_{ij}v_{xj0})\sin\phi_{i0} + ]\Delta\phi_i \\
&\quad + [-g_{ij}\sin\phi_{i0} + b_{ij}(\cos\phi_{i0} - 1)]\Delta v_{xj} + [-b_{ij}\sin\phi_{i0} + g_{ij}(\cos\phi_{i0} - 1)]\Delta v_{yj} \\
\Delta i_{jx} &= [-(g_{ij}v_{xi0} + b_{ij}v_{yi0})\sin\phi_{i0} - (g_{ij}v_{yi0} + b_{ij}v_{xi0})\cos\phi_{i0}]\Delta\phi_i \\
&\quad + [g_{ij}(\cos\phi_{i0} - 1) - b_{ij}\sin\phi_{i0}]\Delta v_{xi} + [b_{ij}(\cos\phi_{i0} - 1) - g_{ij}\sin\phi_{i0}]\Delta v_{yi} \\
\Delta i_{jy} &= [(g_{ij}v_{xi0} + b_{ij}v_{yi0})\cos\phi_{i0} - (g_{ij}v_{yi0} + b_{ij}v_{xi0})\sin\phi_{i0}]\Delta\phi_i \\
&\quad + [g_{ij}\sin\phi_{i0} + b_{ij}(\cos\phi_{i0} - 1)]\Delta v_{xi} + [b_{ij}\sin\phi_{i0} + g_{ij}(\cos\phi_{i0} - 1)]\Delta v_{yi}
\end{aligned}$$

That is

$$\mathbf{\Delta I_{TCPSij}} = \mathbf{b_{TCPSi}}\Delta\phi_i + \mathbf{D_{TCPSi}}\mathbf{\Delta V_{xyij}} \tag{6.77}$$

where

$$\mathbf{\Delta I_{TCPSij}} = \begin{bmatrix} \Delta i_{xi} \\ \Delta i_{yi} \\ \Delta i_{xj} \\ \Delta i_{yj} \end{bmatrix}, \quad \mathbf{\Delta V_{xyij}} = \begin{bmatrix} \Delta v_{xi} \\ \Delta v_{yi} \\ \Delta v_{xj} \\ \Delta v_{yj} \end{bmatrix},$$

$$\mathbf{b_{TCPSi}} = \begin{bmatrix} -(g_{ij}v_{xj0} + b_{ij}v_{yj0})\sin\phi_{i0} + (g_{ij}v_{yj0} + b_{ij}v_{xj0})\cos\phi_{i0} \\ -(g_{ij}v_{xj0} + b_{ij}v_{yj0})\cos\phi_{i0} - (g_{ij}v_{yj0} + b_{ij}v_{xj0})\sin\phi_{i0} \\ -(g_{ij}v_{xi0} + b_{ij}v_{yi0})\sin\phi_{i0} - (g_{ij}v_{yi0} + b_{ij}v_{xi0})\cos\phi_{i0} \\ (g_{ij}v_{xi0} + b_{ij}v_{yi0})\cos\phi_{i0} - (g_{ij}v_{yi0} + b_{ij}v_{xi0})\sin\phi_{i0} \end{bmatrix},$$

$$\mathbf{D_{TCPSi}} = \begin{bmatrix} \mathbf{0} & \mathbf{D_{TCSCi-j}} \\ \mathbf{D_{TCSCi-i}} & \mathbf{0} \end{bmatrix}$$

$$\mathbf{D_{iTCSC-j}} = \begin{bmatrix} g_{ij}(\cos\phi_{i0} - 1) + b_{ij}\sin\phi_{i0} & b_{ij}(\cos\phi_{i0} - 1) + g_{ij}\sin\phi_{i0} \\ -g_{ij}\sin\phi_{i0} + b_{ij}(\cos\phi_{i0} - 1) & -b_{ij}\sin\phi_{i0} + g_{ij}(\cos\phi_{i0} - 1) \end{bmatrix}$$

$$\mathbf{D_{iTCSC-i}} = \begin{bmatrix} g_{ij}(\cos\phi_{i0} - 1) - b_{ij}\sin\phi_{i0} & b_{ij}(\cos\phi_{i0} - 1) - g_{ij}\sin\phi_{i0} \\ g_{ij}\sin\phi_{i0} + b_{ij}(\cos\phi_{i0} - 1) & b_{ij}\sin\phi_{i0} + g_{ij}(\cos\phi_{i0} - 1) \end{bmatrix}$$

Typical application of a TCPS is to regulate power flow. Let the TCPS load flow controller be implemented by a PI control law and a stabilizing control signal, $\Delta\alpha_{stcpsi}$, be added as shown in Fig. 6.10. It can have

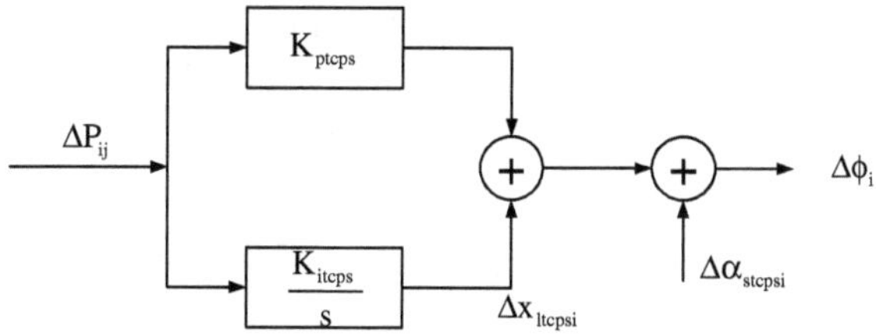


**Fig. 6.10** A PI TCPS load flow controller superimposed with a stabilizing signal

$$\begin{aligned}\Delta\phi_i &= \Delta\alpha_{stcpsi} + K_{ptcps}\Delta P_{ij} + \Delta x_{ltcpsi}\\ \Delta\dot{x}_{ltcpsi} &= K_{itcps}\Delta P_{ij}\end{aligned} \tag{6.78}$$

From Eq. (6.74), it can have

$$\begin{aligned}S^*_{ij} &= \bar{V}_i * \bar{I}_{ij} = \bar{V}^*_i\frac{\bar{V}_i - \bar{V}_j e^{-j\phi_i}}{\bar{z}_{ij}}\\ &= (g_{ij} + jb_{ij})(V_i^2 - \bar{V}_i * \bar{V}_j e^{-j\phi_i})\\ &= (g_{ij} + jb_{ij})V_i^2 - (g_{ij} + jb_{ij})(v_{xi} - jv_{yi})(v_{xj} + jv_{yj})(\cos\phi_i - j\sin\phi_i)\end{aligned} \tag{6.79}$$

Hence,

$$\begin{aligned}P_{ij} = \mathrm{Re}(S^*_{ij}) &= g_{ij}(v_{xi}^2 + v_{yi}^2)\\ &+ g_{ij}[(v_{xi}v_{xj} + v_{yi}v_{yj})\cos\phi_i + (v_{xi}v_{yj} - v_{yi}v_{xj})\sin\phi_i]\\ &- b_{ij}[-(v_{xi}v_{xj} + v_{yi}v_{yj})\sin\phi_i + (v_{xi}v_{yj} - v_{yi}v_{xj})\cos\phi_i]\end{aligned} \tag{6.80}$$

Linearization of Eq. (6.80) is

$$\Delta P_{ij} = -a_{tcps0}\Delta\phi_i + \mathbf{a_{tcpsi}}\Delta\mathbf{V_{xyi}} + \mathbf{a_{tcpsj}}\Delta\mathbf{V_{xyj}} \tag{6.81}$$

where

$$\begin{aligned}a_{tcps0} &= g_{ij}[(v_{xi0}v_{xj0} + v_{yi0}v_{yj0})\sin\phi_{i0} - (v_{xi0}v_{yj0} - v_{yi0}v_{xj0})\cos\phi_{i0}]\\ &+ b_{ij}[-(v_{xi0}v_{xj0} + v_{yi0}v_{yj0})\cos\phi_{i0} + (v_{xi0}v_{yj0} - v_{yi0}v_{xj0})\sin\phi_{i0}]\\ \mathbf{a_{tcpsi}} &= \begin{bmatrix}2g_{ij}v_{xi0} + g_{ij}(v_{xj0}\cos\phi_{i0} + v_{yj0}\sin\phi_{i0}) - b_{ij}(-v_{xj0}\sin\phi_{i0} + v_{yj0}\cos\phi_{i0})\\ 2g_{ij}v_{yi0} + g_{ij}(v_{yj0}\cos\phi_{i0} - v_{xj0}\sin\phi_{i0}) + b_{ij}(v_{yj0}\sin\phi_{i0} + v_{xj0}\cos\phi_{i0})\end{bmatrix}^T\\ \mathbf{a_{tcpsj}} &= \begin{bmatrix}g_{ij}(v_{xi0}\cos\phi_{i0} - v_{yi0}\sin\phi_{i0}) + b_{ij}(v_{xi0}\sin\phi_{i0} + v_{yi0}\cos\phi_{i0})\\ g_{ij}(v_{yi0}\cos\phi_{i0} + v_{xi0}\sin\phi_{i0}) - b_{ij}(-v_{yi0}\sin\phi_{i0} + v_{xi0}\cos\phi_{i0})\end{bmatrix}^T\end{aligned}$$

From Eqs. (6.78) and (6.81), it can have

$$\Delta\phi_i = \frac{1}{1 + a_{tcps0}K_{ptcps}}\Delta\alpha_{stcpsi} + \frac{K_{ptcps}}{1 + a_{tcps0}K_{ptcps}}(\mathbf{a_{tcpsi}}\boldsymbol{\Delta}\mathbf{V_{xyi}} + \mathbf{a_{tcpsj}}\boldsymbol{\Delta}\mathbf{V_{xyj}}) + \frac{1}{1 + a_{tcps0}K_{ptcps}}\Delta x_{ltcpsi} \quad (6.82)$$

$$\begin{aligned}\Delta\dot{x}_{ltcpsi} = & \frac{-a_{tcps0}K_{itcps}}{1 + a_{tcps0}K_{ptcps}}\Delta x_{ltcpsi} + \frac{-a_{tcps0}K_{itcps}}{1 + a_{tcps0}K_{ptcps}}\Delta\alpha_{stcpsi} \\ & + (K_{ptcps} - \frac{a_{tcps0}K_{itcps}K_{ptcps}}{1 + a_{tcps0}K_{itcps}}(\mathbf{a_{tcpsi}}\boldsymbol{\Delta}\mathbf{V_{xyi}} + \mathbf{a_{tcpsj}}\boldsymbol{\Delta}\mathbf{V_{xyj}})\end{aligned} \quad (6.83)$$

Equation (6.83) can be written as

$$\Delta\dot{x}_{ltcpsi} = A_{ltcpsi}\Delta x_{ltcpsi} + \mathbf{B_{LTCPSi}}\boldsymbol{\Delta}\mathbf{V_{xyi}} + \mathbf{B_{LTCPSj}}\boldsymbol{\Delta}\mathbf{V_{xyj}} + b_{ltcpsi}\Delta\alpha_{stcpsi} \quad (6.84)$$

From Eqs. (6.77) and (6.82), it can have

$$\boldsymbol{\Delta}\mathbf{I_{TCPSij}} = \mathbf{C_{TCPSij}}x_{ltcpsi} + \mathbf{b_{TCPSij}}\boldsymbol{\Delta\alpha}_{stcpsi} + \mathbf{D_{TCPSij}}\boldsymbol{\Delta}\mathbf{V_{xyij}} \quad (6.85)$$

where

$$\begin{aligned}\mathbf{b_{TCPSij}} &= \mathbf{C_{TCPSij}} = \frac{1}{1 + a_{tcps0}K_{ptcps}}\mathbf{b_{TCPSi}} \\ \mathbf{D_{TCPSij}} &= \frac{K_{ptcps}}{1 + a_{tcps0}K_{ptcps}}\begin{bmatrix}\mathbf{a_{tcpsi}} & \mathbf{0} \\ \mathbf{0} & \mathbf{a_{tcpsj}}\end{bmatrix} + \mathbf{D_{TCPSi}}\end{aligned} \quad (6.86)$$

Equations (6.84) and (6.85) are the linearized model of the TCPS without the TCPS stabilizer installed. From Eqs. (6.81) and (6.82), the output equation can be obtained as

$$\begin{aligned}\Delta P_{ij} = & \frac{-a_{tcps0}}{1 + a_{tcps0}K_{ptcps}}\Delta\alpha_{stcpsi} + \frac{-a_{tcps0}}{1 + a_{tcps0}K_{ptcps}}\Delta x_{ltcpsi} \\ & + \frac{1}{1 + a_{tcps0}K_{ptcps}}(\mathbf{a_{tcpsi}}\boldsymbol{\Delta}\mathbf{V_{xyi}} + \mathbf{a_{tcpsj}}\boldsymbol{\Delta}\mathbf{V_{xyj}}) \\ = & d_{tcpsi}\Delta\alpha_{stcpsi} + c_{tcpsi}\Delta x_{ltcpsi} + \mathbf{p_{tcpsi}}\boldsymbol{\Delta}\mathbf{V_{xyi}} + \mathbf{p'_{tcpsj}}\boldsymbol{\Delta}\mathbf{V_{xyj}}\end{aligned} \quad (6.87)$$

Let the transfer function of the TCPS load flow controller be $T_{ltcpsi}(s)$. It is assumed that the transfer function of the TCPS stabilizer and feedback signal is $T_{ltcpsi}(s)$ and the integral of deviation of the active power delivered along the transmission line, $\Delta P_{ij}$, where the TCSC is installed. When the TCPS stabilizer is installed, from Fig. 6.10, it can have

$$\Delta\phi_i = \left[T_{ltcpsi}(s) + \frac{1}{s}T_{stcpsi}(s)\right]\Delta P_{ij} = T_{tcpsi}(s)\Delta P_{ij}$$

From Eq. (6.81), it can have

$$\begin{aligned}\Delta\phi_i &= \frac{T_{tcpsi}(s)}{1 + a_{tcps0}T_{tcpsi}(s)}(\mathbf{a}_{tcpsi}\Delta\mathbf{V}_{xyi} + \mathbf{a}_{tcpsj}\Delta\mathbf{V}_{xyj}) \\ &= T'_{tcpsi}(s)(\mathbf{a}_{tcpsi}\Delta\mathbf{V}_{xyi} + \mathbf{a}_{tcpsj}\Delta\mathbf{V}_{xyj})\end{aligned} \tag{6.88}$$

Let the state-space realization of Eq. (6.88) be

$$\dot{\mathbf{X}}_{TCPSi} = \mathbf{A}_{TCPSi}\mathbf{X}_{TCPSi} + \mathbf{B}_{TCPSi}\Delta\mathbf{V}_{xyi} + \mathbf{B}_{TCPSj}\Delta\mathbf{V}_{xyj} \tag{6.89}$$

$$\Delta\phi_i = \mathbf{C}_{TCPSi}\mathbf{X}_{TCPSi} + \mathbf{D}_{TCPS-i}\Delta\mathbf{V}_{xyi} + \mathbf{D}_{TCPS-j}\Delta\mathbf{V}_{xyj} \tag{6.90}$$

By substituting Eq. (6.90) into Eq. (6.77), it can be obtained that

$$\Delta\mathbf{I}_{TCSCij} = \mathbf{C}_{TCSCij}\mathbf{X}_{TCSCi} + \mathbf{D}_{TCSCij}\Delta\mathbf{V}_{xyij} \tag{6.91}$$

where

$$\mathbf{C}_{TCSCij} = \mathbf{b}_{TCPSi}\mathbf{C}_{TCPSi},$$
$$\mathbf{D}_{TCSCij} = \begin{bmatrix}\mathbf{b}_{TCPSi}\mathbf{D}_{TCPS-i} & \mathbf{0} \\ \mathbf{0} & \mathbf{D}_{TCPS-j}\end{bmatrix} + \mathbf{D}_{TCPSi}$$

Equations (6.89) and (6.91) are the linearized model of the TCPS with the stabilizer installed between the ith and jth nodes in the N-machine power system.

#### 6.1.2.5 General Linearized Model of the N-Machine Power System Without the PSSs and Thyristor-Based FACTS Stabilizers Installed

Similar to Eq. (5.52), network equations of M nodes in the N-machine power system installed with thyristor-based FACTS stabilizers can be written as

$$\begin{bmatrix}\mathbf{I}_1 \\ \mathbf{I}_2 \\ \vdots \\ \mathbf{I}_M\end{bmatrix} = \begin{bmatrix}\mathbf{Y}_{11} & \mathbf{Y}_{12} & \cdots & \mathbf{Y}_{1M} \\ \mathbf{Y}_{21} & \mathbf{Y}_{22} & \cdots & \mathbf{Y}_{21} \\ \vdots & \vdots & \vdots & \vdots \\ \mathbf{Y}_{M1} & \mathbf{Y}_{M1} & \cdots & \mathbf{Y}_{MM}\end{bmatrix}\begin{bmatrix}\mathbf{V}_1 \\ \mathbf{V}_2 \\ \vdots \\ \mathbf{V}_M\end{bmatrix} \tag{6.92}$$

where

$$\mathbf{I_i} = \begin{bmatrix} i_{xi} \\ i_{yi} \end{bmatrix}, \mathbf{V_i} = \begin{bmatrix} v_{xi} \\ v_{yi} \end{bmatrix}, \mathbf{Y_{ij}} = \begin{bmatrix} g_{ii} & -b_{ij} \\ b_{ji} & g_{jj} \end{bmatrix}, i, j = 1, 2, \ldots, M \tag{6.93}$$

Let the number of the SVCs, TCSCs, and TCPSs installed in the system be $L_{svc}$, $L_{tcsc}$ and $L_{tcps}$, respectively. Denote

$$\begin{aligned}
\mathbf{\Delta I_{SVC}} &= \begin{bmatrix} \mathbf{\Delta I^T_{SVC1}} & \mathbf{\Delta I^T_{SVC2}} & \cdots & \mathbf{\Delta I^T_{SVCL_{svc}}} \end{bmatrix}^T, \\
\mathbf{\Delta I_{TCSC}} &= \begin{bmatrix} \mathbf{\Delta I^T_{TCSC1j}} & \mathbf{\Delta I^T_{TCSC2j}} & \cdots & \mathbf{\Delta I^T_{TCSCL_{tcsc}j}} \end{bmatrix}^T, \\
\mathbf{\Delta I_{TCPS}} &= \begin{bmatrix} \mathbf{\Delta I^T_{TCPS1j}} & \mathbf{\Delta I^T_{TCPS2j}} & \cdots & \mathbf{\Delta I^T_{TCPSL_{tcps}j}} \end{bmatrix}^T, \\
\mathbf{\Delta V_{SVC}} &= \begin{bmatrix} \mathbf{\Delta V^T_{SVC1}} & \mathbf{\Delta V^T_{SVC2}} & \cdots & \mathbf{\Delta V^T_{SVCL_{svc}}} \end{bmatrix}^T, \\
\mathbf{\Delta V_{TCSC}} &= \begin{bmatrix} \mathbf{\Delta V^T_{TCSC1j}} & \mathbf{\Delta V^T_{TCSC2j}} & \cdots & \mathbf{\Delta V^T_{TCSCL_{tcsc}j}} \end{bmatrix}^T, \\
\mathbf{\Delta V_{TCPS}} &= \begin{bmatrix} \mathbf{\Delta V^T_{TCPS1j}} & \mathbf{\Delta V^T_{TCPS2j}} & \cdots & \mathbf{\Delta V^T_{TCPSL_{tcps}j}} \end{bmatrix}^T
\end{aligned} \tag{6.94}$$

When the SVCs, TCSCs, and TCPSs are not equipped with stabilizers, Eqs. (6.42) and (6.48) are the linearized model of the ith SVC, Eqs. (6.63) and (6.66) are that of the ith TCSC, and Eqs. (6.84) and (6.85) are that of the ith TCPS. Hence, by using the definition of Eq. (6.94), the open-loop (without thyristor-based FACTS stabilizers), state-space representation of $L_{svc}$ SVCs, $L_{tcsc}$ TCSCs, and $L_{tcps}$ TCPSs can be obtained, respectively, from Eqs. (6.42) and (6.48), Eqs. (6.63) and (6.66), and Eqs. (6.84) and (6.85) as

$$\begin{aligned}
\mathbf{\dot{X}_{VSVC}} &= \mathbf{A_{VSVC}X_{VSVC}} + \mathbf{B_{VSVC}\Delta V_{SVC}} \\
\mathbf{\Delta I_{SVC}} &= \mathbf{C_{SVC}X_{VSVC}} + \mathbf{D_{SVC}\Delta V_{SVC}} + \mathbf{b_{SSVC}\Delta\alpha_S}
\end{aligned} \tag{6.95}$$

$$\begin{aligned}
\mathbf{\dot{X}_{LTCSC}} &= \mathbf{A_{LTCSC}X_{LTCSC}} + \mathbf{B_{TCSC}\Delta V_{TCSC}} + \mathbf{b_{LTCSC}\Delta\alpha_{STCSC}} \\
\mathbf{\Delta I_{TCSC}} &= \mathbf{C_{TCSC}X_{LTCSC}} + \mathbf{D_{TCSC}\Delta V_{TCSC}} + \mathbf{b_{TCSC}\Delta\alpha_{STCSC}}
\end{aligned} \tag{6.96}$$

$$\begin{aligned}
\mathbf{\dot{X}_{LTCPS}} &= \mathbf{A_{LTCPS}X_{LTCPS}} + \mathbf{B_{TCPS}\Delta V_{TCPS}} + \mathbf{b_{LTCPS}\Delta\alpha_{STCPS}} \\
\mathbf{\Delta I_{TCPS}} &= \mathbf{C_{TCPS}X_{LTCPS}} + \mathbf{D_{TCPS}\Delta V_{TCPS}} + \mathbf{b_{TCPS}\Delta\alpha_{STCPS}}
\end{aligned} \tag{6.97}$$

where

$$\mathbf{X_{VSVC}} = \begin{bmatrix} \mathbf{X_{VSVC1}} \\ \mathbf{X_{VSVC2}} \\ \vdots \\ \mathbf{X_{VSVCL_{svc}}} \end{bmatrix}, \mathbf{X_{LTCSC}} = \begin{bmatrix} \Delta x_{lt\,csc\,1} \\ \Delta x_{lt\,csc\,2} \\ \vdots \\ \Delta x_{lt\,csc\,1\,L_{t\,csc}} \end{bmatrix}, \mathbf{X_{LTCPS}} = \begin{bmatrix} \Delta x_{lt\,c\,ps1} \\ \Delta x_{lt\,c\,ps1} \\ \vdots \\ x_{ltcpsL_{tcps}} \end{bmatrix},$$

$$\boldsymbol{\Delta\alpha_S} = \begin{bmatrix} \Delta\alpha_{s1} \\ \Delta\alpha_{s2} \\ \vdots \\ \Delta\alpha_{sL_{svc}} \end{bmatrix}, \boldsymbol{\Delta\alpha_{STCSC}} = \begin{bmatrix} \Delta\alpha_{st\,csc\,1} \\ \Delta\alpha_{st\,csc\,2} \\ \vdots \\ \Delta\alpha_{st\,csc\,L_{tcsc}} \end{bmatrix}, \boldsymbol{\Delta\alpha_{STCPS}} = \begin{bmatrix} \Delta\alpha_{stcps1} \\ \Delta\alpha_{stcps2} \\ \vdots \\ \Delta\alpha_{stcpsL_{tcps}} \end{bmatrix}$$

Without loss of generality and for the simplicity of expression, let the order of arranging the network equations of Eq. (6.92) be (1) N nodes of generator terminals; (2) $L_{svc}$ nodes where $L_{svc}$ SVCs are installed; (3) $L_{t\,csc}$ nodes where $L_{t\,csc}$ TCSCs are installed; (4) $L_{tcps}$ nodes where $L_{tcps}$ TCPSs are installed; and (5) other nodes in the system. Linearized network equations of Eq. (6.92) can be written as

$$\begin{bmatrix} \mathbf{\Delta I_{xy}} \\ \mathbf{\Delta I_{SVC}} \\ \mathbf{\Delta I_{TCSC}} \\ \mathbf{\Delta I_{TCPS}} \\ \mathbf{0} \end{bmatrix} = \begin{bmatrix} \mathbf{Y_{gg}} & \mathbf{Y_{gs}} & \mathbf{Y_{gc}} & \mathbf{Y_{gp}} & \mathbf{Y_{go}} \\ \mathbf{Y_{sg}} & \mathbf{Y_{ss}} & \mathbf{Y_{sc}} & \mathbf{Y_{sp}} & \mathbf{Y_{so}} \\ \mathbf{Y_{cg}} & \mathbf{Y_{cs}} & \mathbf{Y_{cc}} & \mathbf{Y_{cp}} & \mathbf{Y_{co}} \\ \mathbf{Y_{pg}} & \mathbf{Y_{ps}} & \mathbf{Y_{pc}} & \mathbf{Y_{pp}} & \mathbf{Y_{po}} \\ \mathbf{Y_{og}} & \mathbf{Y_{os}} & \mathbf{Y_{oc}} & \mathbf{Y_{op}} & \mathbf{Y_{oo}} \end{bmatrix} \begin{bmatrix} \mathbf{\Delta V_{xy}} \\ \mathbf{\Delta V_{SVC}} \\ \mathbf{\Delta V_{TCSC}} \\ \mathbf{\Delta V_{TCPS}} \\ \mathbf{\Delta V_{O}} \end{bmatrix} \tag{6.98}$$

where $\mathbf{V_O}$ is the vector of the voltage at other nodes in the N-machine power system.

The output equation of open-loop system of the SVC stabilizer is Eq. (6.40), that of the TCSC stabilizer is Eq. (6.68), and that of the TCPS stabilizer is Eq. (6.87). According to the notations in Eqs. (6.94)–(6.98), the output equation of open-loop system of $L_{svc}$ SVC stabilizers, $L_{t\,csc}$ TCSC stabilizers, and $L_{tcps}$ TCPS stabilizers can be written, respectively, as

$$\begin{aligned} \mathbf{y_{SVC}} &= \mathbf{p_{SVC}\Delta V_{SVC}} + \mathbf{p_{SVC-O}\Delta V_O} \\ \mathbf{y_{TCSC}} &= \mathbf{d_{TCSC}\Delta\alpha_{STCSC}} + \mathbf{c_{TCSC}\Delta X_{LTCSC}} + \mathbf{p_{TCSC}\Delta V_{TCSC}} \\ \mathbf{y_{TCPS}} &= \mathbf{d_{TCPS}\Delta\alpha_{STCPS}} + \mathbf{c_{TCPS}\Delta X_{LTCPS}} + \mathbf{p_{TCPS}\Delta V_{TCPS}} \end{aligned} \tag{6.99}$$

where

$$\mathbf{y_{SVC}} = \begin{bmatrix}\Delta P_{1j} & \Delta P_{1j} & \cdots & \Delta P_{L_{svc}j}\end{bmatrix}^T,\ \mathbf{y_{TCSC}} = \begin{bmatrix}\Delta P_{1j} & \Delta P_{1j} & \cdots & \Delta P_{L_{tcsc}j}\end{bmatrix}^T,$$

$$\mathbf{y_{TCPS}} = \begin{bmatrix}\Delta P_{1j} & \Delta P_{1j} & \cdots & \Delta P_{L_{cps}j}\end{bmatrix}^T,\ \mathbf{d_{TCSC}} = \mathrm{diag}(d_{tcsc\,i}),$$

$$\mathbf{c_{TCSC}} = \mathrm{diag}(c_{tcsc\,i}),\ \mathbf{d_{TCPS}} = \mathrm{diag}(d_{tcpsi}),\ \mathbf{c_{TCSC}} = \mathrm{diag}(c_{tcpsi}),$$

$$\mathbf{p_{SVC}} = \begin{bmatrix} \mathbf{p_{svc1}} & & & \mathbf{0} \\ & \mathbf{p_{svc2}} & & \\ & & \ddots & \\ \mathbf{0} & & & \mathbf{p_{svcL_{svc}}} \end{bmatrix},\ \mathbf{p_{SVC-O}} = \begin{bmatrix} \mathbf{p'_{svc1}} & & & \mathbf{0} \\ & \mathbf{p'_{svc2}} & & \\ & & \ddots & \\ \mathbf{0} & & & \mathbf{p'_{svcL_{svc}}} \\ & & & \mathbf{0} \\ & & & \vdots \end{bmatrix},$$

$$\mathbf{p_{TCSC}} = \begin{bmatrix} \mathbf{p_{tcsc1}} & & & & & & & \\ & \mathbf{p'_{tcsc1}} & & & & \mathbf{0} & & \\ & & \mathbf{p_{tcsc2}} & & & & & \\ & & & \mathbf{p'_{tcsc2}} & & & & \\ & & & & \ddots & & & \\ & \mathbf{0} & & & & \mathbf{p_{tcscL_{tcsc}}} & & \\ & & & & & & \mathbf{p'_{tcscL_{tcsc}}} \end{bmatrix}$$

$$\mathbf{p_{TCPS}} = \begin{bmatrix} \mathbf{p_{tcps1}} & & & & & & & \\ & \mathbf{p'_{tcps1}} & & & & \mathbf{0} & & \\ & & \mathbf{p_{tcps2}} & & & & & \\ & & & \mathbf{p'_{tcsc2}} & & & & \\ & & & & \ddots & & & \\ & \mathbf{0} & & & & \mathbf{p_{tcpsL_{tcps}}} & & \\ & & & & & & \mathbf{p'_{tcpsL_{tcps}}} \end{bmatrix}$$

Substituting the second equation in Eqs. (6.95), (6.96), (6.97), and Eq. (6.29) into Eq. (6.98), it can be obtained that

$$
\begin{bmatrix} \mathbf{C_{xy}X} + \mathbf{B_\alpha \Delta u} \\ \mathbf{0} \end{bmatrix}
= \begin{bmatrix} \begin{bmatrix} \mathbf{Y_{gg}} - \mathbf{D_{gxy}} & \mathbf{Y_{gs}} & \mathbf{Y_{gc}} & \mathbf{Y_{go}} \\ \mathbf{Y_{sg}} & \mathbf{Y_{ss}} - \mathbf{D_{SVC}} & \mathbf{Y_{sc}} & \mathbf{Y_{sp}} \\ \mathbf{Y_{cg}} & \mathbf{Y_{cs}} & \mathbf{Y_{cc}} - \mathbf{D_{TCSC}} & \mathbf{Y_{cp}} \\ \mathbf{Y_{pg}} & \mathbf{Y_{ps}} & \mathbf{Y_{pc}} & \mathbf{Y_{pp}} - \mathbf{D_{TCPS}} \end{bmatrix} & \begin{bmatrix} \mathbf{Y_{go}} \\ \mathbf{Y_{so}} \\ \mathbf{Y_{co}} \\ \mathbf{Y_{po}} \end{bmatrix} \\ \begin{bmatrix} \mathbf{Y_{og}} & \mathbf{Y_{os}} & \mathbf{Y_{oc}} & \mathbf{Y_{op}} \end{bmatrix} & \mathbf{Y_{oo}} \end{bmatrix} \begin{bmatrix} [\mathbf{\Delta V}] \\ \mathbf{\Delta V_O} \end{bmatrix}
= \begin{bmatrix} \mathbf{Y'_M} & \mathbf{Y_{1o}} \\ \mathbf{Y_{2o}} & \mathbf{Y_{oo}} \end{bmatrix} \begin{bmatrix} [\mathbf{\Delta V}] \\ \mathbf{\Delta V_O} \end{bmatrix}
\tag{6.100}
$$

where

$$
\mathbf{X} = \begin{bmatrix} \mathbf{X_g} \\ \mathbf{X_{VSVC}} \\ \mathbf{X_{LTCSC}} \\ \mathbf{X_{LTCPS}} \end{bmatrix}, \ \mathbf{\Delta V} = \begin{bmatrix} \mathbf{\Delta V_{xy}} \\ \mathbf{\Delta V_{SVC}} \\ \mathbf{\Delta V_{TCSC}} \\ \mathbf{\Delta V_{TCPS}} \end{bmatrix}, \ \mathbf{C_{xy}} = \begin{bmatrix} \mathbf{C_{gxy}} & & & \mathbf{0} \\ & \mathbf{C_{SVC}} & & \\ & & \mathbf{C_{TCSC}} & \\ \mathbf{0} & & & \mathbf{C_{TCPS}} \end{bmatrix},
$$

$$
\mathbf{\Delta u} = \begin{bmatrix} \mathbf{\Delta u_{pss}} \\ \mathbf{\Delta \alpha_S} \\ \mathbf{\Delta \alpha_{STCSC}} \\ \mathbf{\Delta \alpha_{STCPS}} \end{bmatrix}, \ \mathbf{B_\alpha} = \begin{bmatrix} \mathbf{0} & & & \mathbf{0} \\ & \mathbf{b_{SSVC}} & & \\ & & \mathbf{b_{TCSC}} & \\ \mathbf{0} & & & \mathbf{b_{TCPS}} \end{bmatrix}
$$

From Eq. (6.100), it can be obtained that

$$
\begin{aligned} &\mathbf{\Delta V_O} = -\mathbf{Y_{oo}^{-1} Y_{2o} \Delta V} \\ &\mathbf{C_{xy}X} + \mathbf{B_\alpha \Delta u} = (\mathbf{Y'_M} - \mathbf{Y_{1o} Y_{oo}^{-1} Y_{2o}})\mathbf{\Delta V} = \mathbf{Y_M \Delta V} \end{aligned}
\tag{6.101}
$$

By arranging the state equation (the first equation) in Eqs. (6.95), (6.96), (6.97), and (6.29) together according to the notation in Eq. (6.101), it can have

$$
\dot{\mathbf{X}} = \mathbf{A_{xy} X} + \mathbf{B_p \Delta u} + \mathbf{B_{xy} \Delta V}
\tag{6.102}
$$

where

$$
\mathbf{A_{xy}} = \begin{bmatrix} \mathbf{A_{gxy}} & & & \mathbf{0} \\ & \mathbf{A_{VSVC}} & & \\ & & \mathbf{A_{LTCSC}} & \\ \mathbf{0} & & & \mathbf{A_{LTCPS}} \end{bmatrix},
$$

$$
\mathbf{B_p} = \begin{bmatrix} \mathbf{B_g} & & & \mathbf{0} \\ & \mathbf{0} & & \\ & & \mathbf{b_{LTCSC}} & \\ \mathbf{0} & & & \mathbf{b_{LTCPS}} \end{bmatrix}, \ \mathbf{B_{xy}} = \begin{bmatrix} \mathbf{B_{gxy}} & & & \mathbf{0} \\ & \mathbf{B_{VSVC}} & & \\ & & \mathbf{B_{TCSC}} & \\ \mathbf{0} & & & \mathbf{B_{TCPS}} \end{bmatrix}
$$

Deleting $\Delta\mathbf{V}$ in Eq. (6.102) by substituting the second equation of Eq. (6.101) into Eq. (6.102), it can be obtained that

$$\dot{\mathbf{X}} = \mathbf{AX} + \mathbf{B}\Delta\mathbf{u} \tag{6.103}$$

where $\mathbf{A} = \mathbf{A_{xy}} + \mathbf{B_{xy}}\mathbf{Y}_\mathbf{M}^{-1}\mathbf{C_{xy}}$, $\mathbf{B} = \mathbf{B_p} + \mathbf{B_{xy}}\mathbf{Y}_\mathbf{M}^{-1}\mathbf{B}_\alpha$. Equation (6.103) is the state equation of open-loop system without the PSS and thyristor-based FACTS stabilizers installed.

Let the feedback signal of a PSS be the rotor speed of generator where it is installed. Hence, the output equation of PSS can be written as

$$\mathbf{y_{PSS}} = \Delta\boldsymbol{\omega} = \begin{bmatrix}\mathbf{0} & \mathbf{I} & \mathbf{0}\end{bmatrix}\mathbf{X_g} = \mathbf{c_{PSS}}\mathbf{X_g} \tag{6.104}$$

By using the notation of Eq. (6.100), from Eq. (6.99), the first equation in Eq. (6.101) and (6.104), it can have

$$\mathbf{y} = \mathbf{C}'\mathbf{X} + \mathbf{D}'\Delta\mathbf{u} + \mathbf{P_{xy}}\Delta\mathbf{V} \tag{6.105}$$

where

$$\mathbf{y} = \begin{bmatrix}\mathbf{y_{PSS}}\\ \mathbf{y_{SVC}}\\ \mathbf{y_{TCSC}}\\ \mathbf{y_{TCPS}}\end{bmatrix},\quad \mathbf{C}' = \begin{bmatrix}\mathbf{c_{PSS}} & \mathbf{0} & \mathbf{0}\\ \mathbf{0} & \mathbf{0} & \mathbf{0}\\ \mathbf{0} & \mathbf{c_{TCSC}} & \mathbf{0}\\ \mathbf{0} & \mathbf{0} & \mathbf{c_{TCPS}}\end{bmatrix},\quad \mathbf{D}' = \begin{bmatrix}\mathbf{0} & \mathbf{0} & \mathbf{0}\\ \mathbf{0} & \mathbf{0} & \mathbf{0}\\ \mathbf{0} & \mathbf{d_{TCSC}} & \mathbf{0}\\ \mathbf{0} & \mathbf{0} & \mathbf{d_{TCPS}}\end{bmatrix},$$

$$\mathbf{P_{xy}} = \begin{bmatrix}\mathbf{0} & & \mathbf{0} & \\ & \mathbf{p_{SVC}} & & \\ & & \mathbf{p_{TCSC}} & \\ \mathbf{0} & & & \mathbf{p_{TCPS}}\end{bmatrix} + \begin{bmatrix}\mathbf{0}\\ -\mathbf{p_{SVC-o}}\mathbf{Y}_\mathbf{oo}^{-1}\mathbf{Y_{2o}}\\ \mathbf{0}\\ \mathbf{0}\end{bmatrix}$$

Deleting $\Delta\mathbf{V}$ in Eq. (6.105) by substituting the second equation of Eq. (6.101) into Eq. (6.105), the output equation of the system about the PSS and thyristor-based FACTS stabilizers can be obtained as

$$\Delta\mathbf{y} = \mathbf{CX} + \mathbf{D}\Delta\mathbf{u} \tag{6.106}$$

where $\mathbf{C} = \mathbf{C}' + \mathbf{P_{xy}}\mathbf{Y}_\mathbf{M}^{-1}\mathbf{C_{xy}}$, $\mathbf{D} = \mathbf{D}' + \mathbf{P_{xy}}\mathbf{Y}_\mathbf{M}^{-1}\mathbf{B}_\alpha\Delta\mathbf{u}$.

#### 6.1.2.6 General Linearized Model of the N-Machine Power System with the PSSs and Thyristor-Based FACTS Stabilizers Installed

Equations (6.50) and (6.51) are the linearized model of an SVC with stabilizer installed. Equations (6.70) and (6.72) are that of a TCSC with stabilizer installed.

Equations (6.89) and (6.91) are that of a TCPS with stabilizer installed. By using the notation in Eqs. (6.94)–(6.97), the linearized model of $L_{svc}$ SVCs, $L_{tcsc}$ TCSCs, and $L_{tcps}$ TCPSs with stabilizers installed can be obtained from Eqs. (6.50) and (6.51), Eqs. (6.70) and (6.72), and Eqs. (6.89) and (6.91), respectively, as

$$\begin{aligned}\dot{\mathbf{X}}_{\mathbf{SVC}} &= \mathbf{A}_{\mathbf{SVC}}\mathbf{X}_{\mathbf{SVC}} + \mathbf{B}_{\mathbf{SVC}}\mathbf{\Delta V}_{\mathbf{SVC}} \\ \mathbf{\Delta I}_{\mathbf{SVC}} &= \mathbf{C}_{\mathbf{SVC}}\mathbf{X}_{\mathbf{SVC}} + \mathbf{D}_{\mathbf{SVC}}\mathbf{\Delta V}_{\mathbf{SVC}} + \mathbf{D}_{\mathbf{SVC-O}}\mathbf{\Delta V}_{\mathbf{O}}\end{aligned} \tag{6.107}$$

$$\begin{aligned}\dot{\mathbf{X}}_{\mathbf{TCSC}} &= \mathbf{A}_{\mathbf{TCSC}}\mathbf{X}_{\mathbf{TCSC}} + \mathbf{B}_{\mathbf{TCSC}}\mathbf{\Delta V}_{\mathbf{TCSC}} \\ \mathbf{\Delta I}_{\mathbf{TCSC}} &= \mathbf{C}_{\mathbf{TCSC}}\mathbf{X}_{\mathbf{TCSC}} + \mathbf{D}_{\mathbf{TCSC}}\mathbf{\Delta V}_{\mathbf{TCSC}}\end{aligned} \tag{6.108}$$

$$\begin{aligned}\dot{\mathbf{X}}_{\mathbf{TCPS}} &= \mathbf{A}_{\mathbf{TCPS}}\mathbf{X}_{\mathbf{TCPS}} + \mathbf{B}_{\mathbf{TCPS}}\mathbf{\Delta V}_{\mathbf{TCPS}} \\ \mathbf{\Delta I}_{\mathbf{TCPS}} &= \mathbf{C}_{\mathbf{TCPS}}\mathbf{X}_{\mathbf{TCPS}} + \mathbf{D}_{\mathbf{TCPS}}\mathbf{\Delta V}_{\mathbf{TCPS}}\end{aligned} \tag{6.109}$$

where

$$\mathbf{X}_{\mathbf{SVC}} = \begin{bmatrix} \mathbf{X}_{\mathbf{SVC1}} \\ \mathbf{X}_{\mathbf{SVC2}} \\ \vdots \\ \mathbf{X}_{\mathbf{SVCL_{svc}}} \end{bmatrix}, \mathbf{X}_{\mathbf{TCSC}} = \begin{bmatrix} \mathbf{X}_{\mathbf{TCSC1}} \\ \mathbf{X}_{\mathbf{TCSC2}} \\ \vdots \\ \mathbf{X}_{\mathbf{TCSCL_{tcsc}}} \end{bmatrix}, \mathbf{X}_{\mathbf{TCPS}} = \begin{bmatrix} \mathbf{X}_{\mathbf{TCPS1}} \\ \mathbf{X}_{\mathbf{TCPS2}} \\ \vdots \\ \mathbf{X}_{\mathbf{TCPSL_{tcps}}} \end{bmatrix}$$

By substituting the second equation in Eqs. (6.107), (6.108), (6.109), and Eq. (6.33) into Eq. (6.98), it can be obtained that

$$\begin{bmatrix} \mathbf{C_p X} \\ \mathbf{0} \end{bmatrix} = \begin{bmatrix} \mathbf{Y}'_{\mathbf{p}} & \mathbf{Y_{1po}} \\ \mathbf{Y_{2o}} & \mathbf{Y_{oo}} \end{bmatrix} \begin{bmatrix} [\mathbf{\Delta V}] \\ \mathbf{\Delta V_O} \end{bmatrix}$$
$$= \begin{bmatrix} \begin{bmatrix} \mathbf{Y_{gg}} - \mathbf{D_{gp}} & \mathbf{Y_{gs}} & \mathbf{Y_{gc}} & \mathbf{Y_{go}} \\ \mathbf{Y_{sg}} & \mathbf{Y_{ss}} - \mathbf{D_{SVC}} & \mathbf{Y_{sc}} & \mathbf{Y_{sp}} \\ \mathbf{Y_{cg}} & \mathbf{Y_{cs}} & \mathbf{Y_{cc}} - \mathbf{D_{TCSC}} & \mathbf{Y_{cp}} \\ \mathbf{Y_{pg}} & \mathbf{Y_{ps}} & \mathbf{Y_{pc}} & \mathbf{Y_{pp}} - \mathbf{D_{TCPS}} \end{bmatrix} & \begin{bmatrix} \mathbf{Y_{go}} \\ \mathbf{Y_{so}} - \mathbf{D_{SVC-O}} \\ \mathbf{Y_{co}} \\ \mathbf{Y_{po}} \end{bmatrix} \\ [\mathbf{Y_{og}} \quad \mathbf{Y_{os}} \quad \mathbf{Y_{oc}} \quad \mathbf{Y_{op}}] & \mathbf{Y_{oo}} \end{bmatrix} \begin{bmatrix} [\mathbf{\Delta V}] \\ \mathbf{\Delta V_O} \end{bmatrix} \tag{6.110}$$

where

$$\mathbf{X} = \begin{bmatrix} \mathbf{X_g} \\ \mathbf{X_{SVC}} \\ \mathbf{X_{TCSC}} \\ \mathbf{X_{TCPS}} \end{bmatrix}, \mathbf{C_p} = \begin{bmatrix} \mathbf{C_{gp}} & & & \mathbf{0} \\ & \mathbf{C_{SVC}} & & \\ & \mathbf{0} & \mathbf{C_{TCSC}} & \\ & & & \mathbf{C_{TCPS}} \end{bmatrix}$$

From Eq. (6.110), it can be obtained that

$$\begin{aligned} \Delta \mathbf{V_O} &= -\mathbf{Y}_{\mathbf{oo}}^{-1}\mathbf{Y}_{2\mathbf{o}}\Delta \mathbf{V} \\ \mathbf{C_p X} &= (\mathbf{Y}_{\mathbf{p}}^{'} - \mathbf{Y}_{1\mathbf{po}}\mathbf{Y}_{\mathbf{oo}}^{-1}\mathbf{Y}_{2\mathbf{o}})\Delta \mathbf{V} = \mathbf{Y_p}\Delta \mathbf{V} \end{aligned} \quad (6.111)$$

By arranging the state equation (the first equation) in Eqs. (6.107), (6.108), (6.109), and (6.33) together, it can have

$$\dot{\mathbf{X}} = \mathbf{A_p X} + \mathbf{B}_{\mathbf{p}}^{\prime}\Delta \mathbf{V} \quad (6.112)$$

where

$$\mathbf{A_p} = \begin{bmatrix} \mathbf{A_{gp}} & & & \mathbf{0} \\ & \mathbf{A_{SVC}} & & \\ & & \mathbf{A_{TCSC}} & \\ \mathbf{0} & & & \mathbf{A_{TCPS}} \end{bmatrix},$$

$$\mathbf{B}_{\mathbf{p}}^{\prime} = \begin{bmatrix} \mathbf{B_{gp}} & & & \mathbf{0} \\ & \mathbf{B_{SVC}} & & \\ & & \mathbf{B_{TCSC}} & \\ \mathbf{0} & & & \mathbf{B_{TCPS}} \end{bmatrix}$$

From Eqs. (6.111) and (6.112), it can be obtained that

$$\dot{\mathbf{X}} = \mathbf{A_{close}X} \quad (6.113)$$

where $\mathbf{A_{close}} = \mathbf{A_p} + \mathbf{B}_{\mathbf{p}}^{\prime}\mathbf{Y}_{\mathbf{p}}^{-1}\mathbf{C_p}$. Equation (6.113) is the state equation of the closed-loop system with the PSSs and thyristor-based FACTS stabilizers installed in the N-machine power system.

## 6.2 Analysis and Damping Control of Thyristor-Based FACTS Stabilizers Installed in a Multi-machine Power System

### *6.2.1 Damping Torque Analysis in a Multi-machine Power System*

#### 6.2.1.1 Damping Torque Contribution from a Stabilizer Derived from the Heffron–Phillips Model

Equation (5.43) is the Heffron–Phillips model of the N-machine power system with the PSSs installed. As this is a linear system satisfying the principle of superimposition, it only needs to consider the function of one PSS in the discussion. Now, if

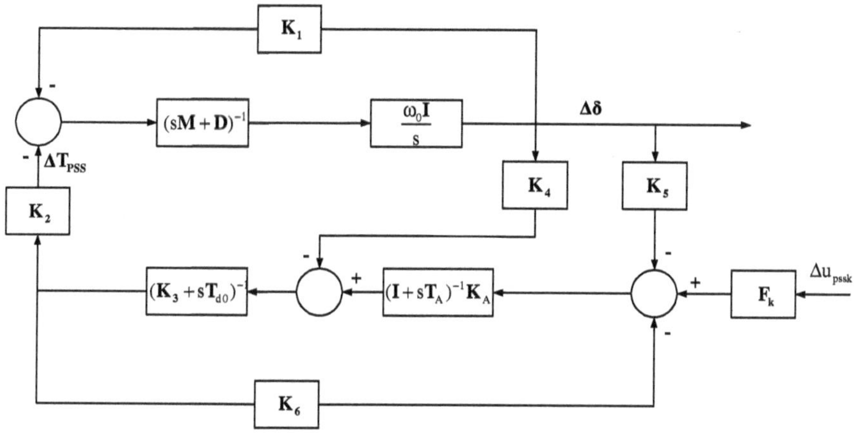


**Fig. 6.11** Heffron–Phillips model of a multi-machine power system installed with a PSS

only the kth PSS is considered, the Heffron–Phillips model of the power system can be obtained from Eq. (5.43) as

$$\begin{aligned} s\Delta\boldsymbol{\delta} &= \omega_0 \mathbf{I}\Delta\boldsymbol{\omega} \\ s\Delta\boldsymbol{\omega} &= \mathbf{M}^{-1}(-\mathbf{K}_1\Delta\boldsymbol{\delta} - \mathbf{K}_2\Delta\mathbf{E}'_q - \mathbf{D}\Delta\boldsymbol{\omega}) \\ s\Delta\mathbf{E}'_q &= \mathbf{T}'^{-1}_{d0}(-\mathbf{K}_3\Delta\mathbf{E}'_q - \mathbf{K}_4\Delta\boldsymbol{\delta} + \Delta\mathbf{E}'_{fd}) \\ s\Delta\mathbf{E}'_{fd} &= -\mathbf{T}_A^{-1}\Delta\mathbf{E}'_{fd} + \mathbf{T}_A^{-1}\mathbf{K}_A(-\mathbf{K}_5\Delta\boldsymbol{\delta} - \mathbf{K}_6\Delta\mathbf{E}'_q + \mathbf{F}_k\Delta u_{pssk}) \end{aligned} \quad (6.114)$$

where $\mathbf{F}_i = [0 \;\; \ldots \;\; 0 \;\; \underset{\text{the kth element}}{\underset{\uparrow}{1}} \;\; 0 \;\; \ldots \;\; 0]^T$ and $\Delta u_{pssk}$ is the output stabilizing signal from the kth PSS. From Eq. (6.114), the Heffron–Phillips model with the kth PSS installed is shown in Fig. 6.11.

Similar to the procedure of damping torque analysis for a PSS installed in a single-machine infinite-bus power system as described in Chap. 2, from Fig. 6.11, the transfer function matrix from the PSS stabilizing signal to the electromechanical oscillation loops of all generators can be obtained as

$$\frac{\Delta\mathbf{T}_{PSS}}{\Delta u_{pssk}} = \mathbf{M}^{-1}\mathbf{K}_2[(\mathbf{I} + s\mathbf{T}_A)(\mathbf{K}_3 + s\mathbf{T}_{d0}) + \mathbf{K}_A\mathbf{K}_6]^{-1}\mathbf{K}_A\mathbf{F}_k = \mathbf{F}_{PSS}(s) \quad (6.115)$$

Let the transfer function and feedback signal of the PSS be $T_{pssk}(s)$ and $\Delta y_k$, respectively, that is

$$\Delta u_{pssk} = T_{pssk}(s)\Delta y_k \quad (6.116)$$

The feedback signal $\Delta y_k$ can always be reconstructed by using $\Delta\omega_j$, $j = 1, 2, \ldots, N$ (see Sect. 6.2.1.2), that is

$$\Delta y_k = \gamma_j(s)\Delta\omega_j, \ j = 1, 2, \ldots, N \tag{6.117}$$

If the low-frequency power oscillation of interests is associated with the ith oscillation mode of the power system, $\bar{\lambda}_i = \xi_i + j\omega_i$, the electric torque contribution of the PSS at the complex angular oscillation frequency $\bar{\lambda}_i = \xi_i + j\omega_i$ can be obtained from Eqs. (6.115) to (6.117) as

$$\mathbf{\Delta T_{PSS}} = \mathbf{\bar{F}_{PSS}}(\bar{\lambda}_i)\bar{T}_{pssk}(\bar{\lambda}_i)\Delta y_k = \begin{bmatrix} \bar{F}_{pss1}(\bar{\lambda}_i)\bar{T}_{pssk}(\bar{\lambda}_i)\bar{\gamma}_1(\bar{\lambda}_i)\Delta\omega_1 \\ \bar{F}_{pss2}(\bar{\lambda}_i)\bar{T}_{pssk}(\bar{\lambda}_i)\bar{\gamma}_2(\bar{\lambda}_i)\Delta\omega_2 \\ \vdots \\ \bar{F}_{pssN}(\bar{\lambda}_i)\bar{T}_{pssk}(\bar{\lambda}_i)\bar{\gamma}_N(\bar{\lambda}_i)\Delta\omega_N \end{bmatrix} \tag{6.118}$$

where $\bar{F}_{pssj}(\bar{\lambda}_i), j = 1, 2, \ldots N$ is the jth element of $\bar{F}_{PSS}(\bar{\lambda}_i)$. Hence, the damping torque contribution from the PSS to the electromechanical oscillation loop of each generator is

$$\Delta T_{dampj} = D_{dampj}\Delta\omega_j = \mathrm{Re}[\tilde{F}_{pssj}(\bar{\lambda}_i)\bar{T}_{pssk}(\bar{\lambda}_i)\bar{\gamma}_j(\bar{\lambda}_i)]\Delta\omega_j \tag{6.119}$$

From the unified Heffron–Phillips model of the N-machine power system installed with a thyristor-based FACTS stabilizer of Fig. 6.5, the transfer function matrix from the FACTS stabilizing signal to the electromechanical oscillation loops of all generators can be obtained as

$$\begin{aligned} \frac{\mathbf{\Delta T_{FACTS}}}{\Delta u_{facts-s}} &= \mathbf{F_{FACTS}}(s) \\ &= \mathbf{M^{-1}K_P} - \mathbf{M^{-1}K_2}[(\mathbf{I} + s\mathbf{T_A})(\mathbf{K_3} + s\mathbf{T_{d0}}) + \mathbf{K_A K_6}]^{-1}[(\mathbf{I} + s\mathbf{T_A})\mathbf{K_q} + \mathbf{K_A K_V}] \end{aligned} \tag{6.120}$$

Similarly, let the transfer function and feedback signal of the thyristor-based FACTS stabilizer be $T_{facts}(s)$ and $\Delta y_k$, respectively, that is

$$\Delta u_{facts-s} = T_{facts}(s)\Delta y_k \tag{6.121}$$

The electric torque contribution of the thyristor-based FACTS stabilizer at the complex angular oscillation frequency $\bar{\lambda}_i = \xi_i + j\omega_i$ can be obtained from Eqs. (6.117), (6.120), and (6.121) as

$$\mathbf{\Delta T_{FACTS}} = \mathbf{\bar{F}_{FACTS}}(\bar{\lambda}_i)\bar{T}_{facts}(\bar{\lambda}_i)\Delta y_k = \begin{bmatrix} \bar{F}_{facts1}(\bar{\lambda}_i)\tilde{T}_{facts}(\bar{\lambda}_i)\bar{\gamma}_1(\bar{\lambda}_i)\Delta\omega_1 \\ \bar{F}_{facts2}(\bar{\lambda}_i)\bar{T}_{facts}(\bar{\lambda}_i)\bar{\gamma}_2(\bar{\lambda}_i)\Delta\omega_2 \\ \vdots \\ \bar{F}_{factsN}(\bar{\lambda}_i)\bar{T}_{facts}(\bar{\lambda}_i)\bar{\gamma}_N(\bar{\lambda}_i)\Delta\omega_N \end{bmatrix} \tag{6.122}$$

where $\bar{F}_{factsj}(\bar{\lambda}_i), j = 1, 2, \ldots, N$ is the jth element of $\bar{\mathbf{F}}_{\mathbf{FACTS}}(\bar{\lambda}_i)$. Hence, the damping torque contribution from the thyristor-based FACTS stabilizer to the electromechanical oscillation loop of each generator is

$$\Delta T_{dampj} = D_{dampj}\Delta\omega_j = \mathrm{Re}[\bar{F}_{factsj}(\bar{\lambda}_i)\bar{T}_{facts}(\bar{\lambda}_i)\bar{\gamma}_j(\bar{\lambda}_i)]\Delta\omega_j \tag{6.123}$$

#### 6.2.1.2 Damping Torque Contribution from a Stabilizer Derived from the General Linearized Model of an N-Machine Power System

The open-loop general linearized model of the N-machine power system is the state equation of Eq. (6.103), that is

$$\dot{\mathbf{X}} = \mathbf{AX} + \mathbf{B}\Delta\mathbf{u} \tag{6.124}$$

where

$$\mathbf{X} = \begin{bmatrix} \mathbf{X_g} \\ \mathbf{X_{VSVC}} \\ \mathbf{X_{LTCSC}} \\ \mathbf{X_{LTCPS}} \end{bmatrix}, \Delta\mathbf{u} = \begin{bmatrix} \Delta\mathbf{u_{pss}} \\ \Delta\boldsymbol{\alpha}_\mathbf{S} \\ \Delta\boldsymbol{\alpha}_\mathbf{STCSC} \\ \Delta\boldsymbol{\alpha}_\mathbf{STCPS} \end{bmatrix}, \Delta\mathbf{u_{pss}} = \begin{bmatrix} \Delta u_{pss1} \\ \Delta u_{pss2} \\ \vdots \\ \Delta u_{pssN} \end{bmatrix},$$

$$\Delta\boldsymbol{\alpha}_\mathbf{S} = \begin{bmatrix} \Delta\alpha_{s1} \\ \Delta\alpha_{s2} \\ \vdots \\ \Delta\alpha_{sL_{svc}} \end{bmatrix}, \Delta\boldsymbol{\alpha}_\mathbf{STCSC} = \begin{bmatrix} \Delta\alpha_{stcsc1} \\ \Delta\alpha_{stcsc2} \\ \vdots \\ \Delta\alpha_{stcscL_{tcsc}} \end{bmatrix}, \Delta\boldsymbol{\alpha}_\mathbf{STCPS} = \begin{bmatrix} \Delta\alpha_{stcps1} \\ \Delta\alpha_{stcps2} \\ \vdots \\ \Delta\alpha_{stcpsL_{tcps}} \end{bmatrix}$$

Equation (6.124) can be rearranged into the following form:

$$\begin{bmatrix} \Delta\dot{\boldsymbol{\delta}} \\ \Delta\dot{\boldsymbol{\omega}} \\ \Delta\dot{\mathbf{Z}} \end{bmatrix} = \begin{bmatrix} \mathbf{0} & \boldsymbol{\omega_0}\mathbf{I} & \mathbf{0} \\ \mathbf{A_{21}} & \mathbf{A_{22}} & \mathbf{A_{23}} \\ \mathbf{A_{31}} & \mathbf{A_{32}} & \mathbf{A_{33}} \end{bmatrix} \begin{bmatrix} \Delta\boldsymbol{\delta} \\ \Delta\boldsymbol{\omega} \\ \Delta\mathbf{Z} \end{bmatrix} + \sum_k \begin{bmatrix} \mathbf{0} \\ \mathbf{B_{2k}} \\ \mathbf{B_{3k}} \end{bmatrix} \Delta u_k \tag{6.125}$$

where $\Delta u_k$ is the output stabilizing signal of the kth stabilizer installed in the power system, which can be a PSS or a thyristor-based FACTS stabilizer, that is

$$\Delta u_k = \Delta u_{pssj},\ j = 1, 2, \ldots, N;\ \Delta u_k = \Delta\alpha_{sj},\ j = 1, 2, \ldots, L_{svc};$$
$$\Delta u_k = \Delta\alpha_{stcscj},\ j = 1,2,\ldots, L_{tcsc};\ \Delta u_k = \Delta\alpha_{stcpsj},\ j = 1, 2, \ldots, L_{tcps}$$

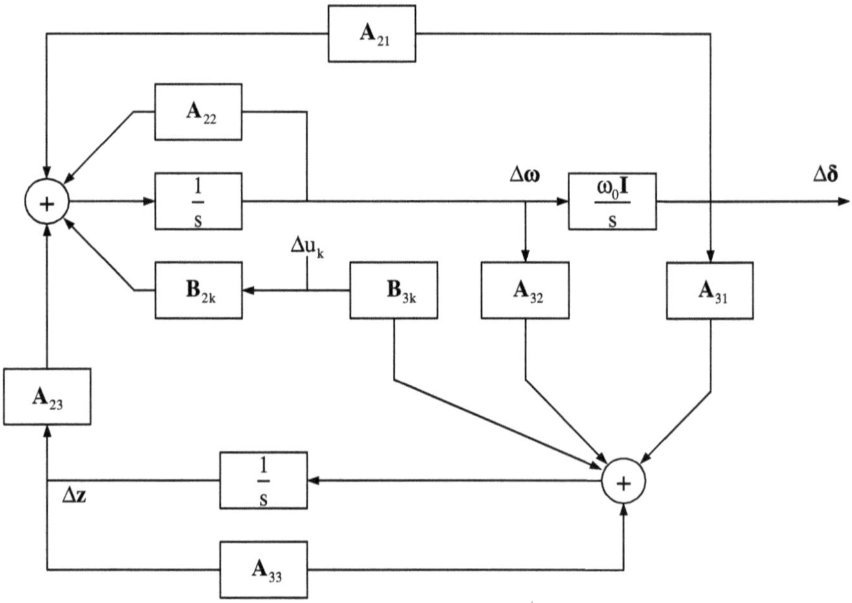


**Fig. 6.12** General linearized model of an N-machine power system

The general linearized model of the N-machine power system of Eq. (6.125) is shown in Fig. 6.12. From the third row of Eq. (6.125), it can be obtained that

$$\Delta \mathbf{Z} = (\mathrm{s}\mathbf{I} - \mathbf{A}_{33})^{-1}\left(\mathbf{A}_{31}\Delta\delta + \mathbf{A}_{32}\Delta\omega + \sum_{k}\mathbf{B}_{3k}\Delta u_k\right) \tag{6.126}$$

Substituting the above equation into Eq. (6.125), it can have

$$\begin{aligned}
\Delta\dot{\delta} &= \omega_0\Delta\omega \\
\Delta\dot{\omega} &= [\mathbf{A}_{21} + \mathbf{A}_{23}(\mathrm{s}\mathbf{I} - \mathbf{A}_{33})^{-1}\mathbf{A}_{31}]\Delta\delta + [\mathbf{A}_{22} + \mathbf{A}_{23}(\mathrm{s}\mathbf{I} - \mathbf{A}_{33})^{-1}\mathbf{A}_{32}]\Delta\omega \\
&\quad + \sum_{k}[\mathbf{B}_{2k} + \mathbf{A}_{23}(\mathrm{s}\mathbf{I} - \mathbf{A}_{33})^{-1}\mathbf{B}_{3k}]\Delta u_k \\
&= \mathbf{A}_{21}(\mathrm{s})\Delta\delta + \mathbf{A}_{22}(\mathrm{s})\Delta\omega + \sum_{k}\mathbf{B}_k(\mathrm{s})\Delta u_k
\end{aligned} \tag{6.127}$$

Hence, the general linearized model of Fig. 6.12 is shown by its compact form in Fig. 6.13.

Let the transfer function and feedback signal of the kth stabilizer be $T_{stk}(s)$ and $\Delta y_k$, respectively, that is

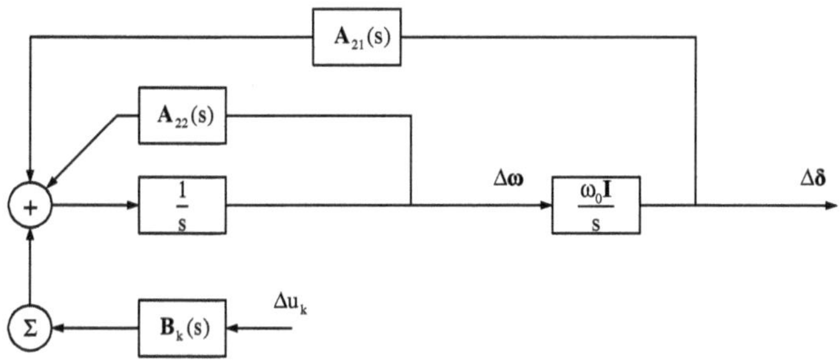


**Fig. 6.13** Compact form of general linearized model of an N-machine power system

$$\Delta u_k = T_{stk}(s)\Delta y_k \tag{6.128}$$

From Fig. 6.13, Eqs. (6.117), and (6.128), the electric torque contribution from the kth stabilizer can be obtained as

$$\mathbf{\Delta T_{ST}} = \bar{\mathbf{B}}_{\mathbf{k}}(\bar{\lambda}_i)\bar{T}_{stk}(\bar{\lambda}_i)\Delta y_k = \begin{bmatrix} \bar{B}_{k1}(\bar{\lambda}_i)\bar{T}_{stk}(\bar{\lambda}_i)\bar{\gamma}_1(\bar{\lambda}_i)\Delta\omega_1 \\ \bar{B}_{k2}(\bar{\lambda}_i)\bar{T}_{stk}(\bar{\lambda}_i)\bar{\gamma}_2(\bar{\lambda}_i)\Delta\omega_2 \\ \vdots \\ \bar{B}_{kN}(\bar{\lambda}_i)\bar{T}_{stk}(\bar{\lambda}_i)\bar{\gamma}_N(\bar{\lambda}_i)\Delta\omega_N \end{bmatrix}$$

where $B_{kj}(\bar{\lambda}_i)$, $j = 1, 2, \ldots, N$ is the jth element of $\mathbf{B_k}(\bar{\lambda}_i)$. Hence, the damping torque contribution from the kth stabilizer to the electromechanical oscillation loop of each generator is (for $j = 1, 2, \ldots, N$)

$$\Delta T_{dampj} = D_{dampj}\Delta\omega_j = \mathrm{Re}[\bar{B}_{kj}(\bar{\lambda}_i)\bar{T}_{stk}(\bar{\lambda}_i)\bar{\gamma}_j(\bar{\lambda}_i)]\Delta\omega_j \tag{6.129}$$

For the kth stabilizer, the general linearized model of the multi-machine power system of Eq. (6.125) can also be rearranged as (for $j = 1, 2, \ldots, N$)

$$\begin{bmatrix} \Delta\dot{\delta}_j \\ \Delta\dot{\omega}_j \\ \mathbf{\Delta\dot{Z}_j} \end{bmatrix} = \begin{bmatrix} 0 & \omega_0 & \mathbf{0} \\ k_j & d_j & \mathbf{A_j} \\ \mathbf{A_{j1}} & \mathbf{A_{j2}} & \mathbf{A_{j3}} \end{bmatrix} \begin{bmatrix} \Delta\delta_j \\ \Delta\omega_j \\ \mathbf{\Delta Z_j} \end{bmatrix} + \begin{bmatrix} 0 \\ b_{jk} \\ \mathbf{B_{jk}} \end{bmatrix} \Delta u_k \tag{6.130}$$

From Eqs. (6.117), (6.128), and (6.130), it can be obtained that

$$\mathbf{\Delta Z_j} = (s\mathbf{I} - \mathbf{A_{j3}})^{-1}\left[\frac{\omega_0}{s}\mathbf{A_{j1}} + \mathbf{A_{j2}} + \mathbf{B_{jk}}T_{stk}(s)\gamma_j(s)\right]\Delta\omega_j \tag{6.131}$$

The output equation of the system about the feedback signal of the stabilizer can be written as

$$\Delta y_k = \begin{bmatrix} c_{1kj} & c_{2kj} & \mathbf{C_{3kj}} \end{bmatrix} \begin{bmatrix} \Delta\delta_j \\ \Delta\omega_j \\ \mathbf{\Delta Z_j} \end{bmatrix} + d_{jk}\Delta u_k \tag{6.132}$$

By using Eqs. (6.117), (6.128), and (6.131), Eq. (6.131) becomes

$$\begin{aligned} \Delta y_k = & \left(\frac{\omega_0}{s} c_{1kj} + c_{2kj}\right)\Delta\omega_j + C_{3kj}(s\mathbf{I} - \mathbf{A_{j3}})^{-1}\left(\frac{\omega_0}{s}\mathbf{A_{j1}} + \mathbf{A_{j2}}\right)\Delta\omega_j \\ & \times [\mathbf{C_{3kj}}(s\mathbf{I} - \mathbf{A_{j3}})^{-1}\mathbf{B_{jk}} + d_{jk}]T_{stk}(s)\gamma_j(s)\Delta\omega_j \end{aligned} \tag{6.133}$$

From Eqs. (6.117) and (6.133), it can have

$$\begin{aligned} \gamma_j(s) = \frac{\Delta y_k}{\Delta\omega_j} = & \left(\frac{\omega_0}{s} c_{1kj} + c_{2kj}\right) + \mathbf{C}_{3kj}(s\mathbf{I} - \mathbf{A_{j3}})^{-1}\left(\frac{\omega_0}{s}\mathbf{A}_{j1} + \mathbf{A}_{j2}\right) \\ & \times [\mathbf{C_{3kj}}(s\mathbf{I} - \mathbf{A_{j3}})^{-1}\mathbf{B_{jk}} + d_{jk}]T_{stk}(s)\gamma_j(s) \end{aligned} \tag{6.134}$$

Hence, finally, it can have

$$\gamma_j(s) = \frac{(\frac{\omega_0}{s} c_{1kj} + c_{2kj}) + \mathbf{C_{3kj}}(s\mathbf{I} - \mathbf{A_{j3}})^{-1}(\frac{\omega_0}{s}\mathbf{A_{j1}} + \mathbf{A_{j2}})}{[\mathbf{C_{3kj}}(s\mathbf{I} - \mathbf{A_{j3}})^{-1}\mathbf{B_{jk}} + d_{jk}]T_{stk}(s)} \tag{6.135}$$

## 6.2.2 *Selection of Installing Location and Feedback Signal of a Stabilizer in a Multi-machine Power System*

### 6.2.2.1 Selection of Installing Location of a PSS by Participation Factor and Sensitivity Index

Selection of installing location and feedback signal of a stabilizer is considered before a stabilizer is installed in a multi-machine power system. At this stage, both the structure and parameters of the stabilizer are not decided. The selection is based on the open-loop linearized model of the power system to predict the effectiveness of the stabilizer. The participation factor, the sensitivity index, the residue, and the damping torque analysis are among the most popular techniques for the selection.

Consider the general linearized open-loop system of Eqs. (6.130) and (6.132). Denote the right and left eigenvectors of the state matrix with respect to the mode $\bar{\lambda}_i$ to be (see Sect. 2.2.1 and 5.2.1.1)

$$\bar{\mathbf{v}}_i = \begin{bmatrix} \bar{v}_{1i} \\ \bar{v}_{2i} \\ \bar{\mathbf{v}}_{3i} \end{bmatrix}, \quad \bar{\mathbf{w}}_i^T = \begin{bmatrix} \bar{w}_{1i} & \bar{w}_{2i} & \bar{\mathbf{w}}_{3i}^T \end{bmatrix} \tag{6.136}$$

The following participation factor (see Eq. (5.60)

$$p_{ji} = \frac{|\bar{w}_{2i}\bar{v}_{2i}|}{|\bar{\mathbf{w}}_i^{\,T}\bar{\mathbf{v}}_i|} \tag{6.137}$$

measures how much the ith oscillation mode and the variation of rotor speed of the jth generator are connected. If the connection is the highest among all generators as measured by the above-defined participation factor, the jth generator is chosen to be the best installing location of a PSS to suppress the power system oscillation associated with the oscillation mode $\bar{\lambda}_i$. The general procedure of the selection is to arrange the state matrix of the system according to Eq. (6.125), that is

$$\mathbf{A} = \begin{bmatrix} \mathbf{0} & \omega_0 I & \mathbf{0} \\ A_{21} & A_{22} & A_{23} \\ A_{31} & A_{32} & A_{33} \end{bmatrix} \tag{6.138}$$

Denote the right and left eigenvectors of the above state matrix with respect to the mode $\bar{\lambda}_i$ to be

$$\bar{\mathbf{v}}_i = \begin{bmatrix} \bar{\mathbf{v}}_{1i} \\ \bar{\mathbf{v}}_{2i} \\ \bar{\mathbf{v}}_{3i} \end{bmatrix}, \ \bar{\mathbf{w}}_i^T = \begin{bmatrix} \bar{\mathbf{w}}_{1i}^T & \bar{\mathbf{w}}_{2i}^T & \bar{\mathbf{w}}_{3i}^T \end{bmatrix} \tag{6.139}$$

The participation factors of all generators can be calculated as

$$p_{ji} = \frac{|\bar{v}_{j2i}\bar{u}_{j2i}|}{|\bar{\mathbf{w}}_{2i}^T\bar{\mathbf{v}}_{2i}|} (\text{for } j = 1, 2, \ldots, N) \tag{6.140}$$

The installing location of the PSS is the generator with the highest participation factor.

The sensitivity index for the selection of installing location of the PSS is computed by adding an increment of artificial damping in the electromechanical oscillation loop of each generator in the power system, $\Delta d_j \Delta\omega_j$ (for $j = 1, 2, \ldots, N$). If the maximum improvement, $\Delta\bar{\lambda}_i$, for the oscillation mode of interests, $\bar{\lambda}_i$, is achieved by the increment on the jth generator, it is selected as the best installing location of the PSS. The physical explanation of the sensitivity index is based on the understanding that the PSS is equivalent to an addition of damping torque in the electromechanical oscillation loop of generators. Hence, if the addition of damping torque on the jth generator achieves the maximum improvement of the oscillation mode of interests, it should be the most effective installing location of the PSS as far as the improvement of the oscillation mode of interests is concerned. The sensitivity index in fact is the sensitivity of the oscillation mode of interests to the coefficient of the added extra damping torque in the electromechanical oscillation mode, $\Delta d_j$. That is

$$\bar{S}_{ij} = \frac{\Delta\bar{\lambda}_i}{\Delta d_j} = \frac{\partial\bar{\lambda}_i}{\partial d_j} \quad (6.141)$$

From Eqs. (6.130) and (6.136), it can have

$$\bar{S}_{ij} = \frac{\partial\bar{\lambda}_i}{\partial d_j} = \bar{\mathbf{w}}_i^T \frac{\partial \mathbf{A}}{\partial d_j} \mathbf{v}_i = \begin{bmatrix} \bar{w}_{1i} & \bar{w}_{2i} & \bar{\mathbf{w}}_{3i}^T \end{bmatrix} \begin{bmatrix} 0 & 0 & \mathbf{0} \\ 0 & 1 & \mathbf{0} \\ \mathbf{0} & \mathbf{0} & \mathbf{0} \end{bmatrix} \begin{bmatrix} \bar{v}_{1i} \\ \bar{v}_{2i} \\ \bar{\mathbf{v}}_{3i} \end{bmatrix} = \bar{w}_{2i}\bar{v}_{2i} \quad (6.142)$$

Equation (6.142) shows that in fact, the participation factor is the normalized sensitivity index. Both methods are equivalent. Application of the sensitivity index often only considers the real part of the oscillation mode of interests, as the objective of installing stabilizer is to improve the damping of the oscillation mode. Let the oscillation mode be $\bar{\lambda}_i = \xi_i + j\omega_i$. The sensitivity index used is

$$S_{ij-\xi} = \frac{\Delta\xi_i}{\Delta d_j} \quad (6.143)$$

Though Eqs. (6.119) and (6.129) show that a PSS supplies damping torque to the electromechanical oscillation loop of all generators in the power system, usually the amount of damping torque contribution to the generator where it is installed is much greater than that of other generators. Hence, participation factor and sensitivity index, though they only tell how much a generator is connected to the oscillation mode of interests, can be used to select the installing location of the PSS. In the case of a thyristor-based FACTS stabilizer, however, they cannot be applied directly. This is because how the thyristor-based FACTS stabilizer distributes its damping torque contribution among all generators also plays a very important role in determining how much it affects the oscillation modes of interest.

#### 6.2.2.2 Selection of Installing Locations and Feedback Signals of a Stabilizer by Damping Torque Analysis [2]

Equations (6.119) and (6.129) demonstrate that a stabilizer (the PSS or thyristor-based FACTS stabilizer) installed in a power system contributes a damping torque, $D_{dampj}\Delta\omega_j$, to the electromechanical oscillation loop of every generator in the system. By using the sensitivity index defined in Eqs. (6.141) or (6.143), the addition of damping torque, $D_{dampj}\Delta\omega_j$ (for $j = 1, 2, \ldots, N$), on each generator affects the oscillation mode of interests. Hence, the total improvement of the oscillation mode of interests $\bar{\lambda}_i$ by the stabilizer can be obtained as

$$\Delta\bar{\lambda}_i = \sum_{j=1}^{N} D_{dampj}\bar{S}_{ij}, \text{ or } \Delta\xi_i = \sum_{j=1}^{N} D_{dampj}S_{ij-\xi} \quad (6.144)$$

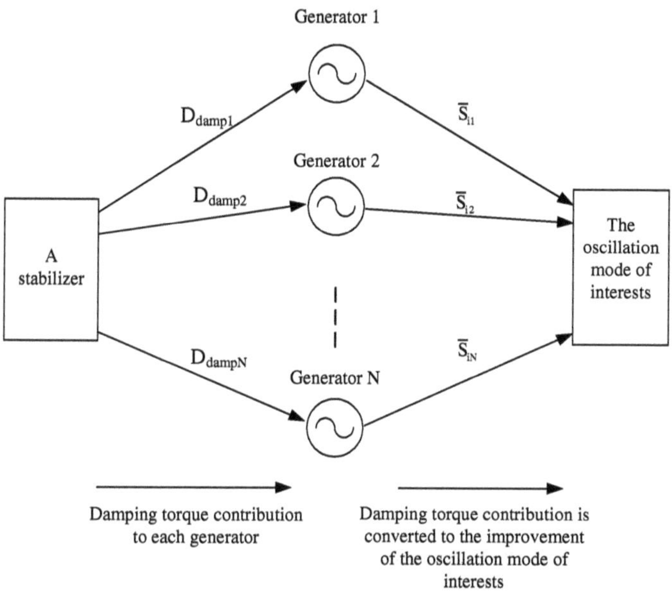


**Fig. 6.14** The oscillation mode of interests as affected by a stabilizer

where $D_{dampj}$ is given by Eq. (6.119) for a PSS and Eq. (6.123) for a thyristor-based FACTS stabilizer based on the Heffron–Phillips model as well as by Eq. (6.129) for the PSS or thyristor-based FACTS stabilizer based on the general linearized model of the N-machine power system.

Equation (6.144) is shown in Fig. 6.14. It presents a clear and full picture on how the stabilizer (the PSS or thyristor-based FACTS stabilizer) contributes the damping torque to every generator and then how the contribution of the damping torque is converted to the influence on the oscillation mode of interests. Damping torque analysis in the multi-machine power system of Eqs. (6.119), (6.123), and (6.129) gives the left-hand half picture of Fig. 6.14 on how the stabilizer distributes the damping torque to all generators. Participation factor or sensitivity index gives equivalently the right-hand half picture of Fig. 6.14 on how the oscillation mode of interests is connected to generators.

Based on Fig. 6.14 or Eq. (6.144), the following indices of the damping torque analysis (DTA) can be proposed for the selection of installing location and feedback signal of the stabilizer:

$$\mathrm{DTA} = \sum_{j=1}^{N} \left|\bar{F}_{pssj}(\bar{\lambda}_i)\bar{\gamma}_j(\bar{\lambda}_i)\bar{S}_{ij}\right| (\text{for a PSS using Heffron–Phillips model}) \quad (6.145)$$

$$\mathrm{DTA} = \sum_{j=1}^{N} \left|\bar{F}_{fsctsj}(\bar{\lambda}_i)\bar{\gamma}_j(\bar{\lambda}_i)\bar{S}_{ij}\right| \quad (6.146)$$

(for a thyristor-based FACTS stabilizer using Heffron–Phillips model)

$$\mathrm{DTA} = \sum_{j=1}^{N} \left|\bar{B}_{kj}(\bar{\lambda}_i)\bar{\gamma}_j(\bar{\lambda}_i)S_{ij}\right| \tag{6.147}$$

(for a PSS or thyristor-based FACTS stabilizer using general linearized model) where $\gamma_j(s) = (\frac{\omega_0}{s}c_{1kj} + c_{2kj}) + \mathbf{C_{3kj}}(s\mathbf{I} - \mathbf{A_{j3}})^{-1}(\frac{\omega_0}{s}\mathbf{A_{j1}} + \mathbf{A_{j2}})$ which is given by Eq. (6.135).

In the above DTA indices, often $\bar{S}_{ij}$ is replaced by $S_{ij-\xi}$ so as to focus the selection on the influence of the stabilizer on the damping of the oscillation mode of interests.

#### 6.2.2.3 Residue Index and Its Connection with the Damping Torque Analysis [3]

Consider the kth stabilizer to be installed in a multi-machine. The general linearized model of the system is (see Eq. (6.125))

$$\begin{bmatrix} \Delta\dot{\boldsymbol{\delta}} \\ \Delta\dot{\boldsymbol{\omega}} \\ \Delta\dot{\mathbf{Z}} \end{bmatrix} = \begin{bmatrix} \mathbf{0} & \omega_0\mathbf{I} & \mathbf{0} \\ \mathbf{A}_{21} & \mathbf{A}_{22} & \mathbf{A}_{23} \\ \mathbf{A}_{31} & \mathbf{A}_{32} & \mathbf{A}_{33} \end{bmatrix} \begin{bmatrix} \Delta\boldsymbol{\delta} \\ \Delta\boldsymbol{\omega} \\ \Delta\mathbf{Z} \end{bmatrix} + \begin{bmatrix} \mathbf{0} \\ \mathbf{B}_{2k} \\ \mathbf{B}_{3k} \end{bmatrix} \Delta u_k$$
$$\Delta y_k = \begin{bmatrix} \mathbf{C}_{1k}^T & \mathbf{C}_{2k}^T & \mathbf{C}_{3k}^T \end{bmatrix} \begin{bmatrix} \Delta\boldsymbol{\delta} \\ \Delta\boldsymbol{\omega} \\ \Delta\mathbf{Z} \end{bmatrix} \tag{6.148}$$

where $\Delta y_k$ is the feedback signal of the stabilizer. If the right and left eigenvectors of the state matrix with respect to the oscillation mode of interests, $\bar{\lambda}_i$, are $\bar{\mathbf{v}}_\mathbf{i}$ and $\bar{\mathbf{w}}_\mathbf{i}^T$, respectively, the controllability, observability, and residue index of the stabilizer are (see Sect. 2.2.1)

$$\bar{b}_{ik} = \bar{\mathbf{w}}_\mathbf{i}^T \begin{bmatrix} \mathbf{0} \\ \mathbf{B}_{2k} \\ \mathbf{B}_{3k} \end{bmatrix}, \ \bar{c}_{ik} = \begin{bmatrix} \mathbf{C}_{1k}^T & \mathbf{C}_{2k}^T & \mathbf{C}_{3k}^T \end{bmatrix} \bar{\mathbf{v}}_i, \ \bar{R}_{ik} = \bar{b}_{ik}\bar{c}_{ik} \tag{6.149}$$

Figure 6.15 shows that $\bar{c}_{ik}$ measures how much the oscillation mode $\bar{\lambda}_i$ is observed in $\Delta y_k$ and $\bar{b}_{ik}$ measures how much the stabilizing signal $\Delta u_k$ affects $\bar{\lambda}_i$. Hence, the residue index $\bar{R}_{ik}$ measures how much the stabilizer is connected with the oscillation mode of interests $\bar{\lambda}_i$. It can be used as an index for the selection of the installing location and feedback signal of the stabilizer.

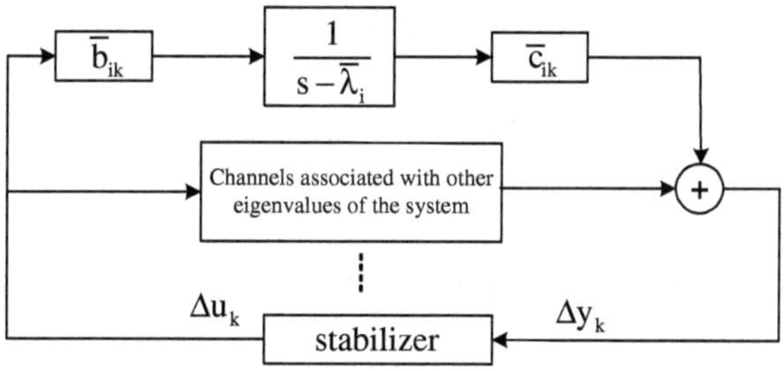


**Fig. 6.15** Configuration of the system in modal decomposition

In fact, the damping torque analysis and the residue introduced above for the selection of installing location and feedback signal of the stabilizer are equivalent. This equivalence can be proved as follows.

From Eq. (2.60), it can have

$$\begin{bmatrix} \boldsymbol{\Delta\delta} \\ \boldsymbol{\Delta\omega} \\ \boldsymbol{\Delta Z} \end{bmatrix} = \mathbf{V} \begin{bmatrix} z_1(0)e^{\lambda_1 t} \\ z_2(0)e^{\lambda_2 t} \\ \vdots \\ z_n(0)e^{\lambda_n t} \end{bmatrix} \tag{6.150}$$

Hence,

$$\Delta\omega_j(s) = \sum_{i=1}^{N} \frac{\bar{v}_{2ij} z_i(0)}{s - \bar{\lambda}_i} \tag{6.151}$$

where $\bar{v}_{2ij}$ is the jth element of $\bar{\mathbf{v}}_{i2}$ in Eq. (6.139). While

$$\Delta y_k = \begin{bmatrix} \mathbf{C}_{1k}^T & \mathbf{C}_{2k}^T & \mathbf{C}_{3k}^T \end{bmatrix} \begin{bmatrix} \boldsymbol{\Delta\delta} \\ \boldsymbol{\Delta\omega} \\ \boldsymbol{\Delta Z} \end{bmatrix} = \begin{bmatrix} \mathbf{C}_{1k}^T & \mathbf{C}_{2k}^T & \mathbf{C}_{3k}^T \end{bmatrix} \mathbf{V} \begin{bmatrix} z_1(0)e^{\lambda_1 t} \\ z_2(0)e^{\lambda_2 t} \\ \vdots \\ z_n(0)e^{\lambda_n t} \end{bmatrix}$$

Hence,

$$\Delta y_k(s) = \sum_{i=1}^{N} \frac{\bar{c}_{ik} z_i(0)}{s - \bar{\lambda}_i} \tag{6.152}$$

From Eqs. (6.150) and (6.152), it can have

$$\frac{\Delta y_k(s)}{\Delta \omega_j(s)} = \frac{\sum_{i=1}^{N} \frac{\bar{c}_{ik} Z_i(0)}{s-\bar{\lambda}_i}}{\sum_{i=1}^{N} \frac{\bar{v}_{2ij} Z_i(0)}{s-\bar{\lambda}_i}} = \frac{(s-\bar{\lambda}_i)\sum_{i=1}^{N} \frac{\bar{c}_{ik} Z_i(0)}{s-\bar{\lambda}_i}}{(s-\bar{\lambda}_i)\sum_{i=1}^{N} \frac{\bar{v}_{2ij} Z_i(0)}{s-\bar{\lambda}_i}} \tag{6.153}$$

In the above equation, by letting $s = \bar{\lambda}_i$ and using Eq. (6.117), it can be obtained that

$$\bar{\gamma}_j(\bar{\lambda}_i) = \frac{\Delta y_k(\bar{\lambda}_i)}{\Delta \omega_j(\bar{\lambda}_i)} = \frac{\bar{c}_{ik} z_i(0)}{\bar{v}_{2ij} z_i(0)} = \frac{\bar{c}_{ik}}{\bar{v}_{2ij}} \tag{6.154}$$

On the other hand, from Eqs. (6.139) and (6.148), it can have

$$\begin{bmatrix} \bar{\mathbf{w}}_{1i}^T & \bar{\mathbf{w}}_{2i}^T & \bar{\mathbf{w}}_{3i}^T \end{bmatrix} \begin{bmatrix} \mathbf{0} & \omega_0 \mathbf{I} & \mathbf{0} \\ \mathbf{A}_{21} & \mathbf{A}_{22} & \mathbf{A}_{23} \\ \mathbf{A}_{31} & \mathbf{A}_{32} & \mathbf{A}_{33} \end{bmatrix} = \bar{\lambda}_i \begin{bmatrix} \bar{\mathbf{w}}_{1i}^T & \bar{\mathbf{w}}_{2i}^T & \bar{\mathbf{w}}_{3i}^T \end{bmatrix} \tag{6.155}$$

The above equation gives

$$\bar{\mathbf{w}}_{2i}^T \mathbf{A}_{23} + \bar{\mathbf{w}}_{3i}^T \mathbf{A}_{33} = \bar{\mathbf{w}}_{3i}^T \bar{\lambda}_i \tag{6.156}$$

Solution of Eq. (6.156) is

$$\bar{\mathbf{w}}_{3i}^T = \bar{\mathbf{w}}_{2i}^T \mathbf{A}_{23} (\bar{\lambda}_i \mathbf{I} - \mathbf{A}_{33})^{-1} \tag{6.157}$$

Hence, from Eqs. (6.148), (6.149), and (6.157), it can have

$$\begin{aligned} \bar{b}_{ik} &= \bar{\mathbf{w}}_i^T \begin{bmatrix} \mathbf{0} \\ \mathbf{B}_{2k} \\ \mathbf{B}_{3k} \end{bmatrix} = \begin{bmatrix} \bar{\mathbf{w}}_{1i}^T & \bar{\mathbf{w}}_{2i}^T & \bar{\mathbf{w}}_{3i}^T \end{bmatrix} \begin{bmatrix} \mathbf{0} \\ \mathbf{B}_{2k} \\ \mathbf{B}_{3k} \end{bmatrix} = \bar{\mathbf{w}}_{2i}^T \mathbf{B}_{2k} + \bar{\mathbf{w}}_{3i}^T \mathbf{B}_{3k} \\ &= \bar{\mathbf{w}}_{2i}^T [\mathbf{B}_{2k} + \mathbf{A}_{23} (\bar{\lambda}_i \mathbf{I} - \mathbf{A}_{33})^{-1} \mathbf{B}_{3k}] = \bar{\mathbf{w}}_{2i}^T \mathbf{B}_k(\bar{\lambda}_i) = \sum_j \bar{w}_{2ij} B_{kj}(\bar{\lambda}_i) \end{aligned} \tag{6.158}$$

From Eqs. (6.139), (6.142), and (6.148), it can have

$$S_{ij} = \frac{\partial \bar{\lambda}_i}{\partial d_j} = \bar{\mathbf{w}}_i^T \frac{\partial \begin{bmatrix} \mathbf{0} & \omega_0 \mathbf{I} & \mathbf{0} \\ \mathbf{A}_{21} & \mathbf{A}_{22} + \mathrm{diag}(d_j) & \mathbf{A}_{23} \\ \mathbf{A}_{31} & \mathbf{A}_{32} & \mathbf{A}_{33} \end{bmatrix}}{\partial d_j} \bar{\mathbf{v}}_i = \bar{w}_{2ij} \bar{v}_{2ij} \tag{6.159}$$

Hence, from Eqs. (6.154), (6.158), and (6.159), the residue index can be obtained as

$$\bar{R}_{ik} = \bar{b}_{ik}\bar{c}_{ik} = \sum_{j} \bar{w}_{2ij}\bar{B}_{kj}(\bar{\lambda}_i)\bar{\gamma}_j(\bar{\lambda}_i)v_{2ij} = \sum_{j} \bar{S}_{ij}\bar{B}_{kj}(\bar{\lambda}_i)\bar{\gamma}_j(\bar{\lambda}_i) \tag{6.160}$$

This proves that the DTA index of Eq. (6.147) for the selection of installing location and feedback signal is equivalent to the residue index $|\bar{R}_{ik}|$.

### *6.2.3 Selection of Robust Installing Locations and Feedback Signals of a Stabilizer by an Eigensolution-Free Method*

#### **6.2.3.1 Robust Selection of Installing Locations and Feedback Signals of a Stabilizer [4]**

A good design of a stabilizer must ensure the maximum effectiveness of the stabilizer. In addition, the robustness of the stabilizer to the variations of power system operation conditions is also an important factor to be considered in the design. In Sect. 4.3.2.3, the issue of robust design of the stabilizer is introduced in the selection of control functions of a UPFC to add the stabilizing control. In this section, it is discussed how the robustness of the stabilizer is considered at the stage of selecting the installing location and feedback signal.

Let the set of the candidate installing locations and feedback signals of the stabilizer in a multi-machine power system be $\Phi(\varphi)$ and that of power system operating conditions $\Omega(\mu)$. The effectiveness of the stabilizer is the function of $\varphi$ and $\mu$, $C(\varphi, \mu)$. If only the maximum effectiveness of the stabilizer to damp a power oscillation is considered, the criterion of selecting the installing location and feedback signal of the stabilizer is

$$\underset{\varphi}{\text{Max}}[C(\varphi, \mu_0)], \quad \mu_0 \in \Omega(\mu) \tag{6.161}$$

In the discussion in previous sections, only the above criterion is considered.

In order to consider the robustness of the stabilizer to the variations of power system operating conditions, the following two criteria should be added.

$$\underset{\varphi}{\text{Max}}[\underset{\mu}{\min} C(\varphi, \mu)] \tag{6.162}$$

$$\underset{\varphi}{\text{Min}}[\underset{\mu}{\max} C(\varphi, \mu) - \underset{\mu}{\min} C(\varphi, \mu)] \tag{6.163}$$

1. The criterion of Eq. (6.162) means that the selection of installing location and feedback signal is made at the operating condition where the stabilizer is least effective. This selection guarantees that when the stabilizer operates at other

operating conditions, it is more effective. Hence, the selection is robust to the variations of system operating conditions.

2. If the damping contribution from the stabilizer changes significantly, on the one hand, with the variations of power system operating conditions, the control could be overstrong at some operating conditions, which would pose much unwanted influence on other modes in the power system. Hence, the criterion of Eq. (6.163) requires that the change of damping contribution from the stabilizer is minimum with the variations of power system operating conditions. However, this criterion should be applied jointly with that of Eq. (6.162), since failing to meet the requirement of the effectiveness obviously is not a proper selection.

Obviously, in order to apply the criteria of Eqs. (6.162) and (6.163), the effectiveness of the stabilizer has to be examined over the range of all system operating conditions. If a method dependent on the eigensolution of system state matrix, such as the residue index, is used, it could mean a heavy computational burden. In the following section, an equivalent residue index is derived which can be used for the eigensolution-free selection of robust installing location and feedback signal of the stabilizer.

#### 6.2.3.2 An Equivalent Residue Index and Eigensolution-Free Selection of Robust Installing Location and Feedback Signal of the Stabilizer

Consider the kth stabilizer to be installed in the N-machine power system. The linearized model of the system is given by Eqs. (6.130) and (6.132) as follows:

$$\begin{bmatrix} \Delta\dot{\delta}_j \\ \Delta\dot{\omega}_j \\ \Delta\dot{\mathbf{Z}}_j \end{bmatrix} = \begin{bmatrix} 0 & \omega_0 & \mathbf{0} \\ k_j & d_j & \mathbf{A}_j \\ \mathbf{A}_{j1} & \mathbf{A}_{j2} & \mathbf{A}_{j3} \end{bmatrix} \begin{bmatrix} \Delta\delta_j \\ \Delta\omega_j \\ \Delta\mathbf{Z}_j \end{bmatrix} + \begin{bmatrix} 0 \\ b_{jk} \\ \mathbf{B}_{jk} \end{bmatrix} \Delta u_k$$
$$\Delta y_k = \begin{bmatrix} c_{1kj} & c_{2kj} & \mathbf{C}_{3kj} \end{bmatrix} \begin{bmatrix} \Delta\delta_j \\ \Delta\omega_j \\ \Delta\mathbf{Z}_j \end{bmatrix} \tag{6.164}$$

Let the right and left eigenvectors of the state matrix, respectively, be

$$\bar{\mathbf{v}}_i = \begin{bmatrix} \bar{v}_{1i} \\ \bar{v}_{2i} \\ \bar{\mathbf{v}}_{3i} \end{bmatrix}, \quad \bar{\mathbf{w}}_i^T = \begin{bmatrix} \bar{w}_{1i} & \bar{w}_{2i} & \bar{\mathbf{w}}_{3i}^T \end{bmatrix} \tag{6.165}$$

According to the definition of the right and left eigenvectors, it can have

$$\begin{gathered}\begin{bmatrix}\bar{w}_{1i} & \bar{w}_{2i} & \bar{\mathbf{w}}_{3i}^T\end{bmatrix}\begin{bmatrix}0 & \omega_0 & \mathbf{0}\\ k_j & d_j & \mathbf{A_j}\\ \mathbf{A_{j1}} & \mathbf{A_{j2}} & \mathbf{A_{j3}}\end{bmatrix} = \bar{\lambda}_i\begin{bmatrix}\bar{w}_{1i} & \bar{w}_{2i} & \bar{\mathbf{w}}_{3i}^T\end{bmatrix}\\ \begin{bmatrix}0 & \omega_0 & \mathbf{0}\\ k_j & d_j & \mathbf{A_j}\\ \mathbf{A_{j1}} & \mathbf{A_{j2}} & \mathbf{A_{j3}}\end{bmatrix}\begin{bmatrix}\bar{v}_{1i}\\ \bar{v}_{2i}\\ \bar{\mathbf{v}}_{3i}\end{bmatrix} = \bar{\lambda}_i\begin{bmatrix}\bar{v}_{1i}\\ \bar{v}_{2i}\\ \bar{\mathbf{v}}_{3i}\end{bmatrix}\end{gathered} \tag{6.166}$$

From Eq. (6.166), it can be obtained that

$$\begin{aligned}\bar{\mathbf{w}}_{i3}^T &= \mathbf{A_j}(\bar{\lambda}_i\mathbf{I} - \mathbf{A_{j3}})^{-1}\bar{w}_{i2}\\ \bar{v}_{i1} &= \frac{\omega_0}{\bar{\lambda}_i}\bar{v}_{i2}\\ \bar{v}_{i3} &= (\bar{\lambda}_i\mathbf{I} - \mathbf{A_{j3}})^{-1}\left(\mathbf{A_{j1}}\frac{\omega_0}{\bar{\lambda}_i} + \mathbf{A_{j2}}\right)\bar{v}_{i2}\end{aligned} \tag{6.167}$$

From Eq. (6.167), controllability and observability index can be obtained, respectively, as

$$\begin{aligned}b_{ik} &= \begin{bmatrix}\bar{w}_{1i} & \bar{w}_{2i} & \bar{\mathbf{w}}_{3i}^T\end{bmatrix}\begin{bmatrix}0\\ b_{jk}\\ \mathbf{B_{jk}}\end{bmatrix}\\ &= [b_{jk} + \mathbf{A_j}(\bar{\lambda}_i\mathbf{I} - \mathbf{A_{j3}})^{-1}\mathbf{B_{jk}}]\bar{w}_{2i} = \bar{K}_{bi}(\bar{\lambda}_i)\bar{w}_{i2}\\ c_{ik} &= \begin{bmatrix}c_{1kj}\ c_{2kj}\ \mathbf{C_{3kj}}\end{bmatrix}\begin{bmatrix}\bar{v}_{1i}\\ \bar{v}_{2i}\\ \bar{\mathbf{v}}_{3i}\end{bmatrix}\\ &= \left[\frac{\omega_0}{\bar{\lambda}_i}c_{1kj} + c_{2kj} + \mathbf{C_{3kj}}(\bar{\lambda}_i\mathbf{I} - \mathbf{A_{j3}})^{-1}(\mathbf{A_{j1}}\frac{\omega_0}{\bar{\lambda}_i} + \mathbf{A_{j2}})\right]\bar{v}_{2i} = \bar{K}_{ci}(\bar{\lambda}_i)\bar{v}_{2i}\end{aligned} \tag{6.168}$$

Hence, the equivalent residue index is

$$\bar{R}_{ik} = \bar{b}_{ik}\bar{c}_{ik} = \bar{K}_{bi}(\bar{\lambda}_i)\bar{K}_{ci}(\bar{\lambda}_i)\bar{v}_{i2}\bar{w}_{i2} \tag{6.169}$$

In the selection of installing location and feedback signal of the stabilizer to be installed in the multi-machine power system, the residue is calculated for comparison among various candidate locations and feedback signals. For example, if there are two candidate installing locations or feedback signals, A and B, and $|\bar{R}_{ikA}| > |\bar{R}_{ikB}|$, then A is considered to be better than B to be the installing location or the feedback signal of the stabilizer. Therefore, it is the ratio, $\frac{|\bar{R}_{ikA}|}{|\bar{R}_{ikB}|}$, not the values of the residue, that determines the selection.

On the other hand, in system state equation of Eq. (6.164), if the installing location or the feedback signal of the stabilizer is different, the control and the output vectors are not the same, but the open-loop state matrix is unchanged, provided that the operating point of the system remains the same. That is,

$$\bar{v}_{i2A} = \bar{v}_{i2B}, \ \bar{w}_{i2A} = \bar{w}_{i2B} \tag{6.170}$$

Therefore, from Eqs. (6.169) and (6.170), it can be obtained that

$$\begin{aligned} \frac{|\bar{R}_{ikA}|}{|\bar{R}_{ikB}|} &= \frac{|\bar{K}_{biA}(\bar{\lambda}_i)\bar{K}_{ciA}(\bar{\lambda}_i)\bar{v}_{i2A}\bar{w}_{i2A}|}{|\bar{K}_{biB}(\bar{\lambda}_i)\bar{K}_{ciB}(\bar{\lambda}_i)\bar{v}_{i2B}\bar{w}_{i2B}|} \\ &= \frac{|\bar{K}_{biA}(\bar{\lambda}_i)\bar{K}_{ciA}(\bar{\lambda}_i)||\bar{v}_{i2A}\bar{w}_{i2A}|}{|\bar{K}_{biB}(\bar{\lambda}_i)\bar{K}_{ciB}(\bar{\lambda}_i)||\bar{v}_{i2B}\bar{w}_{i2B}|} = \frac{|\bar{K}_{biA}(\bar{\lambda}_i)\bar{K}_{ciA}(\bar{\lambda}_i)|}{|\bar{K}_{biB}(\bar{\lambda}_i)\bar{K}_{ciB}(\bar{\lambda}_i)|} \end{aligned} \tag{6.171}$$

Equation (6.171) shows that $|\bar{K}_{bi}(\lambda_i)\bar{K}_{ci}(\lambda_i)|$ can replace $|\bar{R}_{ik}|$ to be the index to measure the effectiveness of the stabilizer so as to select the installing location and feedback signal. Since in most cases, the oscillation mode of interest is lightly damped, i.e. $\bar{\lambda}_i \approx j\omega_i$, so $|\bar{K}_{bi}(\bar{\lambda}_i)\bar{K}_{ci}(\bar{\lambda}_i)| \approx |\bar{K}_{bi}(j\omega_i)\bar{K}_{ci}(j\omega_i)|$, the index can be calculated without knowing the eigensolution of the open-loop state matrix. Therefore, the selection can be made eigensolution-free with the following criteria:

$$\begin{aligned} &\underset{\varphi}{\text{Max}}\left\{\underset{\mu}{\text{Min}}[|\bar{K}_{bi}(j\omega_i)\bar{K}_{ci}(j\omega_i)|], \text{ for all } \mu \in \Omega(\mu)\right\}, \text{ for all } \varphi \in \Phi(\varphi) \\ &\underset{\varphi}{\text{Min}}\left\{\underset{\mu}{\text{Max}}[|\bar{K}_{bi}(j\omega_i)\bar{K}_{ci}(j\omega_i)|] - \underset{\mu}{\text{Min}}[|\bar{K}_{bi}(j\omega_i)\bar{K}_{ci}(j\omega_i)|],\right. \\ &\text{for all } \mu \in \Omega(\mu)\}, \text{ for all } \varphi \in \Phi(\varphi) \end{aligned} \tag{6.172}$$

### 6.2.4 *Stabilizer Design in a Multi-machine Power System Considering Robustness and Interaction of Stabilizers*

#### 6.2.4.1 Selection of an Operating Condition for the Robust Coordinated Design of Multiple Stabilizers in a Multi-machine Power System

Section 4.3.2.3 discusses the selection of an operating condition for the design of a robust stabilizer in a single-machine infinite-bus power system by using the phase compensation method. The selection strategy presented in Sect. 4.3.2.3 is to choose an operating condition where it is predicted that the stabilizer to be installed will be least effective. Hence, once the stabilizer is designed at the selected operating condition to be effective in damping power system oscillation, its effectiveness is ensured when the system operates at other operating conditions.

In a multi-machine power system installed with multiple stabilizers, such a strategy to select an operating condition for the design of robust stabilizers can be similarly applied by using the damping torque analysis as follows.

Consider the kth stabilizer installed in the system to improve the damping of the ith oscillation mode, $\bar{\lambda}_i = \xi_i \pm j\omega_i$, and use the following sensitivity index given by Eq. (6.143)

$$\Delta\xi_i = \sum_{j=1}^{N} D_{dampj} S_{ij-\xi} \tag{6.173}$$

From Eq. (6.147), it can have

$$\Delta\xi_i = \sum_{j=1}^{N} D_{dampj} S_{ij-\xi} = \sum_{j=1}^{N} \mathrm{Re}[\bar{B}_{kj}(\bar{\lambda}_i)\bar{\gamma}_j(\bar{\lambda}_i)] S_{ij-\xi} \tag{6.174}$$

which gives the prediction of the amount of damping provided by the kth stabilizer to the ith oscillation mode, that is, the effectiveness of the stabilizer. Therefore, if it is assumed that the set of all known candidate operating conditions of the multi-machine power system is $\Omega_0$, the following index can be used to predict the effectiveness of the kth stabilizer to damp the ith oscillation mode at an operating condition:

$$W_{ik}(m) = \sum_{j=1}^{N} \mathrm{Re}[\bar{B}_{kj}(\bar{\lambda}_i)\bar{\gamma}_j(\bar{\lambda}_i)] S_{ij-\xi},\ \mu_{ik}(m) \in \Omega_0 \tag{6.175}$$

If $W_{ik}(m_t) = \min\limits_{m}(W_k)$, $\mu_{ik}(m) \in \Omega_0$, it is determined that the kth stabilizer is least effective in damping the ith oscillation mode at the operating condition $\mu_{ik}(m_t) \in \Omega_0$. Hence, $\mu_{ik}(m_t) \in \Omega_0$ can be chosen as the operating condition to design the kth stabilizer. An effective design of the stabilizer at $\mu_{ik}(m_t) \in \Omega_0$ can ensure the effectiveness of the stabilizer at all operating conditions $\mu_{ik}(m_t) \in \Omega_0$.

Obviously, this selection above may result in different operating conditions for different stabilizers and oscillation modes. So, it may be difficult or impossible to select a common operating condition at which all stabilizers can be designed simultaneously in coordination. In other words, if they are designed in coordination at a common operating condition simultaneously, their robustness may not be ensured due to the difference in the changing pattern of their effectiveness to the variations of system operating conditions. This problem can be solved by designing the multiple stabilizers at two stages. The first stage of the design is to set the phase of stabilizers one by one in a sequence according to the selection of robust operating conditions of stabilizers. At the second stage, gains of all stabilizers are tuned simultaneously in coordination to avoid the problem of eigenvalue drift.

For simplicity of expression, the following conventional form of the transfer function of stabilizers is used:

$$T_{stk}(s) = K_k \frac{sT_{wk}}{1 + sT_{wk}} \frac{1}{1 + sT_k} \frac{(1 + sT_{k2})(1 + sT_{k4})}{(1 + sT_{k1})(1 + sT_{k3})} = K_k T_k(s) \tag{6.176}$$

It is assumed that the kth stabilizer is designed mainly to improve the damping of the ith oscillation mode in the power system, $\bar{\lambda}_i = \xi_i \pm j\omega_i$. Without loss of generality, it is assumed that at the operating condition of the kth stabilizer selected by the use of Eq. (6.175), $S_{i1-\xi} \geq S_{i2-\xi} \geq \cdots \geq S_{iN-\xi}$. A compensation angle $\phi_k$ is chosen to satisfy

$$90^\circ > \phi_k - \varphi_{kj} > 0^\circ, \quad j = 1, 2, \ldots, N \tag{6.177}$$

where $\varphi_{kj}$ is the phase of $\bar{B}_{kj}(\bar{\lambda}_i)\bar{\gamma}_j(\bar{\lambda}_i)$. If such a compensation angle cannot be obtained, the selection of the compensation angle $\phi_k$ can be made among the first (N − 1) machines to satisfy

$$90^\circ > \phi_k - \varphi_{kj} > 0^\circ, \quad j = 1, 2, \ldots, N - 1 \tag{6.178}$$

This procedure can be continued until such a compensation angle is obtained. Then, $T_k(s)$ is set to have a phase angle $-\phi_k$, that is, $\bar{T}_k(\bar{\lambda}_i) = |T_k|\angle - \phi_k$. According to Fig. 6.14, as far as the setting of the transfer function of the stabilizer is concerned, the damping contribution by the kth stabilizer to $\bar{\lambda}_i$ can be expressed as

$$\begin{aligned} \Delta\xi_i(k) &= \sum_{j=1}^{N} D_{dampj} S_{ij-\xi} = \sum_{j=1}^{N} \mathrm{Re}[\bar{B}_{kj}(\bar{\lambda}_i)\bar{\gamma}_j(\bar{\lambda}_i)] S_{ij-\xi} \mathrm{Re}[\Delta\bar{T}_{stk}(\bar{\lambda}_i)] \\ &= \sum_{j=1}^{N} S_{ij-\xi} H_{kj} T_k \cos\beta_{kj} \Delta K_k, \quad j = 1, 2, \ldots, N \end{aligned} \tag{6.179}$$

where $\bar{B}_{kj}(\bar{\lambda}_i)\bar{\gamma}_j(\bar{\lambda}_i)\bar{T}_{stk}(\bar{\lambda}_i) = H_{kj}T_k\angle\beta_{kj}, \quad j = 1, 2, \ldots, N$ and $\beta_{kj} = -\phi_k + \varphi_{kj}$. The procedure above in selecting the compensation angle $\phi_k$ ensures that $0 > \beta_{kj} > -90^\circ$ is tenable to the machines which are more sensitive to $\bar{\lambda}_i$ so that they are provided with a positive synchronizing torque. Equation (6.179) shows the damping contribution to the ith oscillation mode by the kth stabilizer set at the selected operating condition. If there are total L stabilizers considered, the total damping contributed from L stabilizers to the ith oscillation mode can be obtained from Eq. (6.179) as

$$\Delta\xi_i = \sum_{k=1}^{L} \Delta\xi_i(k) = \sum_{k=1}^{L}\sum_{j=1}^{N} S_{ij-\xi} H_{kj} T_k \cos\beta_{kj} \Delta K_k, \qquad j = 1, 2, \ldots, N \quad (6.180)$$

Let the target damping of the oscillation modes by the coordinated design of L stabilizers be $\xi_i^*$, $i = 1, 2, \ldots, L$. An objective function can be formed as

$$J(\mathbf{K}) = \sum_{i=1}^{L} Q_i [\xi_i(\mathbf{K}) - \xi_i^*]^2 \quad (6.181)$$

where $\mathbf{K} = [K_1 \quad K_2 \quad \ldots \quad K_L]^T$ and $Q_i$ is the weight for the assignment of the ith oscillation mode. The simultaneous tuning of gains of L stabilizers can be carried out by using a method of optimization to find the solution of the objective function defined by Eq. (6.181). This simultaneous tuning is conducted at the selected operating condition of each stabilizer. At the selected operating condition of the first stabilizer, the initial values of all gains are simply taken to be zero. From the selected operating condition of the second stabilizer on, the gain values set by the previous tuning are taken to be the initial values to start the current tuning. Also in the tuning process, the following constraint is checked:

$$|\xi_i(\mathbf{K})| \geq |\xi_i^*|, \quad i = 1, 2, \ldots, L \quad (6.182)$$

If Eq. (6.182) is satisfied for any $\lambda_l, l = 1, 2, \ldots, L$, which is supposed to be damped by the kth stabilizer, $Q_l = 0$ is set in Eq. (6.181). This arrangement frees the eigenvalue drift towards the "good" direction, the left half complex plane over the target value. Therefore, a unique final solution of gain tuning will be much easier to be found.

#### 6.2.4.2 Non-negatively Interactive Design of a Stabilizer in a Multi-machine Power System [5]

For the design of multiple stabilizers (the PSSs or FACTS-based stabilizers) in a multi-machine power system, the techniques of sequential setting require the stabilizers to be designed one by one in a sequence. However, the stabilizer set later may move the oscillation modes damped by the stabilizers designed previously.

This problem of "eigenvalue drift" caused by the interactions among the multiple stabilizers has motivated the development of the coordinated tuning of multiple stabilizers. By coordinated tuning, the parameters of all interactive stabilizers are set simultaneously to achieve an accurate assignment of all oscillation modes of interest in the power system so that the interactions among the stabilizers are fully taken account of.

However, in the design of multiple stabilizers, only an approximate assignment of oscillation modes is needed as long as the damping of the oscillation modes is improved to meet a certain target value of damping. If the interactions among the stabilizers are positive, in fact, extra benefits of damping provision can be obtained. Hence, the positive interactions should be encouraged instead of being strictly constrained by the coordinated tuning. While what ought to be avoided is the negative interactions among multiple interactive stabilizers, which may result in the provision of negative damping to some oscillation modes, at the same time that positive damping is contributed by the design of the stabilizer. Furthermore, in practice, it is not always the case that when a new stabilizer is put into a power system, all old interactive stabilizers are adjustable so that the coordinated tuning can be implemented. As long as the design of this new stabilizer can ensure no provision of negative damping to all interactive oscillation modes in the power system, it can be set without the coordinated tuning with other old stabilizers. In this section, it is discussed how such a non-negatively interactive stabilizer can be designed.

From Eq. (6.179), the damping provided by a stabilizer (the kth stabilizer) in a multi-machine power system can be denoted as

$$\begin{aligned}\Delta\xi_i(k) &= \sum_{j=1}^{N} \mathrm{Re}[\bar{B}_{kj}(\bar{\lambda}_i)\bar{\gamma}_j(\bar{\lambda}_i)]S_{ij-\xi}\mathrm{Re}[\Delta\bar{T}_{stk}(\bar{\lambda}_i)] \\ &= \mathrm{Re}\left\{\sum_{j=1}^{N}[S_{ij-\xi}\bar{H}_{kij}(\bar{\lambda}_i)]\Delta\bar{T}_{stk}(\bar{\lambda}_i)\right\} = \mathrm{Re}[\bar{C}_{ki}\Delta\bar{T}_{stk}(\bar{\lambda}_i)]\end{aligned} \tag{6.183}$$

where $\bar{C}_{ki} = \sum_{j=1}^{N} S_{ij-\xi}H_{kij}(\bar{\lambda}_i) = |\bar{C}_{ki}|\angle\alpha_{ki}(\bar{\lambda}_i)$. If the target oscillation mode for the installation of the kth stabilizer is $\bar{\lambda}_0 = \xi_0 \pm j\omega_0$ and $\bar{T}_{stk}(\bar{\lambda}_0) = K_k|\bar{T}_k(\bar{\lambda}_0)|\angle\beta_{k0}(\bar{\lambda}_0)$, from Eq. (6.183) the damping improvement of the target oscillation mode by the stabilizer can be obtained as

$$\Delta\xi_0 = \mathrm{Re}[\bar{C}_{k0}\Delta\bar{T}_{stk}(\bar{\lambda}_0)] = |\bar{C}_{k0}|\Delta[K_k|\bar{T}_k(\bar{\lambda}_0)|]\cos[\beta_{k0}(\bar{\lambda}_0) + \alpha_{k0}(\bar{\lambda}_0)] \tag{6.184}$$

For other oscillation modes, $\bar{\lambda}_i = \xi_i \pm j\omega_i$, $i \neq 0$, in the power system, the damping increment of the ith oscillation mode due to the installation of the kth stabilizer is

$$\Delta\xi_i = \mathrm{Re}[\bar{C}_{ki}\Delta\bar{T}_{stk}(\bar{\lambda}_i)] = |\bar{C}_{ki}|\Delta[K_k|\bar{T}_k(\bar{\lambda}_i)|]\cos[\beta_{ki}(\bar{\lambda}_i) + \alpha_{ki}(\bar{\lambda}_i)] \tag{6.185}$$

To ensure a positive damping provision by the stabilizer to the target oscillation mode and at the same time, the imposition of a non-negatively interaction upon the ith oscillation mode, it should have

$$\cos[\beta_{k0}(\bar{\lambda}_0) + \alpha_{k0}(\bar{\lambda}_0)] > 0; \; \cos[\beta_{ki}(\bar{\lambda}_i) + \alpha_{ki}(\bar{\lambda}_i)] \geq 0 \tag{6.186}$$

Therefore, a successful design of non-negatively interactive stabilizer is equivalent to shifting all the vectors, $\bar{C}_{ki}(\bar{\lambda}_i)$, onto the first quadrant in the complex plane, where $\bar{\lambda}_i = \xi_i \pm j\omega_i$, $i = 0, 1, \ldots$ are oscillation modes associated with the interactive stabilizers in the power system. From Eq. (6.184) to (6.186), obviously, it can be seen that the following two factors decide whether the shifting may be successful:

1. the difference between the angles $\alpha_{k0}(\bar{\lambda}_0)$ and $\alpha_{ki}(\bar{\lambda}_i)$;
2. the setting of the phase angle of the stabilizer.

The most successful shifting for $\bar{C}_{ki}(\bar{\lambda}_i)$ is onto the positive real axis so that $\cos[\beta_{ki}(\bar{\lambda}_i) + \alpha_{ki}(\bar{\lambda}_i)] = 1$, $i = 0, 1, \ldots$.

In the case that $\alpha_{ki}(\bar{\lambda}_i) > \alpha_{k0}(\bar{\lambda}_0)$, to achieve the maximum efficiency of the stabilizer to damp its target oscillation mode, according to Eq. (6.184), the phase of the stabilizer could be set to be $\beta_{k0}(\bar{\lambda}_0) = -\alpha_{k0}(\bar{\lambda}_0)$ so that $\bar{C}_{k0}(\bar{\lambda}_0)$ is shifted onto the real axis with $\cos[\beta_{k0}(\bar{\lambda}_0) + \alpha_{k0}(\bar{\lambda}_0)] = 1$. As shown in Fig. 6.16a, $\bar{C}_{k0}(\bar{\lambda}_0)$ may also be moved into the first quadrant. This results in a design of an effective and non-negatively interactive stabilizer.

However, if $\alpha_{ki}(\bar{\lambda}_i) < \alpha_{k0}(\bar{\lambda}_0)$ as shown in Fig. 6.16b, the priority of shifting may have to be given to $\bar{C}_{k0}(\bar{\lambda}_0)$ so as to move it to the first quadrant and in the same time to shift $\bar{C}_{k0}(\bar{\lambda}_0)$ into the first quadrant instead of onto the real axis. Thus, the design may not be most efficient as far as the damping of the target oscillation mode, $\bar{\lambda}_0$, is concerned. However, the stabilizer is positively interactive with $\bar{\lambda}_i$.

In both cases above, if the angle between $\bar{C}_{k0}(\bar{\lambda}_0)$ and $\bar{C}_{ki}(\bar{\lambda}_i)$ is large and even $\alpha_{k0}(\bar{\lambda}_0) - \alpha_{ki}(\bar{\lambda}_i) > 90°$ as shown in Fig. 6.16c, there is the danger that when $\bar{C}_{k0}(\bar{\lambda}_0)$ is shifted into the first quadrant, $\bar{C}_{ki}(\bar{\lambda}_i)$ may not be able to be moved into

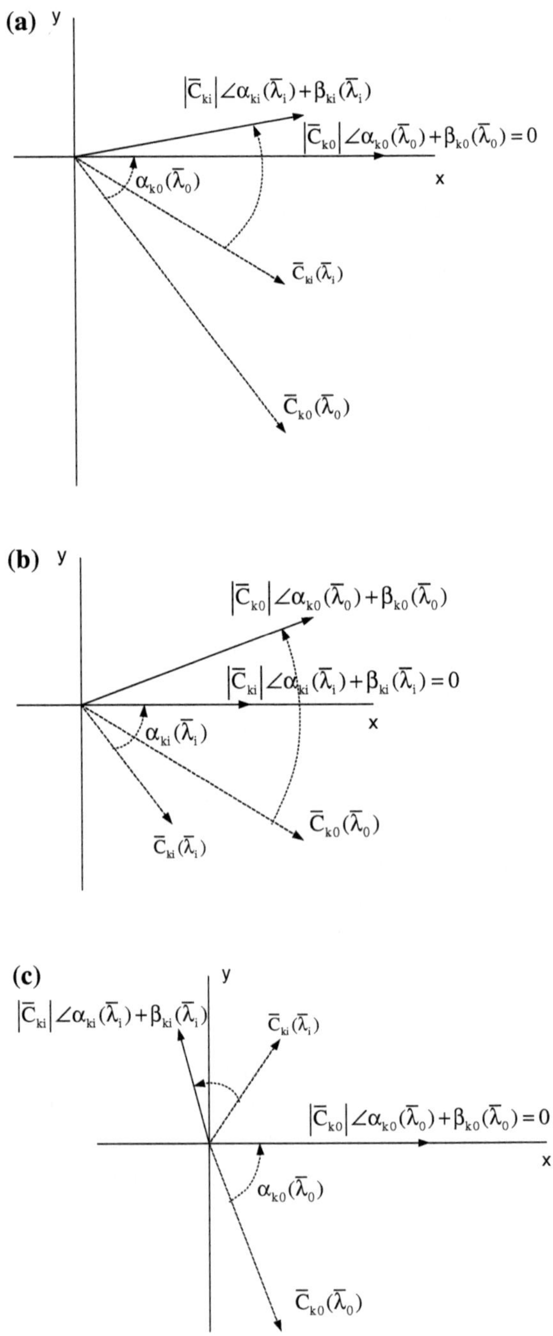


**Fig. 6.16** Non-negatively interactive design of a stabilizer.
**a** $\alpha_{ki}(\bar{\lambda}_i) > \alpha_{k0}(\bar{\lambda}_0)$,
**b** $\alpha_{ki}(\bar{\lambda}_i) < \alpha_{k0}(\bar{\lambda}_0)$,
**c** $\alpha_{k0}(\bar{\lambda}_0) - \alpha_{ki}(\bar{\lambda}_i) > 90°$

the first quadrant in the same time which could result in a negative damping provision to the ith oscillation mode. This means a negative interaction between the stabilizer being designed and the stabilizer which damps the ith oscillation mode.

## 6.3 An Example Two-Area Four-Machine Power System

### *6.3.1 Linearized Model*

#### 6.3.1.1 System Parameters and Operating Conditions

Configuration of a four-machine eleven-node example power system is shown in Fig. 6.17.

Line parameters are given in Table 6.1. Parameters of generators and the AVRs are given in Table 6.2.

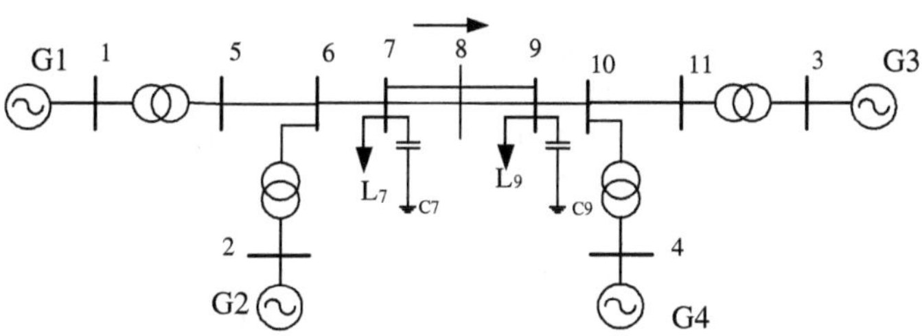


**Fig. 6.17** Configuration of a four-machine example power system

**Table 6.1** Network parameters

| Node | Node | Resistance | Reactance | Susceptance |
|---|---|---|---|---|
| 1 | 5 | 0.0000 | 0.0167 | 0.0000 |
| 2 | 6 | 0.0000 | 0.0167 | 0.0000 |
| 3 | 11 | 0.0000 | 0.0167 | 0.0000 |
| 4 | 10 | 0.0000 | 0.0167 | 0.0000 |
| 5 | 6 | 0.0025 | 0.0250 | 0.04375 |
| 6 | 7 | 0.0010 | 0.0100 | 0.0175 |
| 7 | 8 | 0.0110 | 0.1100 | 0.1925 |
| 7 | 8 | 0.0110 | 0.1100 | 0.1925 |
| 8 | 9 | 0.0110 | 0.1100 | 0.1925 |
| 8 | 9 | 0.0110 | 0.1100 | 0.1925 |
| 9 | 10 | 0.0010 | 0.0100 | 0.0175 |
| 10 | 11 | 0.0025 | 0.0250 | 0.04375 |

**Table 6.2** Generator parameters

| | G1 | G2 | G3 | G4 |
|---|---|---|---|---|
| $T_J$ | 13.0 | 13.0 | 12.35 | 12.35 |
| $X_d$ | 1.8000 | 1.8000 | 1.8000 | 1.8000 |
| $X_q$ | 1.7000 | 1.7000 | 1.7000 | 1.7000 |
| $X_d'$ | 0.3000 | 0.3000 | 0.3000 | 0.3000 |
| $T_{d0}'$ | 8.000 | 8.000 | 8.000 | 8.000 |
| D | 0 | 0 | 0 | 0 |
| M | 6.5 | 6.5 | 6.175 | 6.175 |
| $K_A$ | 50 | 50 | 50 | 50 |
| $T_A$ | 0.1 | 0.1 | 0.1 | 0.1 |

**Table 6.3** Results of load flow computation—node voltage and node injection power

| Node | Magnitude of voltage | Phase of voltage (deg.) | Active power | Reactive power |
|---|---|---|---|---|
| G1 | 1.0300 | 0.5362 | 7.0000 | 2.7299 |
| G2 | 1.0100 | 0.3616 | 7.0000 | 4.4625 |
| G3 | 1.0300 | 0 | 7.3319 | 3.1059 |
| G4 | 1.0100 | −0.1910 | 7.0000 | 5.1643 |
| 5 | 0.9923 | 0.4218 | 0.0000 | 0.0000 |
| 6 | 0.9434 | 0.2388 | 0.0000 | 0.0000 |
| 7 | 0.8989 | 0.0792 | −9.6700 | −1.0000 |
| 8 | 0.8636 | −0.2000 | 0.0000 | 0.0000 |
| 9 | 0.8799 | −0.4834 | −17.6700 | −1.0000 |
| 10 | 0.9320 | −0.3153 | 0.0000 | 0.0000 |
| 11 | 0.9869 | −0.1205 | 0.0000 | 0.0000 |

Initial conditions of the example power system for the load flow computation are

$$V_{G1} = 1.03,\ V_{G2} = 1.01,\ V_{G3} = 1.03\angle 0,\ V_{G4} = 1.03,$$
$$P_{G1} = 7.0,\ P_{G2} = 7.0,\ P_{G4} = 7.0,$$
$$\text{Load}\,L_7 = 9.67 + j1.00,\ \text{Load}\,L_9 = 13.67 + j1.00.$$

Results of load flow computation are given in Tables 6.3 and 6.4. Based on the results of load flow, initial values of state variables can be calculated as follows.

**Table 6.4** Results of load flow computation—line power and power loss

| Starting node (SE) | Ending node (EN) | Power at SN | Power at EN | Power loss |
|---|---|---|---|---|
| 1 | 5 | 7.00 + j2.73 | −7.00 − j1.84 | 0.00 + j0.88 |
| 2 | 6 | 7.00 + j4.46 | −7.00 − j3.33 | 0.00 + j1.12 |
| 3 | 11 | 7.33 + j3.10 | −7.33 − j2.11 | 0.00 + j0.99 |
| 4 | 10 | 7.00 + j5.16 | −7.00 − j3.93 | 0.00 + j1.23 |
| 5 | 6 | 7.00 + j1.82 | −6.86 − j0.57 | 0.13 + j1.25 |
| 6 | 7 | 13.87 + j3.88 | −13.63 − j1.58 | 0.23 + j2.30 |
| 7 | 8 | 1.98 + j0.21 | −1.93 + j0.04 | 0.05 + j0.25 |
| 7 | 8 | 1.98 + j0.21 | −1.93 + j0.04 | 0.05 + j0.25 |
| 8 | 9 | 1.93 − j0.19 | −1.87 + j0.44 | 0.05 + j0.25 |
| 8 | 9 | 1.93 − j0.19 | −1.87 + j0.44 | 0.05 + j0.25 |
| 9 | 10 | −13.93 − j2.04 | 14.18 + j4.57 | 0.25 + j2.53 |
| 10 | 11 | −7.18 − j0.67 | 7.33 + j2.08 | 0.15 + j1.42 |

From the output active and reactive power of each generator, $p_i + jQ_i$, and its terminal voltage, $V_{gi0}$, $i = 1, 2, 3, 4$, output current of each generator can be calculated as

$$I_{gi0} = \frac{P_i - jQ_i}{V^*_{gi0}}, \quad i = 1, 2, 3, 4$$

It gives

$$\begin{aligned}
\bar{I}_{g10} &= 7.0811 + j1.1871 \\
\bar{I}_{g20} &= 7.7126 - j1.2230 \\
\bar{I}_{g30} &= 7.0287 - j2.1249 \\
\bar{I}_{g40} &= 6.2746 - j4.1969
\end{aligned}$$

The imaginary voltage of each generator can be computed as

$$\begin{aligned}
\bar{E}_{Q10} &= \bar{V}_{g10} + jX_{q1}\bar{I}_{g10} = 1.9976\angle 70.7^\circ \\
\bar{E}_{Q20} &= \bar{V}_{g20} + jX_{q2}\bar{I}_{g20} = 2.2619\angle 56.1^\circ \\
\bar{E}_{Q30} &= \bar{V}_{g30} + jX_{q3}\bar{I}_{g30} = 2.0896\angle 40.0^\circ \\
\bar{E}_{Q40} &= \bar{V}_{g40} + jX_{q4}\bar{I}_{g40} = 2.3702\angle 22.6^\circ
\end{aligned}$$

The phase of imaginary voltage is the angular position of each generator, $\delta_{i0}, i = 1, 2, 3, 4$. From $\delta_{i0}, i = 1, 2, 3, 4$, computational results of output current and terminal voltage of each generator can be transformed from the common x–y coordinate to the d–q coordinate as (Eq. (5.49))

**Table 6.5** Initial values of state variables

| | G1 | G2 | G3 | G4 |
|---|---|---|---|---|
| $\delta_{i0}$ (rad.) | 1.2341 | 0.9788 | 0.6990 | 0.3942 |
| $\omega_{i0}$ | 1 | 1 | 1 | 1 |
| $E'_{qi0}$ | 1.0025 | 1.0775 | 1.0181 | 1.1116 |
| $E'_{fdi0}$ | 2.0688 | 2.3456 | 2.1662 | 2.4601 |

$$\begin{bmatrix} i_{di} \\ i_{qi} \end{bmatrix} = \begin{bmatrix} \sin\delta_i & -\cos\delta_i \\ \cos\delta_i & \sin\delta_i \end{bmatrix} \begin{bmatrix} v_{gxi} \\ v_{gyi} \end{bmatrix}, \begin{bmatrix} v_{di} \\ v_{qi} \end{bmatrix} = \begin{bmatrix} \sin\delta_i & -\cos\delta_i \\ \cos\delta_i & \sin\delta_i \end{bmatrix} \begin{bmatrix} v_{gxi} \\ v_{gyi} \end{bmatrix}$$

Since $E'_{fdi0} = 0$, from the third equation of Eq. (5.24), it can have

$$E_{fdi0} = E_{qi0}$$

According to Eq. (5.25), it can have

$$\begin{aligned}
E'_{q10} &= V_{gq10} + X'_{d1} I_{gd10} = 1.0025, E_{fd10} = E'_{q10} + (X_{d1} - X'_{d1}) I_{gd10} = 2.0688 \\
E'_{q20} &= V_{gq20} + X'_{d2} I_{gd20} = 1.0775, E_{fd20} = E'_{q20} + (X_{d2} - X'_{d2}) I_{gd20} = 2.3456 \\
E'_{q30} &= V_{gq30} + X'_{d3} I_{gd30} = 1.0181, E_{fd30} = E'_{q30} + (X_{d3} - X'_{d3}) I_{gd30} = 2.1662 \\
E'_{q40} &= V_{gq40} + X'_{d4} I_{gd40} = 1.1116, E_{fd40} = E'_{q40} + (X_{d4} - X'_{d4}) I_{gd40} = 2.4601
\end{aligned}$$

Computational results of all the above are presented in Table 6.5. Initial values of SVC are given as

$$X_{svcc} = 0.02, X_{svcl} = 0.01, B_{svc0} = 0$$

According to Eq. (6.34), it can have $\alpha_0 = 90°$.

#### 6.3.1.2 Linearized Heffron–Phillips Model

From Tables 6.1 and 6.3, the following network equations can be obtained:

$$\begin{bmatrix} \bar{\mathbf{I}}_g \\ \mathbf{0} \end{bmatrix} = \begin{bmatrix} \bar{\mathbf{Y}}_{11} & \bar{\mathbf{Y}}_{12} \\ \bar{\mathbf{Y}}_{21} & \bar{\mathbf{Y}}_{22} \end{bmatrix} \begin{bmatrix} \bar{\mathbf{E}}_g \\ \bar{\mathbf{V}}_m \end{bmatrix}$$

where

$$\bar{\mathbf{Y}}_{11} = \begin{bmatrix} -j59.9880 & 0 & 0 & 0 \\ 0 & j59.9880 & 0 & 0 \\ 0 & 0 & j59.9880 & 0 \\ 0 & 0 & 0 & j59.9880 \end{bmatrix},$$

$$\bar{\mathbf{Y}}_{12} = \begin{bmatrix} j59.9880 & 0 & 0 & 0 & 0 & 0 & 0 \\ 0 & j59.9880 & 0 & 0 & 0 & 0 & 0 \\ 0 & 0 & 0 & 0 & 0 & 0 & j59.9880 \\ 0 & 0 & 0 & 0 & 0 & j59.9880 & 0 \end{bmatrix},$$

$$\bar{\mathbf{Y}}_{21} = \begin{bmatrix} j59.9880 & 0 & 0 & 0 \\ 0 & j59.9880 & 0 & 0 \\ 0 & 0 & 0 & 0 \\ 0 & 0 & 0 & 0 \\ 0 & 0 & 0 & 0 \\ 0 & 0 & 0 & j59.9880 \\ 0 & 0 & j59.9880 & 0 \end{bmatrix},$$

$$\bar{\mathbf{Y}}_{22} = \begin{bmatrix} \bar{\mathbf{Y}}_{22-1} & \bar{\mathbf{Y}}_{22-2} & \bar{\mathbf{Y}}_{22-3} \end{bmatrix},$$

$$\bar{\mathbf{Y}}_{22-1} = \begin{bmatrix} 3.9604 - j99.5701 & -3.9604 + j39.6040 & 0 \\ -3.9604 + j39.6040 & 13.8614 - j198.5712 & -9.90010 + j99.0100 \\ 0 & -9.9010 + j99.0100 & 23.6692 - j118.0481 \\ 0 & 0 & -1.8002 + j18.0020 \\ 0 & 0 & 0 \\ 0 & 0 & 0 \\ 0 & 0 & 0 \end{bmatrix},$$

$$\bar{\mathbf{Y}}_{222} = \begin{bmatrix} 0 & 0 \\ 0 & 0 \\ -1.8002 + j18.0020 & 0 \\ 3.6004 - j35.6186 & -1.8002 + j18.0020 \\ -1.8002 + j18.0020 & 34.5237 - j118.1020 \\ 0 & -9.9001 + j99.0100 \\ 0 & 0 \end{bmatrix},$$

$$\bar{\mathbf{Y}}_{22-3} = \begin{bmatrix} 0 & 0 \\ 0 & 0 \\ 0 & 0 \\ 0 & 0 \\ -9.9010 + j99.0100 & 0 \\ 13.8614 - j198.5712 & -3.9604 + j39.6040 \\ -3.9604 + j39.6040 & 3.9604 - j99.57010 \end{bmatrix},$$

By keeping only three nodes of generator terminals via deleting all the other nodes in the above network admittance matrix, it can have

$$\bar{\mathbf{Y}}_N = \bar{\mathbf{Y}}_{11} - \bar{\mathbf{Y}}_{12}\bar{\mathbf{Y}}_{22}^{-1}\bar{\mathbf{Y}}_{21}$$
$$= \begin{bmatrix} 1.5150 - j17.7819 & 1.1172 + j15.2884 & 0.1867 + j0.4071 & 0.5275 + j0.9893 \\ 1.1172 + j15.2884 & 5.0840 - j21.9474 & 0.5275 + j0.9893 & 1.4666 + j2.3934 \\ 0.1867 + j0.4071 & 0.5275 + j0.9893 & 1.9728 - j18.0523 & 2.2208 + j14.5440 \\ 0.5275 + j0.9893 & 1.4666 + j\,2.3934 & 2.2208 + j14.5440 & 7.7304 - j23.9732 \end{bmatrix}$$

Thus,

$$\bar{\mathbf{Y}} = \left(\bar{\mathbf{Y}}_N^{-1} + \mathbf{j}\mathbf{x}'_\mathbf{d}\right)^{-1}$$
$$= \begin{bmatrix} 1.2870 - 9.3383i & 1.4623 + 5.9214i & 0.4321 + 0.4042i & 0.6681 + 0.5845i \\ 1.4623 + 5.9214i & 2.4883 - 11.1930i & 0.6681 + 0.5845i & 1.0309 + 0.8430i \\ 0.4321 + 0.4042i & 0.6681 + 0.5845i & 1.7113 - 98981i & 2.0705 + 5.0608i \\ 0.6681 + 0.5845i & 1.0309 + 0.8430i & 2.0705 + 5.0608i & 3.3572 - 12.5138i \end{bmatrix}$$

According to Eq. (5.36), it can have

$$\mathbf{F_{dd}} = \begin{bmatrix} 6.9805 & -5.4977 & -0.6487 & -0.8340 \\ -3.2377 & 5.7984 & -1.0716 & -1.4891 \\ -0.4782 & -0.8486 & 6.8161 & -5.4893 \\ -0.4958 & -1.0508 & -3.4812 & 5.0277 \end{bmatrix},$$
$$\mathbf{F_{qq}} = \begin{bmatrix} 3.2724 & -3.9725 & 0.1877 & 0.5123 \\ -6.6150 & 6.1187 & 0.0321 & 0.4642 \\ -0.4880 & -0.5283 & 3.6934 & -2.6771 \\ -0.9808 & -1.2240 & -5.8082 & 8.0130 \end{bmatrix},$$
$$\mathbf{G_{dd}} = \begin{bmatrix} 9.3383 & -5.3603 & -0.1274 & 0.1073 \\ -6.0988 & 11.1930 & -0.3772 & -0.1341 \\ -0.5680 & -0.7462 & 9.8981 & -4.2062 \\ -0.8876 & -1.2719 & -5.4489 & 12.5138 \end{bmatrix},$$

$$\mathbf{G_{qq}} = \begin{bmatrix} 1.2870 & 2.9102 & 0.5778 & 0.8811 \\ -0.0804 & 2.4883 & 0.8035 & 1.3249 \\ 0.1656 & 0.4807 & 1.7113 & 3.4938 \\ 0.0108 & 0.3945 & 0.4563 & 3.3572 \end{bmatrix},$$
$$\mathbf{H_{dd}} = \begin{bmatrix} 0.2002 & 0.4527 & 0.0899 & 0.1371 \\ -0.0125 & 0.3871 & 0.1250 & 0.2061 \\ 0.0258 & 0.0748 & 0.2662 & 0.5435 \\ 0.0017 & 0.0614 & 0.0710 & 0.5222 \end{bmatrix},$$
$$\mathbf{H_{qq}} = \begin{bmatrix} -1.4526 & 0.8338 & 0.0198 & -0.0167 \\ 0.9487 & -1.7411 & 0.0587 & 0.0209 \\ 0.0884 & 0.1161 & -1.5397 & 0.6543 \\ 0.1381 & 0.1979 & 0.8476 & -1.9466 \end{bmatrix}$$

Thus, according to Eq. (5.38), the following matrices are calculated:

$$\mathbf{F_d} = \begin{bmatrix} 5.9711 & -4.9200 & -0.7327 & -0.3184 \\ -4.2784 & 6.4231 & -1.2632 & -0.8814 \\ -1.0557 & -1.0252 & 6.1938 & -4.1129 \\ -0.9626 & -1.1694 & -4.2732 & 6.4053 \end{bmatrix},$$

$$\mathbf{F_q} = \begin{bmatrix} 0.5778 & -0.9782 & 0.1172 & 0.2833 \\ -2.2267 & 1.8849 & 0.0617 & 0.2801 \\ -0.4224 & -0.2615 & 1.0323 & -0.3484 \\ -0.5768 & -0.4099 & -1.6646 & 2.6513 \end{bmatrix},$$

$$\mathbf{G_d} = \begin{bmatrix} 9.5556 & -4.2382 & 0.2888 & 0.9817 \\ -6.0083 & 11.8998 & -0.0148 & 0.7474 \\ -0.4687 & -0.2367 & 10.3994 & -2.6752 \\ -0.8284 & -0.9169 & -5.1327 & 13.6144 \end{bmatrix},$$

$$\mathbf{G_q} = \begin{bmatrix} 0.5850 & 1.6999 & 0.3921 & 0.6178 \\ 0.1761 & 1.5087 & 0.4498 & 0.7509 \\ 0.1130 & 0.4304 & 0.8213 & 1.8852 \\ 0.0754 & 0.4386 & 0.4397 & 1.7610 \end{bmatrix},$$

From Eqs. (5.41) and (5.42), it can have

$$\begin{aligned}
\mathbf{K}_1 &= \left(\mathbf{V_{td0}} - \mathbf{I}_{q0}\mathbf{X}'_d\right)\mathrm{F_d} + \left(\mathbf{V_{td0}} - \mathbf{I}_{d0}\mathbf{X_q}\right), \\
\mathbf{K}_2 &= \mathbf{I}_{q0} + \left(\mathbf{V_{td0}} - \mathbf{I}_{d0}\mathbf{X_q}\right)\mathbf{G}_d\left(\mathbf{V_{td0}} - \mathbf{I}_{d0}\mathbf{X_q}\right)\mathbf{G_q}, \\
\mathbf{K}_3 &= \mathbf{I} + \left(\mathbf{X}_d - \mathbf{X}'_d\right)\mathbf{G_d}, \\
\mathbf{K}_4 &= \left(\mathbf{X}_d - \mathbf{X}'_d\right)\mathrm{F_d}, \\
\mathbf{K}_5 &= -\mathbf{V}_{t0}^{-1}\mathbf{V}_{q0} - \mathbf{X}'_\mathbf{d}\mathbf{F_d} + \mathbf{V}_{t0}^{-1}\mathbf{V}_{d0}\mathbf{X_q}\mathbf{F_q}, \\
\mathbf{K}_6 &= \mathbf{V}_{t0}^{-1}\mathbf{V}_{q0} - \mathbf{V}_{t0}^{-1}\mathbf{V}_{q0}\mathbf{X}'_\mathbf{d}\mathbf{G_d} + \mathbf{V}_{t0}^{-1}\mathbf{V}_{d0}\mathbf{X_q}\mathbf{G_q}
\end{aligned}$$

Hence, coefficient matrices of the Heffron–Phillips model of the example power system as given by Eq. (5.43) are obtained as

$$\mathbf{K}_1 = \begin{bmatrix} 4.4089 & -4.6359 & -0.1653 & 0.3923 \\ -7.0963 & 7.3556 & -0.4686 & 0.2093 \\ -1.4589 & -1.1060 & 5.5378 & -2.9728 \\ -1.8094 & -1.5087 & -5.9085 & 9.2266 \end{bmatrix},$$

$$\mathbf{K}_2 = \begin{bmatrix} 9.8813 & 1.0858 & 0.9408 & 1.7693 \\ -2.4940 & 12.2358 & 1.0102 & 2.0583 \\ -0.0196 & 0.7701 & 10.9008 & 2.4791 \\ -0.2018 & 0.6184 & -1.3160 & 13.3819 \end{bmatrix},$$

$$\mathbf{K}_3 = \begin{bmatrix} 2.5926 & -0.7064 & 0.0481 & 0.1636 \\ -1.0014 & 2.9833 & -0.0025 & 0.1246 \\ -0.0781 & -0.0395 & 2.7332 & -0.4459 \\ -0.1381 & -0.1528 & -0.8555 & 3.2691 \end{bmatrix},$$

$$\mathbf{K}_4 = \begin{bmatrix} 0.9952 & -0.8200 & -0.1221 & -0.0531 \\ -0.7131 & 1.0705 & -0.2105 & -0.1469 \\ -0.1760 & -0.1709 & 1.0323 & -0.6855 \\ -0.1604 & -0.1949 & -0.7122 & 1.0675 \end{bmatrix},$$

$$\mathbf{K}_5 = \begin{bmatrix} -0.0824 & 0.0069 & 0.0329 & 0.0425 \\ -0.1271 & 0.0315 & 0.0411 & 0.0546 \\ -0.0244 & -0.0056 & -0.0326 & 0.0626 \\ -0.0334 & -0.0103 & -0.0549 & 0.0986 \end{bmatrix},$$

$$\mathbf{K}_6 = \begin{bmatrix} 0.5932 & 0.3146 & 0.0402 & 0.0499 \\ 0.1826 & 0.6569 & 0.0496 & 0.0618 \\ 0.0257 & 0.0583 & 0.5999 & 0.2974 \\ 0.0309 & 0.0712 & 0.1885 & 0.6390 \end{bmatrix}.$$

$$\mathbf{M} = \begin{bmatrix} 6.5 & 0 & 0 & 0 \\ 0 & 6.5 & 0 & 0 \\ 0 & 0 & 6.175 & 0 \\ 0 & 0 & 0 & 6.175 \end{bmatrix}, \mathbf{D} = \begin{bmatrix} 0 & 0 & 0 & 0 \\ 0 & 0 & 0 & 0 \\ 0 & 0 & 0 & 0 \\ 0 & 0 & 0 & 0 \end{bmatrix},$$

$$\mathbf{K_A} = \begin{bmatrix} 50 & 0 & 0 & 0 \\ 0 & 50 & 0 & 0 \\ 0 & 0 & 50 & 0 \\ 0 & 0 & 0 & 50 \end{bmatrix}, \mathbf{T_A} = \begin{bmatrix} 0.1 & 0 & 0 & 0 \\ 0 & 0.1 & 0 & 0 \\ 0 & 0 & 0.1 & 0 \\ 0 & 0 & 0 & 0.1 \end{bmatrix},$$

$$\mathbf{T}'_{\mathbf{do}} = \begin{bmatrix} 8 & 0 & 0 & 0 \\ 0 & 8 & 0 & 0 \\ 0 & 0 & 8 & 0 \\ 0 & 0 & 0 & 8 \end{bmatrix}$$

#### 6.3.1.3 Linearized Heffron–Phillips Model of the System with an SVC Stabilizer to Be Installed

Consider that the SVC stabilizer is installed at node 9 in the example power system. From Fig. 6.2 and Eq. (6.4), it can have

$$\begin{bmatrix} \bar{\mathbf{I}}_g \\ 0 \end{bmatrix} = \begin{bmatrix} \bar{\mathbf{Y}}_{11} & \bar{\mathbf{Y}}_{12} \\ \bar{\mathbf{Y}}_{21} & \bar{\mathbf{Y}}_{22} \end{bmatrix} \begin{bmatrix} \bar{\mathbf{E}}_g \\ \bar{\mathbf{V}}_9 \end{bmatrix}$$

Thus,

$$\bar{\mathbf{I}}_g = \frac{-1}{(\bar{y}_{22} + jb_{svc})} \bar{\mathbf{Y}}_{12}\bar{\mathbf{Y}}_{21}\bar{\mathbf{E}}_g + \bar{\mathbf{Y}}_{11}\bar{\mathbf{E}}_g = [\frac{-1}{(\bar{y}_{22} + jb_{svc})} \bar{\mathbf{Y}}_{12}\bar{\mathbf{Y}}_{21} + \bar{\mathbf{Y}}_{11}]\bar{\mathbf{E}}_g = \bar{\mathbf{Y}}\bar{\mathbf{E}}_g$$

where

$$\bar{\mathbf{Y}}_{11} = \begin{bmatrix} \bar{\mathbf{Y}}_{11-1} & \bar{\mathbf{Y}}_{11-2} \end{bmatrix},$$

$$\bar{\mathbf{Y}}_{11-1} = \begin{bmatrix} -j59.9880 & 0 \\ -8.9964 - j89.9640 & -j59.9880 \\ 33.5910 - j254.3110 & 9.5608 - j92.1942 \\ 45.8239 - j640.5978 & 10.0729 - j231.8408 \end{bmatrix}$$

$$\bar{\mathbf{Y}}_{11-2} = \begin{bmatrix} 0 & 0 \\ 0 & 0 \\ 1.2539 - j20.7674 & 0.0170 + j7.8663 \\ 0.0170 + j7.8663 & 1.2218 - j40.3316 \end{bmatrix},$$

$$\bar{\mathbf{Y}}_{12} = \begin{bmatrix} j59.9880 \\ 8.9931 + j149.9192 \\ -60.1463 + j359.6609 \\ -96.3963 + j907.8644 \end{bmatrix},$$

$$\bar{\mathbf{Y}}_{21} = \begin{bmatrix} -498.2591 + j1004.9460 \\ -169.1906 + j372.5732 \\ -1.2703 + j12.9860 \\ -1.2278 + j32.6445 \end{bmatrix}^T,$$

$$\bar{y}_{22} = 757.0770 - 1400.3458$$

From Eq. (6.9), it can have

$$\mathbf{F}_{\mathrm{dd}} = \begin{bmatrix} 7.0361 & -5.5021 & -0.6541 & -0.8800 \\ -3.4671 & 6.0160 & -1.0512 & -1.4976 \\ -0.5079 & -0.8661 & 6.8227 & -5.4488 \\ -0.6231 & -1.1592 & -3.9462 & 5.7286 \end{bmatrix},$$

$$\mathbf{F}_{\mathrm{qq}} = \begin{bmatrix} 3.2904 & -3.9105 & 0.1630 & 0.4571 \\ -6.2778 & 5.8578 & 0.0239 & 0.3961 \\ -0.4221 & -0.4362 & 3.4441 & -2.5858 \\ -0.8044 & -0.9453 & -4.9335 & 6.6832 \end{bmatrix},$$

$$\mathbf{G}_{\mathrm{dd}} = \begin{bmatrix} 9.3383 & -5.4253 & -0.1673 & -0.0150 \\ -6.0990 & 11.1930 & -0.4145 & -0.2855 \\ -0.5551 & -0.7229 & 9.8981 & -4.4351 \\ -0.8776 & -1.2181 & -5.4050 & 12.5138 \end{bmatrix},$$

$$\mathbf{G}_{\mathrm{qq}} = \begin{bmatrix} 1.2870 & 2.7871 & 0.5675 & 0.8875 \\ 0.0588 & 2.4883 & 0.7849 & 1.3007 \\ 0.2047 & 0.5151 & 1.7113 & 3.1982 \\ 0.1328 & 0.5381 & 0.8275 & 3.3572 \end{bmatrix},$$

$$\mathbf{H}_{\mathrm{dd}} = \begin{bmatrix} 0.2002 & 0.4335 & 0.0883 & 0.1381 \\ 0.0091 & 0.3871 & 0.1221 & 0.2023 \\ 0.0318 & 0.0801 & 0.2662 & 0.4975 \\ 0.0207 & 0.0837 & 0.1287 & 0.5222 \end{bmatrix},$$

$$\mathbf{H}_{\mathrm{qq}} = \begin{bmatrix} -1.4526 & 0.8439 & 0.0260 & 0.0023 \\ 0.9487 & -1.7411 & 0.0645 & 0.0444 \\ 0.0864 & 0.1125 & -1.5397 & 0.6899 \\ 0.1365 & 0.1895 & 0.8408 & -1.9466 \end{bmatrix},$$

$$\mathbf{L}_{\mathrm{dd}} = \begin{bmatrix} 0.1043 \\ 0.2946 \\ 0.6485 \\ 1.6297 \end{bmatrix}, \quad \mathbf{L}_{\mathrm{qq}} = \begin{bmatrix} 0.0788 \\ 0.1519 \\ 0.1730 \\ 0.0638 \end{bmatrix}$$

From Eq. (6.10), it can have

$$\mathbf{F}_{\mathrm{d}} = \begin{bmatrix} 6.1526 & -4.9855 & -0.7232 & -0.4439 \\ -4.4148 & 6.6100 & -1.1952 & -1.0000 \\ -0.9953 & -0.9652 & 6.4009 & -4.4404 \\ -1.0862 & -1.2197 & -4.5385 & 6.8444 \end{bmatrix},$$

$$\mathbf{F}_{\mathrm{q}} = \begin{bmatrix} 0.6169 & -0.9820 & 0.0895 & 0.2756 \\ -2.0932 & 1.7871 & 0.0404 & 0.2658 \\ -0.3695 & -0.2106 & 0.9843 & -0.4041 \\ -0.4845 & -0.3115 & -1.3867 & 2.1827 \end{bmatrix},$$

$$\mathbf{G}_{\mathrm{d}} = \begin{bmatrix} 9.5940 & -4.3354 & 0.2644 & 0.8437 \\ -5.9571 & 11.9432 & -0.0136 & 0.5853 \\ -0.4126 & -0.1860 & 10.4629 & -3.0182 \\ -0.7581 & -0.7453 & -4.9475 & 13.7236 \end{bmatrix},$$

$$\mathbf{G}_{\mathrm{q}} = \begin{bmatrix} 0.6082 & 1.6579 & 0.3971 & 0.6468 \\ 0.2375 & 1.5002 & 0.4534 & 0.7684 \\ 0.1473 & 0.4578 & 0.8635 & 1.7847 \\ 0.1306 & 0.4865 & 0.5748 & 1.7280 \end{bmatrix},$$

$$\mathbf{L}_{\mathrm{d}} = \begin{bmatrix} 0.1662 \\ 0.3477 \\ 0.7077 \\ 1.6776 \end{bmatrix}, \ \mathbf{L}_{\mathrm{q}} = \begin{bmatrix} 0.0604 \\ 0.0793 \\ 0.0886 \\ 0.0548 \end{bmatrix}$$

Substituting into Eqs. (5.40) and (5.41), it can have

$$\begin{aligned}
\mathbf{K}_1 &= \left(\mathbf{V}_{\mathbf{td}0} - \mathbf{I}_{\mathrm{q}0}\mathbf{X}'_{\mathbf{d}}\right)\mathbf{F}_{\mathbf{d}} + \left(\mathbf{V}_{\mathbf{tq}0} - \mathbf{I}_{\mathrm{d}0}\mathbf{X}_{\mathbf{q}}\right)\mathbf{F}_{\mathbf{q}}, \\
\mathbf{K}_2 &= \mathbf{I}_{\mathrm{q}0} + \left(\mathbf{V}_{\mathbf{td}0} - \mathbf{I}_{\mathrm{q}0}\mathbf{X}'_{\mathbf{d}}\right)\mathbf{G}_{\mathbf{d}} + \left(\mathbf{V}_{\mathbf{tq}0} - \mathbf{I}_{\mathrm{d}0}\mathbf{X}_{\mathbf{q}}\right)\mathbf{G}_{\mathbf{q}}, \\
\mathbf{K}_{\mathrm{p}} &= -\left(\left(\mathbf{V}_{\mathrm{td}0} - \mathbf{I}_{\mathrm{qo}}\mathbf{X}'_{\mathrm{d}}\right)\mathbf{L}_{\mathrm{d}} + \left(\mathbf{V}_{\mathrm{td}0} - \mathbf{I}_{\mathrm{do}}\mathbf{X}'_{\mathrm{q}}\right)\mathbf{L}_{\mathrm{q}}\right), \\
\mathbf{K}_3 &= \mathbf{I} + \left(\mathbf{X}_{\mathbf{d}} - \mathbf{X}'_{\mathbf{d}}\right)\mathbf{G}_{\mathbf{d}}, \\
\mathbf{K}_4 &= \left(\mathbf{X}_{\mathbf{d}} - \mathbf{X}'_{\mathbf{d}}\right)\mathbf{F}_{\mathbf{d}}, \\
\mathbf{K}_{\mathrm{q}} &= \left(\mathbf{X}_{\mathbf{d}} - \mathbf{X}'_{\mathbf{d}}\right)\mathbf{L}_{\mathbf{d}}, \\
\mathbf{K}_5 &= -\mathbf{V}_{\mathbf{t}0}^{-1}\mathbf{V}_{\mathrm{q}0}\mathbf{X}'_{\mathbf{d}}\mathbf{F}_{\mathbf{d}} + \mathbf{V}_{\mathrm{t}0}^{-1}\mathbf{V}_{\mathrm{d}0}\mathbf{X}_{\mathbf{q}}\mathbf{F}_{\mathbf{q}}, \\
\mathbf{K}_6 &= \mathbf{V}_{\mathbf{t}0}^{-1}\mathbf{V}_{\mathrm{q}0} - \mathbf{V}_{\mathrm{t}0}^{-1}\mathbf{V}_{\mathrm{q}0}\mathbf{X}'_{\mathbf{d}}\mathbf{G}_{\mathbf{d}} + \mathbf{V}_{\mathrm{t}0}^{-1}\mathbf{V}_{\mathrm{d}0}\mathbf{X}_{\mathrm{q}}\mathbf{G}_{\mathrm{q}}, \\
\mathbf{K}_{\mathrm{v}} &= \mathbf{V}_{\mathbf{t}0}^{-1}\mathbf{V}_{\mathrm{q}0}\mathbf{X}'_{\mathrm{d}}\mathbf{L}_{\mathrm{d}} - \mathbf{V}_{\mathrm{t}0}^{-1}\mathbf{V}_{\mathrm{d}0}\mathbf{X}_{\mathrm{q}}\mathbf{L}_{\mathrm{q}}
\end{aligned}$$

$$\mathbf{K}_1 = \begin{bmatrix} 4.6390 & -4.6990 & -0.2294 & 0.2894 \\ -6.7233 & 7.1872 & -0.5225 & 0.0586 \\ -1.2956 & -0.9680 & 5.6138 & -3.3503 \\ -1.5697 & -1.2865 & -5.2531 & 8.1092 \end{bmatrix},$$

$$\mathbf{K}_2 = \begin{bmatrix} 10.1342 & 0.8111 & 0.9219 & 1.7323 \\ -2.5271 & 12.5659 & 0.9623 & 1.9405 \\ 0.0495 & 0.7864 & 11.4295 & 1.7433 \\ -0.1357 & 0.6002 & -1.4532 & 14.2532 \end{bmatrix},$$

$$\mathbf{K}_3 = \begin{bmatrix} 2.5990 & -0.7226 & 0.0441 & 0.1406 \\ -0.9929 & 2.9905 & -0.0023 & 0.0976 \\ -0.0688 & -0.0310 & 2.7438 & -0.5030 \\ -0.1263 & -0.1242 & -0.8246 & 3.2873 \end{bmatrix},$$

$$\mathbf{K}_4 = \begin{bmatrix} 1.0254 & -0.8309 & -0.1205 & -0.0740 \\ -0.7358 & 1.1017 & -0.1992 & -0.1667 \\ -0.1659 & -0.1609 & 1.0668 & -0.7401 \\ -0.1810 & -0.2033 & -0.7564 & 1.1407 \end{bmatrix},$$

$$\mathbf{K}_5 = \begin{bmatrix} -0.0776 & 0.0029 & 0.0293 & 0.0454 \\ -0.1258 & 0.0326 & 0.0362 & 0.0571 \\ -0.0231 & -0.0035 & -0.0300 & 0.0566 \\ -0.0306 & -0.0063 & -0.0511 & 0.0881 \end{bmatrix},$$

$$\mathbf{K}_6 = \begin{bmatrix} 0.5876 & 0.3150 & 0.0428 & 0.0593 \\ 0.1844 & 0.6494 & 0.0528 & 0.0735 \\ 0.0290 & 0.0633 & 0.5884 & 0.3030 \\ 0.0352 & 0.0778 & 0.1962 & 0.6259 \end{bmatrix},$$

$$\mathbf{K}_\mathrm{P} = \begin{bmatrix} -0.2106 \\ -0.3466 \\ -0.5811 \\ -1.0042 \end{bmatrix}, \mathbf{K}_\mathrm{q} = \begin{bmatrix} -0.0277 \\ -0.0580 \\ -0.1179 \\ -0.2796 \end{bmatrix}, \mathbf{K}_\mathrm{v} = \begin{bmatrix} -0.0033 \\ -0.0000 \\ 0.0059 \\ 0.0364 \end{bmatrix}.$$

## 6.3.2 *Selection of Installing Locations of Stabilizers*

### 6.3.2.1 Selection of the PSS Installing Locations

From the linearized Heffron–Phillips model of the N-machine power system, eigenvalues of the state matrix are calculated as

$$\begin{aligned}
\bar{\lambda}_{1,2} &= -5.1604 \pm j6.0402 \\
\bar{\lambda}_{3,4} &= -5.3163 \pm j4.9927 \\
\bar{\lambda}_{5,6} &= -0.1979 \pm j6.1696 \\
\bar{\lambda}_{7,8} &= -0.1707 \pm j6.3459 \\
\bar{\lambda}_{9,10} &= 0.1641 \pm j3.2658 \\
\bar{\lambda}_{11,12} &= -5.8428 \pm j0.0805 \\
\bar{\lambda}_{13,14} &= -4.1998 \pm j0.0862 \\
\bar{\lambda}_{15,16} &= 0 \pm j0
\end{aligned}$$

According to Eqs. (5.60) and (5.63), the correlation ratio of electromechanical loop can be calculated to identify the electromechanical oscillation modes of the example power system as

$$\begin{aligned}
\sigma_{1,2} &= 0.0080 \\
\sigma_{3,4} &= 0.0503 \\
&\left.\begin{aligned}
\sigma_{5,6} &= 23.5268 \\
\sigma_{7,8} &= 27.3818 \\
\sigma_{9,10} &= 17.0246
\end{aligned}\right\} \quad \text{electromechanical modes} \\
\sigma_{11,12} &= 0.1747 \\
\sigma_{13,14} &= 0.1658
\end{aligned}$$

Hence, three electromechanical oscillation modes of the example power system are

$$\begin{aligned}
\bar{\lambda}_{5,6} &= -0.1979 \pm j6.1696 \\
\bar{\lambda}_{7,8} &= -0.1707 \pm j6.3459 \\
\bar{\lambda}_{9,10} &= 0.1641 \pm j3.2658
\end{aligned}$$

Damp ratio of inter-area oscillation mode $\bar{\lambda}_{9,10}$ is −0.0437. Right and left eigenvectors corresponding to $\bar{\lambda}_9 = 0.1641 + j3.2658$ are calculated as

$$\bar{\mathbf{v}}_9 = \begin{bmatrix} 0.0352 - j0.0797 \\ 0.0312 - j0.0629 \\ -0.1850 + j0.0850 \\ -0.1683 + j0.0753 \\ 0.0007 + j0.0003 \\ 0.0006 + j0.0002 \\ -0.0008 - j0.0016 \\ -0.0007 - j0.0014 \\ 0.0087 - j0.0165 \\ 0.0050 - j0.0249 \\ 0.0052 - j0.0066 \\ 0.0043 - j0.0093 \\ 0.5060 + j0.1325 \\ 0.7359 \\ 0.1030 + j0.1762 \\ 0.1964 + j0.1248 \end{bmatrix}, \quad \bar{\mathbf{w}}_9 = \begin{bmatrix} 0.7994 + j0.5947 \\ 0.7288 + j0.4218 \\ -0.7969 - j0.6005 \\ -0.7313 - j0.4160 \\ 73.0984 - j88.6036 \\ 52.7805 - j81.4813 \\ -73.7524 + j88.2869 \\ -52.1265 + j81.7980 \\ -0.5131 + j1.5515 \\ -0.3709 + j1.8021 \\ 0.5572 - j1.6701 \\ 0.4024 - j1.9651 \\ -0.0002 + j0.0191 \\ 0.0023 + j0.0214 \\ 0.0002 - j0.0206 \\ -0.0026 - j0.0233 \end{bmatrix}$$

1. Selection of installing locations based on the participation factor

From Eq. (6.140), the participation factor of each generator can be calculated to examine how much each generator is involved in the oscillation modes. The results are given in Table 6.6.

For $\bar{\lambda}_9 = 0.1641 + j3.2658$, it can be seen that generator 3 is the best installing location.

2. Selection of installing locations by damping torque analysis

The linearized Heffron–Phillips model for the PSSs to be installed is given in the above section.

$$\begin{aligned} s\boldsymbol{\Delta\delta} &= \boldsymbol{\omega_o I \Delta\omega} \\ s\boldsymbol{\Delta\omega} &= \mathbf{M}^{-1}(-\mathbf{K}_1\boldsymbol{\Delta\delta} - \mathbf{K}_2\boldsymbol{\Delta}\mathbf{E}'_q - \mathbf{D}\Delta\boldsymbol{\omega}) \\ s\boldsymbol{\Delta}\mathbf{E}'_q &= \mathbf{T}'^{-1}_{d0}(-\mathbf{K}_3\boldsymbol{\Delta}\mathbf{E}'_q - \mathbf{K}_4\boldsymbol{\Delta}\bar{\boldsymbol{\delta}} + \boldsymbol{\Delta}\mathbf{E}'_{fd}) \\ s\boldsymbol{\Delta}\mathbf{E}'_{fd} &= -\mathbf{T}_A^{-1}\boldsymbol{\Delta}\mathbf{E}'_{fd} + \mathbf{T}_A^{-1}\mathbf{K}_A(-\mathbf{K}_5\boldsymbol{\Delta\delta} - \mathbf{K}_6\boldsymbol{\Delta}\mathbf{E}'_q + \mathbf{F}_k\Delta u_{pssk}) \end{aligned}$$

**Table 6.6** Participation factor of each generator

| Generator | Electromechanical oscillation modes | | |
|---|---|---|---|
| | −0.1979 ± j6.1696 | −0.1707 ± j6.3459 | 0.1641 ± j3.2658 |
| 1 | 0.2269 | 0.0000 | 0.0868 |
| 2 | 0.2843 | 0.0031 | 0.0591 |
| 3 | 0.0017 | 0.2053 | **0.2032** |
| 4 | 0.0010 | 0.3062 | 0.1551 |

From Eq. (6.145), the DTA can be used to select installing location of the PSSs. If the PSS is installed in generator 3, $\mathbf{F}_3 = [0 \quad 0 \quad 1 \quad 0]^{\mathbf{T}}$. From Eq. (6.115), it can have with $s = \bar{\lambda}_9 = 0.1641 + j3.2658$.

$$\mathbf{F_{PSS}}(\bar{\lambda}_9) = \frac{\mathbf{\Delta T_{PSS}}}{\Delta u_{pss3}} = \mathbf{M}^{-1}\mathbf{K}_2[(\mathbf{I} + s\mathbf{T_A})(\mathbf{K}_3 + s\mathbf{T_{d0}}) + \mathbf{K_A K_6}]^{-1}\mathbf{K_A F_3}$$
$$= \begin{bmatrix} -0.0365 & -j0.0099 \\ -0.0401 & -j0.0102 \\ 0.1147 & -j0.0109 \\ -0.1129 & -j0.0107 \end{bmatrix}$$

For $\bar{\lambda}_9$, the elements of right eigenvector corresponding to $\Delta\omega_j, j = 1, 2, 3, 4$ are

$$\begin{cases} \bar{v}_{95} = 0.0007 + j0.0003 \\ \bar{v}_{96} = 0.0006 + j0.0002 \\ \bar{v}_{97} = -0.0008 - j0.0016 \\ \bar{v}_{98} = -0.0007 - j0.0014 \end{cases}$$

Let the feedback signal be $\Delta\omega_3$, and thus,

$$\Delta\omega_3 = \Delta y = \mathbf{c}_3^{\mathbf{T}}\mathbf{X}$$

where the 7th element of $\mathbf{C}_3$ is 1 and all other elements are zero. From Eq. (6.154), it can have

$$\begin{cases} \bar{\gamma}_1(\bar{\lambda}_9) = \dfrac{\mathbf{c}_3^{\mathbf{T}}\bar{\mathbf{v}}_9}{\bar{v}_{95}} = \dfrac{\bar{v}_{97}}{\bar{v}_{95}} = -1.7522 - j1.5478 \\ \bar{\gamma}_2(\bar{\lambda}_9) = \dfrac{\mathbf{c}_3^{\mathbf{T}}\bar{\mathbf{v}}_9}{\bar{v}_{96}} = \dfrac{\bar{v}_{97}}{\bar{v}_{96}} = -2.2552 - j1.8217 \\ \bar{\gamma}_3(\bar{\lambda}_9) = \dfrac{\mathbf{c}_3^{\mathbf{T}}\bar{\mathbf{v}}_9}{\bar{v}_{97}} = \dfrac{\bar{v}_{97}}{\bar{v}_{97}} = 1 \\ \bar{\gamma}_4(\bar{\lambda}_9) = \dfrac{\mathbf{c}_3^{\mathbf{T}}\bar{\mathbf{v}}_9}{\bar{v}_{98}} = \dfrac{\bar{v}_{97}}{\bar{v}_{98}} = 1.1042 - j0.0115 \end{cases}$$

From Eq. (6.154), sensitive index $\mathbf{S_{ij}}$ can be calculated as

$$\begin{cases} S_{91} = \bar{w}_{291}\bar{v}_{291} = 0.0755 - j0.0427 \\ S_{92} = \bar{w}_{292}\bar{v}_{292} = 0.0493 - j0.0327 \\ S_{93} = \bar{w}_{293}\bar{v}_{293} = 0.1985 + j0.0433 \\ S_{94} = \bar{w}_{294}\bar{v}_{294} = 0.1544 + j0.0150 \end{cases} \tag{6.187}$$

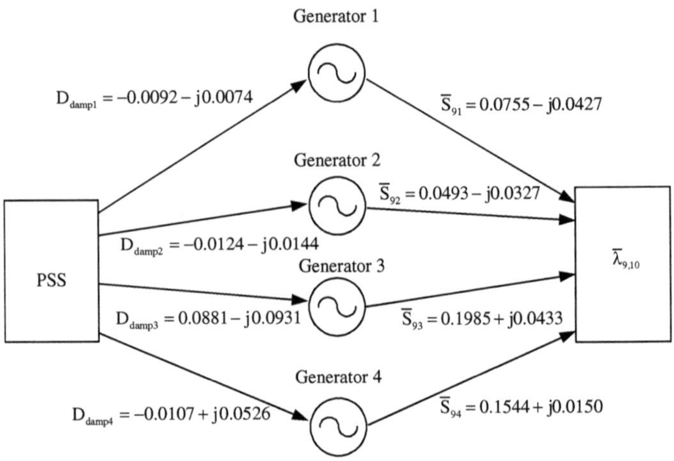


**Fig. 6.18** The PSS to be installed on G3 to affect the inter-area oscillation mode

According to Fig. 6.14 and Eq. (6.145), the DTA can be calculated as

$$\mathrm{DTA}_3 = \sum_{j=1}^{4} \left| \bar{\mathbf{F}}_{\mathbf{pssj}}(\bar{\lambda}_9)\bar{\gamma}_j(\bar{\lambda}_9)\bar{\mathbf{S}}_{9j} \right| = 0.0576$$

where

$$\bar{\mathbf{F}}_{\mathrm{pssj}}(\bar{\lambda}_9)\bar{\gamma}_j(\bar{\lambda}_9) = \begin{bmatrix} -0.0092 & - & j0.0074 \\ -0.0124 & - & j0.0144 \\ 0.0881 & - & j0.0931 \\ -0.0107 & + & j0.0526 \end{bmatrix}$$

Figure 6.18 shows how the PSS to be installed on generator 3 will affect the inter-area oscillation mode. When the PSS is installed on generators 1, 2, or 4, the DTA can be calculated similarly as

$$\mathrm{DTA}_1 = 0.0150, \mathrm{DTA}_2 = 0.0114, \mathrm{DTA}_4 = 0.0354$$

Obviously, generator 3 is the best installing location.

#### 6.3.2.2 Selection of Installing Location of the SVC Stabilizer

From the Heffron–Phillips model of the example power system installed with the SVC, eigenvalues of the state matrix are calculated as

$$\left.\begin{aligned}\bar{\lambda}_{1,2} &= -0.1960 \pm j6.1676\\ \bar{\lambda}_{3,4} &= -0.1607 \pm j6.3093\end{aligned}\right\} \text{electromechanical modes}$$

$$\bar{\lambda}_{5,6} = -5.5111 \pm j5.6757$$

$$\bar{\lambda}_{7,8} = 0.2433 \pm j3.1216 \quad \text{electromechanical mode}$$

$$\bar{\lambda}_{9,10} = -5.9764 \pm j4.4089$$

$$\bar{\lambda}_{11} = -5.8189$$

$$\bar{\lambda}_{12} = -5.5025$$

$$\bar{\lambda}_{13} = -4.9368$$

$$\bar{\lambda}_{14} = -4.1875$$

$$\bar{\lambda}_{15} = -2.4164$$

$$\bar{\lambda}_{16} = 0$$

$$\bar{\lambda}_{17} = 0$$

An SVC stabilizer is to be installed to improve the damping of the inter-area oscillation mode $\bar{\lambda}_7 = 0.2433 + j3.1216$. Right and left eigenvectors corresponding to $\bar{\lambda}_7 = 0.2433 + j3.1216$ are calculated as

$$\bar{\mathbf{v}}_7 = \begin{bmatrix} 0.0624 + j0.0911\\ 0.0563 - j0.0725\\ -0.1749 + j0.0796\\ -0.1611 + j0.0693\\ 0.0008 + j0.0005\\ 0.0006 + j0.0004\\ -0.0008 - j0.0014\\ -0.0007 - 0.0013i\\ 0.0097 - j0.0170\\ 0.0064 - j0.0265\\ 0.0029 - j0.0014\\ 0.0032 - j0.0021\\ 0.5220 + j0.1297\\ 0.7683 + j0.0000\\ -0.0338 + j0.1256\\ 0.0012 + j0.1189\\ -0.0005 + j0.0008 \end{bmatrix}, \quad \bar{\mathbf{w}}_7 = \begin{bmatrix} 0.7677 + j0.5166\\ 0.6952 + j0.3532\\ -0.7436 - j0.4640\\ -0.7193 - j0.4058\\ 69.2025 - j87.3225\\ 48.8981 - j80.1477\\ -62.6622 + j84.9249\\ -55.4385 + j82.5452\\ -0.5691 + j1.5609\\ -0.4629 + j1.9346\\ 0.1767 - j0.8646\\ 0.1625 - j0.8590\\ -0.0010 + j0.0194\\ 0.0014 + j0.0232\\ -0.0010 - j0.0103\\ -0.0011 - j0.0101\\ -3.5217 + j14.0586 \end{bmatrix}$$

From Eqs. (6.36) and (6.37), it can have

$$\begin{cases} \Delta\alpha_{vi} = T_{vsvci}(s)\Delta V_i\\ \Delta b_{svci} = -\frac{1-\cos 2\alpha_{i0}}{\pi X_{svcli}}(\Delta\alpha_{vi} + \Delta\alpha_{si}) \end{cases}$$

Let the transfer function of the SVC voltage controller be

$$T_{vsvci}(s) = K_{vp} + \frac{K_{vi}}{s}$$

where $K_{vp} = 1, K_{vi} = 8$. Let $X_{VSVCi} = \frac{\Delta V_i}{s}$. It can have

$$\begin{cases} \dot{X}_{vsvci} = \Delta V_i \\ \Delta\alpha_{vi} = K_{vi}X_{vsvci} + K_{vp}\Delta V_i \end{cases}$$

According to Eq. (6.42), it can have

$$\begin{cases} \mathbf{A}_{vsvci} = 0, \mathbf{B}_{vsvci} = \begin{bmatrix} \frac{V_{ix0}}{V_{i0}} & \frac{V_{iy0}}{V_{i0}} \end{bmatrix} \\ \mathbf{c}_{vsvci} = K_{vi}, D_{vsvci} = \begin{bmatrix} \frac{V_{ix0}}{V_{i0}} & \frac{V_{iy0}}{V_{i0}} \end{bmatrix} \end{cases}$$

Hence, from Eq. (6.48), it can be obtained that

$$\mathbf{b}_{ssvci} = -\frac{1 - \cos 2\alpha_{i0}}{\pi x_{svcli}} \begin{bmatrix} v_{yi0} \\ -v_{xi0} \end{bmatrix} = \begin{bmatrix} 26.0361 \\ 49.5981 \end{bmatrix}$$

Substituting the above into Eq. (6.103), it can have

$$\mathbf{B} = \begin{bmatrix} \mathbf{0} \\ \mathbf{B}_{2k} \\ \mathbf{B}_{3k} \end{bmatrix} = \begin{bmatrix} 0 \\ 0 \\ 0 \\ 0 \\ 0.0215 \\ 0.0284 \\ 0.0818 \\ 0.0905 \\ 0.0520 \\ 0.0627 \\ 0.0882 \\ 0.0372 \\ 3.0862 \\ 4.3028 \\ 143.3026 \\ 178.5977 \\ -0.5828 \end{bmatrix}$$

The feedback signal of the SVC stabilizer is taken to be the active power, $P_{ij}$, delivered through the node where the SVC is to be installed. Linearization of $P_{ij} = \frac{V_iV_j}{X}\sin(\varphi_i - \varphi_j) = \frac{V_iV_j}{X}\sin\varphi_{ij}$ gives

$$\Delta P_{ij} = \frac{V_i V_j}{X}\sin\varphi_{ij0}\Delta V_i + \frac{V_{i0}V_{j0}}{X}\sin\varphi_{ij0}\Delta V_j + \frac{V_{i0}V_{j0}}{X}\cos\varphi_{ij0}\Delta\varphi_i - \frac{V_{i0}V_{j0}}{X}\cos\varphi_{ij0}\Delta\varphi_j$$

where X is the line reactance between nodes i and j. In the common x–y coordinate, $\Delta V_k$ and $\Delta\varphi_k, k = i, j$ can be written as

$$\Delta V_k = \frac{V_{xk0}}{V_{k0}}\Delta V_{xk} + \frac{V_{yk0}}{V_{k0}}\Delta V_{yk}$$
$$\Delta\varphi_k = \frac{-V_{yk0}}{V_{k0}^2}\Delta V_x + \frac{V_{xk0}}{V_{k0}^2}\Delta V_{yk}$$

Hence,

$$\begin{aligned}\Delta P_{ij} = {} & \frac{V_{i0}V_{j0}}{X}\sin\varphi_{ij0}\frac{V_{xi0}}{V_{i0}}\Delta V_{xi} + \frac{V_{i0}V_{j0}}{X}\sin\varphi_{ij0}\frac{V_{yi0}}{V_{i0}}\Delta V_{yi} \\ & + \frac{V_{i0}V_{j0}}{X}\sin\varphi_{ij0}\frac{V_{xj0}}{V_{j0}}\Delta V_{xj} + \frac{V_{i0}V_{j0}}{X}\sin\varphi_{ij0}\frac{V_{yj0}}{V_{j0}}\Delta V_{yj} \\ & + \frac{V_{i0}V_{j0}}{X}\cos\varphi_{ij0}\frac{-V_{yi0}}{V_{i0}^2}\Delta V_{xi} + \frac{V_{i0}V_{j0}}{X}\cos\varphi_{ij0}\frac{V_{xi0}}{V_{i0}^2}\Delta V_{yi} \\ & - \frac{V_{i0}V_{j0}}{X}\cos\varphi_{ij0}\frac{-V_{yj0}}{V_{j0}^2}\Delta V_{xj} - \frac{V_{i0}V_{j0}}{X}\cos\varphi_{ij0}\frac{V_{xj0}}{V_{j0}^2}\Delta V_{yj}\end{aligned}$$

which can be written as

$$\Delta P_{ij} = a_{11}\Delta V_{xi} + a_{12}\Delta V_{yi} + a_{21}\Delta V_{xj} + a_{22}\Delta V_{yj} \tag{6.188}$$

where

$$\begin{cases} a_{11} = \dfrac{V_{i0}V_{j0}}{X}\sin\varphi_{ij0}\dfrac{V_{xi0}}{V_{i0}} - \dfrac{V_{i0}V_{j0}}{X}\cos\varphi_{ij0}\dfrac{V_{yi0}}{V_{i0}^2} \\ a_{12} = \dfrac{V_{i0}V_{j0}}{X}\sin\varphi_{ij0}\dfrac{V_{yi0}}{V_{i0}} + \dfrac{V_{i0}V_{j0}}{X}\cos\varphi_{ij0}\dfrac{V_{xi0}}{V_{i0}^2} \\ a_{21} = \dfrac{V_{i0}V_{j0}}{X}\sin\varphi_{ij0}\dfrac{V_{xj0}}{V_{j0}} + \dfrac{V_{i0}V_{j0}}{X}\cos\varphi_{ij0}\dfrac{V_{yj0}}{V_{j0}^2} \\ a_{22} = \dfrac{V_{i0}V_{j0}}{X}\sin\varphi_{ij0}\dfrac{V_{yj0}}{V_{j0}} - \dfrac{V_{i0}V_{j0}}{X}\cos\varphi_{ij0}\dfrac{V_{xj0}}{V_{j0}^2} \end{cases}$$

From Eq. (6.101), it can have

$$\mathbf{\Delta V} = \mathbf{Y}_{\mathbf{M}}^{-1}\mathbf{C}_{\mathrm{xy}}\mathbf{\Delta X}$$

Hence, the voltage at any node can always be reconstructed by state variables, that is

$$\begin{aligned} \Delta V_{xi} &= \mathbf{Y_{ix}}^{\mathbf{T}}\mathbf{\Delta X} \\ \Delta V_{yi} &= \mathbf{Y_{iy}}^{\mathbf{T}}\mathbf{\Delta X} \\ \Delta V_{xj} &= \mathbf{Y_{jx}}^{\mathbf{T}}\mathbf{\Delta X} \\ \Delta V_{yj} &= \mathbf{Y_{jy}}^{\mathbf{T}}\mathbf{\Delta X} \end{aligned} \tag{6.189}$$

Because $\Delta y = \Delta P_{ij} = \mathbf{c}_{\mathbf{k}}^{T}\begin{bmatrix} \Delta\delta \\ \Delta\omega \\ \Delta Z \end{bmatrix} = c_k^T\mathbf{\Delta X}$, from Eqs. (6.188) and (6.189) above, it can have

$$\mathbf{c_k} = a_{11}\mathbf{Y}_{ix} + a_{12}\mathbf{Y}_{iy} + a_{21}\mathbf{Y}_{jx} + a_{22}\mathbf{Y}_{jy} \tag{6.190}$$

For the example power system, when the SVC stabilizer is to be installed at node 9, the feedback signal is taken to be $P_{89}$, the active power delivered through node 8 to node 9. From the results of power flow calculation, it can have

$$V_{80} = 0.8636,\ V_{90} = 0.8799,\ \varphi_{80} = -0.2001,\ \varphi_{90} = -0.4834$$

Thus,

$$a_{11} = 0.3999, a_{12} = 0.8355, a_{21} = -0.2098, a_{22} = -0.8856$$

According to Eq. (6.101), for the example power system, it has

$$\mathbf{Y}_{\mathbf{M}}^{-1} = \begin{bmatrix} \mathbf{Y}_{\mathbf{M}}^{-1}{}_{-1} & \mathbf{Y}_{\mathbf{M}}^{-1}{}_{-2} \end{bmatrix}$$

$$\mathbf{Y}_{\mathbf{M}-1}^{-1} = \begin{bmatrix} 7.8484 & 46.0583 & 0.9779 & -15.1586 & 0 & 0 \\ -24.9097 & -4.9233 & 15.1586 & 0.9779 & 0 & 0 \\ 0.9779 & -15.1586 & 15.5048 & 45.9152 & 0 & 0 \\ 15.1586 & 0.9779 & -33.8809 & -6.0720 & 0 & 0 \\ 0 & 0 & 0 & 0 & 13.3470 & 35.9030 \\ 0 & 0 & 0 & 0 & -40.9433 & -10.8393 \\ 0 & 0 & 0 & 0 & 0.0156 & -7.8445 \\ 0 & 0 & 0 & 0 & 7.8445 & 0.0156 \\ 0.1448 & -2.0653 & 0.6714 & -5.1397 & -1.2690 & -12.9499 \\ 2.0653 & 0.1448 & 5.1397 & 0.6714 & 12.9499 & -1.2690 \end{bmatrix}$$

$$\mathbf{Y}_{\mathbf{M}-2}^{-1} = \begin{bmatrix} 0 & 0 & 0.1448 & -2.0653 \\ 0 & 0 & 2.0653 & 0.1448 \\ 0 & 0 & 0.6714 & -5.1397 \\ 0 & 0 & 5.1397 & 0.6714 \\ 0.0156 & -7.8445 & -1.2690 & -12.9499 \\ 7.8445 & 0.0156 & 12.9499 & -1.2690 \\ 10.1321 & 49.4841 & -1.2300 & -32.5540 \\ -66.5814 & -7.7021 & 32.5540 & -1.2300 \\ -1.2300 & -32.5540 & 3.9996 & 65.4289 \\ 32.5540 & -1.2300 & -98.4466 & 48.9889 \end{bmatrix}$$

From Eqs. (6.29) and (6.100), it can be obtained that

$$\mathbf{C}_{xy} = \begin{bmatrix} \mathbf{C}_{xy-1} & \mathbf{C}_{xy-2} \end{bmatrix}$$

$$\mathbf{C}_{xy-1} = \begin{bmatrix} 18.5288 & 0 & 28.8999 & 0 & 0 & 0 & 0 & 0 & 0 & 0 \\ 5.8704 & 0 & -8.0494 & 0 & 0 & 0 & 0 & 0 & 0 & 0 \\ 0 & 0 & 0 & 0 & 18.4483 & 0 & 25.8688 & 0 & 0 & 0 \\ 0 & 0 & 0 & 0 & 3.2065 & 0 & -15.1922 & 0 & 0 & 0 \\ 0 & 0 & 0 & 0 & 0 & 0 & 0 & 0 & 18.5570 & 0 \\ 0 & 0 & 0 & 0 & 0 & 0 & 0 & 0 & -5.6022 & 0 \\ 0 & 0 & 0 & 0 & 0 & 0 & 0 & 0 & 0 & 0 \\ 0 & 0 & 0 & 0 & 0 & 0 & 0 & 0 & 0 & 0 \\ 0 & 0 & 0 & 0 & 0 & 0 & 0 & 0 & 0 & 0 \\ 0 & 0 & 0 & 0 & 0 & 0 & 0 & 0 & 0 & 0 \end{bmatrix}$$

$$\mathbf{C}_{xy-2} = \begin{bmatrix} 0 & 0 & 0 & 0 & 0 & 0 & 0 \\ 0 & 0 & 0 & 0 & 0 & 0 & 0 \\ 0 & 0 & 0 & 0 & 0 & 0 & 0 \\ 0 & 0 & 0 & 0 & 0 & 0 & 0 \\ 18.9260 & 0 & 0 & 0 & 0 & 0 & 0 \\ -23.2767 & 0 & 0 & 0 & 0 & 0 & 0 \\ 0 & 0 & 16.9781 & 0 & 11.7722 & 0 & 0 \\ 0 & 0 & -7.6725 & 0 & -27.5938 & 0 & 0 \\ 0 & 0 & 0 & 0 & 0 & 0 & 201.7248 \\ 0 & 0 & 0 & 0 & 0 & 0 & 398.2669 \end{bmatrix}$$

Thus,

$$\mathbf{Y}_{\mathbf{M}}^{-1} C_{xy} = \begin{bmatrix} \mathbf{Y}_{\mathbf{M}}^{-1}\mathbf{C}_{xy-1} & \mathbf{Y}_{\mathbf{M}}^{-1}\mathbf{C}_{xy-2} & \mathbf{Y}_{\mathbf{M}}^{-1}\mathbf{C}_{xy-3} \end{bmatrix}$$

$$\mathbf{Y}_{\mathbf{M}}^{-1}\mathbf{C}_{\mathrm{xy}-1} = \begin{bmatrix} -0.4930 & 0 & 0.2838 & 0 & -0.1810 & 0 & 0.3392 & 0 \\ 0.5948 & 0 & 0.6333 & 0 & 0.2330 & 0 & 0.0626 & 0 \\ -0.2527 & 0 & 0.1235 & 0 & -0.2656 & 0 & 0.5422 & 0 \\ 0.3017 & 0 & 0.1663 & 0 & 0.5832 & 0 & 0.3967 & 0 \\ -0.0098 & 0 & 0.0107 & 0 & -0.0048 & 0 & 0.0249 & 0 \\ 0.0505 & 0 & 0.0233 & 0 & 0.0571 & 0 & 0.0103 & 0 \\ -0.0006 & 0 & 0.0192 & 0 & 0.0074 & 0 & 0.0343 & 0 \\ 0.0664 & 0 & 0.0341 & 0 & 0.0765 & 0 & 0.0197 & 0 \\ 0.0201 & 0 & 0.0466 & 0 & 0.0382 & 0 & 0.0701 & 0 \\ 0.0878 & 0 & 0.0505 & 0 & 0.1034 & 0 & 0.0355 & 0 \\ -0.3899 & 0 & 0.2505 & 0 & -0.1775 & 0 & 0.3858 & 0 \\ 0.5133 & 0 & 0.4797 & 0 & 0.3233 & 0 & 0.1282 & 0 \\ -0.2474 & 0 & 0.1774 & 0 & -0.1587 & 0 & 0.4654 & 0 \\ 0.3753 & 0 & 0.2541 & 0 & 0.4580 & 0 & 0.2195 & 0 \\ -0.1828 & 0 & 0.1857 & 0 & -0.0946 & 0 & 0.4404 & 0 \\ 0.3638 & 0 & 0.2132 & 0 & 0.4303 & 0 & 0.1542 & 0 \\ -0.0820 & 0 & 0.1175 & 0 & -0.0282 & 0 & 0.2581 & 0 \\ 0.2283 & 0 & 0.1331 & 0 & 0.2697 & 0 & 0.0956 & 0 \\ 0.0090 & 0 & 0.0326 & 0 & 0.0220 & 0 & 0.0522 & 0 \\ 0.0763 & 0 & 0.0422 & 0 & 0.0892 & 0 & 0.0278 & 0 \\ -0.0030 & 0 & 0.0190 & 0 & 0.0051 & 0 & 0.0354 & 0 \\ 0.0613 & 0 & 0.0314 & 0 & 0.0706 & 0 & 0.0179 & 0 \end{bmatrix}$$

$$\mathbf{Y}_{\mathbf{M}}^{-1}\mathbf{C}_{\mathbf{xy}-2} = \begin{bmatrix} 0.0341 & 0 & 0.0519 & 0 & 0.0538 & 0 \\ 0.0111 & 0 & -0.0061 & 0 & 0.0081 & 0 \\ 0.0378 & 0 & 0.0564 & 0 & 0.0591 & 0 \\ 0.0184 & 0 & -0.0075 & 0 & 0.0145 & 0 \\ -0.0050 & 0 & 0.5385 & 0 & 0.0196 & 0 \\ 0.6827 & 0 & 0.4218 & 0 & 0.2397 & 0 \\ 0.0268 & 0 & 0.1369 & 0 & 0.1462 & 0 \\ 0.2835 & 0 & 0.1461 & 0 & 0.5675 & 0 \\ 0.1330 & 0 & 0.1905 & 0 & 0.2048 & 0 \\ 0.2698 & 0 & 0.1561 & 0 & 0.3211 & 0 \\ 0.0393 & 0 & 0.0574 & 0 & 0.0610 & 0 \\ 0.0200 & 0 & -0.0035 & 0 & 0.0176 & 0 \\ 0.0484 & 0 & 0.0660 & 0 & 0.0732 & 0 \\ 0.0325 & 0 & -0.0003 & 0 & 0.0309 & 0 \\ 0.0593 & 0 & 0.0756 & 0 & 0.0875 & 0 \\ 0.0453 & 0 & 0.0047 & 0 & 0.0451 & 0 \\ 0.0973 & 0 & 0.1345 & 0 & 0.1479 & 0 \\ 0.1591 & 0 & 0.0811 & 0 & 0.1849 & 0 \\ 0.0806 & 0 & 0.2184 & 0 & 0.1665 & 0 \\ 0.3305 & 0 & 0.1886 & 0 & 0.3922 & 0 \\ 0.0376 & 0 & 0.4166 & 0 & 0.0745 & 0 \\ 0.5445 & 0 & 0.3213 & 0 & 0.3041 & 0 \end{bmatrix}$$

$$\mathbf{Y}_{\mathbf{M}}^{-1}\mathbf{C}_{\mathbf{xy}-3} = \begin{bmatrix} 0.0701 & 0 & -0.5669 \\ -0.0180 & 0 & 0.8487 \\ 0.0756 & 0 & -0.5768 \\ -0.0251 & 0 & 1.2875 \\ 0.1712 & 0 & -2.2984 \\ 0.0275 & 0 & 4.5241 \\ 0.5319 & 0 & -1.8674 \\ 0.2074 & 0 & 5.6191 \\ 0.2520 & 0 & -1.6762 \\ 0.1031 & 0 & 6.9322 \\ 0.0764 & 0 & -0.5434 \\ -0.0190 & 0 & 1.1876 \\ 0.0857 & 0 & -0.4571 \\ -0.0214 & 0 & 1.6917 \\ 0.0957 & 0 & -0.3264 \\ -0.0208 & 0 & 2.1281 \\ 0.1758 & 0 & -1.0071 \\ 0.0414 & 0 & 4.5797 \\ 0.3350 & 0 & -1.8476 \\ 0.1216 & 0 & 6.1937 \\ 0.2343 & 0 & -2.1590 \\ 0.0690 & 0 & 5.2016 \end{bmatrix}$$

Thus,

$$[\mathbf{Y}_{8x} \quad \mathbf{Y}_{8y} \quad \mathbf{Y}_{9x} \quad \mathbf{Y}_{9y}] = \begin{bmatrix} -0.0650 & 0.2336 & 0.0246 & 0.0886 \\ 0 & 0 & 0 & 0 \\ 0.1252 & 0.1235 & 0.0471 & 0.0467 \\ 0 & 0 & 0 & 0 \\ -0.0082 & 0.2696 & 0.0427 & 0.1022 \\ 0 & 0 & 0 & 0 \\ 0.2609 & 0.0767 & 0.0685 & 0.0288 \\ 0 & 0 & 0 & 0 \\ 0.0972 & 0.1612 & 0.1345 & 0.2724 \\ 0 & 0 & 0 & 0 \\ 0.1353 & 0.0838 & 0.1935 & 0.1609 \\ 0 & 0 & 0 & 0 \\ 0.1476 & 0.1818 & 0.2072 & 0.3158 \\ 0 & 0 & 0 & 0 \\ 0.1755 & 0.0396 & 0.2542 & 0.1013 \\ 0 & 0 & 0 & 0 \\ -0.9722 & 4.6129 & -1.6229 & 6.9396 \end{bmatrix}$$

Finally, according to Eq. (6.190), it is obtained that

$$\mathbf{c_k} = \begin{bmatrix} 0.0856 \\ 0 \\ 0.1020 \\ 0 \\ 0.1225 \\ 0 \\ 0.1285 \\ 0 \\ -0.0959 \\ 0 \\ -0.0589 \\ 0 \\ -0.1122 \\ 0 \\ -0.0398 \\ 0 \\ -2.3398 \end{bmatrix}$$

1. Selection of installing location by using damp torque analysis

From Eq. (6.127), it can have

$$\mathbf{B_k} = \mathrm{B_{2k}} + \mathbf{A_{23}}(\mathbf{sI} - \mathbf{A_{33}})^{-1}\mathbf{B_{3k}} = \begin{bmatrix} 0.0740 + \mathrm{j}0.0070 \\ 0.0102 + \mathrm{j}0.0092 \\ 0.0205 + \mathrm{j}0.0287 \\ 0.0206 + \mathrm{j}0.0374 \end{bmatrix}$$

From Eq. (6.154), it can be calculated that

$$\begin{cases} \bar{\gamma}_1(\bar{\lambda}_7) = \dfrac{\mathbf{c_k^T}\bar{\mathbf{v}}_7}{\bar{v}_{75}} = 25.8592 - \mathrm{j}63.8552 \\ \bar{\gamma}_2(\bar{\lambda}_7) = \dfrac{\mathbf{c_k^T}\bar{\mathbf{v}}_7}{\bar{v}_{76}} = 26.4325 - \mathrm{j}78.5172 \\ \bar{\gamma}_3(\bar{\lambda}_7) = \dfrac{\mathbf{c_k^T}\bar{\mathbf{v}}_7}{\bar{v}_{77}} = 6.2615 + \mathrm{j}39.1056 \\ \bar{\gamma}_4(\bar{\lambda}_7) = \dfrac{\mathbf{c_k^T}\bar{\mathbf{v}}_7}{\bar{v}_{78}} = 7.7533 + \mathrm{j}42.6894 \end{cases}$$

Thus, the following coefficients in Eq. (6.147) are obtained:

$$\bar{B}_{kj}(\bar{\lambda}_i)\bar{\gamma}_j(\bar{\lambda}_i) = \begin{bmatrix} 0.0500 - j0.0515 \\ 0.0474 - j0.0663 \\ -0.1877 + j0.1421 \\ -0.2250 + j0.1462 \end{bmatrix}$$

From Eq. (6.142), the sensitive index $S_{ij}$ can be calculated as

$$\begin{cases} S_{91} = \bar{w}_{291}^T \bar{v}_{291} = 0.0950 - j0.0377 \\ S_{92} = \bar{w}_{291}^T \bar{v}_{292} = 0.0648 - j0.0305 \\ S_{93} = \bar{w}_{291}^T \bar{v}_{293} = 0.1670 + j0.0220 \\ S_{94} = \bar{w}_{291}^T \bar{v}_{294} = 0.1440 + j0.0156 \end{cases}$$

From Eq. (6.147), it can have

$$DTA_9 = 0.3585$$

Figure 6.19 illustrates how the SVC stabilizer to be installed at node 9 contributes the damping to the inter-area oscillation mode.

When the installing location of SVC stabilizer is at node 7 and node 8, the following results can be obtained similarly:

$$DTA_7 = 0.2316, DTA_8 = 0.0478$$

Obviously, the best installing location for the SVC stabilizer is at node 9.

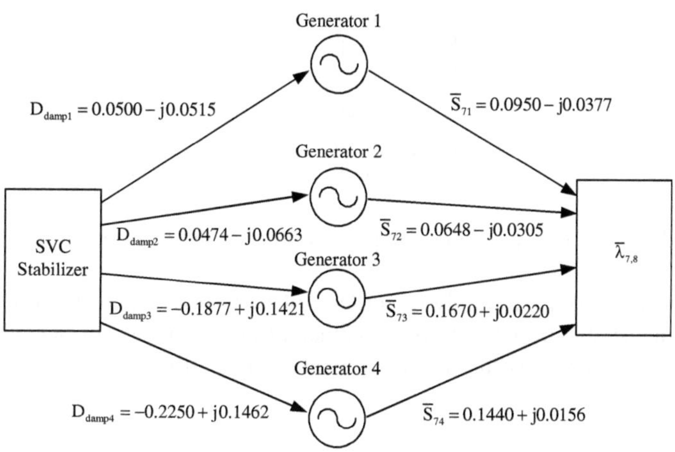


**Fig. 6.19** The SVC stabilizer to be installed at node 9 contributing damping to the inter-area oscillation mode

2. Selection of installing location based on the residue index

Substituting $\mathbf{B_k}$, $\mathbf{c_k}$, $\bar{\mathbf{V}}_9$ and $\bar{\mathbf{W}}_9^T$ into Eq. (6.149), it can have

$$\bar{R}_{ik} = \bar{b}_{ik}\bar{c}_{ik} = (-5.5927 - j0.9465)(0.0498 - j0.0389) = -0.3153 + j0.1705$$

thus, $|\bar{R}_{ik}| = 0.3585$, being equal to $DTA_9 = 0.3585$.
Residue index of node 7 and node 8 is calculated as

Node 7:

$$\bar{R}_{ik} = \bar{b}_{ik}\bar{c}_{ik} = (-1.3986 - j0.8958)(-0.0854 + j0.1103) = 0.2182 - 0.0777i$$
$$|\bar{R}_{ik}| = 0.2316$$

Node 8:

$$\bar{R}_{ik} = \bar{b}_{ik}\bar{c}_{ik} = (-0.4780 + j0.1553)(-0.0791 + j0.0528) = 0.0296 - j0.0375$$
$$|\bar{R}_{ik}| = 0.0478$$

They are equal to $DTA_7 = 0.2316$ and $DTA_8 = 0.0478$, respectively.

#### 6.3.2.3 Robust Installing Locations and Feedback Signals of the SVC Stabilizer

Let six candidate installing locations and feedback signals of the SVC stabilizer in the example power system be

$$\begin{cases} \varphi_1: & \text{node } 7 & \Delta P_{67} \\ \varphi_2: & \text{node } 7 & \Delta P_{78} \\ \varphi_3: & \text{node } 8 & \Delta P_{78} \\ \varphi_4: & \text{node } 8 & \Delta P_{89} \\ \varphi_5: & \text{node } 9 & \Delta P_{89} \\ \varphi_6: & \text{node } 9 & \Delta P_{910} \end{cases}$$

Consider $\Omega(\mu)$ the set of operating conditions of the example power system as given in Table 6.7.

Obviously, in order to apply the criteria of Eqs. (6.162) and (6.163), the effectiveness of the stabilizer has to be examined at ten given system operating conditions. Table 6.8 presents the results of residue calculation.

**Table 6.7** System operating conditions

| MW | Load at node 7 | Load at node 9 |
|---|---|---|
| $\mu_1$ | 927 + j100 | 1767 + j100 |
| $\mu_2$ | 937 + j100 | 1767 + j100 |
| $\mu_3$ | 947 + j100 | 1767 + j100 |
| $\mu_4$ | 957 + j100 | 1767 + j100 |
| $\mu_5$ | 967 + j100 | 1767 + j100 |
| $\mu_6$ | 977 + j100 | 1767 + j100 |
| $\mu_7$ | 987 + j100 | 1767 + j100 |
| $\mu_8$ | 997 + j100 | 1767 + j100 |
| $\mu_9$ | 1007 + j100 | 1767 + j100 |
| $\mu_{10}$ | 1017 + j100 | 1767 + j100 |

**Table 6.8** Residue ($C(\varphi, \mu)$)

| | $\varphi_1$ | $\varphi_2$ | $\varphi_3$ | $\varphi_4$ | $\varphi_5$ | $\varphi_6$ |
|---|---|---|---|---|---|---|
| $\mu_1$ | 0.2263 | 0.2240 | 0.0585 | 0.0534 | 0.3485 | 0.0311 |
| $\mu_2$ | 0.2283 | 0.2260 | 0.0559 | 0.0512 | 0.3509 | 0.0314 |
| $\mu_3$ | 0.2302 | 0.2279 | 0.0532 | 0.0489 | 0.3534 | 0.0317 |
| $\mu_4$ | 0.2319 | 0.2298 | 0.0505 | 0.0465 | 0.3559 | 0.0319 |
| $\mu_5$ | 0.2336 | 0.2316 | 0.0478 | 0.0442 | 0.3585 | 0.0322 |
| $\mu_6$ | 0.2353 | 0.2334 | 0.0451 | 0.0418 | 0.3610 | 0.0325 |
| $\mu_7$ | 0.2369 | 0.2351 | 0.0424 | 0.0394 | 0.3637 | 0.0327 |
| $\mu_8$ | 0.2384 | 0.2368 | 0.0397 | 0.0370 | 0.3663 | 0.0330 |
| $\mu_9$ | 0.2399 | 0.2384 | 0.0371 | 0.0346 | 0.3689 | 0.0333 |
| $\mu_{10}$ | 0.2413 | 0.2400 | 0.0344 | 0.0322 | 0.3716 | 0.0335 |

The following two indices are used to examine the effectiveness of the stabilizer.

$$\begin{cases} 1: \ \min\limits_{\mu} C(\varphi, \mu) \\ 2: \dfrac{\max\limits_{\mu} C(\varphi, \mu) - \min\limits_{\mu} C(\varphi, \mu)}{\bar{C}(\varphi, \mu)} \end{cases}$$

where $\bar{C}(\varphi, \mu)$ is the mean value of the residue ($C(\varphi, \mu)$) at all operating conditions. Two indices above are applied for $\varphi_1 \sim \varphi_6$, and the results of selection are given in Table 6.9.

In order to consider the robustness of the stabilizer to the variations of power system operating conditions, the criteria (6.161), (6.162), and (6.163) should be taken into account. That is

Table 6.9 Results of indices for $\varphi_1 \sim \varphi_6$

| | $\varphi_1$ | $\varphi_2$ | $\varphi_3$ | $\varphi_4$ | $\varphi_5$ | $\varphi_6$ |
|---|---|---|---|---|---|---|
| Index 1 | 0.2263 | 0.2240 | 0.0478 | 0.0442 | 0.3485 | 0.0311 |
| Index 2 | 0.0640 | 0.0689 | 0.5187 | 0.4939 | 0.0642 | 0.0742 |

$$\begin{cases} \underset{\phi}{\mathrm{Max}}[\underset{\mu}{\min} \mathrm{C}(\phi, \mu)] \\ \underset{\phi}{\mathrm{Min}}[\underset{\mu}{\max} \mathrm{C}(\phi, \mu) - \underset{\mu}{\min} \mathrm{C}(\phi, \mu)] \end{cases}$$

Obviously, from Table 6.9, the final selection should be $\varphi_5$. That is, the installing location is at node 9, and the feedback signal is $\Delta P_{89}$.

#### 6.3.2.4 Design of Robust SVC Stabilizer

According to the results obtained in the previous section, the SVC stabilizer is installed at node 9 and the integral of $P_{89}$ is used as the feedback signal. The stabilizer is designed at operating condition $\mu_1$, where it is estimated to be least effective. At $\mu_1$, the inter-area oscillation mode is $\lambda_7 = 0.2473 + j3.0340$. Right and left eigenvectors corresponding to $\bar{\lambda}_7 = 0.2473 + j3.0340$ are calculated as

$$\bar{\mathbf{V}}_9 = \begin{bmatrix} -0.0512 + j0.0923 \\ -0.0466 + j0.0745 \\ 0.1739 - j0.0788 \\ 0.1611 - j0.0682 \\ -0.0008 - j0.0003 \\ -0.0006 - j0.0003 \\ 0.0007 + j0.0013 \\ 0.0006 + j0.0012 \\ -0.0102 + j0.0178 \\ -0.0069 + j0.0278 \\ -0.0033 + j0.0024 \\ -0.0038 + j0.0026 \\ -0.5193 - j0.1312 \\ -0.7712 + j0.0000 \\ 0.0006 - j0.1244 \\ -0.0208 - j0.1235 \\ 0.0006 - j0.0010 \end{bmatrix}, \bar{\mathbf{W}}_9 = \begin{bmatrix} -0.7809 - j0.5431 \\ -0.7129 - j0.3833 \\ 0.7625 + j0.4873 \\ 0.7314 + j0.4390 \\ -75.8300 + j91.8156 \\ -55.1927 + j84.9901 \\ 68.6860 - j90.0933 \\ 62.3367 - j86.7124 \\ 0.6701 - j1.6183 \\ 0.5915 - j2.0830 \\ -0.2084 + j0.8511 \\ -0.2230 + j0.8620 \\ 0.0022 - j0.0204 \\ -0.0002 - j0.0253 \\ 0.0005 + j0.0102 \\ 0.0003 + j0.0104 \\ 4.2900 - j15.2077 \end{bmatrix}$$

Taking the same procedure of computation, it can have

$$
\mathbf{B}_k = \begin{bmatrix} 0 \\ 0 \\ 0 \\ 0 \\ 0.0199 \\ 0.0270 \\ 0.0793 \\ 0.0914 \\ 0.0520 \\ 0.0643 \\ 0.0821 \\ 0.0384 \\ -2.5208 \\ -2.6311 \\ 143.597 \\ 176.7661 \\ -0.5767 \end{bmatrix}, \mathbf{C}_k = \begin{bmatrix} 0.0773 \\ 0 \\ 0.1042 \\ 0 \\ 0.1159 \\ 0 \\ 0.1367 \\ 0 \\ -0.0881 \\ 0 \\ -0.0488 \\ 0 \\ -0.1051 \\ 0 \\ -0.0307 \\ 0 \\ -2.3522 \end{bmatrix}^T
$$

Substituting $\mathbf{B_k}$ $\mathbf{C_k}$ $\bar{\mathbf{V}}_9$ and $\bar{\mathbf{W}}_9^T$ into Eq. (6.149), the residue is calculated as

$$
\bar{R}_{ik} = \bar{b}_{ik}\bar{c}_{ik} = (5.8380 + j1.1365)(-0.0448 + j0.0378) = 0.3485\angle 150.8661^\circ
$$

Let the transfer function of the SVC stabilizer be

$$
T_{svc}(s) = K_{svc}\frac{sT_w}{1 + sT_w}\frac{1}{1 + sT}\frac{(1 + sT_2)(1 + sT_4)}{(1 + sT_1)(1 + sT_3)} = K_{svc}T_k(s)
$$

where $T_1 = 0.79s, T_3 = 0.79s, T = 0.01s, T_w = 10s$.

Firstly, $T_2$ and $T_4$ are set such that the SVC stabilizer to compensate the phase of the residue index to ensure the SVC stabilizer to move the oscillation mode horizontally towards the left on the complex plane. The result is

$$
T_2 = 0.2371s, T_4 = 0.2703s
$$

It gives

$$
T_k(\bar{\lambda}_7) = 1\angle - 150.8661^\circ
$$

To determine the gain value of the SVC stabilizer, $K_{svc}$, a direct searching method in nonlinear programming can be used to move $\bar{\lambda}_7$ to $\bar{\lambda}_7^*$ by minimizing the following objective function:

$$f(K) = \left|Re\left(\bar{\lambda}_7(K)\right) - Re\left(\bar{\lambda}_7^*(K^*)\right)\right|$$

where $Re\left(\bar{\lambda}_7^*(K^*)\right) = -0.45$. The steps of direct searching are as follows:

Step 1: Set the initial searching scale $S = 2$ and minimum searching scale $S_m = 0.0001$. Choose $K^0 = 50$ as the starting point of the searching. The oscillation mode is calculated as $\bar{\lambda}_7 = -0.2031 \pm j4.6030$. Thus, the initial value of objective function is $f(0) = 0.0610$.
Step 2: Search from $K^{(0)}$ to $K^{(0)} + S$, if $f(K^0 + S) < f(K^0)$ let $K^{(0)} \Leftarrow K^0 + S$ and execute Step 2 again; if not, search from $K^{(0)}$ to $K^{(0)} - S$, if $f(K^0 + S) < f(K^0)$ let $K^{(0)} \Leftarrow K^0 - S$ and execute Step 2 again; if not, Step 2 fails and go to Step 3.
Step 3: If searching in Step 2 fails, reduce the searching scale S by half and then go back to Step 2.
Step 4: If $f(K^0) < S_m$, stop searching and take $K^0$ as the solution $K^*$.

For the SVC stabilizer, in 36 steps, the searching stops to have $K^* = 112.5000$ with $f(K^0) = 0$. Thus, the design of the SVC stabilizer is completed. Parameters of SVC stabilizer are set to be

$$K_{svc} = 112.5000, T_2 = 0.2371\ s,\ T_4 = 0.2704\ s$$

With the SVC stabilizer installed, eigenvalues of the state matrix are calculated as

$$\begin{aligned}
&\bar{\lambda}_{1,2} = -49.7492 \pm j85.8280 \\
&\left.\begin{aligned}\bar{\lambda}_{3,4} = -0.1615 \pm j6.3364 \\ \bar{\lambda}_{5,6} = -0.1840 \pm j6.1485\end{aligned}\right\} \quad \text{electromechanical mode} \\
&\bar{\lambda}_{7,8} = -5.7637 \pm j5.5700 \\
&\bar{\lambda}_{9,10} = -0.4500 \pm j4.5671 \quad \text{electromechanical mode} \\
&\bar{\lambda}_{11,12} = -4.3308 \pm j3.9562 \\
&\bar{\lambda}_{13,14} = -0.2099 \pm j22.0032 \\
&\bar{\lambda}_{15} = -6.3962 \\
&\bar{\lambda}_{16} = -3.4632 \\
&\bar{\lambda}_{17,18} = -4.0904 \pm j0.2451 \\
&\bar{\lambda}_{19} = -4.3536 \\
&\bar{\lambda}_{20} = -5.6946 \\
&\bar{\lambda}_{21} = 0 \\
&\bar{\lambda}_{22} = 0
\end{aligned}$$

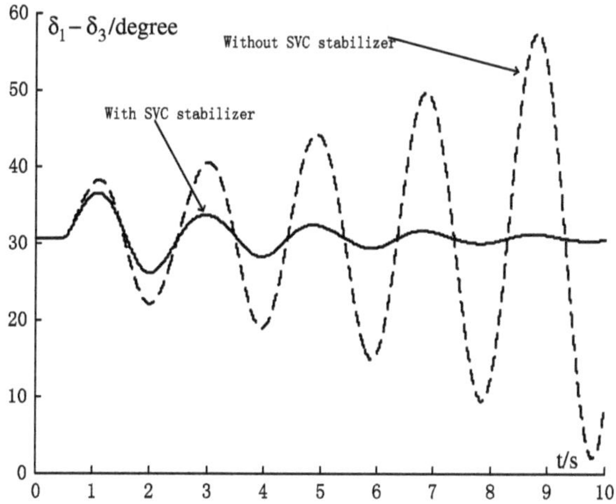


**Fig. 6.20** Simulation result with and without the SVC stabilizer

**Table 6.10** Oscillation mode when the SVC stabilizer is installed and the system is operated at varied operating conditions

| Operating condition | Eigenvalue | Damp ratio |
|---|---|---|
| $\mu_1$ | $-0.4500 \pm j4.5671$ | 0.0983 |
| $\mu_2$ | $-0.4510 \pm j4.5600$ | 0.0984 |
| $\mu_3$ | $-0.4517 \pm j4.5553$ | 0.0987 |
| $\mu_4$ | $-0.4526 \pm j4.5512$ | 0.0990 |
| $\mu_5$ | $-0.4576 \pm j4.5344$ | 0.1004 |
| $\mu_6$ | $-0.4633 \pm j4.5269$ | 0.1018 |
| $\mu_7$ | $-0.4706 \pm j4.5197$ | 0.1036 |
| $\mu_8$ | $-0.4838 \pm j4.4630$ | 0.1078 |
| $\mu_9$ | $-0.4870 \pm j4.4068$ | 0.1098 |
| $\mu_{10}$ | $-0.4901 \pm j4.3511$ | 0.1119 |

It can be observed that the mode of interest is moved to $-0.4500 \pm j4.5671$ which is well damped.

The simulation result of the example power system at operating condition $\mu_1$ without and with the SVC stabilizer installed is presented in Fig. 6.20. At 0.5 s of simulation, a three-phase short-circuit fault occurred at node 8 and was cleared in 0.1 s. From the simulation result, it can be seen that the low-frequency oscillation is well damped.

To examine the robustness of the SVC stabilizer, the oscillation mode of interest is calculated at other operating conditions $\mu_2 - \mu_{10}$ with the SVC stabilizer installed, and the results are given in Table 6.10. From Table 6.10, it can be observed that when the SVC stabilizer is designed at the selected operating condition $\mu_1$ which is predicted to be least effective, its effectiveness is ensured when the system operates at other operating conditions. Hence, the design ensures the robustness of the stabilizer.

## 6.4 Example Three-Machine Power System

### 6.4.1 Dynamic Interactions Among PSSs Installed in Example Three-Machine Power System

In Sect. 5.3, an example three-machine power system is presented. The demonstration of the example is finished with the coordinated design of three PSSs installed in three synchronous generators in Sect. 5.3.2.3, where the forward path of the PSSs before they are designed and installed is calculated as

$$\mathbf{F}_{\mathbf{pss}}(\bar{\lambda}_1^*) = \mathbf{K_2K_A}\left((\mathbf{K_3} + \bar{\lambda}_1^*\mathbf{T'_{d0}})(\mathbf{I} + \bar{\lambda}_1^*\mathbf{T_A}) + \mathbf{K_6K_A}\right)^{-1}$$
$$= \begin{bmatrix} 4.0590\angle -29.34^\circ & 1.7719\angle 153.70^\circ & 1.0376\angle 155.74^\circ \\ 0.6433\angle 113.76^\circ & 6.1802\angle -28.79^\circ & 2.6963\angle 136.09^\circ \\ 1.1801\angle 125.70^\circ & 2.6184\angle 140.40^\circ & 5.3227\angle -31.52^\circ \end{bmatrix}$$

where $\bar{\lambda}_1^* = -0.7094 + j7.0590$ is the target oscillation mode for the coordinated design of the PSSs, i.e. the position of the oscillation mode $\bar{\lambda}_1 = -0.1736 + j7.0590$ to be assigned to by the coordinated design of the PSSs. With the PSSs being designed by the use of the method introduced in Sect. 5.2.2, the oscillation mode is not moved to the exact target position, $\bar{\lambda}_1^*$. The mode with the PSSs installed is in fact moved to $\bar{\lambda}_{pss1} = -0.6094 + j7.0235$. The eigenvalue drift from the target position $\bar{\lambda}_1^*$ to $\bar{\lambda}_{pss1} = -0.6094 + j7.0235$ is due to the dynamic interactions between the PSSs as to be explained as follows.

Consider the example three-machine power system with the PSSs installed on each of the synchronous generators. Transfer function of the PSSs is

$$T_{pssi}(s) = K_{pssi}\frac{(1 + sT_{2i})(1 + sT_{4i})}{(1 + sT_{1i})(1 + sT_{3i})}, T_{1i} = T_{3i} = 0.05, i = 1, 2, 3$$

Parameters of the PSSs are obtained as given in Table 5.6 in Sect. 5.3.2.3. The forward path of the PSSs with $\bar{\lambda}_{pss1} = -0.6094 + j7.0235$ can be obtained to be

$$\mathbf{F}_{\mathbf{pss}}(\bar{\lambda}_{pss1}) = \mathbf{K_2K_A}\left((\mathbf{K_3} + \bar{\lambda}_{pss1}\mathbf{T'_{d0}})(\mathbf{I} + \bar{\lambda}_{pss1}\mathbf{T_A}) + \mathbf{K_6K_A}\right)^{-1}$$
$$= \begin{bmatrix} 4.2427\angle -29.72^\circ & 1.8428\angle 153.34^\circ & 1.0765\angle 155.48^\circ \\ 0.7022\angle 115.88^\circ & 6.4424\angle -29.37^\circ & 2.8752\angle 135.23^\circ \\ 1.2780\angle 125.43^\circ & 2.7740\angle 139.57^\circ & 5.5693\angle -32.24^\circ \end{bmatrix}$$

According to Eq. (6.115), damping torque provided by each PSS to each of the three generators can be calculated as

$$\mathbf{\Delta T_{PSS\mathit{i}}} = \mathrm{Re}[\bar{\mathbf{F}}_{\mathbf{PSS}}(\bar{\lambda}_{pss1})\bar{T}_{pssi}(\bar{\lambda}_{pss1})]\Delta\omega_i = \mathrm{Re}[\mathbf{F}_{\mathbf{pss}}(\bar{\lambda}_{pss1})\mathbf{F_i}\bar{T}_{pssi}(\bar{\lambda}_{pss1})]\Delta\omega_i, i = 1, 2, 3$$

where

$$\mathbf{F_1} = \begin{bmatrix} 1 \\ 0 \\ 0 \end{bmatrix}, \mathbf{F_1} = \begin{bmatrix} 0 \\ 1 \\ 0 \end{bmatrix}, \mathbf{F_1} = \begin{bmatrix} 0 \\ 0 \\ 1 \end{bmatrix}$$

The following results can be obtained according to the above equations:

$$\mathbf{\Delta T_{PSS1}} = \begin{bmatrix} 22.7134\Delta\omega_1 \\ -3.0879\Delta\omega_1 \\ -6.1895\Delta\omega_1 \end{bmatrix}, \mathbf{\Delta T_{PSS2}} = \begin{bmatrix} -3.5277\Delta\omega_2 \\ 12.3402\Delta\omega_2 \\ -5.2042\Delta\omega_2 \end{bmatrix},$$
$$\mathbf{\Delta T_{PSS3}} = \begin{bmatrix} -4.4036\Delta\omega_3 \\ -11.5351\Delta\omega_3 \\ 22.9518\Delta\omega_3 \end{bmatrix}$$

The above results give the damping torque contribution from the ith PSS to the jth generator in the form of $D_{pss-ij}\Delta\omega_i, i, j = 1, 2, 3, i \neq j$. This needs to be converted to the form $D_{pss-ij}\Delta\omega_j, i, j = 1, 2, 3, i \neq j$. The conversion is expressed generally by Eq. (6.117). Without the PSS installed, the conversion coefficient $\bar{\gamma}_j(\bar{\lambda}_i)$ can be calculated as explained from Eqs. (6.128) to (6.135) or from Eqs. (6.150) to (6.154). With the PSSs installed, the conversion can be made as explained as follows.

With the PSSs installed, the closed-loop state equation is

$$s\mathbf{\Delta X} = \mathbf{A_{pss}}\mathbf{\Delta X}$$

which is given at the end of Sect. 5.3.2.3. Hence,

$$\bar{\lambda}_{pss1}\mathbf{\Delta\bar{X}}(\bar{\lambda}_{pss1}) = \mathbf{A_{pss}}\mathbf{\Delta\bar{X}}(\bar{\lambda}_{pss1})$$

Denote the right eigenvector of the state matrix corresponding to $\bar{\lambda}_{pss1}$ to be $\bar{\mathbf{v}}_{\mathbf{pss1}}$; it should have

$$\frac{\Delta\bar{X}_i(\bar{\lambda}_{pss1})}{\Delta\bar{X}_j(\bar{\lambda}_{pss1})} = \frac{\bar{v}_i(\bar{\lambda}_{pss1})}{\bar{v}_j(\bar{\lambda}_{pss1})}$$

where $\Delta\bar{X}_i(\bar{\lambda}_{pss1})$ and $\Delta\bar{X}_j(\bar{\lambda}_{pss1})$ is the ith and jth state variables (element) of $\mathbf{\Delta\bar{X}}(\bar{\lambda}_{pss1})$, and $\bar{v}_i(\bar{\lambda}_{pss1})$ and $\bar{v}_j(\bar{\lambda}_{pss1})$ are the ith and jth elements of $\bar{\mathbf{v}}_{\mathbf{pss1}}$, respectively.

From the closed-loop state matrix obtained in Sect. 5.3.2.3, the right eigenvector corresponding to $\bar{\lambda}_{pss1} = -0.6094 + j7.0235$, $\bar{\mathbf{v}}_{\mathbf{pss1}}$ is calculated as

$$\bar{\mathbf{v}}_{\mathbf{pss1}} = \begin{bmatrix} \bar{\mathbf{v}}_{\mathbf{pss11}} & \bar{\mathbf{v}}_{\mathbf{pss12}} & \bar{\mathbf{v}}_{\mathbf{pss13}} \end{bmatrix}^{T}$$

where

$$\bar{\mathbf{v}}_{\mathbf{pss11}} = \begin{bmatrix} 0.0159 - j0.0289 \\ -0.0554 + j0.0831 \\ -0.0163 + j0.0429 \\ (6.1622 + j4.1263) \times 10^{-4} \\ (-1.7503 - j1.3991) \times 10^{-3} \\ (-9.2666 - j4.4722) \times 10^{-4} \end{bmatrix}, \bar{\mathbf{v}}_{\mathbf{pss12}} = \begin{bmatrix} 0.0049 + j0.0047 \\ -0.0042 - j0.0198 \\ -0.0107 - j0.0087 \\ -0.3141 + j0.2922 \\ 0.7337 + j0 \\ 0.3582 - j0.3680 \end{bmatrix},$$

$$\bar{\mathbf{v}}_{\mathbf{pss13}} = \begin{bmatrix} 0.0005 + j0.0006 \\ 0.0016 + j0.0035 \\ -0.0015 - j0.0020 \\ -0.0015 - j0.0039 \\ -0.0009 - j0.0008 \\ -0.0021 - j0.0035 \end{bmatrix}$$

$\mathbf{\Delta X}$ is defined in Sect. 5.3.2.3 to be

$$\mathbf{\Delta X} = \begin{bmatrix} \mathbf{\Delta X_1} & \mathbf{\Delta X_2} & \mathbf{\Delta X_3} \end{bmatrix}^{T}$$

where

$$\begin{aligned} \mathbf{\Delta X_1} &= \begin{bmatrix} \Delta\delta_1 & \Delta\delta_2 & \Delta\delta_3 & \Delta\omega_1 & \Delta\omega_2 & \Delta\omega_3 \end{bmatrix}^{T} \\ \mathbf{\Delta X_2} &= \begin{bmatrix} \Delta E'_{q1} & \Delta E'_{q2} & \Delta E'_{q3} & \Delta E'_{fd1} & \Delta E'_{fd2} & \Delta E'_{fd3} \end{bmatrix}^{T} \\ \mathbf{\Delta X_3} &= \begin{bmatrix} \Delta X_{pss1} & \Delta U_{pss1} & \Delta X_{pss2} & \Delta U_{pss2} & \Delta X_{pss3} & \Delta U_{pss3} \end{bmatrix}^{T} \end{aligned}$$

Thus, from the last three elements of $\bar{\mathbf{v}}_{\mathbf{pss11}}$, the following results can be obtained as

$$\begin{aligned} \Delta\omega_1 &= \bar{\gamma}_{12}(\bar{\lambda}_{pss1})\Delta\omega_2 = \frac{(6.1622 + j4.1263) \times 10^{-4}}{(-1.7503 - j1.3991) \times 10^{-3}}\Delta\omega_2 = (-0.3298 + j0.0279)\Delta\omega_2, \\ \Delta\omega_1 &= \bar{\gamma}_{12}(\bar{\lambda}_{pss1})\Delta\omega_3 = \frac{(6.1622 + j4.1263) \times 10^{-4}}{(-9.2666 - j4.4722) \times 10^{-4}}\Delta\omega_3 = (-0.7137 - j0.1009)\Delta\omega_3, \\ \Delta\omega_2 &= \bar{\gamma}_{23}(\bar{\lambda}_{pss1})\Delta\omega_3 = \frac{(-1.7503 - j1.3991) \times 10^{-3}}{(-9.2666 - j4.4722) \times 10^{-4}}\Delta\omega_3 = (2.1230 + j0.4853)\Delta\omega_3 \end{aligned}$$

Hence, with $D_{pss-ij}\Delta\omega_i, i, j = 1, 2, 3, i \neq j$ being converted to $D_{pss-ij}\Delta\omega_j, i, j = 1, 2, 3, i \neq j$, the damping torque contributed by each PSS to each of the generators should be calculated as

$$\Delta \mathbf{T}_{\mathbf{PSSi}} = \begin{bmatrix} \mathrm{Re}[\mathbf{F}_{\mathbf{pssi1}}(\bar{\lambda}_{pss1})\bar{T}_{pssi}(\bar{\lambda}_{pss1})\bar{\gamma}_{i1}(\bar{\lambda}_{pss1})]\Delta\omega_1 \\ \mathrm{Re}[\mathbf{F}_{\mathbf{pssi2}}(\bar{\lambda}_{pss1})\bar{T}_{pssi}(\bar{\lambda}_{pss1})\bar{\gamma}_{i2}(\bar{\lambda}_{pss1})]\Delta\omega_2 \\ \mathrm{Re}[\mathbf{F}_{\mathbf{pssi1}}(\bar{\lambda}_{pss1})\bar{T}_{pssi}(\bar{\lambda}_{pss1})\bar{\gamma}_{i3}(\bar{\lambda}_{pss1})]\Delta\omega_3 \end{bmatrix}, i = 1, 2, 3$$

By using the results obtained and equation given above, the damping torque contributed by each PSS to each of the generators is calculated as

$$\Delta \mathbf{T}_{\mathbf{PSS1}} = \begin{bmatrix} 22.7134\Delta\omega_1 \\ 0.8905\Delta\omega_2 \\ 5.9790\Delta\omega_3 \end{bmatrix}, \Delta \mathbf{T}_{\mathbf{PSS2}} = \begin{bmatrix} 11.0506\Delta\omega_1 \\ 12.3402\Delta\omega_2 \\ -14.7900\Delta\omega_3 \end{bmatrix},$$

$$\Delta \mathbf{T}_{\mathbf{PSS1}} = \begin{bmatrix} 4.9442\Delta\omega_1 \\ -3.4809\Delta\omega_2 \\ 22.9518\Delta\omega_3 \end{bmatrix}$$

The above results are summarized in Table 6.11.

In Sect. 5.3.2.3, the required damping torque provided by the PSSs is $[D^*_{pss1} \quad D^*_{pss2} \quad D^*_{pss3}] = \mathbf{p}_{52} = [21.73 \quad 11.84 \quad 21.94]$ to move the oscillation mode from $\bar{\lambda}_1 = -0.1736 + j7.0590$ to $\bar{\lambda}^*_1 = -0.7094 + j7.0590$ by the design of the PSSs. With the PSSs installed, the oscillation mode is only moved to $\bar{\lambda}_{pss1} = -0.6094 + j7.0235$. The reason is clearly indicated by the results in Table 6.11. The actual damping torque provided by the PSSs is $[D_{pss1} \quad D_{pss2} \quad D_{pss3}] = [38.71 \quad 9.75 \quad 14.14]$, not the required damping torque $[D^*_{pss1} \quad D^*_{pss2} \quad D^*_{pss3}] = [21.73 \quad 11.84 \quad 21.94]$ for the PSSs design, because each PSS provides damping torque to all three generators as is shown in Table 6.11.

1. The amount of damping torque provided by each PSS to the generator where it is installed is $[22.7134 \quad 12.3402 \quad 22.9518]$, very close to $[D^*_{pss1} \quad D^*_{pss2} \quad D^*_{pss3}] = [21.73 \quad 11.84 \quad 21.94]$, the required damping torque by the design of the PSS. The difference is caused by the difference between $\bar{\lambda}_{pss1} = -0.6094 + j7.0235$ and $\bar{\lambda}^*_1 = -0.7094 + j7.0590$.

**Table 6.11** The damping torque provided by each PSS

| Damping torque contribution to | By $PSS_1$ | By $PSS_2$ | By $PSS_3$ | Total $D_{pssi}$ |
|---|---|---|---|---|
| G1 | **22.7134** | 11.0506 | 4.9442 | 38.7082 |
| G2 | 0.8905 | **12.3402** | −3.4809 | 9.7498 |
| G3 | 5.9790 | −14.7900 | **22.9518** | 14.1407 |

2. Each of the three PSSs provides positive damping torque to the first generator. Thus, the first generator obtains more damping torque ($38.7082\Delta\omega_1$) than required by the design.
3. PSS installed on the second generator and third generator contributes negative damping torque to the third and second generators, respectively, reducing the actual amount of damping torque obtained by the second and third generators.

### 6.4.2 *Design of Non-negatively Interactive PSSs Installed in the Example Power System*

In Sect. 6.2.4.2, a scheme of non-negatively interactive design of the PSSs is introduced. In this section, the scheme is used to examine whether the coordinated design of three PSSs to be installed in the three-machine power system can be designed to be non-negatively interactive or not as follows.

The state matrix of the example three-machine power system is constructed in Sect. 5.3.1.3 by the following coefficient matrices according to Eq. (5.43) when the PSSs are not installed

$$\mathbf{K}_1 = \begin{bmatrix} 2.3373 & -1.4640 & -0.8733 \\ -1.5290 & 2.2160 & -0.6870 \\ -1.0599 & -0.8201 & 1.8800 \end{bmatrix}, \mathbf{K}_2 = \begin{bmatrix} 2.4574 & -0.9034 & -0.5628 \\ 1.0247 & 3.0671 & -0.4060 \\ 0.3694 & -0.6458 & 2.3234 \end{bmatrix},$$

$$\mathbf{K}_3 = \begin{bmatrix} 1.1889 & -0.1067 & -0.0994 \\ -0.1209 & 3.1565 & -0.7660 \\ -0.4521 & -1.2478 & 3.7141 \end{bmatrix}, \mathbf{K}_4 = \begin{bmatrix} 0.0277 & -0.0008 & -0.0269 \\ -1.2024 & 1.8725 & -0.6701 \\ -1.2795 & -1.0112 & 2.2907 \end{bmatrix},$$

$$\mathbf{K}_5 = \begin{bmatrix} -0.0068 & -0.0076 & 0.0143 \\ -0.0596 & 0.0315 & 0.0281 \\ -00780 & -0.0499 & 0.1278 \end{bmatrix}, \mathbf{K}_6 = \begin{bmatrix} 0.8732 & 0.0711 & 0.0678 \\ 0.4315 & 0.5292 & 0.1730 \\ 0.4691 & 0.1829 & 0.4790 \end{bmatrix},$$

$$\mathbf{M} = \begin{bmatrix} 47.2 & 0 & 0 \\ 0 & 12.8 & 0 \\ 0 & 0 & 6.02 \end{bmatrix}, \mathbf{D} = \begin{bmatrix} 0 & 0 & 0 \\ 0 & 0 & 0 \\ 0 & 0 & 0 \end{bmatrix},$$

$$\mathbf{K_A} = \begin{bmatrix} 200 & 0 & 0 \\ 0 & 200 & 0 \\ 0 & 0 & 200 \end{bmatrix}, \mathbf{T_A} = \begin{bmatrix} 0.02 & 0 & 0 \\ 0 & 0.02 & 0 \\ 0 & 0 & 0.02 \end{bmatrix},$$

$$\mathbf{T}'_{\mathbf{d0}} = \begin{bmatrix} 8.96 & 0 & 0 \\ 0 & 6.00 & 0 \\ 0 & 0 & 5.89 \end{bmatrix}$$

In Sect. 5.3.2.3, it is found that when the damping torque provided by the PSSs installed on the three generators is $\left[\mathrm{D}^*_{\mathrm{pss1}} \;\; \mathrm{D}^*_{\mathrm{pss2}} \;\; \mathrm{D}^*_{\mathrm{pss3}}\right] = \mathbf{p}_{52} = [21.73\; 11.8421.94]$, the electromechanical oscillation mode of interests is moved to $\bar{\lambda}^*_1 = -0.7094 + \mathrm{j}7.0590$. Hence, the state matrix of the example three-machine power system with the PSSs installed (parameters of the PSSs have not been set yet) can be constructed still according to Eq. (5.43) with coefficient matrix **D** being changed to

$$D = \begin{bmatrix} D^*_{pss1} & 0 & 0 \\ 0 & D^*_{pss2} & 0 \\ 0 & 0 & D^*_{pss3} \end{bmatrix} = \begin{bmatrix} 21.73 & 0 & 0 \\ 0 & 11.84 & 0 \\ 0 & 0 & 11.84 \end{bmatrix}$$

Denote the state equation with the PSSs installed (their parameters have not been set yet) to be

$$s\Delta\tilde{\mathbf{X}} = \tilde{\mathbf{A}}_{\mathbf{pss}}\Delta\tilde{\mathbf{X}}$$

where $\Delta\tilde{\mathbf{X}} = [\,\Delta\mathbf{X}_1 \quad \Delta\mathbf{X}_2\,]^T$

$$\Delta\mathbf{X}_1 = [\,\Delta\delta_1 \quad \Delta\delta_2 \quad \Delta\delta_3 \quad \Delta\omega_1 \quad \Delta\omega_2 \quad \Delta\omega_3\,]^T$$
$$\Delta\mathbf{X}_2 = \left[\,\Delta E'_{q1} \quad \Delta E'_{q2} \quad \Delta E'_{q3} \quad \Delta E'_{fd1} \quad \Delta E'_{fd2} \quad \Delta E'_{fd3}\,\right]^T$$

Obviously, $\bar{\lambda}^*_1 = -0.7094 + j7.0590$ is the eigenvalue of state matrix $\tilde{\mathbf{A}}_{\mathbf{pss}}$. Let $\bar{\mathbf{v}}'_{\mathbf{pss1}}$ be the right eigenvector of state matrix $\tilde{\mathbf{A}}_{\mathbf{pss}}$ corresponding to $\bar{\lambda}^*_1$ and $\bar{v}'_j(\bar{\lambda}^*_1)$ be the (j + 3) element of $\bar{\mathbf{v}}'_{\mathbf{pss1}}, j = 1, 2, 3$. It can have

$$\Delta\omega_i = \bar{\gamma}_{ij}(\bar{\lambda}^*_1)\Delta\omega_j = \frac{\bar{v}_i(\bar{\lambda}^*_1)}{\bar{v}_j(\bar{\lambda}^*_1)}\Delta\omega_j$$

The damping torque provided by the ith PSS to the jth generator is

$$\Delta T_{pssij} = \mathrm{Re}[F_{pssij}(\bar{\lambda}^*_1)\bar{\gamma}_{ij}(\bar{\lambda}^*_1)\bar{T}_{pssi}(\bar{\lambda}^*_1)]\Delta\omega_j, i, j = 1, 2, 3$$

$\bar{f}_{pssij} = F_{pssij}(\bar{\lambda}^*_1)\bar{\gamma}_{ij}(\bar{\lambda}^*_1), i, j = 1, 2, 3$ is the forward path from the ith PSS to the electromechanical oscillation loop of the jth generator. A forward path matrix $\bar{\mathbf{F}}(\bar{\lambda}^*_1)$ can be formed with $\bar{f}_{pssij} = F_{pssij}(\bar{\lambda}^*_1)\bar{\gamma}_{ij}(\bar{\lambda}^*_1), i, j = 1, 2, 3$ being the elements. For the example three-machine power system, $\bar{\mathbf{v}}'_{\mathbf{pss1}}$ is calculated as

$$\bar{\mathbf{v}}'_{\mathbf{pss1}} = \left[\,\bar{\mathbf{v}}'_{\mathbf{pss1a}} \quad \bar{\mathbf{v}}'_{\mathbf{pss1b}} \quad \bar{\mathbf{v}}'_{\mathbf{pss1c}}\,\right]^T$$
$$\bar{\mathbf{v}}'_{\mathbf{pss1a}} = [0.0080 - j0.0556 \quad 0.0111 + j0.1722 \quad -0.0188 + j0.0752 \quad -0.0013 - j0.0001]^T$$
$$\bar{\mathbf{v}}'_{\mathbf{pss1b}} = [0.0038 - j0.0006 \quad 0.0017 + j0.0003 \quad -0.0014 + j\,0.0014 \quad 0.0086 - j0.0222]^T$$
$$\bar{\mathbf{v}}'_{\mathbf{pss1c}} = [0.0076 - j0.0021 \quad 0.0958 + j0.0822 \quad -0.9324 + j0 \quad -0.16467 - j0.2189]^T$$

Thus, $\bar{\gamma}_{ij}(\bar{\lambda}_1^*)$ is calculated as

$$\bar{\gamma}_{12} = \bar{v}'_{pss14}/\bar{v}'_{pss15} = (-0.0013 - j0.0001)/(0.0038 - j0.0006) = -0.3187 + j0.0671,$$
$$\bar{\gamma}_{13} = \bar{v}'_{pss14}/\bar{v}'_{pss16} = (-0.0013 - j0.0001)/(0.0017 + j0.0003) = -0.7211 - j0.0733,$$
$$\bar{\gamma}_{21} = 1/\bar{\gamma}_{12} = 1/(-0.3187 + j0.0671) = -3.0048 - j0.6331,$$
$$\bar{\gamma}_{23} = \bar{v}'_{pss15}/\bar{v}'_{pss16} = (0.0038 - j0.0006)/(0.0017 + j0.0003) = 2.1204 + j0.6768,$$
$$\bar{\gamma}_{31} = 1/\bar{\gamma}_{13} = 1/(-0.7211 - j0.0733) = -1.3726 + j0.1395,$$
$$\bar{\gamma}_{32} = 1/\bar{\gamma}_{23} = 1/(2.1204 + j0.6768) = 0.4280 - j0.1366$$

$\bar{\mathbf{F}}(\bar{\lambda}_1^*)$ is calculated as

$$\bar{\mathbf{F}}(\bar{\lambda}_1^*) = \begin{bmatrix} 4.059\angle -29.341^\circ & 5.441\angle -38.198^\circ & 1.4316\angle -18.46^\circ \\ 0.2095\angle -54.347^\circ & 6.1802\angle -28.783^\circ & 1.2114\angle 153.79^\circ \\ 0.8554\angle -60.103^\circ & 5.8279\angle 122.7^\circ & 5.3227\angle -31.527^\circ \end{bmatrix}$$

Three vectors $\bar{f}_{pssij} = F_{pssij}(\bar{\lambda}_1^*)\bar{\gamma}_{ij}(\bar{\lambda}_1^*), j = 1,2,3$ for i = 1,2,3 are plotted in Figs. 6.21, 6.22, and 6.23, respectively.

From Fig. 6.21, it can be seen that if the compensating phase of PSS1 is chosen to be $-\theta_{10} = 29.341^\circ$, it can have $\bar{T}_{pss1}(\bar{\lambda}_1^*) = T_{pss1}\angle\theta_{10} = T_{pss1}\angle 29.341^\circ$. The electric torque provided by PSS1 to each of the three generators is

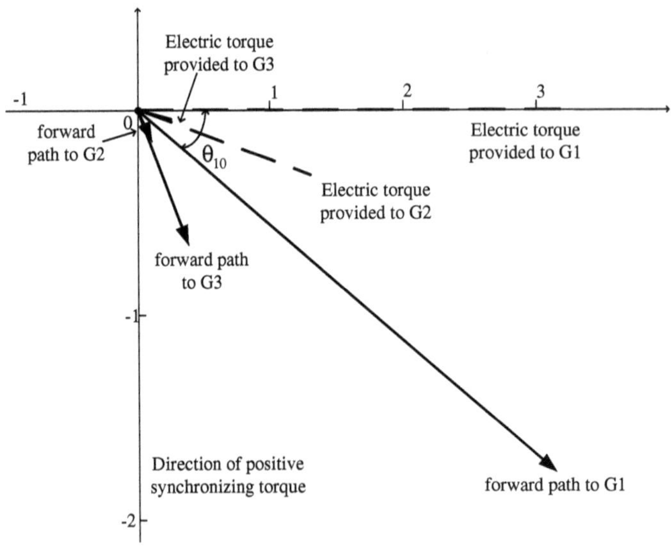


**Fig. 6.21** Forward path of PSS1 to each generator

$$\Delta T_{PSS1-1} = F_{pssi1-1}(\bar{\lambda}_1^*)\bar{\gamma}_{i1}(\bar{\lambda}_1^*)\bar{T}_{pss1}(\bar{\lambda}_1^*)\Delta\omega_1 = 4.059\angle{-29.341^\circ} \times T_{pss1}\angle\theta_{10}\Delta\omega_1 = 4.059T_{pss1}\Delta\omega_1$$
$$\Delta T_{PSS1-2} = F_{pssi1-2}(\bar{\lambda}_1^*)\bar{\gamma}_{i1}(\bar{\lambda}_1^*)\bar{T}_{pss1}(\bar{\lambda}_1^*)\Delta\omega_2 = 0.2095\angle{-54.347^\circ} \times T_{pss1}\angle\theta_{10}\Delta\omega_2$$
$$= 0.2095T_{pss1}\cos(-54.347^\circ + 29.341^\circ)\Delta\omega_2$$
$$\Delta T_{PSS1-3} = F_{pssi1-3}(\bar{\lambda}_1^*)\bar{\gamma}_{i1}(\bar{\lambda}_1^*)\bar{T}_{pss1}(\bar{\lambda}_1^*)\Delta\omega_3 = 0.8554\angle{-60.103^\circ} \times T_{pss1}\angle\theta_{10}\Delta\omega_3$$
$$= 0.8554T_{pss1}\cos(-60.103^\circ + 29.341^\circ)\Delta\omega_3$$

which are all positive. The synchronizing torque provided by PSS1 to each of the generators is

$$\Delta T_{s1-1} = j\mathrm{Im}\{F_{pssi1-1}(\bar{\lambda}_1^*)\bar{\gamma}_{i1}(\bar{\lambda}_1^*)\mathbf{\bar{T}}_{pss1}(\bar{\lambda}_1^*)\}\Delta\omega_1$$
$$= j4.059T_{pss1}\sin(-29.341^\circ + \theta_{10})\Delta\omega_1 = 0$$
$$\Delta T_{s1-2} = j0.2095T_{pss1}\sin(-54.347^\circ + 29.341^\circ)\Delta\omega_1$$
$$\approx j0.2095T_{pss1}\sin(-54.347^\circ + 29.341^\circ)\frac{j\omega_s^*}{\omega_0}\Delta\delta_2$$
$$= -0.2095T_{pss1}\sin(-54.347^\circ + 29.341^\circ)\frac{\omega_s^*}{\omega_0}\Delta\delta_2$$
$$\Delta T_{s1-3} = -0.8554T_{pss1}\sin(-60.103^\circ + 29.341^\circ)\frac{\omega_s^*}{\omega_0}\Delta\delta_3$$

It shows that PSS1 provides positive synchronizing torque to the 2nd and 3rd generators, as the projection of the vector of electric torque provided by PSS1 to G2 and G3 is on the direction of positive synchronizing torque.

From Fig. 6.22, it can be seen that if the compensating phase of PSS2 is chosen to be $-\theta_{20}$, PSS2 will provide positive damping and synchronizing torque to G1 and G2, but negative damping torque to G3. There is no suitable compensating

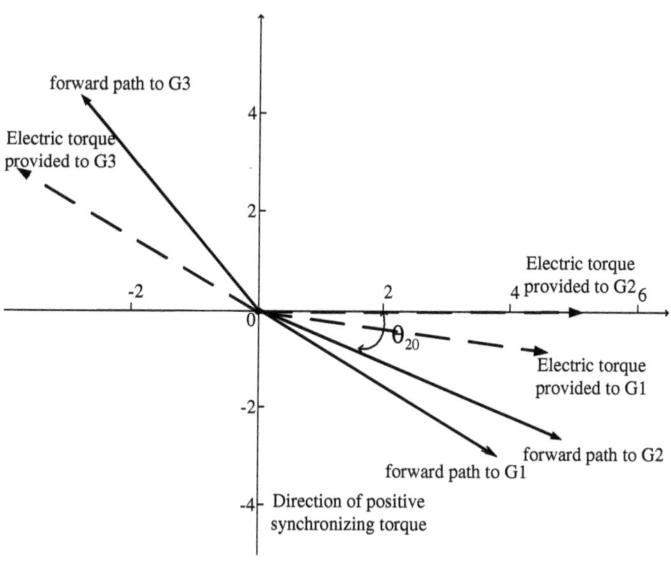


**Fig. 6.22** Forward path of PSS2 to each generator

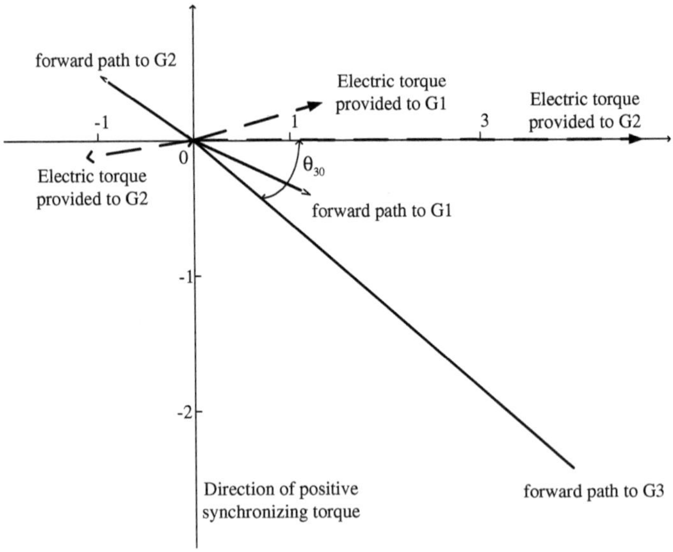


**Fig. 6.23** Forward path of PSS3 to each generator

phase to be selected to ensure the PSS to provide positive damping torque to three generators. Similarly, from Fig. 6.23, it can also be seen that there is no suitable compensating phase available for designing PSS3. Hence, in the example three-machine power system presented in Sect. 5.3, it is inevitable that PSS2 and PSS3 supply negative to other generators.

## References


1. Wang HF, Swift FJ (1998) A unified model for the analysis of FACTS devices in damping power system oscillations part II: multi-machine power systems. IEEE Trans Power Deliv (4)
2. Wang HF, Swift FJ (1997) The indexes for selecting the best locations of PSS or FACTS-based stabilizers in multi-machine power systems: a comparison study. IEE Proc Part C (2)
3. Wang HF, Swift FJ (1997) The connection between modal analysis and electric torque analysis in studying the oscillation stability of multi-machine power systems. Int J Electr Power Energy Syst 19(5):321–330
4. Wang HF (1999) Selection of robust installing locations and feedback signals of FACTS-based stabilizers in multi-machine power systems, IEEE Trans Power Syst (2)
5. Wang HF (1999) Design of non-negatively interactive FACTS-based stabilizers in multi-machine power systems. Int J Power Syst Res 50 (3)

# Chapter 7
# Multi-machine Power Systems Installed with VSC-Based Stabilizers

## 7.1 Mathematical Model of a Multi-machine Power System Installed with VSC-Based Stabilizers

### *7.1.1 Mathematical Model of a Multi-machine Power System Installed with a Shunt VSC-Based Stabilizer*

#### 7.1.1.1 Heffron–Phillips Model of an N-Machine Power System Installed with a Shunt VSC-Based Stabilizer

Without loss of generality, it can be assumed that a shunt VSC is installed at a busbar between nodes 1 and 2 in an N-machine power system, as shown in Fig. 7.1. The shunt VSC can be connected to an energy storage system (ESS), a HVDC, or a static renewable power station, such as a photovoltaic (PV) or fuel cell (FC) power station. It can operate alone as a VSC-based FACTS device, STATCOM (see, Sect. 4.1.1.1).

From Fig. 7.1, it can be obtained by the following:

$$\bar{V}_s = jx_s\bar{I}_s + \bar{V}_c = jx_s(\bar{I}_{1s} + \bar{I}_{s2}) + \bar{V}_c \tag{7.1}$$

Hence,

$$\begin{aligned}\bar{V}_1 &= jx_{1s}\bar{I}_{1s} + \bar{V}_s = jx_{1s}\bar{I}_{1s} + jx_s(\bar{I}_{1s} + \bar{I}_{s2}) + \bar{V}_c = j(x_{1s} + x_s)\bar{I}_{1s} + jx_s\bar{I}_{s2} + \bar{V}_c \\ \bar{V}_2 &= jx_{s2}\bar{I}_{s2} + \bar{V}_s = jx_{s2}\bar{I}_{s2} + jx_s(\bar{I}_{1s} + \bar{I}_{s2}) + \bar{V}_c = jx_s\bar{I}_{1s} + j(x_{s2} + x_s)\bar{I}_{s2} + \bar{V}_c\end{aligned} \tag{7.2}$$

In matrix form, Eq. (7.2) is given as follows:

$$\begin{bmatrix} j(x_{1s} + x_s) & jx_s \\ jx_s & j(x_{s2} + x_s) \end{bmatrix} \begin{bmatrix} \bar{I}_{1s} \\ \bar{I}_{s2} \end{bmatrix} = \begin{bmatrix} \bar{V}_1 \\ \bar{V}_2 \end{bmatrix} - \begin{bmatrix} \bar{V}_c \\ \bar{V}_c \end{bmatrix} \tag{7.3}$$



H. Wang and W. Du, *Analysis and Damping Control of Power System Low-frequency Oscillations*, Power Electronics and Power Systems,
DOI 10.1007/978-1-4899-7696-3_7

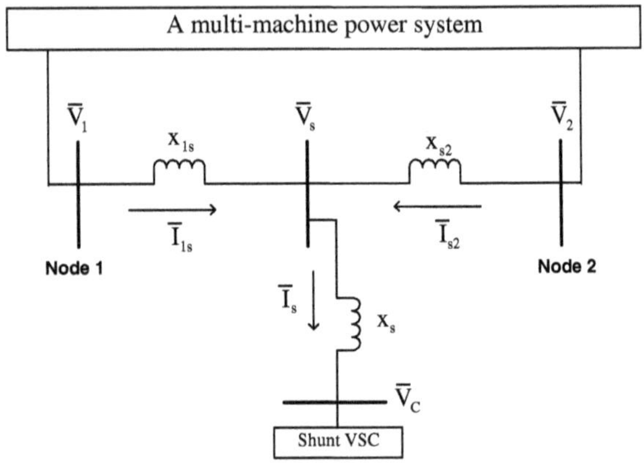


**Fig. 7.1** A multi-machine power system installed with a shunt VSC

That is,

$$\begin{bmatrix}\bar{I}_{1s}\\ \bar{I}_{s2}\end{bmatrix} = \begin{bmatrix} j(x_{1s}+x_s) & jx_s \\ jx_s & j(x_{s2}+x_s)\end{bmatrix}^{-1}\left(\begin{bmatrix}\bar{V}_1\\ \bar{V}_2\end{bmatrix} - \begin{bmatrix}\bar{V}_c\\ \bar{V}_c\end{bmatrix}\right) \tag{7.4}$$

For the N-machine power system, firstly the set of network equations with N generator's internal nodes, nodes 1 and 2 left (without node s where the shunt VSC is installed) can be established as follows:

$$\begin{bmatrix}0\\0\\ \bar{\mathbf{I}}_{\mathbf{g}}\end{bmatrix} = \begin{bmatrix}\bar{y}_{11} & \bar{y}_{12} & \bar{\mathbf{Y}}_{\mathbf{13}}\\ \bar{y}_{21} & \bar{y}_{22} & \bar{\mathbf{Y}}_{\mathbf{23}}\\ \bar{\mathbf{Y}}_{\mathbf{31}} & \bar{\mathbf{Y}}_{\mathbf{32}} & \bar{\mathbf{Y}}_{\mathbf{33}}\end{bmatrix}\begin{bmatrix}\bar{V}_1\\ \bar{V}_2\\ \bar{\mathbf{E}}_{\mathbf{g}}\end{bmatrix} \tag{7.5}$$

Secondly, the network equations of (7.5) can be modified with node s added to be

$$\begin{aligned}\begin{bmatrix}0\\0\end{bmatrix} &= \begin{bmatrix}\bar{y}'_{11} & 0\\ 0 & \bar{y}'_{22}\end{bmatrix}\begin{bmatrix}\bar{V}_1\\ \bar{V}_2\end{bmatrix} + \begin{bmatrix}\bar{I}_{1s}\\ \bar{I}_{s2}\end{bmatrix} + \begin{bmatrix}\bar{\mathbf{Y}}_{\mathbf{13}}\\ \bar{\mathbf{Y}}_{\mathbf{23}}\end{bmatrix}\bar{\mathbf{E}}_{\mathbf{g}}\\ \bar{\mathbf{I}}_{\mathbf{g}} &= \begin{bmatrix}\bar{\mathbf{Y}}_{31} & \bar{\mathbf{Y}}_{32}\end{bmatrix}\begin{bmatrix}\bar{V}_1\\ \bar{V}_2\end{bmatrix} + \bar{\mathbf{Y}}_{33}\bar{\mathbf{E}}_{\mathbf{g}}\end{aligned} \tag{7.6}$$

where $\bar{y}'_{11}$ and $\bar{y}'_{22}$ are obtained by excluding $x_{12} = x_{1s} + x_{s2}$ from $\bar{y}_{11}$ and $\bar{y}_{22}$ in Eq. (7.5), respectively.

Finally from Eqs. (7.4) and (7.6), it can have the following:

$$\begin{aligned}\begin{bmatrix}0\\0\end{bmatrix} &= \begin{bmatrix}\bar{y}'_{11} & 0\\ 0 & \bar{y}'_{22}\end{bmatrix}\begin{bmatrix}\bar{V}_1\\ \bar{V}_2\end{bmatrix}\\ &+ \begin{bmatrix} j(x_{1s}+x_s) & jx_s \\ jx_s & j(x_{s2}+x_s)\end{bmatrix}^{-1}\left(\begin{bmatrix}\bar{V}_1\\ \bar{V}_2\end{bmatrix} - \begin{bmatrix}\bar{V}_c\\ \bar{V}_c\end{bmatrix}\right) + \begin{bmatrix}\bar{\mathbf{Y}}_{\mathbf{13}}\\ \bar{\mathbf{Y}}_{\mathbf{23}}\end{bmatrix}\bar{\mathbf{E}}_{\mathbf{g}}\end{aligned}$$

That is.

$$\begin{aligned}\begin{bmatrix}\bar{V}_1\\ \bar{V}_2\end{bmatrix} &= \left(\mathbf{I}+\begin{bmatrix}j(x_{1s}+x_s) & jx_s\\ jx_s & j(x_{s2}+x_s)\end{bmatrix}\begin{bmatrix}\bar{y}'_{11} & 0\\ 0 & \bar{y}'_{22}\end{bmatrix}\right)^{-1}\\ &\times\left(\begin{bmatrix}\bar{V}_c\\ \bar{V}_c\end{bmatrix}-\begin{bmatrix}j(x_{1s}+x_s) & jx_s\\ jx_s & j(x_{s2}+x_s)\end{bmatrix}\begin{bmatrix}\bar{\mathbf{Y}}_{13}\\ \bar{\mathbf{Y}}_{23}\end{bmatrix}\bar{\mathbf{E}}_g\right)\\ &= \bar{\mathbf{Y}}_S^{-1}\left(\begin{bmatrix}\bar{V}_c\\ \bar{V}_c\end{bmatrix}-\begin{bmatrix}j(x_{1s}+x_s) & jx_s\\ jx_s & j(x_{s2}+x_s)\end{bmatrix}\begin{bmatrix}\bar{\mathbf{Y}}_{13}\\ \bar{\mathbf{Y}}_{23}\end{bmatrix}\bar{\mathbf{E}}_g\right)\end{aligned} \tag{7.7}$$

Substituting Eq. (7.7) into (7.6), it can be obtained that

$$\begin{aligned}\bar{\mathbf{I}}_g &= \bar{\mathbf{C}}\bar{V}_c+\bar{\mathbf{Y}}\bar{\mathbf{E}}_g = \begin{bmatrix}\bar{\mathbf{Y}}_{31} & \bar{\mathbf{Y}}_{32}\end{bmatrix}\bar{\mathbf{Y}}_s^{-1}\begin{bmatrix}1\\ 1\end{bmatrix}\bar{V}_c\\ &+\left(\bar{\mathbf{Y}}_{33}-\begin{bmatrix}\bar{\mathbf{Y}}_{31} & \bar{\mathbf{Y}}_{32}\end{bmatrix}\bar{\mathbf{Y}}_s^{-1}\begin{bmatrix}j(x_{1s}+x_s) & jx_s\\ jx_s & j(x_{s2}+x_s)\end{bmatrix}\begin{bmatrix}\bar{\mathbf{Y}}_{13}\\ \bar{\mathbf{Y}}_{23}\end{bmatrix}\right)\bar{\mathbf{E}}_g\end{aligned} \tag{7.8}$$

From Eqs. (7.4) and (7.7), it can have the following:

$$\begin{bmatrix}\bar{I}_{1s}\\ \bar{I}_{s2}\end{bmatrix} = \begin{bmatrix}j(x_{1s}+x_s) & jx_s\\ jx_s & j(x_{s2}+x_s)\end{bmatrix}^{-1}\left[\left(\bar{\mathbf{Y}}_S^{-1}-\mathbf{I}\right)\begin{bmatrix}\bar{V}_c\\ \bar{V}_c\end{bmatrix}-\bar{\mathbf{Y}}_S^{-1}\begin{bmatrix}\bar{\mathbf{Y}}_{13}\\ \bar{\mathbf{Y}}_{23}\end{bmatrix}\bar{\mathbf{E}}_g\right] \tag{7.9}$$

Hence from Fig. 7.1 and Eq. (7.9), it can have the following:

$$\bar{I}_s = i_{sx}+ji_{sy} = \bar{I}_{1s}+\bar{I}_{s2} = \begin{bmatrix}1 & 1\end{bmatrix}\begin{bmatrix}\bar{I}_{1s}\\ \bar{I}_{s2}\end{bmatrix} = \bar{y}_s\bar{V}_c+\sum_{j=1}^{N}\bar{y}_{sj}\bar{E}_{gj} \tag{7.10}$$

where $\bar{E}_{gj}$ is the jth element of $\bar{\mathbf{E}}_g$. According to Eq. (5.10), $\bar{E}_{gj}$ is given as follows:

$$\bar{E}_{gj} = E'_{qj}e^{j\delta_j}+(x_{qj}-x'_{dj})i_{qj}e^{i(\delta_j-90^\circ)} \tag{7.11}$$

From Eqs. (7.1) and (7.10), it can have the following:

$$\bar{V}_s = jx_s\bar{I}_s+\bar{V}_c = (jx_s+\bar{y}_s)\bar{V}_c+\sum_{j=1}^{N}\bar{y}_{sj}\bar{E}_{gj} \tag{7.12}$$

In x–y coordinate of the power system, it can have the following:

$$\bar{V}_c = mkV_{dc}(\cos\gamma+j\sin\gamma) = mkV_{dc}\angle\gamma \tag{7.13}$$

where $\gamma$ is the angle between $\bar{V}_c$ and x-axis. Active power received by the capacitor of the VSC from the power system is given as:

$$V_{dc}I_{dc1} = i_{sx}v_{cx} + i_{sy}v_{cy} = i_{sx}mkV_{dc}\cos\gamma + i_{sy}mkV_{dc}\sin\gamma$$

Hence,

$$I_{dc1} = i_{sx}mk\cos\gamma + i_{sy}mk\sin\gamma \tag{7.14}$$

In this section, for the simplicity of discussion, effect of dynamic and control functions associated with $I_{dc2}$ (see, Fig. 4.1) is not considered, and the VSC adopts the PWM (pulse width modulation) algorithm. Hence, the dynamic equation on the DC side of the VSC is given as follows [1]:

$$\dot{V}_{dc} = \frac{1}{C_{dc}}I_{dc1} = \frac{1}{C_{dc}}(i_{sx}mk\cos\gamma + i_{sy}mk\sin\gamma) \tag{7.15}$$

AC and DC voltage control functions implemented via the DC/AC converter are as follows:

$$\begin{aligned} m &= m_0 + T_m(s)(V_s - V_{sref}) + u_{vsc-pss} \\ \phi &= \phi_0 + T_{dc}(s)(V_{dc} - V_{dcref}) \end{aligned} \tag{7.16}$$

where $u_{vsc-pss}$ is the stabilizing control signal, and $\phi$ is the angle between $\bar{V}_s$ and $\bar{V}_c$

$$\mathrm{ac}\tan^{-1}\frac{v_{sy}}{v_{sx}} = \phi + \gamma \tag{7.17}$$

$T_m(s)$ and $T_{dc}(s)$ are the transfer functions of AC and DC voltage control function of the VSC, respectively.

Denote the elements of matrix $\bar{\mathbf{C}}$ and $\bar{\mathbf{Y}}$ in Eq. (7.8) as

$$\bar{\mathbf{C}} = \lceil \bar{c}_i \rceil, \quad \bar{\mathbf{Y}} = \left\lceil \bar{y}_{ij} \right\rceil \tag{7.18}$$

Similar to Eqs. (6.6) and (6.7), from Eqs. (7.8) and (7.18) it can have the following:

$$\begin{aligned} \bar{I}_{gi} &= i_{di} + ji_{qi} = (i_{xi} + ji_{yi})e^{j(90^\circ - \delta_i)} \\ &= e^{j}(90^\circ - \delta_i)\left(\bar{c}_i\bar{V}_c + \sum_{j=1}^{N}\bar{y}_{ij}\left[E'_{qj}e^{j\delta_j} + \left(x_{qj} - x'_{dj}\right)i_{qj}e^{j(\delta_j - 90^\circ)}\right]\right) \\ &= e^{j(90^\circ - \delta_i)}\bar{c}_i\bar{V}_c + \sum_{j=1}^{N}\bar{y}_{ij}\left[E'_{qj}e^{j(90^\circ + \delta_j - \delta_i)} + \left(x_{qj} - x'_{dj}\right)i_{qj}e^{j(\delta_j - \delta_i)}\right] \\ &= c_i mkV_{dc}e^{j(90^\circ - \delta_i + \gamma + \beta_i)} + \sum_{j=1}^{N}y_{ij}\left[E'_{qj}e^{j\left(90^\circ + \alpha_{ij} + \delta_j - \delta_i\right)} + \left(x_{qj} - x'_{dj}\right)i_{qj}e^{j(\alpha_{ij} + \delta_j - \delta_i)}\right] \end{aligned} \tag{7.19}$$

where $\bar{c}_i = c_i e^{j\beta_i}, \bar{y}_{ij} = y_{ij} e^{j\alpha_{ij}}$. From Eq. (7.19), it can be obtained that

$$\begin{aligned} i_{di} &= -c_i mkV_{dc} \sin(\gamma + \beta_i - \delta_i) \\ &\quad + \sum_{j=1}^{N} y_{ij} \left[ -E'_{qj} \sin(\alpha_{ij} - \delta_{ij}) + \left( x_{qj} - x'_{dj} \right) i_{qj} \cos(\alpha_{ij} - \delta_{ij}) \right] \\ i_{qi} &= c_i mkV_{dc} \cos(\gamma + \beta_i - \delta_i) \\ &\quad + \sum_{j=1}^{N} y_{ij} \left[ E'_{qj} \cos(\alpha_{ij} - \delta_{ij}) + \left( x_{qj} - x'_{dj} \right) i_{qj} \sin(\alpha_{ij} - \delta_{ij}) \right] \end{aligned} \tag{7.20}$$

Linearization of Eq. (7.20) can be written in the matrix form to be

$$\begin{aligned} \mathbf{\Delta I_d} &= \mathbf{F_{dd}\Delta\delta} + \mathbf{G_{dd}\Delta E'_q} + \mathbf{H_{dd}\Delta I_q} + \mathbf{L_{dd-dc}}\Delta V_{dc} + \mathbf{L_{dd-m}}\Delta m + \mathbf{L_{dd-g}}\Delta\gamma \\ \mathbf{\Delta I_q} &= \mathbf{F_{qq}\Delta\delta} + \mathbf{G_{qq}\Delta E'_q} + \mathbf{H_{qq}\Delta I_q} + \mathbf{L_{qq-dc}}\Delta V_{dc} + \mathbf{L_{qq-m}}\Delta m + \mathbf{L_{qq-g}}\Delta\gamma \end{aligned} \tag{7.21}$$

That can give the following:

$$\begin{aligned} \mathbf{\Delta I_d} &= \mathbf{F_d\Delta\delta} + \mathbf{G_d\Delta E'_q} + \mathbf{L_{d-dc}}\Delta V_{dc} + \mathbf{L_{d-m}}\Delta m + \mathbf{L_{d-g}}\Delta\gamma \\ \mathbf{\Delta I_q} &= \mathbf{F_q\Delta\delta} + \mathbf{G_q\Delta E'_q} + \mathbf{L_{q-dc}}\Delta V_{dc} + \mathbf{L_{q-m}}\Delta m + \mathbf{L_{q-g}}\Delta\gamma \end{aligned} \tag{7.22}$$

Substituting Eq. (7.22) into Eqs. (5.40) and (5.41), it can have the following:

$$\begin{aligned} \mathbf{\Delta\dot{\delta}} &= \omega_0 \mathbf{\Delta\omega} \\ \mathbf{\Delta\dot{\omega}} &= \mathbf{M}^{-1}(-\mathbf{K_1\Delta\delta} - \mathbf{K_2\Delta E'_q} - \mathbf{D\Delta\omega} + \mathbf{K_{p-dc}}\Delta V_{dc} + \mathbf{K_{p-m}}\Delta m + \mathbf{K_{p-g}}\Delta\gamma) \\ \mathbf{\Delta\dot{E}'_q} &= \mathbf{T'^{-1}_{d0}}(-\mathbf{K_3\Delta E'_q} - \mathbf{K_4\Delta\delta} + \mathbf{\Delta E'_{fd}} + \mathbf{K_{q-dc}}\Delta V_{dc} + \mathbf{K_{q-m}}\Delta m + \mathbf{K_{q-g}}\Delta\gamma) \\ \mathbf{\Delta\dot{E}'_{fd}} &= -\mathbf{T_A^{-1}\Delta E'_{fd}} + \mathbf{T_A^{-1}K_A}(-\mathbf{K_5\Delta\delta} - \mathbf{K_6\Delta E'_q} + \mathbf{K_{v-dc}}\Delta V_{dc} + \mathbf{K_{v-m}}\Delta m + \mathbf{K_{v-g}}\Delta\gamma) \end{aligned} \tag{7.23}$$

Linearization of Eqs. (7.10) and (7.11) is given as follows:

$$\begin{aligned} \Delta i_{sx} &= a_{x-dc}\Delta V_{dc} + a_{x-m}\Delta m + a_{x-g}\Delta\gamma + \mathbf{a_{x-d}\Delta\delta} + \mathbf{a_{x-e}\Delta E'_q} \\ \Delta i_{sy} &= a_{y-dc}\Delta V_{dc} + a_{y-m}\Delta m + a_{y-g}\Delta\gamma + \mathbf{a_{y-d}\Delta\delta} + \mathbf{a_{y-e}\Delta E'_q} \end{aligned} \tag{7.24}$$

By using Eq. (7.24), linearization of Eq. (7.15) can be obtained as

$$\Delta\dot{V}_{dc} = \mathbf{K_7\Delta\delta} + \mathbf{K_8\Delta E'_q} + K_9\Delta V_{dc} + K_{10}\Delta m + K_{11}\Delta\gamma \tag{7.25}$$

Linearization of Eqs. (7.11) and (7.12) can give the following:

$$\Delta v_{sx} = b_{x-dc}\Delta V_{dc} + b_{x-m}\Delta m + b_{x-g}\Delta\gamma + \mathbf{b_{x-d}}\boldsymbol{\Delta\delta} + \mathbf{b_{x-e}}\boldsymbol{\Delta E'_q}$$
$$\Delta v_{sy} = b_{y-dc}\Delta V_{dc} + b_{y-m}\Delta m + b_{y-g}\Delta\gamma + \mathbf{b_{y-d}}\boldsymbol{\Delta\delta} + \mathbf{b_{y-e}}\boldsymbol{\Delta E'_q}$$

Hence, linearization of Eq. (7.16) is given as

$$\begin{aligned}\Delta m &= T_m(s)\Delta V_s + \Delta u_{vsc-pss}\\ &= T_m(s)(b_{dc}\Delta V_{dc} + b_m\Delta m + b_g\Delta\gamma + \mathbf{b_d}\boldsymbol{\Delta\delta} + \mathbf{b_e}\boldsymbol{\Delta E'_q}) + \Delta u_{vsc-pss}\\ \Delta\phi &= T_{dc}(s)\Delta V_{dc}\end{aligned} \tag{7.26}$$

That is,

$$\begin{aligned}\Delta m &= T_m(s)\Delta V_s + \Delta u_{vsc-pss}\\ &= \frac{T_m(s)}{1 - b_m T_m(s)}(b_{dc}\Delta V_{dc} + b_g\Delta\gamma + \mathbf{b_d}\boldsymbol{\Delta\delta} + \mathbf{b_e}\boldsymbol{\Delta E'_q}) + \Delta u_{vsc-pss}\\ \Delta\phi &= T_{dc}(s)\Delta V_{dc}\end{aligned} \tag{7.27}$$

By using Eq. (7.26), linearization of Eq. (7.17) can be obtained as

$$\Delta\gamma = \mathbf{K_{gd}}\boldsymbol{\Delta\delta} + \mathbf{K_{gq}}\boldsymbol{\Delta E'_q} + K_{gdc}\Delta V_{dc} + K_{gm}\Delta m + K_{gf}\Delta\phi \tag{7.28}$$

By substituting Eq. (7.28) into Eqs. (7.23), (7.25), and (7.27), Heffron–Phillips model of the power system installed with the shunt VSC-based stabilizer can be obtained as

$$\begin{aligned}\boldsymbol{\Delta\dot\delta} &= \omega_0\boldsymbol{\Delta\omega}\\ \boldsymbol{\Delta\dot\omega} &= \mathbf{M}^{-1}(-\mathbf{K'_1}\boldsymbol{\Delta\delta} - \mathbf{K'_2}\boldsymbol{\Delta E'_q} - \mathbf{D}\boldsymbol{\Delta\omega} + \mathbf{K'_{p-dc}}\Delta V_{dc} + \mathbf{K'_{p-m}}\Delta m + \mathbf{K'_{p-g}}\Delta\phi)\\ \boldsymbol{\Delta\dot E'_q} &= \mathbf{T'^{-1}_{d0}}(-\mathbf{K'_3}\boldsymbol{\Delta E'_q} - \mathbf{K'_4}\boldsymbol{\Delta\delta} + \boldsymbol{\Delta E'_{fd}} + \mathbf{K'_{q-dc}}\Delta V_{dc} + \mathbf{K'_{q-m}}\Delta m + \mathbf{K'_{q-g}}\Delta\phi)\\ \boldsymbol{\Delta\dot E'_{fd}} &= -\mathbf{T_A^{-1}}\boldsymbol{\Delta E'_{fd}} + \mathbf{T_A^{-1}}\mathbf{K_A}(-\mathbf{K'_5}\boldsymbol{\Delta\delta} - \mathbf{K'_6}\boldsymbol{\Delta E'_q} + \mathbf{K'_{v-dc}}\Delta V_{dc} + \mathbf{K'_{v-m}}\Delta m + \mathbf{K'_{v-g}}\Delta\phi)\\ \Delta\dot V_{dc} &= \mathbf{K'_7}\boldsymbol{\Delta\delta} + \mathbf{K'_8}\boldsymbol{\Delta E'_q} + K'_9\Delta V_{dc} + K'_{10}\Delta m + K'_{11}\Delta\phi\\ \Delta m &= \frac{T_m(s)}{1 - b_m T_m(s)}(b'_{dc}\Delta V_{dc} + b'_g\Delta\phi + \mathbf{b'_d}\boldsymbol{\Delta\delta} + \mathbf{b'_e}\boldsymbol{\Delta E'_q}) + \Delta u_{vsc-pss}\\ \Delta\phi &= T_{dc}(s)\Delta V_{dc}\end{aligned} \tag{7.29}$$

The model is shown in Figs. 7.2 and 7.3 [2, 3].

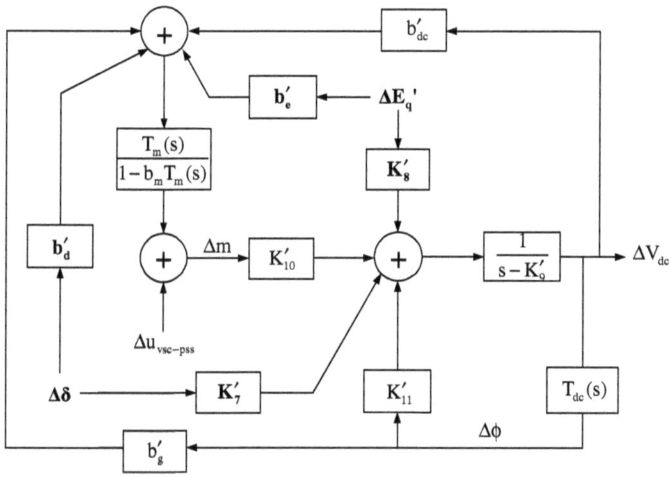


Fig. 7.2 Heffron–Phillips model—part of shunt VSC

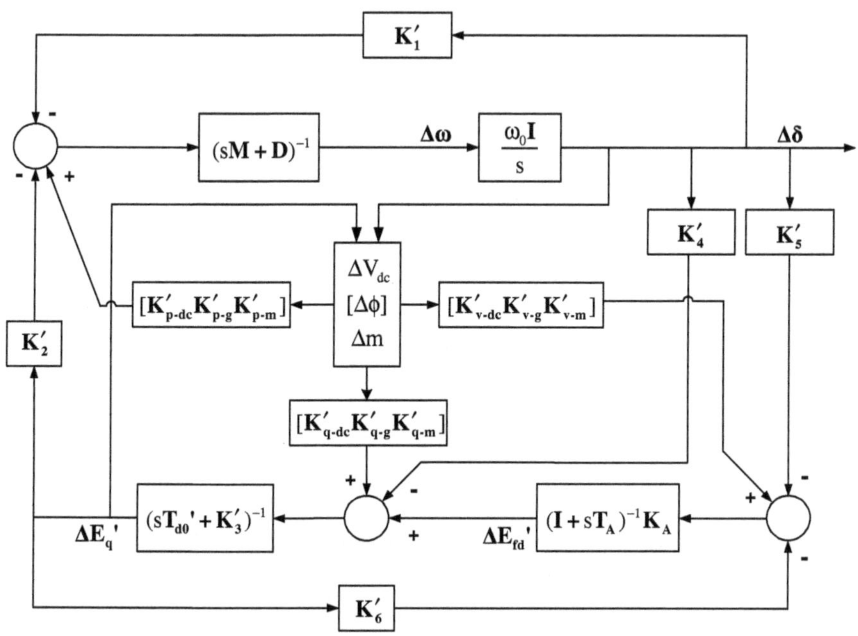


Fig. 7.3 Heffron–Phillips model—part of power system

#### 7.1.1.2 General Linearized Model of an N-Machine Power System Installed with a Shunt VSC-Based Stabilizer

Let $\mathbf{\Delta V_{sxy}} = \begin{bmatrix} \Delta v_{sx} \\ \Delta v_{sy} \end{bmatrix}$. Linearization of Eq. (7.17) can be obtained as

$$\Delta\gamma = -\Delta\phi + \mathbf{a}_s^T \mathbf{\Delta V_{sxy}} \tag{7.30}$$

Equation (7.1) can be written as

$$v_{sx} + jv_{sy} = jx_s(i_{sx} + ji_{sy}) + mkV_{dc}\cos\gamma + jmkV_{dc}\cos\gamma \tag{7.31}$$

By using Eq. (7.30), linearization of Eq. (7.31) can give the following:

$$\begin{aligned} \Delta i_{sx} &= c_{x-dc}\Delta V_{dc} + c_{x-m}\Delta m + c_{x-f}\Delta\phi + \mathbf{a}_{sx}^T \mathbf{\Delta V_{sxy}} \\ \Delta i_{sy} &= c_{y-dc}\Delta V_{dc} + c_{y-m}\Delta m + c_{y-f}\Delta\phi + \mathbf{a}_{sy}^T \mathbf{\Delta V_{sxy}} \end{aligned} \tag{7.32}$$

Hence, by using Eq. (7.32), linearization of Eq. (7.15) can be obtained as

$$\Delta\dot{V}_{dc} = c_{d-dc}\Delta V_{dc} + c_{d-m}\Delta m + c_{d-f}\Delta\phi + \mathbf{a}_{sdc}^T \mathbf{\Delta V_{sxy}} \tag{7.33}$$

Because

$$\Delta V_s = \begin{bmatrix} \frac{V_{sx0}}{V_{s0}} & \frac{V_{sy0}}{V_{s0}} \end{bmatrix} \mathbf{\Delta V_{sxy}}, \tag{7.34}$$

the state-space realization of the AC controller of the shunt VSC (the first equation in Eq. (7.26)) can be written as

$$\begin{aligned} \dot{\mathbf{X}}_{AC} &= \mathbf{A}_{AC}\mathbf{X}_{AC} + \mathbf{B}_{AC}\mathbf{\Delta V_{sxy}} \\ \Delta m &= \mathbf{C}_{AC}\mathbf{X}_{AC} + \mathbf{D}_{AC}\mathbf{\Delta V_{sxy}} + \Delta u_{vsc-pss} \end{aligned} \tag{7.35}$$

Let the state-space realization of the DC controller of the shunt VSC (the second equation in Eq. (7.26)) be

$$\begin{aligned} \dot{\mathbf{X}}_{DC} &= \mathbf{A}_{DC}\mathbf{X}_{DC} + \mathbf{B}_{DC}\Delta V_{dc} \\ \Delta\phi &= \mathbf{C}_{DC}\mathbf{X}_{DC} + \mathbf{D}_{DC}\Delta V_{dc} \end{aligned} \tag{7.36}$$

From Eqs. (7.33), (7.35), and (7.36), it can have the following:

$$\begin{aligned} \Delta\dot{V}_{dc} = {} & (c_{d-dc} + c_{d-f}D_{DC})\Delta V_{dc} + c_{d-m}\mathbf{C}_{AC}\mathbf{X}_{AC} + c_{d-f}\mathbf{C}_{DC}\mathbf{X}_{DC} \\ & + (c_{d-m}\mathbf{D}_{AC} + \mathbf{a}_{sdc}^T)\mathbf{\Delta V_{sxy}} + c_{d-m}\Delta u_{vsc-pss} \end{aligned} \tag{7.37}$$

From Eqs. (7.35)–(7.37), the overall state equation of the shunt VSC dynamics and control functions can be written as

$$\dot{\mathbf{X}}_{\mathrm{VSC}} = \mathbf{A}_{\mathrm{VSC}}\mathbf{X}_{\mathrm{AC}} + \mathbf{B}_{\mathrm{VSC}}\mathbf{\Delta V}_{\mathbf{sxy}} + \mathbf{B}_{\mathbf{vsc-pss}}\Delta \mathrm{u}_{\mathrm{vsc-pss}} \tag{7.38}$$

where

$$\mathbf{X}_{\mathrm{AC}} = \begin{bmatrix} \Delta \mathrm{V}_{\mathrm{dc}} \\ \mathbf{X}_{\mathrm{AC}} \\ \mathbf{X}_{\mathrm{DC}} \end{bmatrix}, \quad \mathbf{A}_{\mathrm{VSC}} = \begin{bmatrix} \mathrm{c}_{\mathrm{d-dc}} + \mathrm{c}_{\mathrm{d-f}}\mathrm{D}_{\mathrm{DC}} & \mathrm{c}_{\mathrm{d-m}}\mathbf{C}_{\mathrm{AC}} & \mathrm{c}_{\mathrm{d-f}}\mathbf{C}_{\mathbf{DC}} \\ \mathbf{0} & \mathbf{A}_{\mathrm{AC}} & \mathbf{0} \\ \mathbf{B}_{\mathrm{DC}} & \mathbf{0} & \mathbf{A}_{\mathrm{DC}} \end{bmatrix},$$

$$\mathbf{B}_{\mathrm{VSC}} = \begin{bmatrix} \mathrm{c}_{\mathrm{d-m}}\mathbf{D}_{\mathbf{AC}} + \mathbf{a}_{\mathbf{sdc}}^{\mathrm{T}} \\ \mathbf{B}_{\mathrm{AC}} \\ \mathbf{0} \end{bmatrix}, \quad \mathbf{B}_{\mathbf{vsc-pss}} = \begin{bmatrix} \mathrm{c}_{\mathrm{d-m}} \\ \mathbf{0} \\ \mathbf{0} \end{bmatrix}$$

By using Eqs. (7.35) and (7.36), Eq. (7.32) becomes as

$$\begin{aligned} \Delta \mathrm{i}_{\mathrm{sx}} &= (\mathrm{c}_{\mathrm{x-dc}} + \mathrm{c}_{\mathrm{x-f}}\mathrm{D}_{\mathrm{DC}})\Delta \mathrm{V}_{\mathrm{dc}} + \mathrm{c}_{\mathrm{x-m}}\mathbf{C}_{\mathrm{AC}}\mathbf{X}_{\mathrm{AC}} + \mathrm{c}_{\mathrm{x-f}}\mathbf{C}_{\mathbf{DC}}\mathbf{X}_{\mathbf{DC}} \\ &\quad + (\mathrm{c}_{\mathrm{x-m}}\mathbf{D}_{\mathrm{AC}} + \mathbf{a}_{\mathbf{sx}}^{\mathrm{T}})\mathbf{\Delta V}_{\mathbf{sxy}} + \mathrm{c}_{\mathrm{x-m}}\Delta \mathrm{u}_{\mathrm{vsc-pss}} \\ \Delta \mathrm{i}_{\mathrm{sy}} &= (\mathrm{c}_{\mathrm{y-dc}} + \mathrm{c}_{\mathrm{y-f}}\mathrm{D}_{\mathrm{DC}})\Delta \mathrm{V}_{\mathrm{dc}} + \mathrm{c}_{\mathrm{y-m}}\mathbf{C}_{\mathrm{AC}}\mathbf{X}_{\mathrm{AC}} + \mathrm{c}_{\mathrm{y-f}}\mathbf{C}_{\mathbf{DC}}\mathbf{X}_{\mathbf{DC}} \\ &\quad + (\mathrm{c}_{\mathrm{y-m}}\mathbf{D}_{\mathrm{AC}} + \mathbf{a}_{\mathbf{sx}}^{\mathrm{T}})\mathbf{\Delta V}_{\mathbf{sxy}} + \mathrm{c}_{\mathrm{y-m}}\Delta \mathrm{u}_{\mathrm{vsc-pss}} \end{aligned} \tag{7.39}$$

Let $\mathbf{\Delta I}_{\mathbf{sxy}} = \begin{bmatrix} \Delta \mathrm{i}_{\mathrm{sx}} \\ \Delta \mathrm{i}_{\mathrm{sy}} \end{bmatrix}$. From Eq. (7.39), it can have the following:

$$\mathbf{\Delta I}_{\mathbf{sxy}} = \mathbf{C}_{\mathrm{VSC}}\mathbf{X}_{\mathrm{VSC}} + \mathbf{D}_{\mathrm{VSC}}\mathbf{\Delta V}_{\mathbf{sxy}} + \mathbf{D}_{\mathbf{vsc-pss}}\Delta \mathrm{u}_{\mathrm{vsc-pss}} \tag{7.40}$$

where

$$\mathbf{C}_{\mathrm{VSC}} = \begin{bmatrix} \mathrm{c}_{\mathrm{x-dc}} + \mathrm{c}_{\mathrm{x-f}}\mathrm{D}_{\mathrm{DC}} & \mathrm{c}_{\mathrm{x-m}}\mathbf{C}_{\mathrm{AC}} & \mathrm{c}_{\mathrm{x-f}}\mathbf{C}_{\mathbf{DC}} \\ \mathrm{c}_{\mathrm{y-dc}} + \mathrm{c}_{\mathrm{y-f}}\mathrm{D}_{\mathrm{DC}} & \mathrm{c}_{\mathrm{y-m}}\mathbf{C}_{\mathrm{AC}} & \mathrm{c}_{\mathrm{y-f}}\mathbf{C}_{\mathbf{DC}} \end{bmatrix},$$

$$\mathbf{D}_{\mathrm{VSC}} = \begin{bmatrix} \mathrm{c}_{\mathrm{x-m}}\mathbf{D}_{\mathrm{AC}} + \mathbf{a}_{\mathbf{sx}}^{\mathrm{T}} \\ \mathrm{c}_{\mathrm{y-m}}\mathbf{D}_{\mathrm{AC}} + \mathbf{a}_{\mathbf{sy}}^{\mathrm{T}} \end{bmatrix}, \quad \mathbf{D}_{\mathbf{vsc-pss}} = \begin{bmatrix} \mathrm{c}_{\mathrm{x-m}} \\ \mathrm{c}_{\mathrm{y-m}} \end{bmatrix}$$

If it is assumed that the feedback signal of the shunt VSC-based stabilizer is the active power, $\mathrm{P}_{\mathrm{s2}}$, delivered along the transmission line where the shunt VSC is installed, similar to Eq. (6.40), it can have the following:

$$\Delta \mathrm{P}_{\mathrm{s2}} = \mathbf{p}_{\mathbf{s-vsc}}\mathbf{\Delta V}_{\mathbf{sxy}} + \mathbf{p}_{\mathbf{2-vsc}}\mathbf{\Delta V}_{\mathbf{xy2}} \tag{7.41}$$

where $\Delta \mathbf{V}_{\mathbf{xy2}} = \begin{bmatrix} \Delta v_{x2} \\ \Delta v_{y2} \end{bmatrix}$ and $v_{x2} + jv_{y2} = \bar{V}_2$. Similar to Eq. (6.98), let the linearized network equation of the N-machine power system with the shunt VSC-based stabilizer installed be (only with N nodes of generators, node s and 2 of system of Fig. 7.1 left) the following:

$$\begin{bmatrix} \Delta \mathbf{I}_{xy} \\ \Delta \mathbf{I}_{sxy} \\ \mathbf{0} \end{bmatrix} = \begin{bmatrix} \mathbf{Y}_{gg} & \mathbf{Y}_{gs} & \mathbf{Y}_{g2} \\ \mathbf{Y}_{sg} & \mathbf{Y}_{ss} & \mathbf{Y}_{s2} \\ \mathbf{Y}_{2g} & \mathbf{Y}_{2s} & \mathbf{Y}_{22} \end{bmatrix} \begin{bmatrix} \Delta \mathbf{V}_{xy} \\ \Delta \mathbf{V}_{sxy} \\ \Delta \mathbf{V}_{xy2} \end{bmatrix} \tag{7.42}$$

By substituting Eq. (6.29) and (7.40) into Eq. (7.42), it can have the following:

$$\begin{bmatrix} \mathbf{C}_{gxy}\mathbf{X}_{g} \\ \mathbf{C}_{VSC}\mathbf{X}_{VSC} + \mathbf{D}_{vsc-pss}\Delta u_{vsc-pss} \\ \mathbf{0} \end{bmatrix} = \begin{bmatrix} \mathbf{Y}_{gg} - \mathbf{D}_{gxy} & \mathbf{Y}_{gs} & \mathbf{Y}_{g2} \\ \mathbf{Y}_{sg} & \mathbf{Y}_{ss} - \mathbf{D}_{VSC} & \mathbf{Y}_{s2} \\ \mathbf{Y}_{2g} & \mathbf{Y}_{2s} & \mathbf{Y}_{22} \end{bmatrix} \begin{bmatrix} \Delta \mathbf{V}_{xy} \\ \Delta \mathbf{V}_{sxy} \\ \Delta \mathbf{V}_{xy2} \end{bmatrix} \tag{7.43}$$

From Eq. (7.43), it can be obtained that

$$\begin{aligned} \Delta \mathbf{V}_{2xy} &= -\mathbf{Y}_{22}^{-1}[\mathbf{Y}_{2g} \quad \mathbf{Y}_{2s}] \begin{bmatrix} \Delta \mathbf{V}_{xy} \\ \Delta \mathbf{V}_{sxy} \end{bmatrix} \\ &\quad \times \begin{bmatrix} \mathbf{C}_{gxy} & \mathbf{0} \\ \mathbf{0} & \mathbf{C}_{VSC} \end{bmatrix} \begin{bmatrix} \mathbf{X}_{g} \\ \mathbf{X}_{VSC} \end{bmatrix} + \begin{bmatrix} \mathbf{0} \\ \mathbf{D}_{vsc-pss} \end{bmatrix} \Delta u_{vsc-pss} \\ &= \left( \begin{bmatrix} \mathbf{Y}_{gg} - \mathbf{D}_{gxy} & \mathbf{Y}_{gs} \\ \mathbf{Y}_{sg} & \mathbf{Y}_{ss} - \mathbf{D}_{VSC} \end{bmatrix} - \begin{bmatrix} \mathbf{Y}_{g2} \\ \mathbf{Y}_{s2} \end{bmatrix} \mathbf{Y}_{22}^{-1}[\mathbf{Y}_{2g} \quad \mathbf{Y}_{2s}] \right) \begin{bmatrix} \Delta \mathbf{V}_{xy} \\ \Delta \mathbf{V}_{sxy} \end{bmatrix} \\ &= \mathbf{Y}_{gsxy} \begin{bmatrix} \Delta \mathbf{V}_{xy} \\ \Delta \mathbf{V}_{sxy} \end{bmatrix} \end{aligned} \tag{7.44}$$

By arranging the state equation of the generators of Eq. (6.29) and that of the shunt VSC of Eq. (7.38) together, it can have the following:

$$\begin{bmatrix} \dot{\mathbf{X}}_{g} \\ \dot{\mathbf{X}}_{VSC} \end{bmatrix} = \begin{bmatrix} \mathbf{A}_{gxy} & \mathbf{0} \\ \mathbf{0} & \mathbf{A}_{VSC} \end{bmatrix} \begin{bmatrix} \mathbf{X}_{g} \\ \mathbf{X}_{VSC} \end{bmatrix} + \begin{bmatrix} \mathbf{B}_{gxy} & \mathbf{0} \\ \mathbf{0} & \mathbf{B}_{VSC} \end{bmatrix} \begin{bmatrix} \Delta \mathbf{V}_{xy} \\ \Delta \mathbf{V}_{sxy} \end{bmatrix} + \begin{bmatrix} \mathbf{0} \\ \mathbf{B}_{vsc-pss} \end{bmatrix} \Delta u_{vsc-pss} \tag{7.45}$$

By substituting Eq. (7.44) into (7.45), the open-loop state equation of the system without the shunt VSC-based stabilizer installed can be obtained as

$$\dot{\mathbf{X}} = \mathbf{A}\mathbf{X} + \mathbf{B}\Delta u_{vsc-pss} \tag{7.46}$$

where

$$\mathbf{X} = \begin{bmatrix} \mathbf{X_g} \\ \mathbf{X_{VSC}} \end{bmatrix}, \quad \mathbf{B} = \begin{bmatrix} \mathbf{0} \\ \mathbf{B_{vsc-pss}} \end{bmatrix} + \begin{bmatrix} \mathbf{B_{gxy}} & \mathbf{0} \\ \mathbf{0} & \mathbf{B_{VSC}} \end{bmatrix} \mathbf{Y}_{\mathbf{gsxy}}^{-1} \begin{bmatrix} \mathbf{0} \\ \mathbf{D_{vsc-pss}} \end{bmatrix}$$

$$\mathbf{A} = \begin{bmatrix} \mathbf{A_{gxy}} & \mathbf{0} \\ \mathbf{0} & \mathbf{A_{VSC}} \end{bmatrix} + \begin{bmatrix} \mathbf{B_{gxy}} & \mathbf{0} \\ \mathbf{0} & \mathbf{B_{VSC}} \end{bmatrix} \mathbf{Y}_{\mathbf{gsxy}}^{-1} \begin{bmatrix} \mathbf{C_{gxy}} & \mathbf{0} \\ \mathbf{0} & \mathbf{C_{VSC}} \end{bmatrix}$$

From Eqs. (7.41) and (7.44), it can have the following:

$$\begin{aligned} \Delta P_{s2} &= \left( \begin{bmatrix} \mathbf{0} & \mathbf{p_{s-vsc}} \end{bmatrix} - \mathbf{p_{2-vsc}} \mathbf{Y}_{22}^{-1} \begin{bmatrix} \mathbf{Y_{2g}} & \mathbf{Y_{2s}} \end{bmatrix} \right) \begin{bmatrix} \Delta \mathbf{V_{xy}} \\ \Delta \mathbf{V_{sxy}} \end{bmatrix} \\ &= \left( \begin{bmatrix} \mathbf{0} & \mathbf{p_{s-vsc}} \end{bmatrix} - \mathbf{p_{2-vsc}} \mathbf{Y}_{22}^{-1} \begin{bmatrix} \mathbf{Y_{2g}} & \mathbf{Y_{2s}} \end{bmatrix} \right) \mathbf{Y}_{\mathbf{gsxy}}^{-1} \\ &\quad \times \left( \begin{bmatrix} \mathbf{C_{gxy}} & \mathbf{0} \\ \mathbf{0} & \mathbf{C_{VSC}} \end{bmatrix} \begin{bmatrix} \mathbf{X_g} \\ \mathbf{X_{VSC}} \end{bmatrix} + \begin{bmatrix} \mathbf{0} \\ \mathbf{D_{vsc-pss}} \end{bmatrix} \right) \Delta u_{vsc-pss} \end{aligned} \tag{7.47}$$

Hence, the output equation of the system is given as

$$\Delta y = \mathbf{CX} + \mathbf{D} \Delta u_{vsc-pss} \tag{7.48}$$

where

$$\mathbf{C} = \left( \begin{bmatrix} \mathbf{0} & \mathbf{p_{s-vsc}} \end{bmatrix} - \mathbf{p_{2-vsc}} \mathbf{Y}_{22}^{-1} \begin{bmatrix} \mathbf{Y_{2g}} & \mathbf{Y_{2s}} \end{bmatrix} \right) \mathbf{Y}_{\mathbf{gsxy}}^{-1} \begin{bmatrix} \mathbf{C_{gxy}} & \mathbf{0} \\ \mathbf{0} & \mathbf{C_{VSC}} \end{bmatrix}$$

$$\mathbf{D} = \left( \begin{bmatrix} \mathbf{0} & \mathbf{p_{s-vsc}} \end{bmatrix} - \mathbf{p_{2-vsc}} \mathbf{Y}_{22}^{-1} \begin{bmatrix} \mathbf{Y_{2g}} & \mathbf{Y_{2s}} \end{bmatrix} \right) \mathbf{Y}_{\mathbf{gsxy}}^{-1} \begin{bmatrix} \mathbf{0} \\ \mathbf{D_{vsc-pss}} \end{bmatrix}$$

## 7.1.2 Mathematical Model of a Multi-machine Power System Installed with a UPFC-Based Stabilizer

### 7.1.2.1 Heffron–Phillips Model of an N-Machine Power System Installed with a UPFC-Based Stabilizer [4, 5]

Without loss of generality, it can be assumed that a UPFC is installed at a busbar between nodes 1 and 2 in an N-machine power system, as shown in Fig. 7.4.

From Fig. 7.4, it can be obtained (see, Sect. 4.3.1.1) as

$$\bar{V}_{Et} = jx_E \bar{I}_E + \bar{V}_E, \quad \bar{V}_{Bt} = jx_B \bar{I}_{E2} + \bar{V}_B \tag{7.49}$$

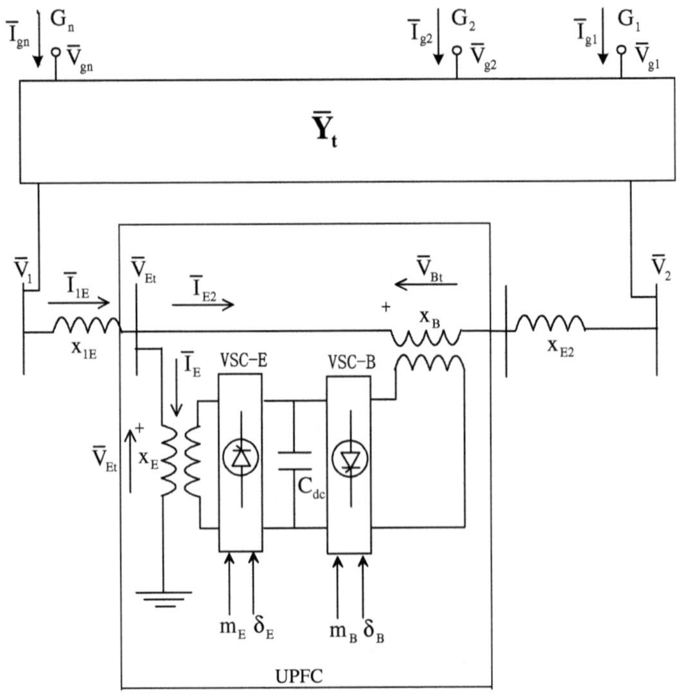


**Fig. 7.4** A multi-machine power system installed with a UPFC

and

$$\begin{aligned}\bar{V}_1 &= jx_{1E}\bar{I}_{1E} + \bar{V}_{Et}\\ \bar{V}_{Et} &= jx_{E2}\bar{I}_{E2} + \bar{V}_{Bt} + \bar{V}_2\\ \bar{I}_E &= \bar{I}_{1E} - \bar{I}_{E2}\end{aligned} \tag{7.50}$$

Substituting Eq. (7.49) into (7.50), it can have the following:

$$\begin{bmatrix}\bar{I}_{1E}\\ \bar{I}_{E2}\end{bmatrix} = \frac{1}{x_\Sigma}\begin{bmatrix}-j(x_E + x_{E2} + x_B) & jx_E\\ -jx_E & j(x_{1E} + x_E)\end{bmatrix}\begin{bmatrix}\bar{V}_1\\ \bar{V}_2\end{bmatrix} + \frac{1}{x_\Sigma}\begin{bmatrix}j(x_{E2} + x_B) & jx_E\\ -jx_E & j(x_{1E} + x_E)\end{bmatrix}\begin{bmatrix}\bar{V}_E\\ \bar{V}_B\end{bmatrix} \tag{7.51}$$

where $x_\Sigma = (x_{1E} + x_E)(x_E + x_{E2} + x_B) - x_E^2$. Similar to Eq. (7.6), it should have the following:

$$\begin{aligned}\begin{bmatrix}0\\ 0\end{bmatrix} &= \begin{bmatrix}\bar{y}'_{11} & 0\\ 0 & \bar{y}'_{22}\end{bmatrix}\begin{bmatrix}\bar{V}_1\\ \bar{V}_2\end{bmatrix} + \begin{bmatrix}\bar{I}_{1E}\\ \bar{I}_{E2}\end{bmatrix} + \begin{bmatrix}\bar{\mathbf{Y}}_{13}\\ \bar{\mathbf{Y}}_{23}\end{bmatrix}\bar{\mathbf{E}}_{\mathbf{g}}\\ \bar{\mathbf{I}}_{\mathbf{g}} &= \begin{bmatrix}\bar{\mathbf{Y}}_{31} & \bar{\mathbf{Y}}_{32}\end{bmatrix}\begin{bmatrix}\bar{V}_1\\ \bar{V}_2\end{bmatrix} + \bar{\mathbf{Y}}_{33}\bar{\mathbf{E}}_{\mathbf{g}}\end{aligned} \tag{7.52}$$

where $\bar{y}'_{11}$ and $\bar{y}'_{22}$ are obtained by excluding $x_{12} = x_{1E} + x_{E2}$ from $\bar{y}_{11}$ and $\bar{y}_{22}$ in Eq. (7.5). Substituting Eq. (7.51) into (7.52), it can be obtained that

$$\begin{aligned}\begin{bmatrix}0\\0\end{bmatrix} = &\left(\begin{bmatrix}\bar{y}'_{11} & 0\\ 0 & \bar{y}'_{22}\end{bmatrix} + \frac{1}{x_{\Sigma}}\begin{bmatrix}-j(x_E + x_{E2} + x_B) & jx_E\\ -jx_E & j(x_{1E} + x_E)\end{bmatrix}\right)\begin{bmatrix}\bar{V}_1\\ \bar{V}_2\end{bmatrix}\\ &+ \frac{1}{x_{\Sigma}}\begin{bmatrix}j(x_{E2} + x_B) & jx_E\\ -jx_E & j(x_{1E} + x_E)\end{bmatrix}\begin{bmatrix}\bar{V}_E\\ \bar{V}_B\end{bmatrix} + \begin{bmatrix}\bar{\mathbf{Y}}_{\mathbf{13}}\\ \bar{\mathbf{Y}}_{\mathbf{23}}\end{bmatrix}\bar{\mathbf{E}}_{\mathbf{g}}\end{aligned} \tag{7.53}$$

The above equation gives the following:

$$\begin{bmatrix}\bar{V}_1\\ \bar{V}_2\end{bmatrix} = -\bar{\mathbf{Y}}_{\mathbf{E}}^{-1}\left(\frac{1}{x_{\Sigma}}\begin{bmatrix}j(x_{E2} + x_B) & jx_E\\ -jx_E & j(x_{1E} + x_E)\end{bmatrix}\begin{bmatrix}\bar{V}_E\\ \bar{V}_B\end{bmatrix} + \begin{bmatrix}\bar{\mathbf{Y}}_{\mathbf{13}}\\ \bar{\mathbf{Y}}_{\mathbf{23}}\end{bmatrix}\bar{\mathbf{E}}_{\mathbf{g}}\right) \tag{7.54}$$

where $\bar{\mathbf{Y}}_{\mathbf{E}} = \begin{bmatrix}\bar{y}'_{11} & 0\\ 0 & \bar{y}'_{22}\end{bmatrix} + \frac{1}{x_{\Sigma}}\begin{bmatrix}-j(x_E + x_{E2} + x_B) & jx_E\\ -jx_E & j(x_{1E} + x_E)\end{bmatrix}$

Substituting Eq. (7.54) into (7.52), it can be obtained that

$$\begin{aligned}\bar{\mathbf{I}}_{\mathbf{g}} &= \bar{\mathbf{C}}_E\bar{V}_E + \bar{\mathbf{C}}_{\mathbf{B}}\bar{V}_B + \bar{\mathbf{Y}}\bar{\mathbf{E}}_{\mathbf{g}}\\ &= -\begin{bmatrix}\bar{\mathbf{Y}}_{\mathbf{31}} & \bar{\mathbf{Y}}_{\mathbf{32}}\end{bmatrix}\bar{\mathbf{Y}}_{\mathbf{E}}^{-1}\frac{1}{x_{\Sigma}}\begin{bmatrix}j(x_{E2} + x_B) & jx_E\\ -jx_E & j(x_{1E} + x_E)\end{bmatrix}\begin{bmatrix}\bar{V}_E\\ \bar{V}_B\end{bmatrix}\\ &\quad + \left(\bar{\mathbf{Y}}_{\mathbf{33}} - \begin{bmatrix}\bar{\mathbf{Y}}_{\mathbf{31}} & \bar{\mathbf{Y}}_{\mathbf{32}}\end{bmatrix}\bar{\mathbf{Y}}_{\mathbf{E}}^{-1}\begin{bmatrix}\bar{\mathbf{Y}}_{\mathbf{13}}\\ \bar{\mathbf{Y}}_{\mathbf{23}}\end{bmatrix}\right)\bar{\mathbf{E}}_{\mathbf{g}}\end{aligned} \tag{7.55}$$

From Eqs. (7.51) and (7.54), it can have the following:

$$\begin{aligned}\begin{bmatrix}\bar{I}_{1E}\\ \bar{I}_{E2}\end{bmatrix} = &\left(-\frac{1}{x_{\Sigma}}\begin{bmatrix}-j(x_E + x_{E2} + x_B) & jx_E\\ -jx_E & j(x_{1E} + x_E)\end{bmatrix}\bar{\mathbf{Y}}_{\mathbf{E}}^{-1} + \mathbf{I}\right)\\ &\times\frac{1}{x_{\Sigma}}\begin{bmatrix}j(x_{E2} + x_B) & jx_E\\ -jx_E & j(x_{1E} + x_E)\end{bmatrix}\begin{bmatrix}\bar{V}_E\\ \bar{V}_B\end{bmatrix}\\ &-\frac{1}{x_{\Sigma}}\begin{bmatrix}-j(x_E + x_{E2} + x_B) & jx_E\\ -jx_E & j(x_{1E} + x_E)\end{bmatrix}\bar{\mathbf{Y}}_{\mathbf{E}}^{-1}\begin{bmatrix}\bar{\mathbf{Y}}_{\mathbf{13}}\\ \bar{\mathbf{Y}}_{\mathbf{23}}\end{bmatrix}\bar{\mathbf{E}}_{\mathbf{g}}\end{aligned} \tag{7.56}$$

Hence from Fig. 7.4 and Eq. (7.56), it can have the following:

$$\begin{aligned}\bar{I}_E &= i_{Ex} + ji_{Ey} = \bar{I}_{1E} - \bar{I}_{E2} = \begin{bmatrix}1 & -1\end{bmatrix}\begin{bmatrix}\bar{I}_{1E}\\ \bar{I}_{E2}\end{bmatrix} = \bar{y}_E\bar{V}_E + \bar{y}_B\bar{V}_B + \sum_{j=1}^{N}\bar{y}_{Ej}\bar{E}_{gj}\\ \bar{I}_{E1} &= i_{E1x} + ji_{E1y} = \begin{bmatrix}1 & 0\end{bmatrix}\begin{bmatrix}\bar{I}_{1E}\\ \bar{I}_{E2}\end{bmatrix} = \bar{y}_{E1}\bar{V}_E + \bar{y}_{B1}\bar{V}_B + \sum_{j=1}^{N}\bar{y}_{E1j}\bar{E}_{gj}\\ \bar{I}_{E2} &= i_{E2x} + ji_{E2y} = \begin{bmatrix}0 & 1\end{bmatrix}\begin{bmatrix}\bar{I}_{1E}\\ \bar{I}_{E2}\end{bmatrix} = \bar{y}_{E2}\bar{V}_E + \bar{y}_{B2}\bar{V}_B + \sum_{j=1}^{N}\bar{y}_{E2j}\bar{E}_{gj}\end{aligned} \tag{7.57}$$

where $\bar{E}_{gj}$ is the jth element of $\bar{\mathbf{E}}_{\mathbf{g}}$, which is given by Eq. (7.11). From Eqs. (7.49) and (7.57), it can have the following:

$$\begin{aligned}\bar{V}_{Et} &= (jx_E\bar{y}_E + 1)\bar{V}_E + jx_E\bar{y}_B\bar{V}_B + jx_E\sum_{j=1}^{N}\bar{y}_{Ej}\bar{E}_{gj}\\ \bar{V}_{Bt} &= jx_B\bar{y}_{E2}\bar{V}_E + (jx_B\bar{y}_{B2} + 1)\bar{V}_B + jx_B\sum_{j=1}^{N}\bar{y}_{E2j}\bar{E}_{gj}\end{aligned} \tag{7.58}$$

In x–y coordinate of the power system, let the following be denoted

$$\begin{aligned}\bar{V}_E &= m_E k_E V_{dc}(\cos\gamma_E + j\sin\gamma_E) = m_E k_E V_{dc}\angle\gamma_E\\ \bar{V}_B &= m_B k_B V_{dc}(\cos\gamma_B + j\sin\gamma_B) = m_B k_B V_{dc}\angle\gamma_B\end{aligned} \tag{7.59}$$

Linearization of Eqs. (7.11) and (7.12) can give the following:

$$\begin{aligned}\Delta v_{Etx} &= b_{Etx-dc}\Delta V_{dc} + b_{Etx-mE}\Delta m_E + b_{Etx-gE}\Delta\gamma_E\\ &\quad + b_{Etx-mB}\Delta m_B + b_{Etx-gB}\Delta\gamma_B + \mathbf{b}_{\mathbf{Etx-d}}\boldsymbol{\Delta\delta} + \mathbf{b}_{\mathbf{Etx-e}}\boldsymbol{\Delta}\mathbf{E}'_{\mathbf{q}}\\ \Delta v_{Ety} &= b_{Ety-dc}\Delta V_{dc} + b_{Ety-mE}\Delta m_E + b_{Ety-gE}\Delta\gamma_E\\ &\quad + b_{Ety-mB}\Delta m_B + b_{Ety-gB}\Delta\gamma_B + \mathbf{b}_{\mathbf{Ety-d}}\boldsymbol{\Delta\delta} + \mathbf{b}_{\mathbf{Ety-e}}\boldsymbol{\Delta}\mathbf{E}'_{\mathbf{q}}\\ \Delta v_{Btx} &= b_{Btx-dc}\Delta V_{dc} + b_{Btx-mE}\Delta m_E + b_{Btx-gE}\Delta\gamma_E\\ &\quad + b_{Btx-mB}\Delta m_B + b_{Btx-gB}\Delta\gamma_B + \mathbf{b}_{\mathbf{Btx-d}}\boldsymbol{\Delta\delta} + \mathbf{b}_{\mathbf{Btx-e}}\boldsymbol{\Delta}\mathbf{E}'_{\mathbf{q}}\\ \Delta v_{Bty} &= b_{Bty-dc}\Delta V_{dc} + b_{Bty-mE}\Delta m_E + b_{Bty-gE}\Delta\gamma_E\\ &\quad + b_{Bty-mB}\Delta m_B + b_{Bty-gB}\Delta\gamma_B + \mathbf{b}_{\mathbf{Bty-d}}\boldsymbol{\Delta\delta} + \mathbf{b}_{\mathbf{Bty-e}}\boldsymbol{\Delta}\mathbf{E}'_{\mathbf{q}}\end{aligned} \tag{7.60}$$

In x–y coordinate of the power system, denote the angle between $\bar{V}_{Et}$ and $\bar{V}_E$ to be $\delta_E$ and that between $\bar{V}_{Bt}$ and $\bar{V}_B$ to be $\delta_B$; that is,

$$\begin{aligned}\text{ac}\ \tan^{-1}\frac{v_{Ety}}{v_{Etx}} &= \delta_E + \gamma_E\\ \text{ac}\ \tan^{-1}\frac{v_{Bty}}{v_{Btx}} &= \delta_B + \gamma_B\end{aligned} \tag{7.61}$$

From Eq. (7.60) and linearization of Eq. (7.61), it can be obtained that

$$\begin{aligned}\Delta\gamma_E &= \mathbf{K}_{\mathbf{gEd}}\boldsymbol{\Delta\delta} + \mathbf{K}_{\mathbf{gEq}}\boldsymbol{\Delta}\mathbf{E}'_{\mathbf{q}} + K_{gEdc}\Delta V_{dc}\\ &\quad + K_{gEmE}\Delta m_E + K_{gEdE}\Delta\delta_E + K_{gEmB}\Delta m_B + K_{gEdB}\Delta\delta_B\\ \Delta\gamma_B &= \mathbf{K}_{\mathbf{gBd}}\boldsymbol{\Delta\delta} + \mathbf{K}_{\mathbf{gBq}}\boldsymbol{\Delta}\mathbf{E}'_{\mathbf{q}} + K_{gBdc}\Delta V_{dc}\\ &\quad + K_{gBmE}\Delta m_E + K_{gBdE}\Delta\delta_E + K_{gBmB}\Delta m_B + K_{gBdB}\Delta\delta_B\end{aligned} \tag{7.62}$$

Denote the elements of matrix, $\bar{\mathbf{C}}_{\mathbf{E}}$, $\bar{\mathbf{C}}_{\mathbf{B}}$, and $\bar{\mathbf{Y}}$ in Eq. (7.55) as

$$\bar{\mathbf{C}}_{\mathbf{E}} = \lceil \bar{c}_{Ei} \rceil, \quad \bar{\mathbf{C}}_{\mathbf{B}} = \lceil \bar{c}_{Bi} \rceil, \quad \bar{\mathbf{Y}} = \left\lceil \bar{y}_{ij} \right\rceil \tag{7.63}$$

From Eqs. (6.7) and (7.55), it can have the following:

$$\begin{aligned}
\bar{I}_{gi} &= i_{di} + j i_{qi} = (i_{xi} + j i_{yi}) e^{j(90^\circ - \delta_i)} \\
&= e^{j(90^\circ - \delta_i)} (\bar{c}_{Ei} \bar{V}_E + \bar{c}_{Bi} \bar{V}_B + \sum_{j=1}^{N} \bar{y}_{ij} [E'_{qj} e^{j\delta_j} + (x_{qj} - x'_{dj}) i_{qj} e^{j(\delta_j - 90^\circ)}]) \\
&= e^{j(90^\circ - \delta_i)} (\bar{c}_{Ei} \bar{V}_E + \bar{c}_{Bi} \bar{V}_B) + \sum_{j=1}^{N} \bar{y}_{ij} [E'_{qj} e^{j(90^\circ + \delta_j - \delta_i)} + (x_{qj} - x'_{dj}) i_{qj} e^{j(\delta_j - \delta_i)}] \\
&= c_{Ei} m_E k_E V_{dc} e^{j(90^\circ - \delta_i + \gamma_E + \beta_{Ei})} + c_{Bi} m_B k_B V_{dc} e^{j(90^\circ - \delta_i + \gamma_B + \beta_{Bi})} \\
&\quad + \sum_{j=1}^{N} y_{ij} [E'_{qj} e^{j(90^\circ + \alpha_{ij} + \delta_j - \delta_i)} + (x_{qj} - x'_{dj}) i_{qj} e^{j(\alpha_{ij} + \delta_j - \delta_i)}]
\end{aligned} \tag{7.64}$$

where $\bar{c}_{Ei} = c_{Ei} e^{j\beta_{Ei}}, \bar{c}_{Bi} = c_{Bi} e^{j\beta_{Bi}}, \bar{y}_{ij} = y_{ij} e^{j\alpha_{ij}}$. From Eq. (7.64) it can be obtained that

$$\begin{aligned}
i_{di} &= -c_{Ei} m_E k_E V_{dc} \sin(\gamma_E + \beta_{Ei} - \delta_i) - c_{Bi} m_B k_B V_{dc} \sin(\gamma_B + \beta_{Bi} - \delta_i) \\
&\quad + \sum_{j=1}^{N} y_{ij} [-E'_{qj} \sin(\alpha_{ij} - \delta_{ij}) + (x_{qj} - x'_{dj}) i_{qj} \cos(\alpha_{ij} - \delta_{ij})] \\
i_{qi} &= c_{Ei} m_E k_E V_{dc} \cos(\gamma_E + \beta_{Ei} - \delta_i) + c_{Bi} m_B k_B V_{dc} \cos(\gamma_B + \beta_{Bi} - \delta_i) \\
&\quad + \sum_{j=1}^{N} y_{ij} [E'_{qj} \cos(\alpha_{ij} - \delta_{ij}) + (x_{qj} - x'_{dj}) i_{qj} \sin(\alpha_{ij} - \delta_{ij})]
\end{aligned} \tag{7.65}$$

Linearization of Eq. (7.65) is given as

$$\begin{aligned}
\mathbf{\Delta I_d} &= \mathbf{F_{dd} \Delta \delta} + \mathbf{G_{dd} \Delta E'_q} + \mathbf{H_{dd} \Delta I_q} + \mathbf{L_{dd-dc}} \Delta V_{dc} \\
&\quad + \mathbf{L_{dd-mE}} \Delta m_E + \mathbf{L_{dd-gE}} \Delta \gamma_E + \mathbf{L_{dd-mB}} \Delta m_B + \mathbf{L_{dd-gB}} \Delta \gamma_B \\
\mathbf{\Delta I_q} &= \mathbf{F_{qq} \Delta \delta} + \mathbf{G_{qq} \Delta E'_q} + \mathbf{H_{qq} \Delta I_q} + \mathbf{L_{qq-dc}} \Delta V_{dc} \\
&\quad + \mathbf{L_{qq-mE}} \Delta m_E + \mathbf{L_{qq-gE}} \Delta \gamma_E + \mathbf{L_{qq-mB}} \Delta m_B + \mathbf{L_{qq-gB}} \Delta \gamma_B
\end{aligned} \tag{7.66}$$

By using Eq. (7.62), from above equation it can have the following:

$$\begin{aligned}\Delta \mathbf{I_d} &= \mathbf{F_d}\Delta\boldsymbol{\delta} + \mathbf{G_d}\Delta\mathbf{E'_q} + \mathbf{L_{d-dc}}\Delta \mathrm{V_{dc}} \\ &\quad + \mathbf{L_{d-mE}}\Delta \mathrm{m_E} + \mathbf{L_{d-dE}}\Delta\delta_E + \mathbf{L_{d-mB}}\Delta \mathrm{m_B} + \mathbf{L_{d-dB}}\Delta\delta_B \\ \Delta \mathbf{I_q} &= \mathbf{F_q}\Delta\boldsymbol{\delta} + \mathbf{G_q}\Delta\mathbf{E'_q} + \mathbf{L_{q-dc}}\Delta \mathrm{V_{dc}} \\ &\quad + \mathbf{L_{q-mE}}\Delta \mathrm{m_E} + \mathbf{L_{q-dE}}\Delta\delta_E + \mathbf{L_{q-mB}}\Delta \mathrm{m_B} + \mathbf{L_{q-dB}}\Delta\delta_B\end{aligned} \tag{7.67}$$

Substituting Eq. (7.67) into Eqs. (5.40) and (5.41), it can be obtained that

$$\begin{aligned}\Delta\dot{\boldsymbol{\delta}} &= \omega_0\Delta\boldsymbol{\omega} \\ \Delta\dot{\boldsymbol{\omega}} &= \mathbf{M}^{-1}(-\mathbf{K_1}\Delta\boldsymbol{\delta} - \mathbf{K_2}\Delta\mathbf{E'_q} - \mathbf{D}\Delta\boldsymbol{\omega} + \mathbf{K_{p-dc}}\Delta \mathrm{V_{dc}} \\ &\quad + \mathbf{K_{p-mE}}\Delta \mathrm{m_E} + \mathbf{K_{p-dE}}\Delta\delta_E + \mathbf{K_{p-mB}}\Delta \mathrm{m_B} + \mathbf{K_{p-dB}}\Delta\delta_B) \\ \Delta\dot{\mathbf{E}}'_{\mathbf{q}} &= \mathbf{T'^{-1}_{d0}}(-\mathbf{K_3}\Delta\mathbf{E'_q} - \mathbf{K_4}\Delta\boldsymbol{\delta} + \Delta\mathbf{E'_{fd}} + \mathbf{K_{q-dc}}\Delta \mathrm{V_{dc}} \\ &\quad + \mathbf{K_{q-mE}}\Delta \mathrm{m_E} + \mathbf{K_{q-dE}}\Delta\delta_E + \mathbf{K_{q-mB}}\Delta \mathrm{m_B} + \mathbf{K_{q-dB}}\Delta\delta_B) \\ \Delta\dot{\mathbf{E}}'_{\mathbf{fd}} &= -\mathbf{T_A^{-1}}\Delta\mathbf{E'_{fd}} + \mathbf{T_A^{-1}}\mathbf{K_A}(-\mathbf{K_5}\Delta\boldsymbol{\delta} - \mathbf{K_6}\Delta\mathbf{E'_q} + \mathbf{K_{v-dc}}\Delta \mathrm{V_{dc}} \\ &\quad + \mathbf{K_{v-mE}}\Delta \mathrm{m_E} + \mathbf{K_{v-dE}}\Delta\delta_E + \mathbf{K_{v-mB}}\Delta \mathrm{m_B} + \mathbf{K_{v-dB}}\Delta\delta_B)\end{aligned} \tag{7.68}$$

Dynamic equation of the VSCs of the UPFC on the DC side is given by the following equation (see Eq. (4.73)):

$$\dot{V}_{dc} = \frac{1}{C_{dc}}\left(\frac{m_E k_E V_{Et}}{x_E}\sin\delta_E + \frac{m_B k_B V_{Bt}}{x_B}\sin\delta_B\right) \tag{7.69}$$

By using Eqs. (7.60) and (7.62), linearization of Eq. (7.69) can be obtained as

$$\begin{aligned}\Delta\dot{V}_{dc} &= \mathbf{K_7}\Delta\boldsymbol{\delta} + \mathbf{K_8}\Delta\mathbf{E'_q} + K_9\Delta V_{dc} \\ &\quad + K_{10-E}\Delta m_E + K_{11-E}\Delta\delta_E + K_{10-B}\Delta m_B + K_{11-B}\Delta\delta_B\end{aligned} \tag{7.70}$$

Without loss of generality and for the purpose of demonstration, it is assumed that the shunt VSC of the UPFC implements the AC and DC voltage control functions, the series VSC realizes the line active and reactive power regulation, and the damping control is added on the AC voltage control function as follows:

$$\begin{aligned}m_E &= m_{E0} + T_{ac}(s)(V_{Et} - V_{Etref}) + u_{UPFC-pss} \\ \delta_E &= \delta_{E0} + T_{dc}(s)(V_{dc} - V_{dcref}) \\ m_B &= m_{B0} + T_{ap}(s)(P_{Et} - P_{Etref}) \\ \delta_B &= \delta_{B0} + T_{rp}(s)(Q_{Et} - Q_{Etref})\end{aligned} \tag{7.71}$$

where $u_{UPFC-pss}$ is the damping control signal, $T_{ac}(s)$, $T_{dc}(s)$,$T_{ap}(s)$, and $T_{rp}(s)$ is the transfer function of the AC voltage, DC voltage, active, and reactive power controller of the UPFC, respectively. Linearization of Eq. (7.71) is given as

$$\begin{aligned}\Delta m_E &= T_{ac}(s)\Delta V_{Et} + \Delta u_{UPFC-pss}\\ \Delta\delta_E &= T_{dc}(s)\Delta V_{dc}\\ \Delta m_B &= T_{ap}(s)\Delta P_{Et}\\ \Delta\delta_B &= T_{rp}(s)\Delta Q_{Et}\end{aligned} \tag{7.72}$$

By using Eqs. (7.60) and (7.62), it can be obtained that

$$\begin{aligned}\Delta V_{Et} &= b_{Et-dc}\Delta V_{dc} + b_{Et-mE}\Delta m_E + b_{Et-dE}\Delta\delta_E\\ &\quad + b_{Et-mB}\Delta m_B + b_{Et-dB}\Delta\delta_B + \mathbf{b_{Et-d}\Delta\delta} + \mathbf{b_{Et-e}\Delta E'_q}\end{aligned} \tag{7.73}$$

From Eqs. (7.57) and (7.58), it can have the following:

$$\begin{aligned}P_{Et} + jQ_{Et} = \bar{V}_{Et}\bar{I}^*_{E1} &= \left[(jx_E\bar{y}_E + 1)\bar{V}_E + jx_E\bar{y}_B\bar{V}_B + jx_E\sum_{j=1}^{N}\bar{y}_{Ej}\bar{E}_{gj}\right]\\ &\quad \times\left[\bar{y}^*_{E1}\bar{V}^*_E + \bar{y}^*_{B1}\bar{V}^*_B + \sum_{j=1}^{N}\bar{y}^*_{E1j}\bar{E}^*_{gj}\right]\end{aligned} \tag{7.74}$$

By using Eq. (7.62), linearization of the above equation can be obtained as

$$\begin{aligned}\Delta P_{Et} &= b_{p-dc}\Delta V_{dc} + b_{p-mE}\Delta m_E + b_{p-dE}\Delta\delta_E\\ &\quad + b_{p-mB}\Delta m_B + b_{p-dB}\Delta\delta_B + \mathbf{b_{p-d}\Delta\delta} + \mathbf{b_{p-e}\Delta E'_q}\\ \Delta Q_{Et} &= b_{q-dc}\Delta V_{dc} + b_{q-mE}\Delta m_E + b_{q-dE}\Delta\delta_E\\ &\quad + b_{q-mB}\Delta m_B + b_{q-dB}\Delta\delta_B + \mathbf{b_{q-d}\Delta\delta} + \mathbf{b_{q-e}\Delta E'_q}\end{aligned} \tag{7.75}$$

Substituting Eqs. (7.73) and (7.75) into Eq. (7.72) gives the following:

$$\begin{aligned}\Delta m_E &= \frac{T_{ac}(s)}{1 - b_{Et-mE}T_{ac}(s)}(b_{Et-dc}\Delta V_{dc} + b_{Et-dE}\Delta\delta_E + b_{Et-mB}\Delta m_B\\ &\quad + b_{Et-dB}\Delta\delta_B + \mathbf{b_{Et-d}\Delta\delta} + \mathbf{b_{Et-e}\Delta E'_q}) + \frac{1}{1 - b_{Et-mE}T_{ac}(s)}\Delta u_{pss}\\ \Delta\delta_E &= T_{dc}(s)\Delta V_{dc}\\ \Delta m_B &= \frac{T_{ap}(s)}{1 - b_{p-mB}T_{ap}(s)}(b_{p-dc}\Delta V_{dc} + b_{p-mE}\Delta m_E\\ &\quad + b_{p-dE}\Delta\delta_E + b_{p-dB}\Delta\delta_B + \mathbf{b_{p-d}\Delta\delta} + \mathbf{b_{p-e}\Delta E'_q})\\ \Delta\delta_B &= \frac{T_{rp}(s)}{1 - b_{q-dB}T_{rp}(s)}(b_{q-dc}\Delta V_{dc} + b_{q-mE}\Delta m_E\\ &\quad + b_{q-dE}\Delta\delta_E + b_{q-mB}\Delta m_B + \mathbf{b_{q-d}\Delta\delta} + \mathbf{b_{q-e}\Delta E'_q})\end{aligned} \tag{7.76}$$

Equation (7.68), (7.70), and (7.76) are the Heffron–Phillips model of the power system installed with the UPFC.

#### 7.1.2.2 General Linearized Model of an N-Machine Power System Installed with a UPFC-Based Stabilizer

Let

$$\boldsymbol{\Delta V}_{\mathbf{Etxy}} = \begin{bmatrix} \Delta v_{Etx} \\ \Delta v_{Ety} \end{bmatrix}, \quad \boldsymbol{\Delta V}_{\mathbf{Btxy}} = \begin{bmatrix} \Delta v_{Btx} \\ \Delta v_{Bty} \end{bmatrix}, \quad \boldsymbol{\Delta V}_{\mathbf{E2xy}} = \begin{bmatrix} \Delta v_{x2} \\ \Delta v_{y2} \end{bmatrix},$$
$$\boldsymbol{\Delta I}_{\mathbf{Exy}} = \begin{bmatrix} \Delta i_{Ex} \\ \Delta i_{Ey} \end{bmatrix}, \quad \boldsymbol{\Delta I}_{\mathbf{E2xy}} = \begin{bmatrix} \Delta i_{E2x} \\ \Delta i_{E2y} \end{bmatrix}$$

From Eqs. (7.49) and (7.50), it can have the following:

$$\bar{V}_{Et} = jx_E\bar{I}_E + \bar{V}_E, \quad \bar{V}_{Et} = j(x_{E2} + x_B)\bar{I}_{E2} + \bar{V}_B + \bar{V}_2$$

that is,

$$\begin{aligned} \bar{I}_E &= \frac{1}{jx_E}(\bar{V}_{Et} - \bar{V}_E) \\ \bar{I}_{E2} &= \frac{1}{j(x_{E2} + x_B)}(\bar{V}_{Et} - \bar{V}_B - \bar{V}_2) \end{aligned} \tag{7.77}$$

Substituting Eq. (7.77) into (7.49), it can be obtained that

$$\bar{V}_{Bt} = \frac{x_B}{(x_{E2} + x_B)}(\bar{V}_{Et} - \bar{V}_B - \bar{V}_2) + \bar{V}_B \tag{7.78}$$

By using Eq. (7.59), linearization of Eqs. (7.77) and (7.78) becomes the following:

$$\begin{aligned} \boldsymbol{\Delta I}_{\mathbf{Exy}} &= a_{E-dc}\Delta V_{dc} + a_{E-m}\Delta m_E + a_{E-d}\Delta\delta_E + \mathbf{a}^{\mathbf{T}}_{\mathbf{E-Et}}\boldsymbol{\Delta V}_{\mathbf{Etxy}} \\ \boldsymbol{\Delta I}_{\mathbf{E2xy}} &= a_{E2-dc}\Delta V_{dc} + a_{E2-m}\Delta m_B + a_{E2-d}\Delta\delta_B \\ &\quad + \mathbf{a}^{\mathbf{T}}_{\mathbf{E2-Et}}\boldsymbol{\Delta V}_{\mathbf{Etxy}} + \mathbf{a}^{\mathbf{T}}_{\mathbf{E2-2}}\boldsymbol{\Delta V}_{\mathbf{E2xy}} \end{aligned} \tag{7.79}$$

$$\begin{aligned} \boldsymbol{\Delta V}_{\mathbf{Btxy}} &= b_{Bt-dc}\Delta V_{dc} + b_{Bt-m}\Delta m_B + b_{Bt-d}\Delta\delta_B \\ &\quad + \mathbf{b}^{\mathbf{T}}_{\mathbf{Bt-Et}}\boldsymbol{\Delta V}_{\mathbf{Etxy}} + \mathbf{b}^{\mathbf{T}}_{\mathbf{Bt-2}}\boldsymbol{\Delta V}_{\mathbf{E2xy}} \end{aligned} \tag{7.80}$$

By using Eqs. (7.79) and (7.80), linearizing Eq. (4.73) gives the following:

$$\begin{aligned}\Delta\dot{V}_{dc} &= c_{dc}\Delta V_{dc} + c_{mE}\Delta m_E + c_{dE}\Delta\delta_E + c_{mB}\Delta m_B \\ &\quad + c_{dB}\Delta\delta_B + \mathbf{c}_{Et}^T\boldsymbol{\Delta}\mathbf{V}_{Etxy} + \mathbf{c}_2^T\boldsymbol{\Delta}\mathbf{V}_{E2xy}\end{aligned} \tag{7.81}$$

From Eq. (7.74), it can have the following:

$$P_{Et} + jQ_{Et} = \bar{V}_{Et}\bar{I}_{E1}^{*} = \bar{V}_{Et}(\bar{I}_E + \bar{I}_{E2})^{*} \tag{7.82}$$

Hence by using Eq. (7.79), linearization of Eq. (7.82) can give the following:

$$\begin{aligned}\Delta P_{Et} &= c_{p-dc}\Delta V_{dc} + c_{p-mE}\Delta m_E + c_{p-dE}\Delta\delta_E + c_{p-mB}\Delta m_B \\ &\quad + c_{p-dB}\Delta\delta_B + \mathbf{c}_{p-Et}^T\boldsymbol{\Delta}\mathbf{V}_{Etxy} + \mathbf{c}_{p-2}^T\boldsymbol{\Delta}\mathbf{V}_{Exy2} \\ \Delta Q_{Et} &= c_{q-dc}\Delta V_{dc} + c_{q-mE}\Delta m_E + c_{q-dE}\Delta\delta_E + c_{q-mB}\Delta m_B \\ &\quad + c_{q-dB}\Delta\delta_B + \mathbf{c}_{q-Et}^T\boldsymbol{\Delta}\mathbf{V}_{Etxy} + \mathbf{c}_{q-2}^T\boldsymbol{\Delta}\mathbf{V}_{Exy2}\end{aligned} \tag{7.83}$$

Because

$$\Delta V_{Et} = \begin{bmatrix} \frac{V_{Etx0}}{V_{Et0}} & \frac{V_{Ety0}}{V_{Et0}} \end{bmatrix}\boldsymbol{\Delta}\mathbf{V}_{Etxy}, \tag{7.84}$$

the state-space realization of four UPFC controllers of Eq. (7.72) can be written as follows:

$$\begin{aligned}\dot{\mathbf{X}}_{UPFC-AC} &= \mathbf{A}_{UPFC-AC}\mathbf{X}_{UPFC-AC} + \mathbf{B}_{UPFC-AC}\boldsymbol{\Delta}\mathbf{V}_{Etxy} \\ \Delta m_E &= \mathbf{C}_{UPFC-AC}\mathbf{X}_{UPFC-AC} + \mathbf{D}_{UPFC-AC}\boldsymbol{\Delta}\mathbf{V}_{Etxy} + \Delta u_{pss}\end{aligned} \tag{7.85}$$

$$\begin{aligned}\dot{\mathbf{X}}_{UPFC-DC} &= \mathbf{A}_{UPFC-DC}\mathbf{X}_{UPFC-DC} + \mathbf{B}_{UPFC-DC}\Delta V_{dc} \\ \Delta\delta_E &= \mathbf{C}_{UPFC-DC}\mathbf{X}_{UPFC-DC} + \mathbf{D}_{UPFC-DC}\Delta V_{dc}\end{aligned} \tag{7.86}$$

$$\begin{aligned}\dot{\mathbf{X}}_{UPFC-P} &= \mathbf{A}_{UPFC-P}\mathbf{X}_{UPFC-P} + \mathbf{B}_{UPFC-P}\Delta P_{Et} \\ \Delta m_B &= \mathbf{C}_{UPFC-P}\mathbf{X}_{UPFC-P} + \mathbf{D}_{UPFC-P}\Delta P_{Et}\end{aligned} \tag{7.87}$$

$$\begin{aligned}\dot{\mathbf{X}}_{UPFC-Q} &= \mathbf{A}_{UPFC-Q}\mathbf{X}_{UPFC-Q} + \mathbf{B}_{UPFC-Q}\Delta Q_{Et} \\ \Delta\delta_B &= \mathbf{C}_{UPFC-Q}\mathbf{X}_{UPFC-AC} + \mathbf{D}_{UPFC-Q}\Delta Q_{Et}\end{aligned} \tag{7.88}$$

Denote

$$\mathbf{X}_{UPFC} = \begin{bmatrix} \Delta V_{dc} & \mathbf{X}_{UPFC-AC}^T & \mathbf{X}_{UPFC-DC}^T & \mathbf{X}_{UPFC-P}^T & \mathbf{X}_{UPFC-Q}^T \end{bmatrix}^T \tag{7.89}$$

By using Eqs. (7.83), (7.85), and (7.86), Eqs. (7.87) and (7.88) can be written as

$$\begin{aligned}\dot{\mathbf{X}}_{\mathrm{UPFC-P}} &= \mathbf{A}_{\mathrm{UPFC-PP}}\mathbf{X}_{\mathrm{UPFC}} + \mathbf{B}_{\mathrm{UPFC-PEt}}\Delta\mathbf{V}_{\mathrm{Etxy}} + \mathbf{B}_{\mathrm{UPFC-P2}}\Delta\mathbf{V}_{\mathrm{E2xy}} + \mathbf{B}_{\mathrm{UPFC-PPSS}}\Delta u_{\mathrm{pss}} \\ \Delta m_{\mathrm{B}} &= \mathbf{C}_{\mathrm{UPFC-PP}}\mathbf{X}_{\mathrm{UPFC}} + \mathbf{D}_{\mathrm{UPFC-PEt}}\Delta\mathbf{V}_{\mathrm{Etxy}} + \mathbf{D}_{\mathrm{UPFC-P2}}\Delta\mathbf{V}_{\mathrm{E2xy}} + \mathbf{D}_{\mathrm{UPFC-PPSS}}\Delta u_{\mathrm{pss}}\end{aligned} \tag{7.90}$$

$$\begin{aligned}\dot{\mathbf{X}}_{\mathrm{UPFC-Q}} &= \mathbf{A}_{\mathrm{UPFC-QQ}}\mathbf{X}_{\mathrm{UPFC}} + \mathbf{B}_{\mathrm{UPFC-QEt}}\Delta\mathbf{V}_{\mathrm{Etxy}} + \mathbf{B}_{\mathrm{UPFC-Q2}}\Delta\mathbf{V}_{\mathrm{E2xy}} + \mathbf{B}_{\mathrm{UPFC-QPSS}}\Delta u_{\mathrm{pss}} \\ \Delta \delta_{\mathrm{B}} &= \mathbf{C}_{\mathrm{UPFC-QQ}}\mathbf{X}_{\mathrm{UPFC}} + \mathbf{D}_{\mathrm{UPFC-QEt}}\Delta\mathbf{V}_{\mathrm{Etxy}} + \mathbf{D}_{\mathrm{UPFC-Q2}}\Delta\mathbf{V}_{\mathrm{E2xy}} + \mathbf{D}_{\mathrm{UPFC-QPSS}}\Delta u_{\mathrm{pss}}\end{aligned} \tag{7.91}$$

By substituting the second equation of Eqs. (7.85), (7.86), (7.90), and (7.91) into Eq. (7.81), it can have the following:

$$\Delta\dot{V}_{\mathrm{dc}} = \mathbf{c}_{\mathrm{DC-UPFC}}\mathbf{X}_{\mathrm{UPFC}} + \mathbf{c}_{\mathrm{DC-Et}}^{\mathrm{T}}\Delta\mathbf{V}_{\mathrm{Etxy}} + \mathbf{c}_{\mathrm{DC-2}}^{\mathrm{T}}\Delta\mathbf{V}_{\mathrm{E2xy}} + c_{\mathrm{DC-PSS}}\Delta u_{\mathrm{pss}} \tag{7.92}$$

From the first equation of Eqs. (7.85), (7.86), and (7.90)–(7.92), the overall state equation of the UPFC dynamic and control functions can be written as

$$\dot{\mathbf{X}}_{\mathrm{UPFC}} = \mathbf{A}_{\mathrm{UPFC}}\mathbf{X}_{\mathrm{UPFC}} + \mathbf{B}_{\mathrm{UPFC-Et}}\Delta\mathbf{V}_{\mathrm{Etxy}} + \mathbf{B}_{\mathrm{UPFC-2}}\Delta\mathbf{V}_{\mathrm{E2xy}} + \mathbf{B}_{\mathrm{UPFC-pss}}\Delta u_{\mathrm{pss}} \tag{7.93}$$

Let

$$\Delta\mathbf{V}_{\mathrm{UPFC}} = \begin{bmatrix}\Delta\mathbf{V}_{\mathrm{Etxy}}^{\mathrm{T}} & \Delta\mathbf{V}_{\mathrm{E2xy}}^{\mathrm{T}}\end{bmatrix}^{\mathrm{T}}, \quad \Delta\mathbf{I}_{\mathrm{UPFC}} = \begin{bmatrix}\Delta\mathbf{I}_{\mathrm{Exy}}^{\mathrm{T}} & \Delta\mathbf{I}_{\mathrm{E2xy}}^{\mathrm{T}}\end{bmatrix}^{\mathrm{T}}$$

From Eqs. (7.79) and (7.93), it can have the following:

$$\begin{aligned}\dot{\mathbf{X}}_{\mathrm{UPFC}} &= \mathbf{A}_{\mathrm{UPFC}}\mathbf{X}_{\mathrm{UPFC}} + \mathbf{B}_{\mathrm{UPFC}}\Delta\mathbf{V}_{\mathrm{UPFC}} + \mathbf{B}_{\mathrm{UPFC-pss}}\Delta u_{\mathrm{pss}} \\ \Delta\mathbf{I}_{\mathrm{UPFC}} &= \mathbf{C}_{\mathrm{UPFC}}\mathbf{X}_{\mathrm{UPFC}} + \mathbf{D}_{\mathrm{UPFC}}\Delta\mathbf{V}_{\mathrm{UPFC}} + \mathbf{D}_{\mathrm{UPFC-pss}}\Delta u_{\mathrm{pss}}\end{aligned} \tag{7.94}$$

Similar to Eq. (6.98) and (7.42), let the linearized network equation of the N-machine power system with the UPFC-based stabilizer installed be

$$\begin{bmatrix}\Delta\mathbf{I}_{\mathrm{xy}} \\ \Delta\mathbf{I}_{\mathrm{UPFC}} \\ \mathbf{0}\end{bmatrix} = \begin{bmatrix}\mathbf{Y}_{\mathrm{gg}} & \mathbf{Y}_{\mathrm{gUPFC}} & \mathbf{Y}_{\mathrm{g2}} \\ \mathbf{Y}_{\mathrm{UPFCg}} & \mathbf{Y}_{\mathrm{UPFC}} & \mathbf{Y}_{\mathrm{UPFC-O}} \\ \mathbf{Y}_{\mathrm{Og}} & \mathbf{Y}_{\mathrm{O-UPFC}} & \mathbf{Y}_{\mathrm{O}}\end{bmatrix}\begin{bmatrix}\Delta\mathbf{V}_{\mathrm{xy}} \\ \Delta\mathbf{V}_{\mathrm{UPFC}} \\ \Delta\mathbf{V}_{\mathrm{O}}\end{bmatrix} \tag{7.95}$$

Similar to the derivation from Eqs. (7.43)–(7.46), from Eqs. (7.94) and (7.95) the open-loop, state equation of the system without the UPFC stabilizer installed can be obtained. The output equation can be obtained by using Eqs. (7.83), (7.85), (7.86), (7.90), and (7.91).

## 7.2 Design of a Shunt VSC-Based Stabilizer by Localized Phase Compensation Method to Suppress Inter-area Line Power Oscillations

In Sect. 2.3.1.3, the phase compensation method is introduced for the design of a PSS on the basis of the Heffron–Phillips model. In Sect. 4.2.2, it is demonstrated that when a general linearized model is used, a SSSC stabilizer can be designed by the use of the phase compensation method. Those are examples of applying the phase compensation method in a single-machine infinite-bus power system. They are the cases of suppressing a local-mode power oscillation along a transmission line, where the power oscillation is closely related to the rotor motion of the synchronous generator which is connected at one end of the transmission line. Acceleration and deceleration of the generator absorbs and releases the active power in responding to the power oscillation dynamically. Hence if a stabilizer (the PSS or FACTS-based stabilizer) is designed to supply positive damping torque to damp the oscillation of the rotor motion of the generator, it thus can effectively suppress the local-mode power oscillation.

As it is introduced in Chaps. 5 and 6, a stabilizer in a multi-machine power system contributes the damping torque to the electromechanical oscillation loop of every generator. Hence, the application of the phase compensation method involves the dynamic interactions of all generators, as it is shown in Sect. 6.2.4.2. The main reason is that the phase compensation method has been introduced on the basis of the damping torque analysis. In addition, like the modal analysis, the damping torque analysis (hence the phase compensation method) needs the global information of the multi-machine power system to establish the linearized model of the whole power system, which in practice may not always be readily available and is difficult to be validated when the system is large and complex.

This section introduces a different phase compensation method for the design of a VSC-based stabilizer to suppress an inter-area line power oscillation. It is established straight on the damping analysis of the inter-area mode line power oscillation along the transmission line where the VSC-based stabilizer locates, rather than relying on the concept of the damping torque analysis. Its application considers the direct contribution from the stabilizer to the damping of the inter-area line power oscillation and does not involve in explicit consideration of interactive dynamics of generators. Moreover, it does not need to obtain and validate the complete power system information for applying the phase compensation method to design the VSC-based stabilizer, because the design only requires the locally available parameters of the transmission lines and the VSC-based unit. Hence, it is named as the localized phase compensation method.

### *7.2.1 Localized Small-Signal Model of a VSC-Based Unit in a Multi-machine Power System*

#### 7.2.1.1 Line Power Small-Signal Oscillation [6]

Figure 7.5 shows a VSC-based unit installed in a multi-machine power system. The unit has a DC/DC converter behind the DC capacitor, $C_{dc}$, which can be connected to an energy storage system. It can also be connected to a power generation plant, such as a fuel cell (FC) or photovoltaic (PV) power plant. Without the DC/DC converter ($I_{dc2} = 0$), the unit is a STATCOM. From Eq. (7.13), it can have the following:

$$\bar{V}_c = mk_c V_{dc}\angle\gamma = mk_c V_{dc}\angle\varphi + \phi \tag{7.96}$$

where the modulation phase, $\phi$, is the phase difference between $\bar{V}_c = V_c\angle\gamma$ and $\bar{V}_s = V_s\angle\varphi$ at node s, i.e. $\phi = \gamma - \varphi$. From Fig. 7.5, it can have the following:

$$\begin{aligned}\bar{V}_s &= jx_{sB}\bar{I}_{sB} + \bar{V}_B = jX_{sB}\left[\bar{I}_{As} - \left(\frac{\bar{V}_S - \bar{V}_C}{jx_s}\right)\right] + \bar{V}_B \\ &= jx_{sB}\bar{I}_{As} - \frac{x_{sB}}{x_s}\bar{V}_s + \frac{x_{sB}}{x_s}\bar{V}_c + \bar{V}_B\end{aligned} \tag{7.97}$$

The above equation gives the following:

$$\bar{V}_s = \frac{jx_{sB}}{1 + \frac{x_{sB}}{x_s}}\bar{I}_{As} + \frac{x_{sB}}{x_s(1 + \frac{x_{sB}}{x_s})}\bar{V}_c + \frac{\bar{V}_B}{1 + \frac{x_{sB}}{x_s}} \tag{7.98}$$

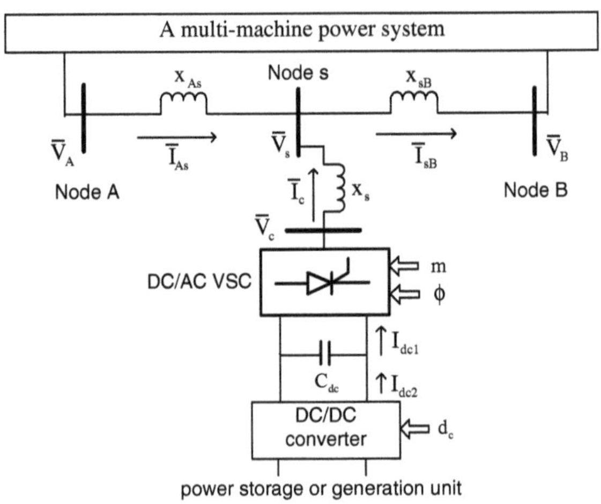


**Fig. 7.5** A VSC-based unit in a multi-machine power system

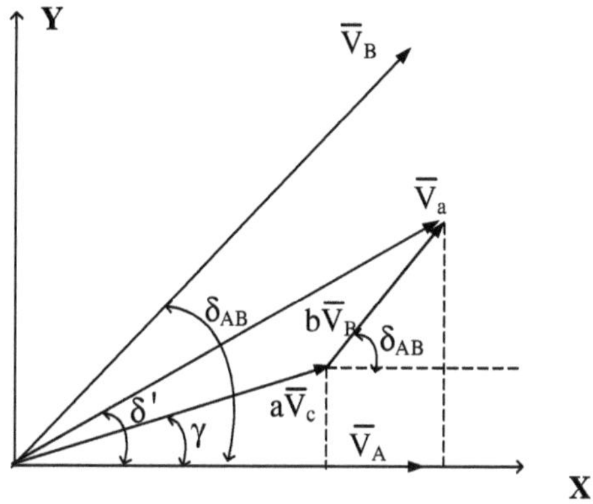


**Fig. 7.6** Phasor diagram in the x–y coordinate

Hence, it can be obtained that

$$\begin{aligned} \bar{V}_A &= jx_{As}\bar{I}_{As} + \bar{V}_s = j(x_{As} + \frac{x_s x_{sB}}{x_s + x_{sB}})\bar{I}_{As} \\ &+ \frac{x_{sB}}{x_s + x_{sB}}\bar{V}_c + \frac{x_s}{x_s + x_{sB}}\bar{V}_B = jx_\Sigma\bar{I}_{As} + \bar{V}_a \end{aligned} \tag{7.99}$$

where

$$x_\Sigma = (x_{As} + \frac{x_s x_{sB}}{x_s + x_{sB}}), \quad \bar{V}_a = \frac{x_{sB}}{x_s + x_{sB}}\bar{V}_c + \frac{x_s}{x_s + x_{sB}}\bar{V}_B = a\bar{V}_c + b\bar{V}_B$$

From Eq. (7.99) and the phasor diagram of Fig. 7.6, the active power delivered along the transmission line from node A to s in Fig. 7.5 can be obtained as

$$\begin{aligned} P_A &= \frac{V_a V_A}{x_\Sigma}\sin\delta' = \frac{V_A}{x_\Sigma}(bV_B \sin\delta_{AB} + akmV_{dc}\sin\gamma) \\ &= \frac{V_A}{x_\Sigma}bV_B\sin\delta_{AB} + \frac{V_A}{x_\Sigma}akmV_{dc}\sin(\varphi + \phi) \end{aligned} \tag{7.100}$$

Hence, the small-signal variation of line power is given as

$$\begin{aligned} \Delta P_A &= h_{AB}\Delta\delta_{AB} + h_A\Delta V_A + h_B\Delta V_B + h_\varphi\Delta\varphi \\ &+ h_{dc}\Delta V_{dc} + h_m\Delta m + h_f\Delta\phi \end{aligned} \tag{7.101}$$

where

$$h_{AB} = \frac{\partial P_A}{\partial \delta_{AB}}, \quad h_A = \frac{\partial P_A}{\partial V_A}, \quad h_B = \frac{\partial P_A}{\partial V_B}, \quad h_\varphi = \frac{\partial P_A}{\partial \varphi},$$
$$h_{dc} = \frac{\partial P_A}{\partial V_{dc}} = \frac{V_{A0}}{x_\Sigma} akm_0 \sin(\varphi_0 + \phi_0)\Delta V_{dc}$$
$$h_m = \frac{\partial P_A}{\partial m} = \frac{V_{A0}}{x_\Sigma} akV_{dc0} \sin(\varphi_0 + \phi_0)$$
$$h_f = \frac{\partial P_A}{\partial \phi} = \frac{V_{A0}}{x_\Sigma} akm_0 V_{dc0} \cos(\varphi_0 + \phi_0)$$

Equation (7.101) indicates that the small-signal variation of line power is directly affected by the variation of line terminal variables, $\Delta\delta_{AB}$, $\Delta V_A$, $\Delta V_B$, $\Delta\varphi$ and the internal variables of the VSC-based unit, $\Delta V_{dc}$, $\Delta m$, $\Delta\phi$.

#### 7.2.1.2 Control Function Model of the VSC

The VSC is controlled by the DC/AC conversion to regulate active and reactive power exchange between the VSC-based unit and the rest of the power system. Two most popular control algorithms are the pulse width modulation (PWM) and the pulse amplitude modulation (PAM). Figure 7.7 shows the configuration of a voltage control type of the PWM, where m and $\phi$ are the modulation ratio and phase of the PWM, respectively.

From Fig. 7.7, the linearized model of the PWM VSC control functions can be written as

$$\begin{aligned} \Delta m &= -\left(K_{pac} + \frac{K_{ac}}{s}\right)\Delta V_s + \Delta u_{pss} \\ \Delta\phi &= -\left(K_{pdc} + \frac{K_{dc}}{s}\right)\Delta V_{dc} + \Delta u_{pss} \end{aligned} \tag{7.102}$$

where $K_{pac}$, $K_{ac}$, $K_{pdc}$, and $K_{dc}$ is the gain of the AC and DC voltage proportional and integral (PI) controllers. Obviously, if the PWM VSC controllers are not PI

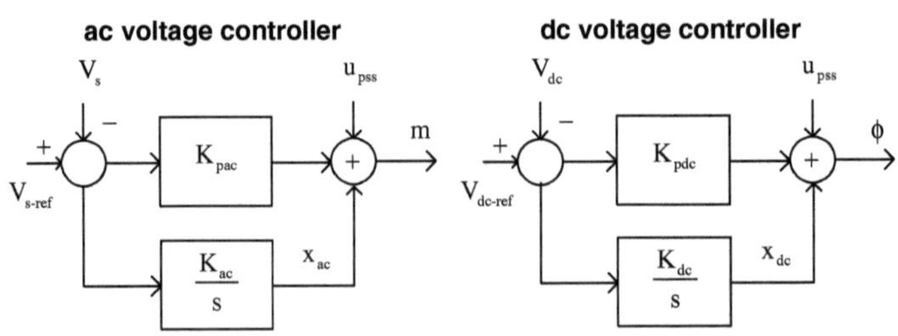


**Fig. 7.7** Configuration of PWM VSC controllers (voltage control type)

controllers, then small-signal model of the VSC control can always be expressed by the following general state-space representation as

$$\begin{aligned} s\mathbf{X}_{\mathbf{PWM}} &= \mathbf{A}_{\mathbf{PWM}}\mathbf{X}_{\mathbf{PWM}} + \mathbf{b}_{\mathbf{PWM-s}}\Delta V_s + \mathbf{b}_{\mathbf{PWM-dc}}\Delta V_{dc} \\ \Delta m &= \mathbf{C}_{AC}\mathbf{X}_{\mathbf{PWM}} + d_{AC}\Delta V_s + \Delta u_{pss} \\ \Delta\phi &= \mathbf{C}_{DC}\mathbf{X}_{\mathbf{PWM}} + d_{DC}\Delta V_{dc} + \Delta u_{pss} \end{aligned} \tag{7.103}$$

where $\mathbf{X}_{\mathbf{PWM}} = \begin{bmatrix} x_{ac} & x_{dc} \end{bmatrix}^T$ is the vector of state variables of the PWM PI controllers as shown in Eq. (7.103) and in Fig. 7.7.

Figure 7.8 shows the configuration of the voltage control type of the PAM VSC control. From Fig. 7.8, the small-signal model of the PAM VSC control can be written as

$$\Delta\phi = \left(K_{ppam} + \frac{K_{pam}}{s}\right)\left[\frac{K_{ac}}{s}(-\Delta V_s + \Delta u_{pss}) - \Delta V_{dc}\right] \tag{7.104}$$

where $K_p$, $K_{ppam}$, and $K_{pam}$ is the gain of corresponding proportional or integral controllers in Fig. 7.8. The state-space representation of Eq. (7.104) can be written as

$$\begin{aligned} s\mathbf{X}_{\mathbf{PAM}} &= \mathbf{A}_{\mathbf{PAM}}\mathbf{X}_{\mathbf{PAM}} + \mathbf{b}_{\mathbf{PAM}}\Delta V_s + \mathbf{b}_{\mathbf{pam-pss}}\Delta u_{pss} \\ \Delta\phi &= \mathbf{C}_{\mathbf{PAM}}\mathbf{X}_{\mathbf{PAM}} + d_{DC}\Delta V_{dc} \end{aligned} \tag{7.105}$$

where $\mathbf{X}_{\mathbf{PAM}} = \begin{bmatrix} x_{pam1} & x_{pam2} \end{bmatrix}^T$ the vector of state variables of the PAM PI controllers as shown in Eq. (7.104) and in Fig. 7.8.

The VSC controllers can also be the current control type, where the limit of current can be easily applied for the protection of the VSC-based unit. Figure 7.9 shows the configuration of the current control type of the VSC PWM controllers. As the difference in the control type (voltage or current) does not affect the following discussion, the voltage control type of the VSC controllers is used as the representative in this section to demonstrate how the localized small-signal model of the VSC-based unit can be established.

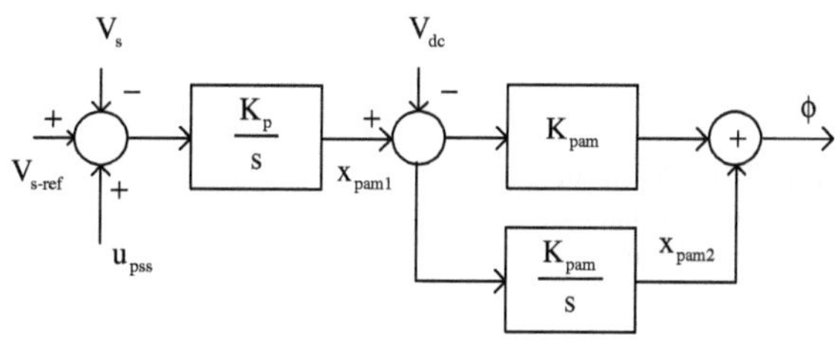


**Fig. 7.8** Configuration of PAM VSC controller (voltage control type)

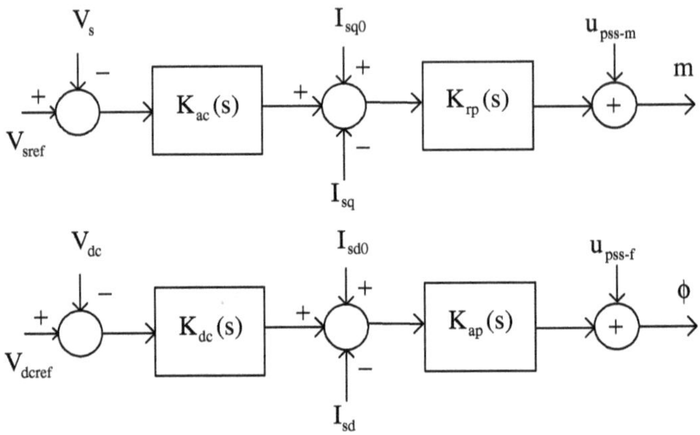


**Fig. 7.9** Configuration of PWM VSC controllers (current control type)

#### 7.2.1.3 Dynamic Equation of the VSC

From Fig. 7.5, it can have $\bar{V}_c = jx_s\bar{I}_s + \bar{V}_s$. Hence,

$$i_{sd} = \frac{mkV_{dc}\sin\phi}{x_s}, \quad i_{sq} = \frac{V_s - mkV_{dc}\cos\phi}{x_s} \tag{7.106}$$

where subscripts d and q denote the d and q components of the corresponding variable, respectively, in the d–q coordinate of the VSC-based unit, which is shown in Fig. 7.10. Linearization of the above equations is given as

$$\begin{aligned}\Delta i_{sd} &= b_{1d}\Delta V_{dc} + b_{2d}\Delta m + b_{3d}\Delta\phi \\ \Delta i_{sq} &= b_{1q}\Delta V_{dc} + b_{2q}\Delta m + b_{3q}\Delta\phi + b_{4q}\Delta V_s\end{aligned} \tag{7.107}$$

The active power exchange between the VSC-based unit and the rest of the power system is (see Eq. (7.96) given as $v_{cd} = mkV_{dc}\cos\phi$, $v_{cq} = mkV_{dc}\sin\phi$)

$$V_{dc}I_{dc1} = i_{sd}v_{cd} + i_{sq}v_{cq} = i_{sd}mkV_{dc}\cos\phi + i_{sq}mkV_{dc}\sin\phi$$

**Fig. 7.10** Phasor diagram in the d–q coordinate of VSC-based unit

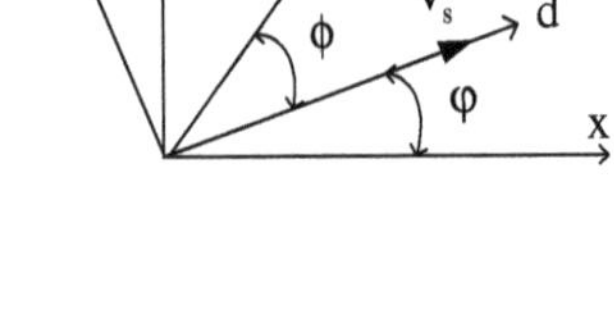

Hence

$$I_{dc1} = i_{sd} mk\cos\phi + i_{sq} mk\sin\phi \tag{7.108}$$

Thus, linearization of Eq. (7.108) can be obtained from Eq. (7.107) as

$$\Delta I_{dc1} = b_{d1}\Delta V_{dc} + b_{d2}\Delta m + b_{d3}\Delta\phi + b_{d4}\Delta V_s \tag{7.109}$$

The small-signal model of a power storage or generation unit can always be expressed by the following state-space representation:

$$\begin{aligned} s\mathbf{X_P} &= \mathbf{A_P X_P} + \mathbf{b_P}\Delta V_{dc} + \mathbf{b_{vsc}}\Delta u_{vsc} \\ \Delta I_{dc2} &= \mathbf{C_P X_P} + d_P \Delta V_{dc} \end{aligned} \tag{7.110}$$

where $\mathbf{X_P}$ and $\Delta u_{vsc}$ are the state variable vector and input of the power storage or generation unit, respectively. In Sect. 7.3, a FC (fuel cell) power plant is used as an example to demonstrate the procedure to derive the small-signal model of Eq. (7.110). A similar procedure to what is demonstrated in Sect. 7.3 can be applied to derive the model of Eq. (7.110) for the energy storage system.

Dynamic equation of the VSC is given as

$$\dot{V}_{dc} = \frac{1}{C_{dc}}(-I_{dc1} + I_{dc2}) \tag{7.111}$$

Hence by using Eqs. (7.109)–(7.111), linearization of Eq. (7.111) can be obtained as

$$s\Delta V_{dc} = b_1\Delta V_{dc} + b_2\Delta m + b_3\Delta\phi + b_4\Delta V_s + \mathbf{b_{dc} X_P} \tag{7.112}$$

#### 7.2.1.4 Localized Small-Signal Model of the VSC-Based Unit

From Fig. 7.5 and Eq. (7.99), it can have the following:

$$\begin{aligned} \bar{V}_A &= jx_\Sigma(i_{Asx} + ji_{Asy}) + \bar{V}_a \\ &= -x_\Sigma i_{Asy} + V_a\cos\delta' + jx_\Sigma i_{Asx} + jV_a\sin\delta' \end{aligned} \tag{7.113}$$

where subscripts x and y denote the x and y components of the corresponding variable, respectively, in the x–y coordinate of the power system as shown in Fig. 7.6. In x–y coordinate, Eq. (7.113) is given as

$$i_{Asx} = -\frac{V_a\sin\delta'}{x_\Sigma}, \quad i_{Asy} = \frac{V_a\cos\delta' - V_A}{x_\Sigma} \tag{7.114}$$

because

$$\begin{aligned}\bar{V}_A &= jx_{As}\bar{I}_{As} + \bar{V}_s \\ &= jx_{As}(i_{Asx} + ji_{Asy}) - X_{As}i_{AsY} + v_{sx} + jv_{sy}\end{aligned} \tag{7.115}$$

it can be obtained from Eqs. (7.114) and (7.115) that

$$\begin{aligned}v_{sx} &= V_A + x_{As}i_{Asy} = V_A + x_{As}\frac{V_a\cos\delta' - V_A}{x_\Sigma} \\ &= \frac{x_{As}}{x_\Sigma}V_a\cos\delta' + \left(\frac{x_\Sigma - x_{As}}{x_\Sigma}\right)V_A \\ &= \frac{x_{As}}{x_\Sigma}(bV_B\cos\delta_{AB} + akmV_{dc}\cos\gamma) + \left(\frac{x_\Sigma - x_{As}}{x_\Sigma}\right)V_A \\ v_{sy} &= -x_{As}I_{Asx} = x_{As}\frac{V_a\sin\delta'}{x_\Sigma} \\ &= \frac{x_{As}}{x_\Sigma}(bV_B\sin\delta_{AB} + akmV_{dc}\sin\gamma)\end{aligned} \tag{7.116}$$

Linearization of Eq. (7.116) gives the following:

$$\begin{aligned}\Delta v_{sx} &= a'_{1x}\Delta V_{dc} + a'_{2x}\Delta m + a'_{3x}\Delta\gamma + a'_{4x}\Delta\delta_{AB} + a'_{5x}\Delta V_A + a'_{6x}\Delta V_B \\ \Delta v_{sy} &= a'_{1y}\Delta V_{dc} + a'_{2y}\Delta m + a'_{3y}\Delta\gamma + a'_{4y}\Delta\delta_{AB} + a'_{5y}\Delta V_A + a'_{6y}\Delta V_B\end{aligned} \tag{7.117}$$

However, the modulation phase, $\phi$, is the angle between $\bar{V}_s$ and $\bar{V}_c$, as given in Eq. (7.96), that is

$$\gamma = \phi + \varphi = \phi + \arctan^{-1}\frac{v_{sy}}{v_{sx}} \tag{7.118}$$

from Eqs. (7.117) and (7.118), it can have the following:

$$\Delta\gamma = \Delta\phi + a'_1\Delta V_{dc} + a'_2\Delta m + a'_3\Delta\gamma + a'_4\Delta\delta_{AB} + a'_5\Delta V_A + a'_6\Delta V_B \tag{7.119}$$

The above equation gives the following:

$$\Delta\gamma = a_g\Delta\phi + a_{g1}\Delta V_{dc} + a_{g2}\Delta m + a_{g4}\Delta\delta_{AB} + a_{g5}\Delta V_A + a_{g6}\Delta V_B \tag{7.120}$$

By substituting Eq. (7.120) into (7.117), it can be obtained that

$$\begin{aligned}\Delta v_{sx} &= a_{1x}\Delta V_{dc} + a_{2x}\Delta m + a_{3x}\Delta\phi + a_{4x}\Delta\delta_{AB} + a_{5x}\Delta V_A + a_{6x}\Delta V_B \\ \Delta v_{sy} &= a_{1y}\Delta V_{dc} + a_{2y}\Delta m + a_{3y}\Delta\phi + a_{4y}\Delta\delta_{AB} + a_{5y}\Delta V_A + a_{6y}\Delta V_B\end{aligned} \tag{7.121}$$

In x–y coordinate of the power system, $\bar{V}_s = v_{sx} + jv_{sy}$. Hence, the linearization of $V_s$ can be obtained from Eq. (7.121) as

$$\Delta V_s = \frac{v_{sx0}}{V_{s0}} \Delta v_{sx} + \frac{v_{sy0}}{V_{s0}} \Delta v_{sy} = a_{1f} \Delta V_{dc} + a_{2f} \Delta m + a_{3f} \Delta \phi + a_{4f} \Delta \delta_{AB} + a_{5f} \Delta V_A + a_{6f} \Delta V_B \quad (7.122)$$

Finally from Eqs. (7.101), (7.102), (7.104), (7.110), (7.112), and (7.122), the localized small-signal model of the VSC-based unit can be shown by the block diagram in Fig. 7.11 when the VSC adopts the PWM control algorithm, and Fig. 7.12 when the PAM is adopted by the VSC. To distinguish the stabilizer added on the AC and DC voltage control loops of the PWM VSC (see Fig. 7.7) as well as on the control function of the PAM VSC (see Fig. 7.8), in Figs. 7.10 and 7.11, the stabilizing signal is denoted as $\Delta u_{pss}(m)$, $\Delta u_{pss}(\phi)$, and $\Delta u_{pss}(pam)$, respectively. In the following section, the corresponding stabilizer is named as the PWM reactive power stabilizer, the PWM active power stabilizer, and the PAM stabilizer. Obviously, the establishment of the localized small-signal model of the VSC-based unit only requires the locally available data and parameters associated with the transmission lines and the VSC-based unit. The global information of the whole power system is not needed.

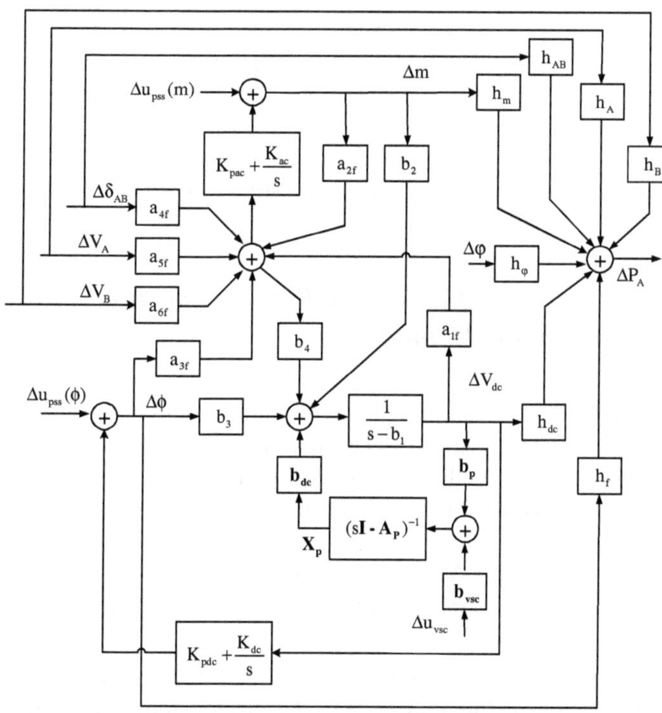


**Fig. 7.11** Block diagram of localized small-signal model of VSC-based unit adopting the PWM control

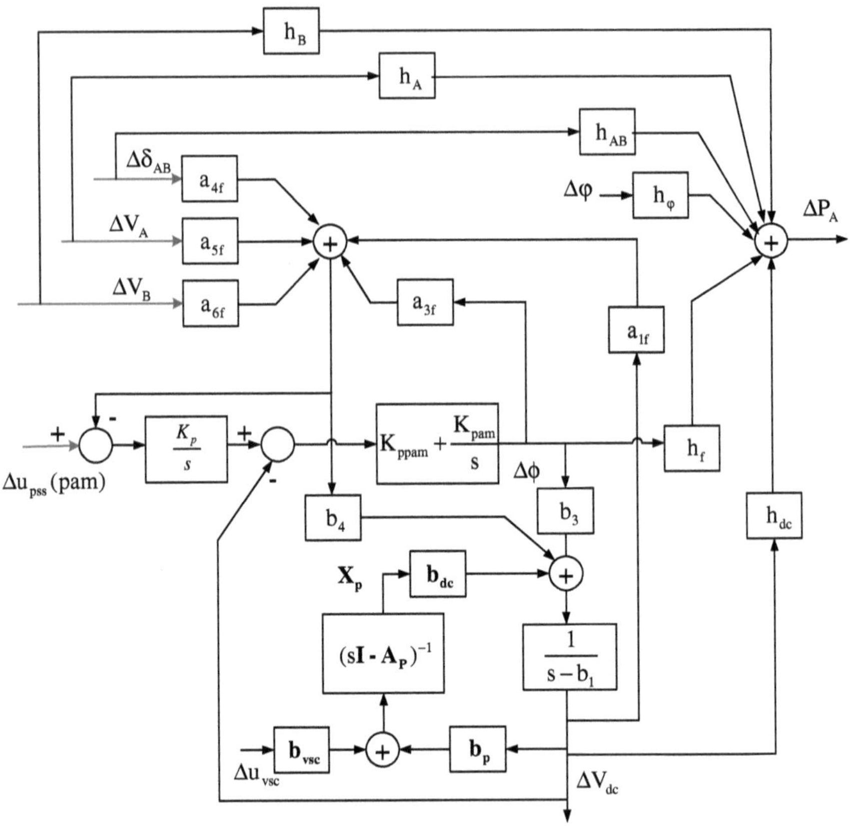


**Fig. 7.12** Block diagram of localized small-signal model of VSC-based unit adopting the PAM control

## 7.2.2 *Design of VSC-Based Stabilizer by Localized Phase Compensation Method*

In the Heffron–Phillips model of a single-machine system installed with a PSS of Fig. 2.10, the electric torque to the electromechanical oscillation loop is contributed to the deviation of rotor angular position $\Delta\delta$ and PSS stabilizing signal $\Delta u_{pss}$. The principle of superimposition of linear systems is given as

$$\begin{gathered}\Delta T = \Delta T(\Delta\delta) + \Delta T(\Delta u_{pss}) \\ \Delta T(\Delta\delta) = F_{delta}(s)\Delta\delta, \Delta T(\Delta u_{pss}) = F_{pss}(s)\Delta u_{pss}\end{gathered} \tag{7.123}$$

where $F_{delta}(s)$ and $F_{pss}(s)$ are the transfer functions of the forward path from $\Delta\delta$ and the stabilizing signal of the PSS to $\Delta T$, the electric torque contribution to the electromechanical oscillation loop, respectively. The damping torque analysis is the decomposition of the electric torque on $\Delta\delta - \Delta\omega$ coordinate into a synchronizing torque (proportional to $\Delta\delta$) and damping torque (proportional to $\Delta\omega$),

$$\begin{aligned} \Delta T(\Delta\delta) &= F_{delta}(j\omega_s)\Delta\delta = C_{delta}\Delta\delta + D_{delta}\Delta\omega \\ \Delta T(\Delta u_{pss}) &= F_{pss}(j\omega_s)\Delta u_{pss} = C_{pss}\Delta\delta + D_{pss}\Delta\omega \end{aligned} \tag{7.124}$$

Since only the damping torque contribution affects the damping of the rotor motion, the phase compensation method is to ensure that the direct contribution from the PSS is only the damping torque; that is, $\Delta T(\Delta u_{pss}) = D_{pss}\Delta\omega$. In Eq. (7.124), $\Delta T(\Delta\delta) = C_{delta}\Delta\delta + D_{delta}\Delta\omega$ represents the effect of the variable (i.e. $\Delta\delta$) external to the PSS on the damping of power oscillation. It is not considered when the PSS is designed (though it is a part of dynamic of the power system).

As it has been explained in Sect. 2.3.2.2, in fact, the damping torque analysis converts the problem of suppression of the local-mode line power oscillation (i.e. $\Delta P_t$) into that of the damping of the rotor motion (i.e. $\Delta\delta$). That is why the damping torque analysis in a multi-machine power system involves all generators. However, if the damping of the local-mode line power oscillation, $\Delta P_t$, is considered directly, from Fig. 2.10, Eqs. (7.123) and (7.124), it can have the following:

$$\begin{aligned} \Delta P_t &= \Delta P_t(\Delta\delta) + \Delta P_t(\Delta u_{pss}) \\ \Delta P_t(\Delta\delta) &= K_1\Delta\delta + C_{delta}\Delta\delta + D_{delta}\Delta\omega \\ \Delta P_t(\Delta u_{pss}) &= C_{pss}\Delta\delta + D_{pss}\Delta\omega \end{aligned} \tag{7.125}$$

According to the damping torque analysis, only the component in $\Delta P_t$ proportional to $\Delta\omega$ determines the damping of line power oscillation. The phase compensation method is to ensure that the PSS only contributes the damping component. The localized phase compensation method introduced below is based on the establishment of Eq. (7.125) in a multi-machine power system.

According to the principle of superimposition of linear systems, from Figs. 7.11 and 7.11, it can have the following:

$$\begin{aligned} \Delta P_A &= \Delta P_A(\Delta E) + \Delta P_A(\Delta u_{pss}) \\ \Delta P_A(\Delta \mathbf{E}) &= [F_{AB}(s) \quad F_A(s) \quad F_B(s) \quad F_\varphi(s) \quad F_{vsc}(s)]\Delta\mathbf{E} \\ \Delta P_A(\Delta u_{pss}) &= F_{pss}(s)\Delta u_{pss} \end{aligned} \tag{7.126}$$

where $\mathbf{\Delta E} = [\Delta\delta_{AB} \quad \Delta V_A \quad \Delta V_B \quad \Delta\varphi \quad \Delta u_{vsc}]^T$; $\Delta u_{pss}$ is $\Delta u_{pss}(m)$, $\Delta u_{pss}(\phi)$ or $\Delta u_{pss}(pam)$; $F_{AB}(s)$, $F_A(s)$, $F_B(s)$,$F_\varphi(s)$, $F_{vsc}(s)$, and $F_{pss}(s)$ is the transfer function of the forward path from $\Delta\delta_{AB}$, $\Delta V_A$, $\Delta V_B$, $\Delta\varphi$, $\Delta u_{vsc}$, and $\Delta u_{pss}$ to $\Delta P_A$ in Figs. 7.11 and 7.12, respectively.

In fact, by denoting $\mathbf{X_{VSC}} = \begin{bmatrix} \Delta V_{dc} & \mathbf{X_{PWM}^T} & \mathbf{X_P^T} \end{bmatrix}^T$, from Eqs. (7.101), (7.103), (7.105), (7.110), (7.112), and (7.126), it can be obtained that

$$\begin{aligned} s\dot{\mathbf{X}}_{\mathbf{VSC}} &= \mathbf{A_{VSC}X_{VSC}} + \mathbf{B_{VSC}\Delta E} + \mathbf{B_{PSS}}\Delta u_{pss} \\ \Delta P_A &= \mathbf{C_{VSC}X_{VSC}} + \mathbf{D_{VSC}\Delta E} + d_{PSS}\Delta u_{pss} \end{aligned} \tag{7.127}$$

Hence comparing Eqs. (7.126) and (7.127), it can have

$$\Delta P_A(\mathbf{\Delta E}) = [\mathbf{C_{SVC}}(s\mathbf{I} - \mathbf{A_{SVC}})^{-1}\mathbf{B_{SVC}} + \mathbf{D_{SVC}}]\mathbf{\Delta E} \tag{7.128}$$

$$\Delta P_A(\Delta u_{pss}) = [\mathbf{C_{SVC}}(s\mathbf{I} - \mathbf{A_{SVC}})^{-1}\mathbf{B_{PSS}} + d_{pss}]\Delta u_{pss} \tag{7.129}$$

That is,

$$[F_{AB}(s) \quad F_A(s) \quad F_B(s) \quad F_\varphi(s) \quad F_{vsc}(s)] = [\mathbf{C_{SVC}}(s\mathbf{I} - \mathbf{A_{SVC}})^{-1}\mathbf{B_{SVC}} + \mathbf{D_{SVC}}] \tag{7.130}$$

$$F_{pss}(s) = [\mathbf{C_{SVC}}(s\mathbf{I} - \mathbf{A_{SVC}})^{-1}\mathbf{B_{PSS}} + d_{pss}]\Delta u_{pss} \tag{7.131}$$

In Eq. (7.129), $\Delta P_A(\Delta u_{pss})$ is the direct contribution from a VSC-based stabilizer to the line power variation (oscillation), $\Delta P_A$. Obviously, Eqs. (7.128) and (7.129) are the extensions of Eq. (7.125) in the multi-machine power system.

Main Conclusion: If the stabilizer is designed to ensure $\Delta P_A(\mathbf{\Delta u_{pss}})$ to be proportional to the time derivative of $\Delta P_A$, that is

$$\Delta P_A(\Delta u_{pss}) = D_{pss}\Delta\dot{P}_A(D_{pss} > 0) \tag{7.132}$$

the stabilizer will supply positive damping to suppress the inter-area line power oscillation.

The above main conclusion can be established by using the graphical explanation (similar to the introduction in Sect. 2.3.2.2) based on the linearized $P - \delta$ curve and equal-area criterion as follows.

Figure 7.13 shows the linearized $P_A - \delta_{AB}$ curve, where $(\delta_{AB0}, P_{A0})$ is the operating point of the power system at steady state. It is assumed that the small-signal oscillation of $\Delta P_A$ starts from point "a" in Fig. 7.13 with the operating point moving down. In $\delta_{AB} - \Delta\dot{\delta}_{AB}$ coordinate, $\Delta P_A(\mathbf{\Delta E})$ can always be decomposed to be

$$\Delta P_A(\mathbf{\Delta E}) = C_{other}\Delta\delta_{AB} + D_{other}\Delta\dot{\delta}_{AB} \tag{7.133}$$

where $C_{other}$ and $D_{other}$ are two constants. Without affecting the establishment of the main conclusion, it is assumed that $D_{other} > 0$. When there is no stabilizing control ($\Delta P_A(\mathbf{\Delta u_{pss}}) = 0$), $\Delta P_A = \Delta P_A(\mathbf{\Delta E}) = C_{other}\Delta\delta_{AB} + D_{other}\Delta\dot{\delta}_{AB}$ which is shown by the dashed curve in Fig. 7.13. This is because when the operating point moves

**Fig. 7.13** Graphical explanation of the main conclusion

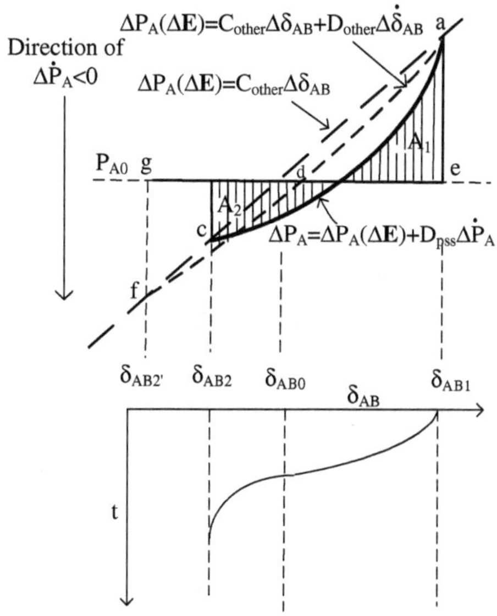


down, $\Delta\dot{\delta}_{AB} < 0$. Hence, $D_{other}\Delta\dot{\delta}_{AB} < 0$ is added on the line $\Delta P_A(\mathbf{\Delta E}) = C_{other}\Delta\delta_{AB}$. When the operating point arrives at point "f", it stops moving ($D_{other}\Delta\dot{\delta}_{AB} = 0$). Hence, it should be on the line $\Delta P_A = \Delta P_A(\mathbf{\Delta E}) = C_{other}\Delta\delta_{AB}$. According to the equal-area criterion, the area "ade" is equal to that of "dgf".

When the stabilizing control is added and is set according to Eq. (7.132), $D_{pss}\Delta\dot{P}_A < 0$ ($\Delta\dot{P}_A < 0$) will be added on the dashed curve $\Delta P_A(\mathbf{\Delta E}) = C_{other}\Delta\delta_{AB} + D_{other}\Delta\dot{\delta}_{AB}$ as shown in Eq. (7.126). Hence, when the operating point moves down, it should move below the dashed curve along the highlighted trajectory in Fig. 7.13. When the operating point stops on the line $\Delta P_A(\mathbf{\Delta E}) = C_{other}\Delta\delta_{AB}$, the area $A_1$ is equal to area $A_2$ at point "c", where $\Delta\dot{P}_A = 0$ and $D_{other}\Delta\dot{\delta}_{AB} = 0$. Obviously, $\delta_{AB1} - \delta_{AB0} > \delta_{AB0} - \delta_{AB2}$, indicating that extra positive damping is provided by the stabilizing control to the power oscillation. A similar analysis can be carried out to examine the case when the operating point moves up from point "c". Thus, the above main conclusion is established.

Hence, if $\Delta P_A$ is used as the feedback signal of the VSC-based stabilizer and the transfer function of stabilizer is constructed by a lead-lag, then it can have the following:

$$\Delta u_{pss} = K_{pss}\frac{sT_w}{1 + sT_w}\frac{1 + sT_2}{1 + sT_1}\frac{1 + sT_4}{1 + sT_3}\Delta P_A \tag{7.134}$$

According to Eqs. (7.129), (7.132), and (7.134), parameters of the stabilizer can be set at the angular oscillation frequency, $\omega_s$, to satisfy the following equation

$$F_{pss}(s)K_{pss}\frac{sT_w}{1+sT_w}\frac{1+sT_2}{1+sT_1}\frac{1+sT_4}{1+sT_3}=D_{pss}s \tag{7.135}$$

with $s = j\omega_s$. That is, the stabilizer is set with a phase to compensate the phase of $\frac{F_{pss}(S)}{S}$ such that a pure positive damping is provided to the line power oscillation.

### *7.2.3 Robustness of an ESS-Based Stabilizer to Variation of Line-Loading Conditions*

#### 7.2.3.1 Localized Small-Signal Model of an Energy Storage System [7]

In Fig. 7.5, the shunt-connected VSC-based unit can be an energy storage system (ESS). At the steady-state operation of the power system, there is no exchange of active power between the ESS and the rest of power system, i.e. $\phi_0 = 0$. Hence from Eq. (7.106), it can have $i_{sd0} = 0$ and

$$\begin{aligned}\Delta i_{sd} &= \Delta\frac{mkV_{dc}\sin\phi}{x_s} = \frac{k\sin\phi_0}{x_s}(V_{dc0}\Delta m + m_0\Delta V_{dc})\\ &+ \frac{m_0kV_{dc0}\cos\phi_0}{x_s}\Delta\phi = \frac{m_0kV_{dc0}}{x_s}\Delta\phi = \frac{V_{c0}}{x_s}\Delta\phi\\ i_{sq0} &= \frac{V_{s0}-V_{c0}}{x_s}\end{aligned} \tag{7.136}$$

From the above equation, linearization of Eq. (7.108) can be obtained as

$$\begin{aligned}\Delta I_{dc1} &= i_{sd0}(k\cos\phi_0\Delta m - m_0k\sin\phi_0\Delta\phi)\\ &\quad + i_{sq0}(k\sin\phi_0\Delta m + m_0k\cos\phi_0\Delta\phi)\\ &\quad + m_0k\cos\phi_0\Delta i_{sd} + m_0k\sin\phi_0\Delta i_{sq}\\ &= i_{sq0}m_0k\Delta\phi + m_0k\frac{V_{c0}}{x_s}\Delta\phi\end{aligned} \tag{7.137}$$

Hence, linearization of Eq. (7.111) for the ESS is given as

$$s\Delta V_{dc} = -k_{dc}\Delta\phi + \frac{\Delta I_{dc2}}{C_{dc}} \tag{7.138}$$

where $k_{dc}$ is a constant.

Denote the transfer function of the AC and DC voltage controllers of the VSC adopting the PWM algorithm (see Fig. 7.7 and Eq. (7.102) where the controllers adopt the PI control law) generally to be $K_{ac}(s)$ and $K_{dc}(s)$, the supplementary damping control signal superimposed on the AC and DC voltage control functions, respectively, to be $u_{pss-m}$ and $u_{pss-f}$, that is

$$\begin{aligned} m &= m_0 + K_{ac}(s)(V_{sref} - V_s) + u_{pss-m} \\ \phi &= \phi_0 + K_{dc}(s)(V_{dcref} - V_{dc}) + u_{pss-f} \end{aligned} \tag{7.139}$$

Because the ESS supplementary damping controller associated with $u_{pss-m}$ and $u_{pss-f}$ is implemented through regulating, respectively, the exchange of reactive and active power between the ESS and the rest of the power system, it is named as ESS reactive power PSS (power system stabilizer) and active power PSS correspondingly.

Similarly, denote the transfer functions of the voltage controllers of the VSC adopting the PAM algorithm generally to be $K_1(s)$ and $K_2(s)$ (see Fig. 7.8 and Eq. (7.104) where the controllers adopt the I and PI control law, respectively), the added damping control signal to be $u_{pss-pam}$, that is,

$$\phi = \phi_0 + K_1(s)\left[\frac{1}{k}K_2(s)(V_{sref} - V_s + u_{pss-pam}) + \frac{1}{k}V_{sref} - V_{dc}\right] \tag{7.140}$$

The damping controller associated with $u_{pss-pam}$ is called the PAM PSS. By using Eq. (7.122), linearization of Eqs. (7.138)–(7.140) can be obtained as

$$\begin{aligned} \Delta m &= -\frac{K_{ac}(s)(a_{1f}\Delta V_{dc} + a_{3f}\Delta\phi + a_{4f}\Delta\delta_{AB} + a_{5f}\Delta V_A + a_{6f}\Delta V_B)}{1 + a_{2f}K_{ac}(s)} \\ &\quad + \frac{1}{1 + a_{2f}K_{ac}(s)}\Delta u_{pss-m} \\ \Delta\phi &= -K_{dc}(s)\Delta V_{dc} + \Delta u_{pss-f} \end{aligned} \tag{7.141}$$

$$\begin{aligned} \Delta\phi = &\frac{-K_1(s)K_2(s)(a_{4f}\Delta\delta_{AB} + a_{5f}\Delta V_A + a_{6f}\Delta V_B)}{k + a_{3f}K_1(s)K_2(s)} \\ &+ \frac{K_1(s)K_2(s)}{k + a_{3f}K_1(s)K_2(s)}\Delta u_{pss-pam} - \frac{K_1(s)K_2(s)a_{1f} + kK_1(s)}{k + a_{3f}K_1(s)K_2(s)}\Delta V_{dc} \end{aligned} \tag{7.142}$$

From Eqs. (7.138), (7.141), and (7.142), localized small-signal model of the ESS is shown in Figs. 7.14 and 7.15, respectively, for the VSC adopting the PWM and PAM, where $\Delta P_A(\mathbf{u}_{ESS}) = h_{dc}\Delta V_{dc} + h_m\Delta m + h_f\Delta\phi$, $\mathbf{u}_{ESS} = [\Delta V_{dc} \quad \Delta m \quad \Delta\phi]^T$.

#### 7.2.3.2 Robustness of the ESS-Based Stabilizer to the Variations of Line-Loading Conditions

In Fig. 7.13, when the operating point moves along the linearized $P_A - \delta_{AB}$ curve, sign of $\Delta\dot{\delta}_{AB}$ and $\Delta\dot{P}_A$ is always same and $\Delta\dot{\delta}_{AB} = 0$ when $\Delta\dot{P}_A = 0$. Hence, it can

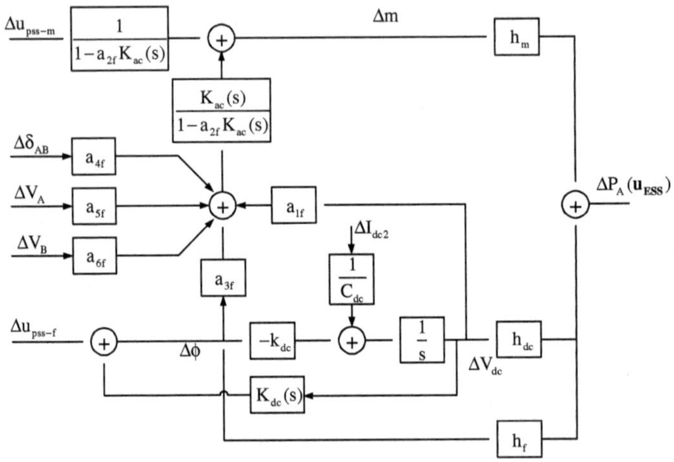


**Fig. 7.14** Localized small-signal model of the ESS with the VSC adopting the PWM algorithm

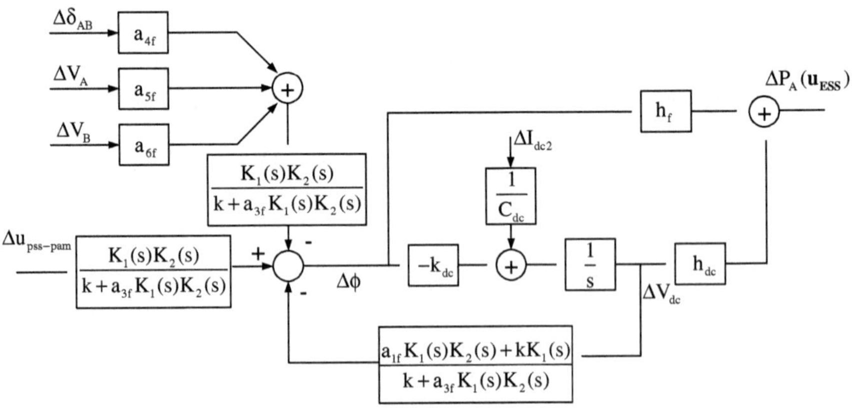


**Fig. 7.15** Localized small-signal model of the ESS with the VSC adopting the PAM algorithm

be easily concluded that in Eq. (7.126), if the decomposition of $\Delta P_A(\Delta u_{pss})$ in $\Delta\delta_{AB} - \Delta\dot{\delta}_{AB}$ coordinate is

$$\Delta P_A(\Delta u_{pss}) = C_{PSS}\Delta\delta_{AB} + D_{PSS}\Delta\dot{\delta}_{AB} \tag{7.143}$$

the stabilizer associated with $\Delta u_{pss}$ can suppress the line power oscillation if and only if $D_{PSS} > 0$. The bigger $D_{PSS} > 0$ is, the more damping is provided by the stabilizer.

Figures 7.16 and 7.17 show the block diagram of the direct damping contribution from the ESS reactive power PSS and active power PSS to the small-signal oscillation of line power, $\Delta P_A(\Delta u_{pss-m})$ and $\Delta P_A(\Delta u_{pss-f})$, respectively, when the

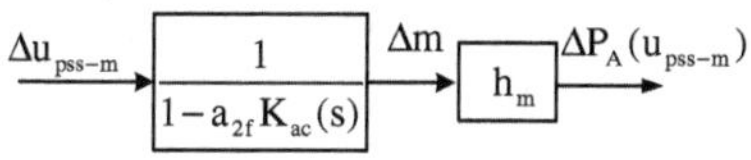


**Fig. 7.16** Forward path of ESS reactive power PSS

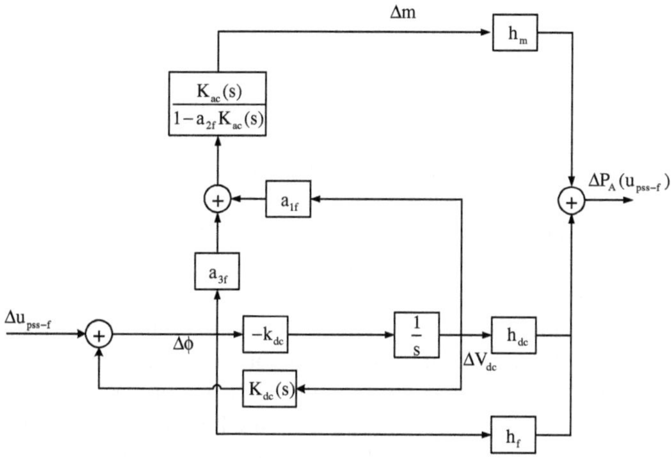


**Fig. 7.17** Forward path of ESS active power PSS

PWM algorithm is used. In Fig. 7.16, if $\Delta m$ is decomposed in $\Delta\delta_{AB} - \Delta\dot{\delta}_{AB}$ coordinate to be

$$\Delta m = C'_{pss-m}\Delta\delta_{AB} + D'_{pss-m}\Delta\dot{\delta}_{AB} \tag{7.144}$$

from Fig. 7.15, Eqs. (7.101), (7.126), (7.143), and (7.144), it can have ($\phi_0 = 0$)

$$D_{PSS} = h_m D'_{pss-m} = \frac{V_A}{x_\Sigma} akV_{dc0}\sin\varphi_0 D'_{pss-m} \tag{7.145}$$

At a higher line-loading condition, $\varphi_0$ and hence $D_{PSS}$ is bigger. This means the ESS reactive power PSS is more effective to suppress the line power oscillation at a higher line-loading condition.

In the case of the ESS active power PSS when the PWM control is used as shown in Fig. 7.17, if the decomposition of $\Delta m$, $\Delta V_{dc}$, and $\Delta\phi$ in $\Delta\delta_{AB} - \Delta\dot{\delta}_{AB}$ coordinate, respectively, is

$$\begin{aligned} \Delta m &= C_{pss-fm}\Delta\delta_{AB} + D_{pss-fm}\Delta\dot{\delta}_{AB} \\ \Delta V_{dc} &= C_{pss-fdc}\Delta\delta_{AB} + D_{pss-fdc}\Delta\dot{\delta}_{AB} \\ \Delta\phi &= C_{pss-ff}\Delta\delta_{AB} + D_{pss-ff}\Delta\dot{\delta}_{AB} \end{aligned} \tag{7.146}$$

from the above decomposition, Fig. 7.17, Eqs. (7.101), (7.126), (7.143) and (7.144), then it can have ($\phi_0 = 0$)

$$\begin{aligned} D_{PSS} &= h_m D_{pss-fm} + h_{dc} D_{pss-fm} + h_f D_{pss-ff} \\ &= \frac{V_A}{x_\Sigma} ak(V_{dc0} \sin\varphi_0 D_{pss-fm} + m_0 \sin\varphi_0 D_{pss-fdc} + m_0 V_{dc0} \cos\varphi_0 D_{pss-ff}) \end{aligned} \tag{7.147}$$

When the line-loading condition increases, $\sin\varphi_0$ increases but $\cos\varphi_0$ decreases in Eq. (7.147). Hence, $D_{PSS}$ does not change as much as it does in the case of the ESS reactive power PSS. This means that the robustness of the ESS active power PSS to the variations of line-loading conditions is expected to be better than that of the ESS reactive power PSS. The same conclusion can be drawn from Fig. 7.18 for the ESS PAM PSS when the PAM control is used.

The same conclusion has been established in Sect. 4.1.2.2 in the case of a single-machine infinite-bus power system. Function of the ESS reactive power stabilizer involves the regulation of magnitude of the terminal voltage of the VSC. Hence, in essence it works via controlling the exchange of reactive power between the ESS and the rest of the power system. Both the ESS active power stabilizer and the PAM stabilizer function through the regulation of phase of the terminal voltage of the ESS, that is, the exchange of the active power between the ESS and the rest of the power system. Therefore, the ESS active power stabilizer and PAM stabilizer suppress the line power oscillation directly by absorbing active power from or injecting it into the power system when they "see" the excess or lack of active power delivered along the transmission line. Hence, their capability to damp the power oscillation is less affected by the change of system loading conditions. The ESS reactive power stabilizer affects indirectly the variation of active power delivered along the transmission line through the regulation of injection or absorbing of reactive power from the ESS. At a lower loading condition, regulation of reactive power exchange is related less closely with the variation of line active power. Hence, the stabilizer is less effective in damping the power oscillation.

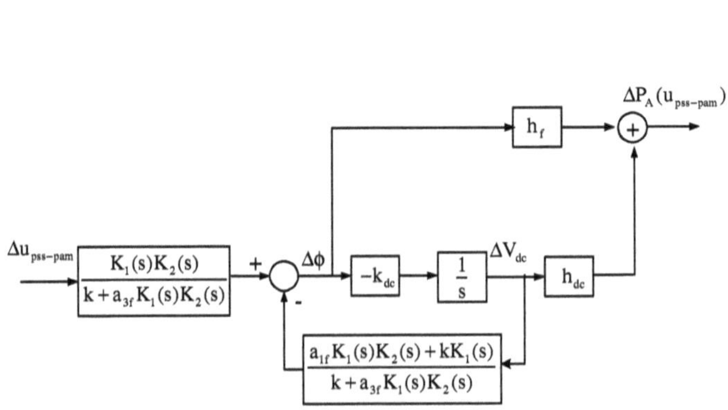


**Fig. 7.18** Forward path of ESS PAM PSS

# 7.3 An Example of Multi-machine Power System with a Grid-Connected FC Power Plan

Fuel cell (FC) power generation is one of the most promising technologies of clean energy for the future. Among several types of the FC, the high-temperature solid oxide fuel cell (SOFC) is considered to be the most attractive FC technology for grid-connected power generation. It is foreseen that, in future power systems, gird-connected high-capacity SOFC power plants could play certain important role as one of the main clean power generation sources.

This section presents an example of a multi-machine power system with a SOFC power plant which is integrated into the power system via a shunt VSC. The objective is to demonstrate how the linearized mathematical model of the power system with the shunt VSC-based SOFC power plant can be established for power system small-signal stability analysis and control. It shows how the localized small-signal model of the SOFC power plant can be developed and a stabilizer attached to the VSC of the SOFC power plant can be effectively designed by using the localized phase compensation method introduced in Sect. 7.2.

## *7.3.1 Mathematical Model of a Multi-machine Power System with a Grid-Connected FC Power Plant*

### 7.3.1.1 Dynamic Model of the SOFC Power Plant

Power generation by a SOFC is the conversion of chemical energy in hydrogen ($H_2$) and oxygen ($O_2$) into electric energy through the following basic reaction at two nodes of the SOFC:

$$\text{Anode:}\quad H_2 + O^{=} \rightarrow H_2O + 2e^{-}$$

$$\text{Cathode:}\quad \frac{O_2}{2} + 2e^{-} \rightarrow O^{-}$$

DC current $I_{fc}$ is generated from the releasing of electrons ($2e^{-}$) at the anode due to the above chemical reaction. A dynamic model of the SOFC power plant suitable for use in power system studies is shown in Fig. 7.19. In Fig. 7.19, electrical dynamic describes the chemical reaction to restore the charge that has been quickly drained by the load. This is modelled by a first-order transfer function. To ensure the FC operation within the safe operating area, current control reference, $I_{fc-ref}$, is limited by the following boundaries:

$$I_{fc-ref-max} = \frac{U_{max}}{2K_r} q_{h2-in},\quad I_{fc-ref-min} = \frac{U_{min}}{2K_r} q_{h2-in},\quad K_r = \frac{N_0}{4F}$$

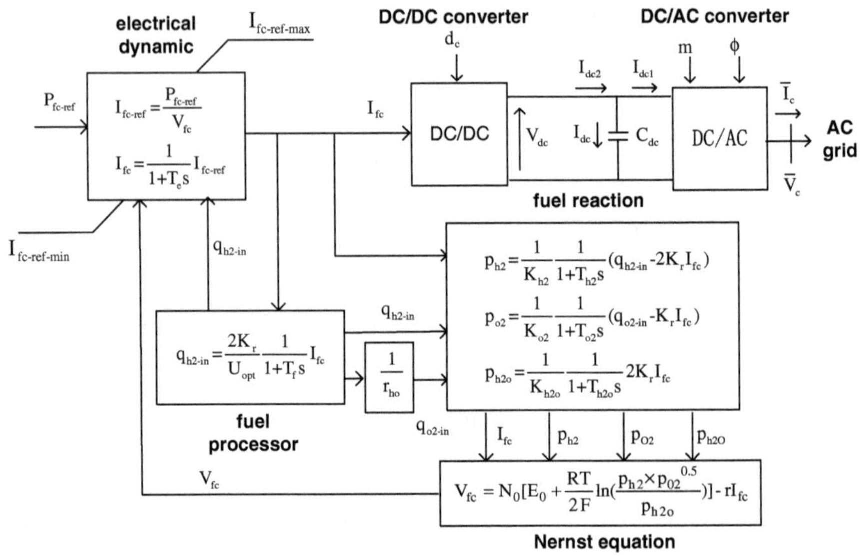


**Fig. 7.19** Dynamic model of a grid-connected SOFC power plant

where $U_{max}$ and $U_{min}$ are the maximum and minimum fuel unitization, respectively, $N_0$ the number of cells in series in the FC stack, $q_{h2-in}$ the hydrogen input flow rate, and F the Faraday constant.

Fuel processor depicts the dynamics of fuel supply which is a process to reform the natural gas supply to the hydrogen-rich fuel. This can be simply represented by a first-order lag model of time constant $T_f$, where $U_{opt}$ is the optimal fuel utilization. The chemical process of fuel reaction inside the FC stack is described by three first-order transfer functions for hydrogen, oxygen, and water, respectively, where $r_{ho}$ is the ratio of hydrogen to oxygen, $K_{h2}$, $K_{o2}$, and $K_{h2o}$ the valve molar constant, $T_{h2}$, $K_{o2}$, and $T_{h2o}$ the time constant and $p_{h2}$, $p_{o2}$, and $p_{h2o}$ the partial pressure for hydrogen, oxygen, and water flow, respectively. The internal EMF generated by the SOFC stack is calculated by the well-known Nernst equation, where $E_0$ is the ideal standard potential, *R* the universal gas constant, T the absolute temperature, and r the ohmic loss.

Grid connection of the SOFC power plant can be implemented by a two-stage configuration of power electronics interface as shown in Fig. 7.19. In order to avoid the underused or overused fuel, the DC/DC converter can be controlled to ensure a constant current output from the fuel cell stack. That is,

$$d_c = d_{c0} + T_{dc}(s)(I_{fc-ref} - I_{fc}) \tag{7.148}$$

where $d_c$ is the duty cycle and $T_{dc}(s)$ the transfer function of the FC current controller.

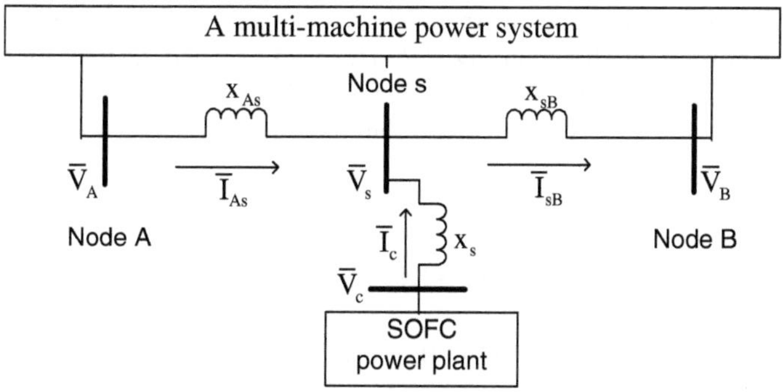


**Fig. 7.20** A multi-machine power system with a SOFC power plant connected at node s

Figure 7.20 shows the SOFC power plant connected to a multi-machine power system at node s via a VSC. Voltage at the AC terminal of the DC/AC converter of the SOFC power plant is (see Eq. (7.96)) given as

$$\bar{V}_c = mkV_{dc}\angle\gamma = mkV_{dc}\angle\varphi + \phi \tag{7.149}$$

If the DC/AC converter is controlled by the PWM control, m and $\phi$ are the modulation ratio and phase of the PWM, respectively. If the control algorithm of the DC/AC converter is the PAM, then m = 1.

From Eq. (7.108), it can have the following:

$$I_{dc1} = i_{sd}mk\cos\phi + i_{sq}mk\sin\phi \tag{7.150}$$

where subscripts d and q denote the d and q components of the corresponding variable, respectively, in the d–q coordinate of the SOFC power plant as shown in Fig. 7.20. Because

$$P_{fc} = I_{fc}V_{fc} = I_{dc2}V_{dc}, \quad V_{dc} = \frac{V_{fc}}{1 - d_c}, \tag{7.151}$$

it can have $I_{dc2} = (1 - d_c)I_{fc}$ such that the dynamics of the DC/AC converter can be expressed as

$$\begin{aligned} \dot{V}_{dc} &= \frac{1}{C_{dc}}(-I_{dc1} + I_{dc2}) \\ &= \frac{1}{C_{dc}}[-i_{sd}mk\cos\phi - i_{sq}mk\sin\phi + (1 - d_c)I_{fc}] \end{aligned} \tag{7.152}$$

From the point of view of power system operation, node s can be treated as a PV bus. Active power output from the SOFC power plant is kept to be $p_{fc\text{-}ref}$ because of the current controller implemented in the DC/DC converter. Voltage control scheme applied by the DC/AC converter will be very useful as far as the power system operation control is concerned. This can be achieved by either the PWM or the PAM algorithm as shown in Fig. 7.22 (see Figs. 7.7 and 7.8), where $K_{ac}(s)$ and $K_{dc}(s)$ are the two transfer functions of the PWM- and PAM-controlled DC/AC converter, and $u_{pss}$ is a supplementary damping control signal to improve power system stability.

#### 7.3.1.2 Linearized Model of the SOFC Power Plant

Similar to Eq. (7.106), from Fig. 7.20, it can have the following:

$$i_{sd} = \frac{mkV_{dc}\sin\phi}{x_s}, \quad i_{sq} = \frac{V_s - mkV_{dc}\cos\phi}{x_s} \tag{7.153}$$

In the common x–y coordinate of the power system, $\bar{V}_s = v_{sx} + jv_{sy}$. Hence, linearization of $V_s$ can be obtained as

$$\Delta V_s = \mathbf{v}_{s0}^{T}\Delta\mathbf{V}_{sxy} \tag{7.154}$$

where $\mathbf{v}_{s0}^{T} = \begin{bmatrix} \frac{V_{sx0}}{V_{s0}} & \frac{V_{sy0}}{V_{s0}} \end{bmatrix}$, $\Delta\mathbf{V}_{sxy} = \begin{bmatrix} \Delta v_{sx} \\ \Delta v_{sy} \end{bmatrix}$. From Eqs. (7.153) and (7.154), it can have the following:

$$\begin{aligned} \Delta i_{sd} &= b_{1d}\Delta V_{dc} + b_{2d}\Delta m + b_{3d}\Delta\phi \\ \Delta i_{sq} &= b_{1q}\Delta V_{dc} + b_{2q}\Delta m + b_{3q}\Delta\phi + \mathbf{b}_{4q}^{T}\Delta\mathbf{V}_{sxy} \end{aligned} \tag{7.155}$$

If the linearization of Eq. (7.148) is $\Delta d_c = -T_{dc}(s)\Delta I_{fc}$ and the state-space realization of transfer function $T_{dc}(s)$ is $(\mathbf{A_d}, \mathbf{B_d}, \mathbf{C_d}, \mathbf{D_d})$, then it can have the following:

$$\begin{aligned} \dot{\mathbf{X}}_{\mathbf{d}} &= \mathbf{A_d X_d} + \mathbf{B_d}\Delta I_{fc} \\ \Delta d_c &= -\mathbf{C_d X_d} - D_d\Delta I_{fc} \end{aligned} \tag{7.156}$$

If the state-space realization of transfer function $K_{ac}(s)$ and $K_{dc}(s)$ is $(\mathbf{A_{AC}}, \mathbf{B_{AC}}, \mathbf{C_{AC}}, \mathbf{D_{AC}})$ and $(\mathbf{A_{DC}}, \mathbf{B_{DC}}, \mathbf{C_{DC}}, \mathbf{D_{DC}})$, respectively, then from Fig. 7.22 it can have that for the PWM control

$$\begin{aligned} \dot{\mathbf{X}}_{AC} &= \mathbf{A}_{AC}\mathbf{X}_{AC} + \mathbf{B}_{AC}\Delta\mathbf{V}_{sxy} \\ \Delta m &= \mathbf{C}_{AC}\mathbf{X}_{AC} + \mathbf{D}_{AC}\Delta\mathbf{V}_{sxy} + \Delta u_{pss} \end{aligned} \tag{7.157}$$

$$\begin{aligned} \dot{\mathbf{X}}_{DC} &= \mathbf{A}_{DC}\mathbf{X}_{DC} + \mathbf{B}_{DC}\Delta V_{dc} \\ \Delta\phi &= \mathbf{C}_{DC}\mathbf{X}_{DC} + D_{DC}\Delta V_{dc} + \Delta u_{pss} \end{aligned} \tag{7.158}$$

and for the PAM control

$$\begin{aligned}
\dot{\mathbf{X}}_{AC} &= \mathbf{A}_{AC}\mathbf{X}_{AC} + \mathbf{B}_{AC}\mathbf{\Delta V_{sxy}} + \mathbf{b_{pam-pss}}\Delta u_{pss} \\
\dot{\mathbf{X}}_{\mathbf{DC}} &= \mathbf{A_{DC}}\mathbf{X_{DC}} + \mathbf{B_{DC}}\Delta V_{dc} \\
\Delta\phi &= \mathbf{C_{DC}}\mathbf{X_{DC}} + D_{DC}\mathbf{C}_{AC}\mathbf{X}_{AC} + D_{DC}\mathbf{D}_{AC}\mathbf{\Delta V_{sxy}} - D_{DC}\Delta V_{dc} + d_{pam-pss}\Delta u_{pss}
\end{aligned} \tag{7.159}$$

By using Eqs. (7.155)–(7.159), linearization of Eq. (7.152) can be obtained as

$$\begin{aligned}
\Delta\dot{V}_{dc} &= a_1 V_{dc} + a_2\Delta I_{fc} + \mathbf{a_3^T X_d} + \mathbf{a_4^T X_{AC}} \\
&\quad + \mathbf{a_5^T X_{DC}} + \mathbf{a_6^T \Delta V_{sxy}} + a_7\Delta u_{pss}
\end{aligned} \tag{7.160}$$

where $\Delta u_{pss}$ can be $\Delta u_{m-pss}$, $\Delta u_{\phi-pss}$ or $\Delta u_{pam-pss}$ with different coefficient $a_7$.

Linearization of dynamic model of the SOFC power plant of Fig. 7.19 can be obtained as

$$\begin{aligned}
\Delta I_{fc-ref} &= -\frac{P_{fc-ref0}}{V_{fc0}^2}\Delta V_{fc}, \quad \Delta I_{fc} = \frac{1}{1+T_e s}\Delta I_{fc-ref} \\
\Delta q_{h2-in} &= \frac{2K_r}{U_{opt}}\frac{1}{1+T_f s}\Delta I_{fc}, \quad \Delta q_{o2-in} = \frac{1}{r_{ho}}\Delta q_{h2-in} \\
\Delta p_{h2} &= \frac{1}{K_{h2}}\frac{1}{1+T_{h2}s}(\Delta q_{h2-in} - 2K_r\Delta I_{fc}) \\
\Delta p_{o2} &= \frac{1}{K_{o2}}\frac{1}{1+T_{o2}s}(\Delta q_{o2-in} - K_r\Delta I_{fc}) \\
\Delta p_{h2o} &= \frac{1}{K_{h2o}}\frac{1}{1+T_{h2o}s}2K_r\Delta I_{fc} \\
\Delta V_{fc} &= c_1\Delta p_{h2} + c_2\Delta p_{02} + c_3\Delta p_{h2o} + c_4\Delta I_{fc}
\end{aligned} \tag{7.161}$$

Rearranging Eq. (7.161) can give the following:

$$\begin{aligned}
\Delta\dot{I}_{fc} &= d_1\Delta p_{h2} + d_2\Delta p_{02} + d_3\Delta p_{h2o} + d_4\Delta I_{fc} \\
\Delta\dot{q}_{h2-in} &= d_5\Delta q_{h2-in} + d_6\Delta I_{fc} \\
\Delta\dot{p}_{h2} &= d_7\Delta q_{h2-in} + d_8\Delta p_{h2} + d_9\Delta I_{fc} \\
\Delta\dot{p}_{o2} &= d_{10}\Delta q_{h2-in} + d_{11}\Delta p_{o2} + d_{12}\Delta I_{fc} \\
\Delta p_{h2o} &= d_{13}\Delta p_{h2o} + d_{14}\Delta I_{fc}
\end{aligned} \tag{7.162}$$

From Eqs. (7.156)–(7.160) and (7.162), the state-space equation of the SOFC power plant can be obtained as

$$\dot{\mathbf{X}}_{\mathbf{SOFC}} = \mathbf{A_{SOFC}}\mathbf{X_{SOFC}} + \mathbf{B_{SOFC}}\mathbf{\Delta V_{sxy}} + \mathbf{B_{PSS}}\Delta u_{pss} \tag{7.163}$$

Fig. 7.21 Phasor diagram in the d–q coordinate of the SOFC power plant

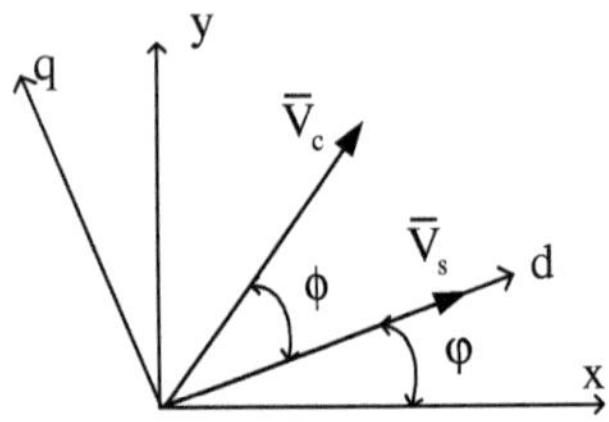


where

$$\mathbf{X_{SOFC}} = \begin{bmatrix} \Delta V_{dc} & \Delta I_{fc} & \Delta q_{h2-in} & \Delta p_{h2} & \Delta p_{o2} & \Delta p_{h2o} & \mathbf{X_d^T} & \mathbf{X_{AC}^T} & \mathbf{X_{DC}^T} \end{bmatrix}^T$$

Since $\varphi = \mathrm{ac}\ \tan^{-1}\frac{V_{sy}}{V_{sx}}$, it can have the following:

$$\Delta\varphi = \mathbf{a_0^T}\boldsymbol{\Delta}\mathbf{V_{sxy}} \tag{7.164}$$

where $\mathbf{a_0^T} = \begin{bmatrix} \frac{-V_{sy0}}{V_{sx0}^2 + V_{sy0}^2} & \frac{V_{sx0}}{V_{sx0}^2 + V_{sy0}^2} \end{bmatrix}$. From Fig. 7.21, it can have the following:

$$\begin{bmatrix} i_{sx} \\ i_{sy} \end{bmatrix} = \begin{bmatrix} \cos\varphi & -\sin\varphi \\ \sin\varphi & \cos\varphi \end{bmatrix} \begin{bmatrix} i_{sd} \\ i_{sq} \end{bmatrix} \tag{7.165}$$

Linearization of Eq. (7.165) is given as

$$\begin{bmatrix} \Delta i_{sx} \\ \Delta i_{sy} \end{bmatrix} = \begin{bmatrix} \cos\varphi_0 & -\sin\varphi_0 \\ \sin\varphi_0 & \cos\varphi_0 \end{bmatrix} \begin{bmatrix} \Delta i_{sd} \\ \Delta i_{sq} \end{bmatrix} + \begin{bmatrix} -\sin\varphi_0 & -\cos\varphi_0 \\ \cos\varphi_0 & -\sin\varphi_0 \end{bmatrix} \begin{bmatrix} i_{sd0} \\ i_{sq0} \end{bmatrix} \Delta\varphi \tag{7.166}$$

From Eqs. (7.155), (7.164), and (7.166), it can be obtained that

$$\boldsymbol{\Delta}\mathbf{I_{sxy}} = \mathbf{b_1}\Delta V_{dc} + \mathbf{b_2}\Delta m + \mathbf{b_3}\Delta\phi + \mathbf{b_4^T}\boldsymbol{\Delta}\mathbf{V_{sxy}} \tag{7.167}$$

where $\boldsymbol{\Delta}\mathbf{I_{sxy}} = \begin{bmatrix} \Delta i_{sx} \\ \Delta i_{sy} \end{bmatrix}$. By using Eqs. (7.157)–(7.159) and (7.163), Eq. (7.167) can be expressed as

$$\boldsymbol{\Delta}\mathbf{I_{sxy}} = \mathbf{C_{SOFC}}\mathbf{X_{SOFC}} + \mathbf{D_{SOFC}}\boldsymbol{\Delta}\mathbf{V_{sxy}} + \mathbf{D_{PSS}}\Delta u_{pss} \tag{7.168}$$

#### 7.3.1.3 Linearized Model of Multi-machine Power System with SOFC Power Plant

Similar to the derivation in Sect. 6.1, it is assumed that there are N generators and M nodes in the multi-machine power system with the gird-connected SOFC power plant. In the common x–y coordinate of the power system, node voltage and current are denoted as $v_{xj} + jv_{yj}$ and $i_{xj} + ji_{yj}$, $j = 1, 2, \ldots, M$, respectively. Without the loss of generality, it can be assumed that node 1 to N is generator node and node s in Fig. 7.20 is node (N + 1). Defining (see Eq. (5.51))

$$\begin{aligned}\mathbf{\Delta V_{xy}} &= \begin{bmatrix}\Delta v_{gx1} & \Delta v_{gy1} & \Delta v_{gx2} & \Delta v_{gy2} & \cdots & \cdots & \Delta v_{gxN} & \Delta v_{gyN}\end{bmatrix}^T \\ \mathbf{\Delta I_{xy}} &= \begin{bmatrix}\Delta i_{x1} & \Delta i_{y1} & \Delta i_{x2} & \Delta i_{y2} & \cdots & \cdots & \Delta i_{xN} & \Delta i_{yN}\end{bmatrix}^T\end{aligned}$$

Linearized model of generators is (see Eq. (6.33)) given as

$$\begin{aligned}\dot{\mathbf{X}}_{\mathbf{gp}} &= \mathbf{A_{gp}X_{gp}} + \mathbf{B_{gp}\Delta V_{xy}} \\ \mathbf{\Delta I_{xy}} &= \mathbf{C_{gp}X_{gp}} + \mathbf{D_{gp}\Delta V_{xy}}\end{aligned} \tag{7.169}$$

where $\mathbf{X_{gp}}$ is the state variable vector of generator dynamics, including state variables of the AVRs and the PSSs. Denote the voltage at other nodes by the following vector:

$$\mathbf{\Delta V_{oxy}} = \begin{bmatrix}\Delta V_{(N+2)x} & \Delta V_{(N+2)y} & \cdots & \Delta V_{Mx} & \Delta V_{Mx}\end{bmatrix}^T$$

The network equations can be established as

$$\begin{bmatrix}\mathbf{\Delta I_{xy}} \\ \mathbf{\Delta I_{sxy}} \\ \mathbf{0}\end{bmatrix} = \begin{bmatrix}\mathbf{Y_{gg}} & \mathbf{Y_{gs}} & \mathbf{Y_{go}} \\ \mathbf{Y_{sg}} & \mathbf{Y_{ss}} & \mathbf{Y_{so}} \\ \mathbf{Y_{og}} & \mathbf{Y_{os}} & \mathbf{Y_{oo}}\end{bmatrix}\begin{bmatrix}\mathbf{\Delta V_{xy}} \\ \mathbf{\Delta V_{sxy}} \\ \mathbf{\Delta V_{oxy}}\end{bmatrix} \tag{7.170}$$

Substituting Eqs. (7.163), (7.168), and (7.169) into Eq. (7.170), it can have the following:

$$\begin{bmatrix}\mathbf{C_{gp}X_{gp}} \\ \mathbf{C_{SOFC}X_{SOFC}} + \mathbf{D_{PSS}}\Delta u_{pss} \\ \mathbf{0}\end{bmatrix} = \begin{bmatrix}\mathbf{Y_{gg}} - \mathbf{D_{gp}} & \mathbf{Y_{gs}} & \mathbf{Y_{go}} \\ \mathbf{Y_{sg}} & \mathbf{Y_{ss}} - \mathbf{D_{SOFC}} & \mathbf{Y_{so}} \\ \mathbf{Y_{og}} & \mathbf{Y_{os}} & \mathbf{Y_{2o}}\end{bmatrix}\begin{bmatrix}\mathbf{\Delta V_{xy}} \\ \mathbf{\Delta V_{sxy}} \\ \mathbf{\Delta V_{oxy}}\end{bmatrix} \tag{7.171}$$

Equation (7.171) can give the following:

$$\begin{aligned}\mathbf{\Delta V_{oxy}} &= -\mathbf{Y_{oo}^{-1}}\begin{bmatrix}\mathbf{Y_{og}} & \mathbf{Y_{os}}\end{bmatrix}\begin{bmatrix}\mathbf{\Delta V_{xy}} \\ \mathbf{\Delta V_{sxy}}\end{bmatrix} \\ \begin{bmatrix}\mathbf{\Delta V_{xy}} \\ \mathbf{\Delta V_{sxy}}\end{bmatrix} &= \mathbf{Y_{gsxy}^{-1}}\left(\begin{bmatrix}\mathbf{C_{gp}} & \mathbf{0} \\ \mathbf{0} & \mathbf{C_{SOFC}}\end{bmatrix}\begin{bmatrix}\mathbf{X_{gp}} \\ \mathbf{X_{SOFC}}\end{bmatrix} + \begin{bmatrix}\mathbf{0} \\ \mathbf{D_{PSS}}\end{bmatrix}\Delta u_{pss}\right)\end{aligned} \tag{7.172}$$

where $\mathbf{Y_{gsxy}} = \begin{bmatrix}\mathbf{Y_{gg}} - \mathbf{D_{gp}} & \mathbf{Y_{gs}} \\ \mathbf{Y_{sg}} & \mathbf{Y_{ss}} - \mathbf{D_{SOFC}}\end{bmatrix} - \begin{bmatrix}\mathbf{Y_{go}} \\ \mathbf{Y_{so}}\end{bmatrix}\mathbf{Y_{oo}^{-1}}\begin{bmatrix}\mathbf{Y_{og}} & \mathbf{Y_{os}}\end{bmatrix}$

Arranging the first equation in Eqs. (7.169) and (7.163) together, it can have the following:

$$\begin{bmatrix} \dot{\mathbf{X}}_{\mathbf{gp}} \\ \dot{\mathbf{X}}_{\mathbf{SOFC}} \end{bmatrix} = \begin{bmatrix} \mathbf{A}_{\mathbf{gp}} & \mathbf{0} \\ \mathbf{0} & \mathbf{A}_{\mathbf{SOFC}} \end{bmatrix} \begin{bmatrix} \mathbf{X}_{\mathbf{gp}} \\ \mathbf{X}_{\mathbf{SOFC}} \end{bmatrix} + \begin{bmatrix} \mathbf{B}_{\mathbf{gp}} & \mathbf{0} \\ \mathbf{0} & \mathbf{B}_{\mathbf{SOFC}} \end{bmatrix} \begin{bmatrix} \Delta\mathbf{V}_{\mathbf{xy}} \\ \Delta\mathbf{V}_{\mathbf{sxy}} \end{bmatrix} + \begin{bmatrix} \mathbf{0} \\ \mathbf{B}_{\mathbf{PSS}} \end{bmatrix} \Delta u_{pss} \tag{7.173}$$

Substituting Eq. (7.172) into (7.173), system state equation can be obtained as

$$\mathbf{X} = \mathbf{AX} + \mathbf{B}\Delta u_{pss} \tag{7.174}$$

where

$$\mathbf{X} = \begin{bmatrix} \mathbf{X}_{\mathbf{gp}} \\ \mathbf{X}_{\mathbf{SOFC}} \end{bmatrix}$$

$$\mathbf{A} = \begin{bmatrix} \mathbf{A}_{\mathbf{gp}} & \mathbf{0} \\ \mathbf{0} & \mathbf{A}_{\mathbf{SOFC}} \end{bmatrix} + \begin{bmatrix} \mathbf{B}_{\mathbf{gp}} & \mathbf{0} \\ \mathbf{0} & \mathbf{B}_{\mathbf{SOFC}} \end{bmatrix} \mathbf{Y}_{\mathbf{gsxy}}^{-1} \begin{bmatrix} \mathbf{C}_{\mathbf{gp}} & \mathbf{0} \\ \mathbf{0} & \mathbf{C}_{\mathbf{SOFC}} \end{bmatrix},$$

$$\mathbf{B} = \begin{bmatrix} \mathbf{0} \\ \mathbf{B}_{\mathbf{PSS}} \end{bmatrix} + \begin{bmatrix} \mathbf{B}_{\mathbf{gp}} & \mathbf{0} \\ \mathbf{0} & \mathbf{B}_{\mathbf{SOFC}} \end{bmatrix} \mathbf{Y}_{\mathbf{gsxy}}^{-1} \begin{bmatrix} \mathbf{0} \\ \mathbf{D}_{\mathbf{PSS}} \end{bmatrix}$$

System small-signal stability can be improved by the supplementary stabilizing control superimposed on the normal control functions of the DC/AC converter (see Fig. 7.22). If the feedback signal of the stabilizer is the deviation of active power delivered along the transmission line from node A to s in Fig. 7.20, $\Delta P_A$, then it can have the following:

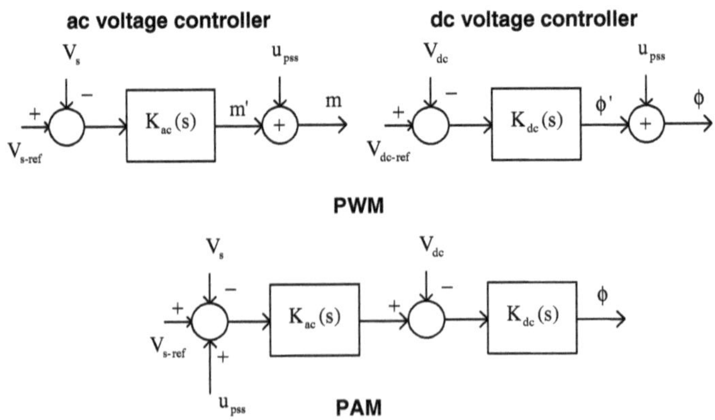


**Fig. 7.22** Control of DC/AC converter either by the PWM or by the PAM algorithm

$$P_A = \mathrm{Re}\left(\frac{V_A - V_s}{Z_{As}} V_A^*\right) = \mathrm{Re}[(g_{As} + jb_{As})(V_A V_A^* - V_s V_A^*)]$$
$$= g_{As}\left(v_{Ax}^2 + v_{Ay}^2\right) + b_{As}(v_{Ax} v_{sy} - v_{Ay} v_{sx})$$

From the above equation, it can have the following:

$$\Delta P_A = \begin{bmatrix} 2g_{As} v_{Ax0} + b_{As} v_{sy0} & 2g_{As} v_{Ay0} - b_{As} v_{sx0} \end{bmatrix} \begin{bmatrix} \Delta v_{Ax} \\ \Delta v_{Ay} \end{bmatrix}$$
$$+ \begin{bmatrix} b_{As} v_{Ax0} & -b_{As} v_{Ay0} \end{bmatrix} \begin{bmatrix} \Delta v_{sx} \\ \Delta v_{sy} \end{bmatrix}$$
$$= \mathbf{p_s} \Delta \mathbf{V_{sxy}} + \mathbf{p'_s} \Delta \mathbf{V_{oxy}} \tag{7.175}$$

From Eqs. (7.172) and (7.175), it can be obtained that

$$\Delta P_A = \left(\begin{bmatrix} \mathbf{0} & \mathbf{p_s} \end{bmatrix} - \mathbf{p'_s} \mathbf{Y_{oo}^{-1}} \begin{bmatrix} \mathbf{Y_{og}} & \mathbf{Y_{os}} \end{bmatrix}\right) \mathbf{Y_{gsxy}^{-1}}$$
$$\times \left(\begin{bmatrix} \mathbf{C_{gp}} & \mathbf{0} \\ \mathbf{0} & \mathbf{C_{SOFC}} \end{bmatrix} \begin{bmatrix} \mathbf{X_{gp}} \\ \mathbf{X_{SOFC}} \end{bmatrix} + \begin{bmatrix} \mathbf{0} \\ \mathbf{D_{PSS}} \end{bmatrix} \Delta u_{pss}\right)$$
$$= \mathbf{CX} + \mathbf{D} \Delta u_{pss} \tag{7.176}$$

Equation (7.176) is the output equation of state-space representation of the power system. By denoting $y = \Delta P_A$, the linearized model of the multi-machine power system with the gird-connected SOFC power plant is obtained as (Eqs. (7.174) and (7.176))

$$\begin{aligned} \dot{\mathbf{X}} &= \mathbf{AX} + \mathbf{B} \Delta u_{pss} \\ y &= \mathbf{CX} + \mathbf{D} \Delta u_{pss} \end{aligned} \tag{7.177}$$

This is the linearized mathematical model of the whole power system. It can be used for system small-signal stability analysis and design of the stabilizer attached to the VSC of the SOFC power plant.

## 7.3.2 *Design of a Stabilizer Attached to the VSC of FC Power Plant by Localized Phase Compensation Method*

#### 7.3.2.1 Localized Design of the Stabilizer

Based on the linearized model of Eq. (7.177), power system small-signal stability can be examined by modal analysis, i.e. computation of system oscillation modes and associated eigenvectors. For the design of the supplementary stabilizer, various methods proposed for the design of the PSS or FACTS stabilizers introduced in Chap. 5 and 6 can be used. However, even the stabilizer is to suppress a locally observed oscillation, such as the power oscillation along the transmission line from node A to s in Fig. 7.20, and the mathematical model of the whole power system of Eq. (7.177) needs to be established. This increases significantly not only the complicity of the design, but also the possibility of error in obtaining system model of Eq. (7.177), as it is difficult to gain and validate the accurate system parameters and operational data when the scale of the power system is large. In Sect. 7.2, a much simpler method to design the stabilizer by using a localized linearized model of the VSC-based unit is introduced. The stabilizer is designed by a localized phase compensation method to suppress the power oscillation along the transmission line where the VSC-based unit is installed. In this section, a stabilizer attached to the VSC of the grid-connected SOFC power plant is used as an example to demonstrate the application of the localized phase compensation method as follows.

By substituting Eq. (7.121) into (7.163), the following localized state-space equation of the SOFC power plant dynamics can be obtained as

$$s\dot{\mathbf{X}}_{\mathbf{SOFC}} = \mathbf{A}_{\mathbf{L-SOFC}}\mathbf{X}_{\mathbf{SOFC}} + \mathbf{B}_{\mathbf{E-SOFC}}\boldsymbol{\Delta}\mathbf{E} + \mathbf{B}_{\mathbf{PSS}}\Delta u_{pss} \tag{7.178}$$

By using Eqs. (7.157)–(7.159), (7.164), and (7.121), Eq. (7.101) can be written as

$$\Delta P_A = \mathbf{C}_{\mathbf{L-SOFC}}\mathbf{X}_{\mathbf{SOFC}} + \mathbf{C}_{\mathbf{E-SOFC}}\boldsymbol{\Delta}\mathbf{E} + d_{SOFC}\Delta u_{pss} \tag{7.179}$$

From Eqs. (7.178) and (7.179), it can have the following:

$$\Delta P_A = \Delta P_A(\boldsymbol{\Delta}\mathbf{E}) + \Delta P_A(\Delta u_{pss}) \tag{7.180}$$

where

$$\Delta P_A(\boldsymbol{\Delta}\mathbf{E}) = [\mathbf{C}_{\mathbf{L-SOFC}}(s\mathbf{I} - \mathbf{A}_{\mathbf{L-SOFC}})^{-1}\mathbf{B}_{\mathbf{E-SOFC}} + \mathbf{C}_{\mathbf{E-SOFC}}]\boldsymbol{\Delta}\mathbf{E}$$
$$\Delta P_A(\Delta u_{pss}) = [\mathbf{C}_{\mathbf{L-SOFC}}(s\mathbf{I} - \mathbf{A}_{\mathbf{L-SOFC}})^{-1}\mathbf{B}_{\mathbf{PSS}} + d_{SOFC}]\Delta u_{pss}$$

$\Delta P_A(\Delta u_{pss})$ is the direct contribution from the stabilizer to the variation (oscillation) of line active power, $\Delta P_A$. If $\Delta P_A$ is used as the feedback signal of the stabilizer, then the transfer function of stabilizer is constructed by a lead-lag network

$$\Delta u_{pss} = K_{pss}\frac{sT_w}{1+sT_w}\frac{1+sT_2}{1+sT_1}\frac{1+sT_4}{1+sT_3}\Delta P_A \tag{7.181}$$

Denote

$$F_{SOFC}(s) = \mathbf{C}_{\mathbf{L-SOFC}}(s\mathbf{I} - \mathbf{A}_{\mathbf{L-SOFC}})^{-1}\mathbf{B}_{\mathbf{PSS}} + d_{SOFC} \tag{7.182}$$

According to Sect. 7.2.2, parameters of the stabilizer can be set at the angular oscillation frequency, $\omega_s$, to satisfy the following equation:

$$F_{SOFC}(s)K_{pss}\frac{sT_w}{1+sT_w}\frac{1+sT_2}{1+sT_1}\frac{1+sT_4}{1+sT_3} = D_{pss}s \tag{7.183}$$

with $s = j\omega_s$.

The stabilizer will supply positive damping to suppress the line power oscillation effectively.

#### 7.3.2.2 An Example

Figure 7.23 shows the configuration of a four-machine two-area power system integrated with a SOFC power plant. This is the modified example power system used in Sect. 6.3. The system has a poorly damped inter-area oscillation mode. Parameters of the SOFC power plant are as follows:

$$T = 1273,\ F = 96487,\ R = 8.314,\ E_0 = 1.18,\ N_0 = 384,\ K_r = 0.966e{-6},$$
$$U_{max} = 0.9,\ U_{min} = 0.8,\ U_{opt} = 0.85,\ K_{h2} = 8.43e{-4},\ K_{h2o} = 2.81e{-4},$$
$$K_{o2} = 2.52e{-3},\ T_{h2} = 26.1,\ T_{h2o} = 78.3,\ T_{o2} = 2.91,$$
$$r = 0.126,\ T_f = 5,\ T_e = 0.08,\ r_{ho} = 1.145,\ C_{dc} = 1.0$$

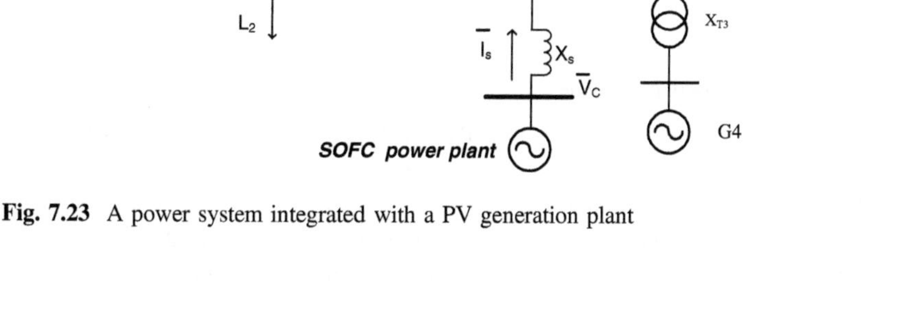


**Fig. 7.23** A power system integrated with a PV generation plant

To enhance the damping of the inter-area oscillation mode, a stabilizer can be added onto the DC/AC VSC of the SOFC power plant as discussed in the above section. For the purpose of demonstration, the following three cases are studied:

1. The PWM is adopted and the stabilizer is superimposed on the AC voltage controller (see Fig. 7.22). This is the PWM reactive power PSS.
2. The stabilizer is added on the DC voltage controller (see Fig. 7.22) of the PWM. This is the PWM active power PSS.
3. The PAM algorithm is used (see Fig. 7.22). The added stabilizer is named the PAM stabilizer.

Three types of stabilizers are designed at the line-loading condition $P_{A0} = 500$ MW by using the method introduced in Sect. 7.2.2 based on the localized model of the SOFC power plant of Eqs. (7.178) and (7.179). To confirm the effectiveness of the stabilizers designed, results of modal analysis obtained by using the complete mathematical model of the power system of Eq. (7.177) are given in Table 7.1 and simulation results using the nonlinear mathematical model of the example power system are shown in Fig. 7.24. In the nonlinear simulation, the example power system is subject to a disturbance of 1 % increase of mechanical power input to G4 for 100 ms at 0.2 s of simulation. From Table 7.1 and Fig. 7.24, it can be seen that the stabilizers are designed successfully by using the localized phase compensation method to effectively suppress the inter-area power oscillation.

Three types of stabilizers are tested at a lower line-loading condition of the example power system with $P_{A0} = 300$ MW. Results of modal analysis are given in Table 7.1. Figure 7.25 presents the results of nonlinear simulation (subject to the same disturbance as that in Fig. 7.24). From Table 7.1 and Fig. 7.25, it can be seen that at the lower line-loading condition, PWM reactive power stabilizer designed at the higher line-loading condition with $P_{A0} = 500$ MW cannot suppress the inter-area oscillation effectively. The robustness of the PWM active power and PAM stabilizer is better as they can still function to damp the inter-area power oscillation when the line-loading conditions of the power system change.

Similar to the case discussed in Sect. 7.2.3 about the ESS-based stabilizer, the PWM active power and PAM stabilizer function through regulating the phase of terminal voltage of the SOFC power plant, that is, the exchange of active power between the SOFC power plant and the rest of the power system. Therefore, they suppress the inter-area power oscillation directly by absorbing active power from or injecting it into the power system when they "see" the excess or lack of active

**Table 7.1** Inter-area oscillation mode

| | $P_{A0} = 500$ MW | $P_{A0} = 300$ MW |
|---|---|---|
| Without any stabilizer | $-0.103 \pm j3.471$ | $-0.883 \pm j3.602$ |
| With PWM reactive power stabilizer | $-0.355 \pm j3.169$ | $-0.221 \pm j3.331$ |
| With PWM active power stabilizer | $-0.405 \pm j3.365$ | $-0.320 \pm j3.454$ |
| With PAM stabilizer | $-0.529 \pm j3.363$ | $-0.398 \pm j3.419$ |

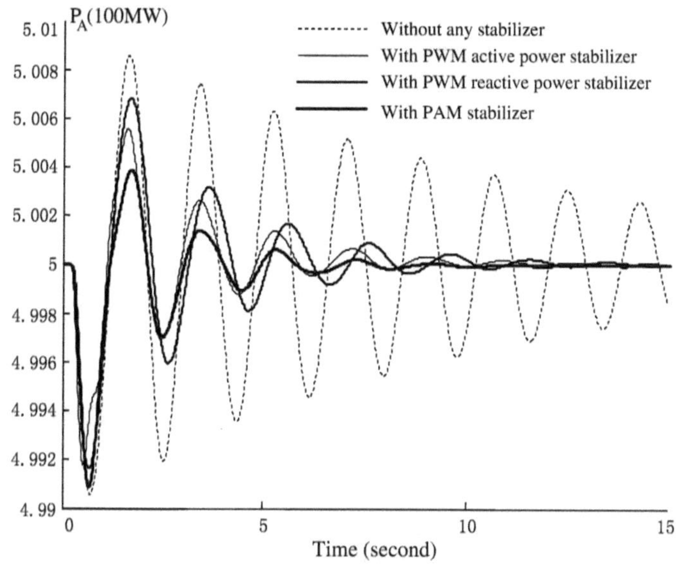


**Fig. 7.24** Results of nonlinear simulation when the system is subject to a small disturbance ($P_{A0}$ = 500 MW)

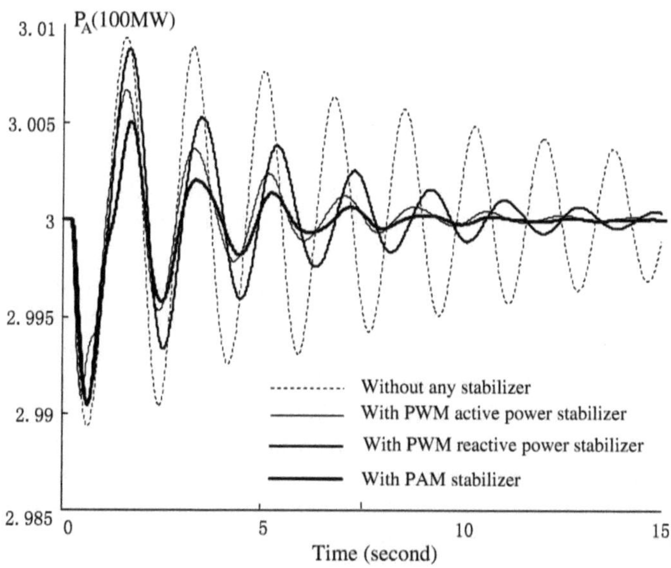


**Fig. 7.25** Results of nonlinear simulation when the system is subject to a small disturbance ($P_{A0}$ = 300 MW)

power delivered along the transmission line. Hence, their capability to damp the power oscillation is less affected by the change of system loading conditions. Function of the PWM reactive power stabilizer involves the regulation of magnitude of terminal voltage of the SOFC power plant. Hence, in essence it works via controlling the exchange of reactive power between the SOFC power plant and the rest of the power system. The stabilizer affects indirectly the variation of active power delivered along the transmission line through the regulation of injection into or absorbing of reactive power from the SOFC power plant. At a lower line-loading condition, regulation of reactive power exchange is related less closely with the variation of line active power. Hence, the stabilizer is less effective.

## 7.4 Damping of Multi-mode Oscillations by Multiple Stabilizers Attached to a Single UPFC

To suppress power system multi-mode oscillations, the commonly used schemes are to deploy multiple decentralized PSS or/and FACTS stabilizers. Usually, a stabilizer is designed to improve the damping of one selected oscillation mode and hence damping of multi-mode oscillations is achieved by multiple stabilizers jointly. Under these schemes, parameters of multiple stabilizers are set either individually or simultaneously. In engineering practice nowadays, strategy of individual setting is often used because the setting can be easily tuned in field. In many cases, it is effective and acceptable for the multiple stabilizers to provide reasonable damping to multi-mode oscillations. Extensive research in the last three decades has demonstrated that the coordinated design to simultaneously set multiple stabilizers can provide better results to effectively suppress power system multi-mode oscillations as introduced in Sects. 5.2.2 and 6.2.4.

However, from the point of view of engineering practice, at the stage of field tuning after parameters of multiple stabilizers are set in coordination, it could be a difficult and challenging task to finally implement the schemes of simultaneous setting of multiple stabilizers. This is because often parameters of stabilizers set in coordination need to be further tuned carefully in the field to provide required control performance. However, those stabilizers are often at different geographical locations in a power system. Cross-location and simultaneous field tuning of parameters of multiple stabilizers is practically very difficult. In this case, if parameters of stabilizers are tuned in field one by one in a sequence, the overall strategy of deploying multiple stabilizers will be the combination of simultaneous setting and individual field tuning. It is unlikely that the optimality of stabilizer's design by setting their parameters simultaneously can be achieved eventually by implementing individual parameters tuning in the field.

Hence, this section demonstrates a scheme of attaching multiple stabilizers to a single UPFC to damp power system multi-mode oscillations. The method takes the full advantage of the UPFC, which can perform multiple control functions of active, reactive power, or/and voltage regulation. Multiple stabilizers are superimposed on different control functions of a single multifunctional UPFC. Each stabilizer is set to specially meet the need of damping improvement of one selected oscillation mode and multiple stabilizers are designed in coordination with damp multi-mode oscillations. This arrangement of multiple stabilizers is as same as that of conventional schemes of coordinated design of multiple stabilizers introduced in Sects. 5.2.2 and 6.2.4. Hence, the optimal design to suppress multi-mode oscillation can be achieved. Moreover, being attached to one UPFC, multiple stabilizers are at a same geographical location in the power system. Therefore, problem of cross-location field tuning of stabilizer's parameters is effectively avoided.

### *7.4.1 Coordinated Design of Multiple Stabilizers by Artificial Fish Swarm Algorithm (AFSA) [8]*

Methods of optimal coordinated design of multiple stabilizers can be classified under two categories: mathematical optimization and non-mathematical optimization. In Sect. 5.2.2, a representative mathematical method based on the nonlinear programming is introduced. One of the most well-known methods of non-mathematical optimization is the genetic algorithm (GA) which has been tested for the coordinated design of multiple stabilizers in power systems. In this section, a different method of non-mathematical optimization—artificial fish swarm algorithm (AFSA)—is presented for the design of multiple stabilizers attached to a single UPFC to suppress multi-mode power system oscillation as follows.

#### 7.4.1.1 Objective Function

The coordinated design of multiple stabilizers is based on the following objective function

$$J = f(x) = \min\{\varsigma_i,\ i = 1, 2, \ldots, m\} \tag{7.184}$$

where $\varsigma_i$ is the damping of the ith oscillation mode and m the number of oscillation modes to be damped by m stabilizers. Hence, the objective function J is the damping of the oscillation mode of the power system that is of the least damping. The coordinated design is to search for the maximum of J = f(x).

Transfer function of the jth stabilizer superimposed on a selected primary control function of the UPFC is given as

$$K_j \frac{sT_{wj}}{1 + sT_{wj}} \frac{1 + sT_{2j}}{1 + sT_{1j}} \frac{1 + sT_{4j}}{1 + sT_{3j}}$$

Parameters of the stabilizer, $T_{2j}$, $T_{4j}$, and $T_{wj}$ are predetermined, and $T_{1j}$, $T_{3j}$, and gain $K_j$ are to be set by using the AFSA. In the parameter space of J, parameters of the jth stabilizer involved in the coordinated design are subject to the following constrains:

$$K_{jmin} \leq K_j \leq K_{jmax};$$
$$T_{1jmin} \leq T_{1j} \leq T_{1jmax}; T_{3jmin} \leq T_{3j} \leq T_{3jmax}$$

The AFSA is an algorithm based on the population evolution similar to the GA. The basic argument of the AFSA is that in a water area, the place where there are most fish is where there is most food. The optimal search imitates the behaviours of fish to seek food. In the n-dimensional parameter space of the objective function $J = f(x)$ defined by Eq. (7.184), the algorithm starts with constructing k artificial fish and the position of the ith fish is denoted by vector $\mathbf{x_i} = \begin{bmatrix} x_{i1} & x_{i2} & \cdots & x_{in} \end{bmatrix}$. The amount of food at present position is denoted by $Y_i = f(\mathbf{x_i})$. Distance between artificial fish is $d_{ij} = \|\mathbf{x_i} - \mathbf{x_j}\|$ and the visual distance of artificial fish is denoted by Visual. Crowding index is CI and maximal moving step is Step. The AFSA is implemented by three main behaviours of artificial fish as introduced in the following section.

#### 7.4.1.2 The Artificial Fish Swarm Algorithm

*Preying behaviour*

It is assumed that the present position of artificial fish is $\mathbf{x_i}(t)$ and the amount of food at the position is $Y_i(t)$. Within the visual distance of artificial fish, i.e. $d_{ij} <$ Visual, a new position $\mathbf{x_j}(t+1)$ is randomly chosen and $Y_i(t+1)$ is calculated. If $Y_i(t+1) > Y_i(t)$, then the artificial fish moves one step forwards. Otherwise, a new position is randomly chosen.

*Swarming behaviour*

This is to count the number of fish within the visual distance, $n_f$ and to calculate the central position $\mathbf{x_c}$ of this group of fish. The amount of food at the central position is $Y_c(t)$. If $Y_c(t)/n_f > CIY_i(t)$, this means that there is more food at the central position and the position is not very crowded. Hence, the artificial fish moves one step towards the central position. Otherwise, it performs preying.

*Following behaviour*

If within the visual distance of an artificial fish, there is most amount of food $Y_{max}(t)$ at position $\mathbf{x_{max}}$. Using $\mathbf{x_{max}}$ as the centre, the number of fish $n_{max}$ within its visual distance ($d_{ij} <$ Visual) is counted. If $Y_{max}(t)/n_{max} > CIY_i(t)$, then there is more food at $\mathbf{x_{max}}$ and it is not very crowded there. The artificial fish moves one step towards $\mathbf{x_{max}}$. Otherwise, it performs preying.

In the AFSA, preying behaviour guarantees the convergence of the algorithm. The introduction of crowding index CI is to prevent that a local optima is found as

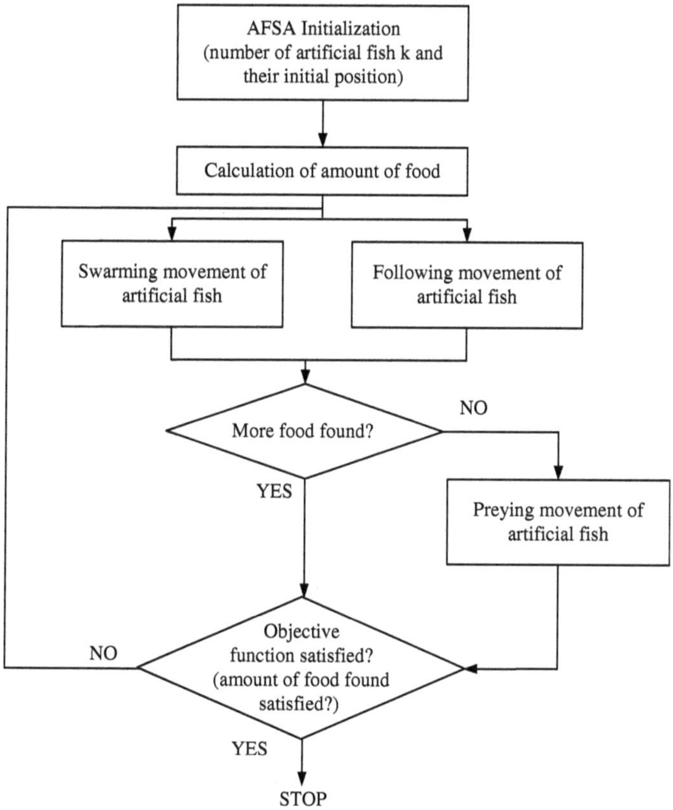


**Fig. 7.26** Flow chart of the AFSA

the final solution. The swarming behaviours are to ensure the stability of the algorithm and global optimization. Following behaviours are to accelerate the convergence of the algorithm towards the global solution. A flow chart of implementing the AFSA is shown in Fig. 7.26. The AFSA essentially is a random global optimization algorithm. It is of many attractive features of the genetic algorithm, but it does not need to perform crossover and mutation. Hence, it is relatively easier and simpler.

## *7.4.2 Examples*

### 7.4.2.1 An Example of Damping Two-Mode Oscillations by Stabilizers Added on a UPFC

The 4-machine 2-area power system of Fig. 7.24 is used as an example, where the SOFC power plant is replaced by a UPFC and G1 and G2 are not installed with the

PSS. The UPFC employs four control functions as defined by Eq. (7.71). In the system, there are three oscillation modes. One local mode related to G3 and G4 is well damped. An inter-area mode (G1 and G2 against G3 and 4) and a local mode (G1 against G2) are poorly damped. Two stabilizers are superimposed on two normal control functions of the UPFC to suppress the poorly damped local-mode and inter-area mode oscillation. To select the control functions to be attached with a stabilizer, the controllability index is calculated and presented in Table 7.2. From Table 7.2, it can be seen that in order to suppress two-mode oscillations, the first stabilizer should be added to the function of the AC voltage regulation to damp the inter-area oscillation mode and the second stabilizer to DC voltage regulation to damp the local oscillation mode.

Two stabilizers superimposed on the functions of the AC and DC voltage regulation of the UPFC are designed in coordination based on the linearized model of the power system presented in Sect. 7.2. Feedback signal of the first stabilizer is the deviation of active power delivered from node 7–8 and that of the second stabilizer is the difference between the deviation of active power from node 5–7 and that from node 2–7. Results of coordinated design of two stabilizers by using the AFSA are given in Table 7.3. From the third, fourth, and fifth rows of Table 7.4, it can be seen that two stabilizers can enhance the damping of two oscillation modes effectively.

**Table 7.2** Results of controllability computation of the UPFC normal control functions

| Control function to attach stabilizer | Inter-area mode | Local mode |
|---|---|---|
| Active power regulation | 3.77 | 0.10 |
| Reactive power regulation | 5.85 | 0.11 |
| AC voltage regulation | 49.29 | 1.41 |
| DC voltage regulation | 138.26 | 7.43 |

**Table 7.3** Results of coordinated design ($T_2 = T_4 = 0.05$, $T_w = 5$)

| Parameters | The first stabilizer | The second stabilizer |
|---|---|---|
| $T_1$ | 0.01 | 0.37 |
| $T_3$ | 0.12 | 0.07 |
| K | 0.25 | 28.36 |

**Table 7.4** Oscillation modes in the example power system with the UPFC

| | Inter-area mode | Local mode |
|---|---|---|
| Without any stabilizer | $-0.12062 \pm j3.4785$ | $-0.20745 \pm j5.8104$ |
| With the first stabilizer | $-0.70652 \pm j3.0758$ | $-0.21014 \pm j5.8047$ |
| With the second stabilizer | $-0.05775 \pm j3.3794$ | $-1.1759 \pm j5.7957$ |
| With two stabilizers | $-0.6812 \pm j3.0858$ | $-1.2403 \pm j5.6547$ |

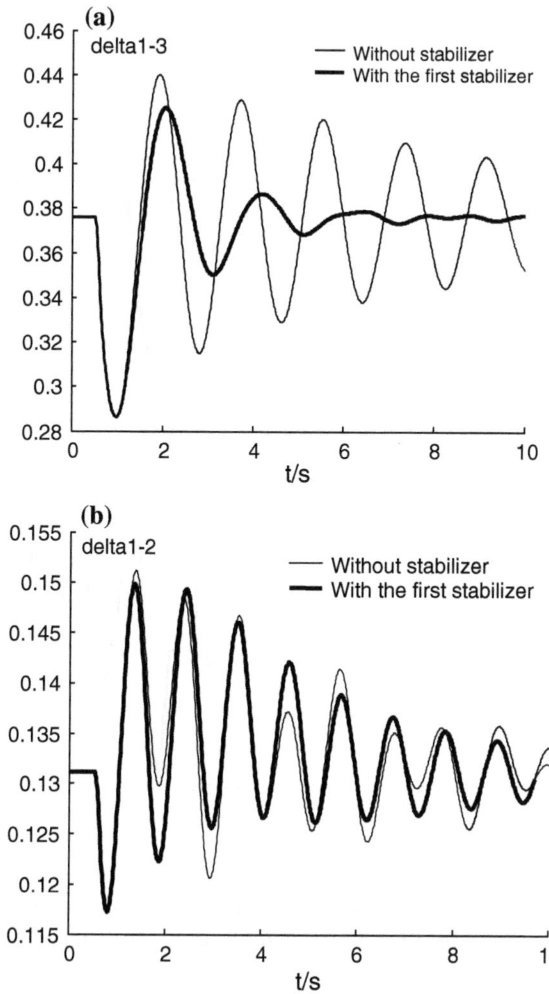


**Fig. 7.27** Simulation results when the first stabilizer is installed. **a** Inter-area mode oscillation and **b** local-mode oscillation

Figures 7.27 and 7.28 present the results of nonlinear simulation when the example power system is subject to a three-phase to-earth fault at node 8 for 100 ms. Figure 7.27 is that when only the first stabilizer is installed. Because the first stabilizer is set to damp inter-area mode, in Fig. 7.27 the local-mode oscillation is not suppressed effectively. From Fig. 7.28, it can be seen that with two stabilizers being superimposed on the AC and DC voltage regulator of the UPFC, respectively, two-mode oscillations of the example power system are damped successfully. Hence, this example demonstrates that it is feasible to suppress multi-mode oscillations by arranging two stabilizers on a single UPFC.

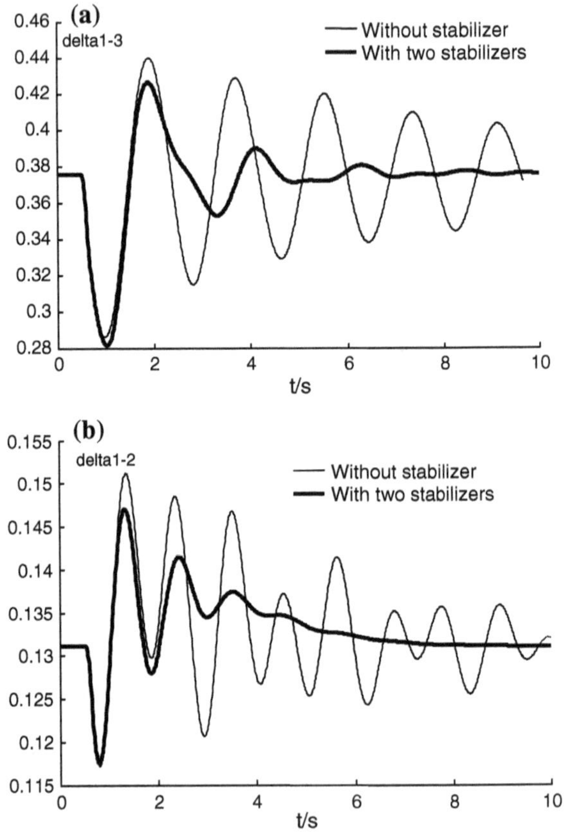


**Fig. 7.28** Simulation results when two stabilizers are installed. **a** Inter-area mode oscillation and **b** local-mode oscillation

#### 7.4.2.2 An Example of Damping Three-Mode Oscillations by Stabilizers Added on a Multi-terminal UPFC

In order to show that the example presented in the above section is not the unique case, the configuration of the example power system of Fig. 7.24 is modified to form a 6-machine power system as shown in Fig. 7.29. In this example power system, there are three weakly damped oscillation modes as shown in the second row of Table 7.5. They are: (1) the first inter-area oscillation mode (first mode) of group of machines 1 and 2 against that of machines 3, 4, 5, and 6; (2) the second inter-area oscillation mode (second mode) of group of machines 3 and 4 against that of machines 5 and 6; (3) a local oscillation mode (local mode) of machine 1 against machine 2. Two other local oscillation modes are very well damped and hence are not listed.

In the example power system, a multi-terminal UPFC (MUPFC) is installed at node 7. The MUPFC is of one shunt and two series VSCs and its mathematical model is given as:

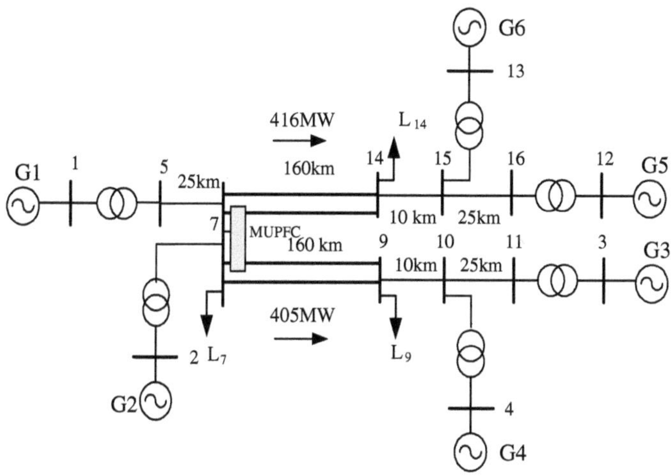


**Fig. 7.29** A 6-machine power system installed with a MUPFC

**Table 7.5** Oscillation modes in the example power system of Fig. 7.28

| | First mode | Second mode | Local mode |
|---|---|---|---|
| Without any stabilizer | −0.115 ± j4.296 | −0.113 ± j3.338 | −0.230 ± j5.845 |
| With the first stabilizer | −0.677 ± j3.987 | −0.059 ± j3.395 | −0.234 ± j5.825 |
| With the first and second stabilizers | −0.827 ± j3.806 | −0.557 ± j3.588 | −0.234 ± j5.828 |
| With three stabilizers | −0.635 ± j3.792 | −0.610 ± j3.663 | −0.933 ± j5.623 |

$$\frac{dv_{dc}}{dt} = \frac{3m_E}{4C_{dc}}[\cos\delta_E \quad \sin\delta_E]\begin{bmatrix} i_{Ed} \\ i_{Eq} \end{bmatrix} + \frac{3m_{B\_a}}{4C_{dc}}[\cos\delta_{B\_a} \quad \sin\delta_{B\_a}]\begin{bmatrix} i_{Bd\_a} \\ i_{Bq\_a} \end{bmatrix} + \frac{3m_{B\_b}}{4C_{dc}}[\cos\delta_{B\_b} \quad \sin\delta_{B\_b}]\begin{bmatrix} i_{Bd\_b} \\ i_{Bq\_b} \end{bmatrix}$$

$$\begin{bmatrix} v_{Btd\_a} \\ v_{Btq\_a} \end{bmatrix} = \begin{bmatrix} 0 & -x_{B\_a} \\ x_{B\_a} & 0 \end{bmatrix}\begin{bmatrix} i_{Bd\_a} \\ i_{Bq\_a} \end{bmatrix} + \begin{bmatrix} \dfrac{m_{B\_a}\cos\delta_{B\_a}v_{dc}}{2} \\ \dfrac{m_{B\_a}\sin\delta_{B\_a}v_{dc}}{2} \end{bmatrix};$$

$$\begin{bmatrix} v_{Btd\_b} \\ v_{Btq\_b} \end{bmatrix} = \begin{bmatrix} 0 & -x_{B\_b} \\ x_{B\_b} & 0 \end{bmatrix}\begin{bmatrix} i_{Bd\_b} \\ i_{Bq\_b} \end{bmatrix} + \begin{bmatrix} \dfrac{m_{B\_b}\cos\delta_{B\_b}v_{dc}}{2} \\ \dfrac{m_{B\_b}\sin\delta_{B\_b}v_{dc}}{2} \end{bmatrix};$$

$$\begin{bmatrix} v_{Etd} \\ v_{Etq} \end{bmatrix} = \begin{bmatrix} 0 & -x_E \\ x_E & 0 \end{bmatrix}\begin{bmatrix} i_{Ed} \\ i_{Eq} \end{bmatrix} + \begin{bmatrix} \dfrac{m_E\cos\delta_E v_{dc}}{2} \\ \dfrac{m_E\sin\delta_E v_{dc}}{2} \end{bmatrix}$$

where $m_E, m_{B\_a}, m_{B\_b}$ and $\delta_E, \delta_{B\_a}, \delta_{B\_b}$ are the modulation ratios and phase angles of shunt and series VSCs, respectively. The MUPFC is to perform the primary control functions to regulate the voltage at node 7 and power flow along line 7–9 and line 7–14 with the following six PI controllers.

DC voltage regulator: $\delta_E = \delta_{E0} + (K_{dcp} + \frac{K_{dci}}{S})(v_{dcref} - v_{dc})$

AC voltage regulator: $m_E = m_{E0} + (K_{vp} + \frac{K_{vi}}{S})(V_{ref} - V_7)$

Active and reactive power regulators for line 7–9:

$$\delta_{B\_a} = \delta_{B\_a0} + \left(K_{Bapp} + \frac{K_{Bapi}}{S}\right)(P_{7-9ref} - P_{7-9})$$

$$m_{B\_a} = m_{B\_a0} + \left(K_{Baqp} + \frac{K_{Baqi}}{S}\right)(Q_{7-9ref} - Q_{7-9})$$

Active and reactive power regulators for line 7–14:

$$\delta_{B\_b} = \delta_{B\_b0} + \left(K_{Bbpp} + \frac{K_{Bbpi}}{S}\right)(P_{7-14ref} - P_{7-14})$$

$$m_{B\_b} = m_{B\_b0} + \left(K_{Bbqp} + \frac{K_{Bbqi}}{S}\right)(Q_{7-14ref} - Q_{7-14})$$

In order to improve the damping of those three oscillation modes, three stabilizers are to be added on three of six MUPFC normal control functions. A linearized model of the system can be established by using the similar procedure presented in Sect. 7.2. Based on the linearized model, controllability index to superimpose damping function on the normal control functions of the MUPFC is calculated and presented in Table 7.6. Maximal controllability predicts the best damping effect (bold figures in Table 7.6). Hence, the AC voltage control function is selected to add the first stabilizer and the reactive power control function for line 7–9 is to add the second stabilizer to damp the first and second inter-area oscillation mode, respectively. DC voltage control function is chosen to add the third stabilizer to suppress the local oscillation mode. Locally available signals of active power are used as the feedback signals and those three stabilizers are set in coordination using the AFSA. Results of the coordinated design are given in Table 7.7.

**Table 7.6** Controllability index of the MUPFC normal control functions

| Control functions | 1st mode | 2nd mode | Local mode |
|---|---|---|---|
| AC voltage | **62.06** | 6.93 | 9.93 |
| Reactive power of line 7–9 | 9.86 | **19.52** | 0.86 |
| Reactive power of line 7–14 | 9.04 | 17.85 | 0.91 |
| DC voltage | 72.59 | 5.47 | **16.07** |
| Active power of line 7–9 | 0.89 | 0.98 | 0.18 |
| Active power of line 7–14 | 0.88 | 0.99 | 0.18 |

**Table 7.7** Results of coordinated design ($T_2 = T_4 = 0.05$, $T_w = 5$)

| Parameters | The first stabilizer | The second stabilizer | The third stabilizer |
|---|---|---|---|
| $T_1$ | 0.05 | 0.5 | 0.3 |
| $T_3$ | 0.3 | 0.1 | 0.05 |
| K | 0.6 | 2.0 | 20.0 |

Success of coordinated design is confirmed by the results of computation of oscillation modes presented in Table 7.5. From the third, fourth, and fifth rows of Table 7.5, it can be seen that those three stabilizers are set to improve the damping of three oscillations of the example power system separately and the multi-mode oscillations are damped effectively by their joint actions.

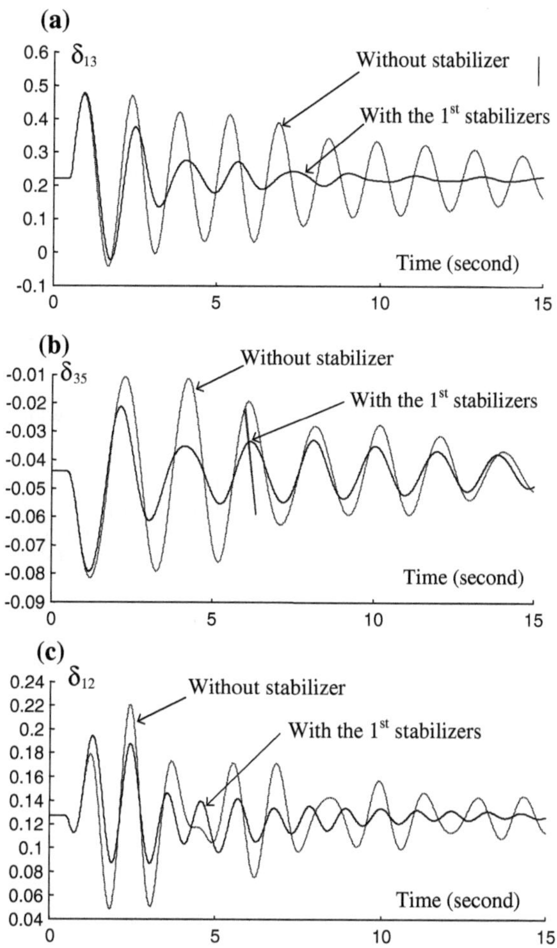


**Fig. 7.30** Simulation results when only the first stabilizer is installed to damp the first inter-area oscillation mode. **a** The first mode inter-area oscillation, **b** the second mode inter-area oscillation, and **c** local-mode oscillation

Figures 7.30, 7.31, and 7.32 give the results of nonlinear simulation of the example power system when it is subject to a three-phase to-earth fault at node 7 for 100 ms. They confirm the results of eigensolution in Table 7.5 that each stabilizer is designed to damp the targeted oscillation mode and with three stabilizers three-mode oscillations are suppressed successfully.

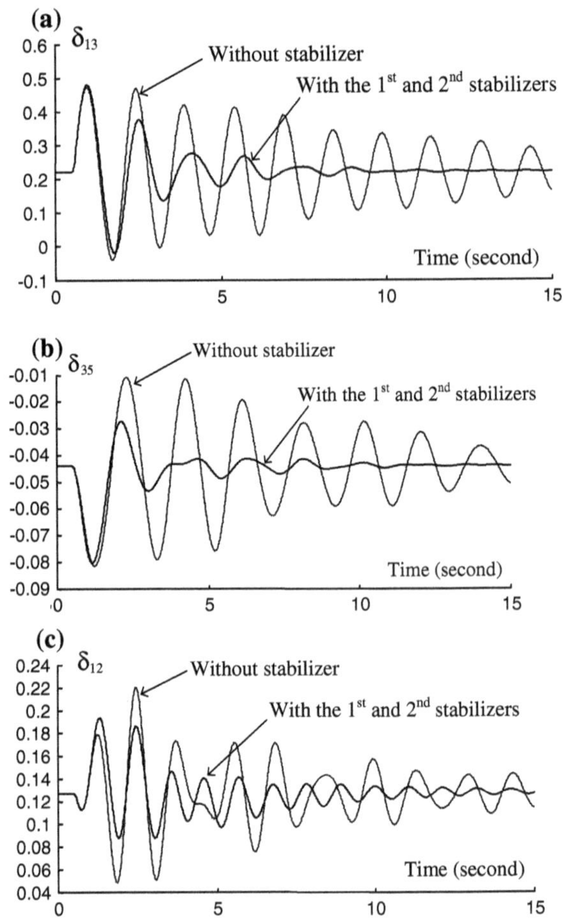


**Fig. 7.31** Simulation results when only the first and second stabilizers are installed to damp the first and second inter-area oscillation modes. **a** The first mode inter-area oscillation, **b** the second mode inter-area oscillation, and **c** local-mode oscillation

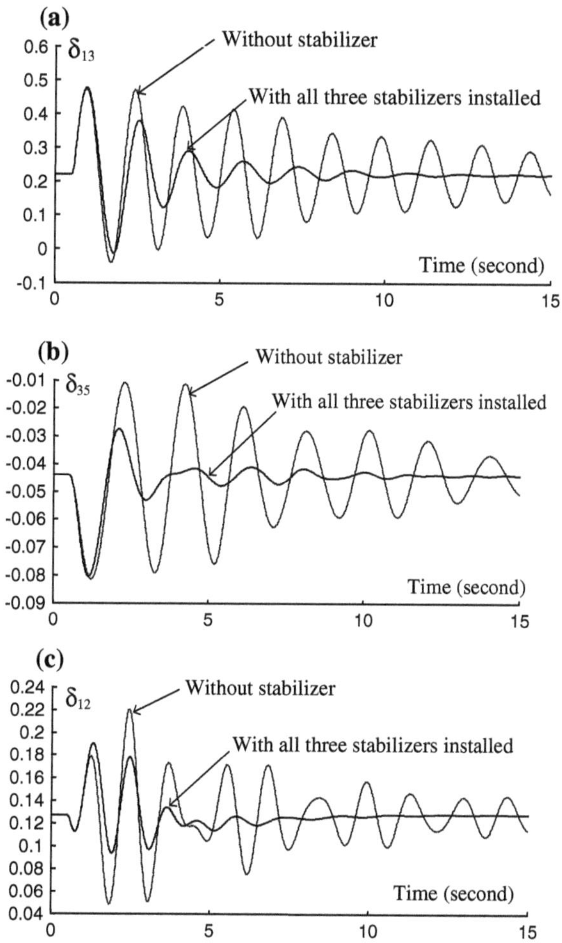


**Fig. 7.32** Simulation when all three stabilizers are installed. **a** The first mode inter-area oscillation, **b** the second mode inter-area oscillation, and **c** local-mode oscillation

# References

1. CIGRE TF 30-01-08 Report (1999) Modelling of power electronics equipment (FACTS) in load flow and stability programs
2. Wang HF (1999) Phillips-Heffron model of power systems installed with STATCOM and applications. IEE Proc Part C 146(5):521–527
3. Du W (2009) Power system small signal oscillation stability as affected by static synchronous compensator (STATCOM) and energy storage system (ESS). PhD thesis, University of Bath, UK
4. Wang HF (1999) Applications of modelling UPFC into multi-machine power systems. IEE Proc Part C 3
5. Wang HF (2000) A unified model for the analysis of FACTS devices in damping power system oscillations part III: unified power flow controller. IEEE Trans Power Del 3

6. Du W, Wang HF, Cao J, Xiao LY (2012) Application of the phase compensation method for the design a DC/AC converter based stabilizer to damp power system inter-area oscillation. IEEE Trans Power Syst 27(3):1302–1310
7. Du W, Wang HF, Cao J, Xiao LY (2012) Robustness of an energy storage system based stabilizer to suppress inter-area oscillation in a multi-machine power system. IET Proc Gener Transm Distrib 6(4):339–351
8. Du W, Wu X, Wang HF, Dunn R (2010) Feasibility study to damp power system multi-mode oscillations by using a single FACTS device. Int J Electr Power Energy Syst 32(6):645–655

# Index




H. Wang and W. Du, *Analysis and Damping Control of Power System Low-frequency Oscillations*, Power Electronics and Power Systems,
DOI 10.1007/978-1-4899-7696-3

MIX
Papier aus verantwortungsvollen Quellen
Paper from responsible sources
FSC® C105338
www.fsc.org

If you have any concerns about our products,
you can contact us on
**ProductSafety@springernature.com**

In case Publisher is established outside the EU,
the EU authorized representative is:
**Springer Nature Customer Service Center GmbH**
**Europaplatz 3, 69115 Heidelberg, Germany**

Printed by Libri Plureos GmbH
in Hamburg, Germany